Control of Power Electronic Converters
with Microgrid Applications

IEEE Press
445 Hoes Lane
Piscataway, NJ 08854

Control of Power Electronic Converters with Microgrid Applications

Arindam Ghosh PhD
Curtin University
Australia

Firuz Zare PhD
Queensland University of Technology
Australia

Published by John Wiley & Sons, Inc., Hoboken, New Jersey.
Published simultaneously in Canada.

Library of Congress Cataloging-in-Publication Data Applied for

Hardback: 9781119815433

Cover Design: Wiley
Cover Image: © LPETTET/Getty Images

Set in 9.5/12.5pt STIXTwoText by Straive, Pondicherry, India

Dedicated to my wife, Supriya, and my son, Aviroop, and in loving memory of my parents.

Arindam Ghosh

Dedicated to my lovely family, my wife, Leila, and my son, Farzan.

Firuz Zare

Contents

Author Biographies

Arindam Ghosh

Arindam Ghosh is a Research Academic Professor at Curtin University, Perth, Australia. He obtained his PhD from the University of Calgary, Canada. He was with the Indian Institute of Technology Kanpur from 1985 to 2006 and a Research Capacity Building Professor at Queensland University of Technology, Brisbane, Australia from 2006 to 2013. He was a Fulbright Scholar in 2003. He is a Fellow of the Indian National Academy of Engineering: INAE (2005) and a Fellow of the Institute of Electrical and Electronics Engineers: IEEE (2006). He was conferred the IEEE PES Nari Hingorani Custom Power Award in 2019. He has published over 450 peer reviewed journal and conference articles and has authored 2 books.

Firuz Zare

Professor Firuz Zare is a Fellow of the IEEE and Head of the School of Electrical Engineering and Robotics at Queensland University of Technology in Australia. He has over 20 years of experience in academia, industry, and international standardization committees, including eight years in two large R&D centers working on grid-connected inverters, energy conversion systems, and power quality projects. He has been a very active member and leader in IEC, Danish, and Australian standardization committees and has been a Task Force Leader (International Project Manager) of Active Infeed Converters to develop the first international standard IEC 61000-3-16 within the IEC standardization SC77A. Professor Zare has received several awards, such as an Australian Future Fellowship, John Madsen Medal, Symposium Fellowship, and early career excellence research award. He was awarded a technology leadership program by the Danish Innovation Council to attend a one-year leadership program delivered by Harvard Business School in Boston, USA in 2015. Professor Zare is a Senior Editor of the *IEEE Access* journal, a Guest Editor and Associate Editor of the *IEEE Journal of Emerging and Selected Topics in Power Electronics*, and an Editorial board member of several international journals.

Preface

Power converter applications in power systems have a long history. One of the first installations of high-voltage direct current (HVDC) transmission systems was on the Swedish island of Gotland in 1954. Mercury arc valves were used in the project. These were replaced by thyristor valves in 1967. Since then, other thyristor-based devices like the static var compensator (SVC), the thyristor-controlled series compensator (TCSC), etc., started finding applications in power transmission systems. However, with the advance of insulated-gate bipolar transistor (IGBT) technology, voltage source converters (VSCs) have started gaining prominence in power system applications. Currently, several VSC-based devices have been used in power transmission applications, such as in VSC-HVDC, flexible alternating current transmission systems (FACTS) devices, etc. At the same time, VSC applications in power distribution systems have been gaining prominence in custom power technologies and in microgrids.

With increased concerns about climate change, there has been an increased application of power electronic converters in power systems and an increase in the use of solar photovoltaic (PV) or wind power generation. Since these renewable generators are intermittent in nature, energy storage devices (predominantly battery energy storages) are being used for both storing energy and smoothing power fluctuations. Since VSCs generate harmonics, they are equipped with output passive filters. These filters can cause resonance with the rest of the system. Therefore, the control of power electronic devices has gained prominence in recent times. A very large number of publications have appeared in different IEEE Transactions about converter controls and their usages.

The concept of a microgrid has gained much attention in recent times. Microgrids are small power systems that have distributed generators (DGs), battery storage units, and customer loads located in close proximity. They can either be connected to the utility grids or be operated independently in an autonomous mode. They can provide fuel diversity and can increase the reliability and resilience of power delivery systems. Microgrids have been installed in communities, university campuses, hospitals, manufacturing sites, as well as in military installations. Moreover, remote area microgrids have the potential of providing reliable power to locations that are far away from power lines. Even though small or medium-sized diesel or gas-fired generators can be used in a microgrid, power-converter-interfaced generators are most prevalent as they interconnect renewable generators and battery storages. Therefore, power converter control is a very critical issue for microgrid applications as well.

The aim of this book is twofold: to review the control theories used for smart power converter control and to review the applications of these control concepts in power electronic converters used in power distribution systems. A voltage source converter can have several different control aspects that depend on its application. However, the basic principles are somewhat common. Therefore, a systematic approach has been taken for the application-specific converter control design in the book.

Three chapters in the book cover control theory. Most of the materials that are presented in these chapters can be used for a senior level undergraduate course or a junior level graduate course. There are several worked examples and design tips that can be used in MATLAB®, a product of MathWorks. The advantage of using MATLAB® is that complex control algorithms can easily be tested and verified using this software. In this book, MATLAB® has also been used for power converter controller design, while the design concepts have been verified through the Manitoba HVDC Research Center's EMTDC/PSCAD simulation package.

The book is organized in 11 chapters. Chapter 1 introduces the book. This chapter presents a basic introduction to power electronic components and power converter modes of operation and topologies. The need for harmonic filtering is also discussed briefly. Since most of the power converters can be modeled as piece-wise linear circuits, they need be linearized for feedback control design. This is also discussed in this chapter.

The methods of analysis of AC signals are presented in Chapter 2. Topics such as symmetrical components (phasor and instantaneous), Clarke and Park transforms, and the principle and use of phase locked loop (PLL) are covered in this chapter.

Chapter 3 provides an in-depth review of the classical control for single-input, single-output (SISO) systems. Since most classical control analysis and design approaches are similar for both continuous-time and discrete-time systems, more focus has been given to continuous-time systems in the book. Topics such as Routh–Hurwitz's criterion, root locus, frequency response methods, Nyquist stability criterion, relative stability, compensator design, and the PID controller and its tuning are covered along with several numerical examples. At the end of the chapter, discrete-time representation and z-transform are discussed.

Power converter control design in classical domain is discussed in Chapter 4. Specifically, DC-DC converters, such as buck and boost converters, are analyzed in detail. The process of deriving models of these converters using averaging methods and then designing classical controllers using these linearized models are explained. It also shows that a simple output voltage control is not sufficient for a boost converter since it has a right-half s-plane zero. A two-loop control design is also presented.

State space analysis and control design in both continuous- and discrete-time domains are presented in Chapter 5. Different topics such as the representation of a SISO system in state space domain, solutions of state equations, eigenvalues, and eigenvectors are covered in this chapter. Also, modal analysis using diagonalization, controllability, and observability are discussed. A state feedback control design using pole placement and a linear quadratic regulator is explained. The process of eliminating any steady state error using an integral control action is also described. At the end of the chapter, the process of deriving a DC-DC boost converter model using state space averaging as well as designing a controller that has a much superior performance are demonstrated.

Chapter 6 discusses control system design in the discrete-time domain, where prediction-based controllers are explained. Topics that are covered in this chapter include minimum variance prediction and control, pole placement in the polynomial domain, generalized predictive control, and self-tuning adaptive control that combines recursive parameter estimation with control design. A numerical example of the self-tuning control of a boost converter is also presented.

The open-loop control of DC-AC converters is covered in Chapter 7, where hysteretic current control and sinusoidal pulse with modulation (SPWM) for both bipolar and unipolar modulations are discussed. The concept of space vectors and space vector pulse width modulation (SVPWM) are also presented in this chapter. It also discusses how the performance of SPWM can be improved through a third harmonic injection. Different multilevel converters – such as diode-clamped, flying capacitor, cascaded, and modular – are also discussed in this chapter, along with the SPWM methods that can be used in multilevel converter output voltage modulation.

Chapter 8 presents several techniques of closed-loop control of DC-AC converters, and discusses both voltage and current controllers. To eliminate the harmonics generated by voltage source converters, they are equipped by output passive LC or LCL filters. First, a typical filter design principle is discussed. This is followed by a discussion of the state feedback based PWM and SVPWM voltage control of VSCs and sliding mode voltage control. Current control, using both state feedback and output feedback, is also discussed.

Power conditioning devices that are used for power quality improvements in power distribution networks use DC-AC converters that need to be controlled in some specific manner to achieve their goals. Such devices are discussed in Chapter 9, where, in particular, the structure and operating principles of a distribution static compensator (DSTATCOM) are presented. The chapter demonstrates that this device can be used for both voltage control, where a distribution bus voltage can be controlled against the load harmonics and unbalance, and for current control for load compensation. The associated converter control method is also presented.

Chapter 10 discusses microgrids. Both DC and AC microgrids are considered. The primary control applications in these microgrids are in the form of droop controllers, which are covered in detail in this chapter. Examples of different converter control principles that can be used for renewable energy integration are included in this chapter as well as the evolving smart power distribution systems that may contain several microgrids. Some of the possible connection and operating principles of microgrid networks are discussed. Specifically, the power exchange between the connected microgrid through a dedicated feeder is discussed in detail.

With the increased usage of power converters in power systems, higher-frequency harmonics have been causing concerns for the operational health of power components and appliances. In Chapter 11, some of the aspects of harmonic analysis and the harmonic propagation aspects in distribution system are highlighted. Furthermore, the standards that are evolving to tackle the harmonic problem are also presented.

Arindam Ghosh
Firuz Zare

Acknowledgments

I thank two of my best friends and collaborators – Professor Gerard Ledwich and Professor Avinash Joshi – for the many hours of discussions that I have had with them over the last three decades. Many concepts in this book have been formulated or clarified through such discussions. The book is also a product of my long-time friendship and collaboration with Firuz and the enthusiasm that we both have about the applications of power electronics in power systems. I also thank Professor Saikat Chakrabarti for being a source of encouragement and support over the last ten years.

I have been very blessed to have some outstanding PhD students. The critical discussions that I have had with them have enriched my knowledge in the diverse areas covered in this book. In particular, I thank Professor Mahesh Mishra, Professor Rajesh Gupta, Professor Anshuman Shukla, Dr Amit Jindal, Dr Ritwik Majumder, Dr Manjula Dewadasa, Dr Alireza Nami, Associate Professor Pooya Davari, Dr Megha Goyal, Dr Ehsan Pashajavid, Dr Amit Datta, and Dr Blessy John for helping to clarify doubts and for their contributions in the formulation of several concepts that have been included in the book.

I thank my wife, Supriya, for carefully proofreading the entire manuscript and my son, Aviroop, for making critical comments about several technical elements in the book. I also thank them for providing me with mental and moral support during the stressful times in the process of writing this book.

Arindam Ghosh

I know many people in industry, academia and standardization committees who have contributed to the development and the creation of knowledge in power electronics, harmonics, and power quality standards. I would like to start by thanking my PhD supervisor and colleague Professor Gerard Ledwich for his advice and contribution during my PhD program and later as a collaborator. I have known Professor Frede Blaabjerg since 2001 when I moved to Denmark. I would like to thank him for his contribution and technical discussions on several joint projects.

Many thanks to my post-docs and PhD students. We have worked together on different topics and industry-based projects. In particular, I would like to thank to Dr Alireza Nami, Dr Jafar Adabi, Dr Pooya Davari, Dr Jalil Yaghoobi, Dr Abdulrahman Alduraibi, Dr Davood Solati Alkaran, Dr Hamid Soltani, Mr. Arash Moradi, Mr. Amir Ganjavi, and Mr. Kiarash Gharani Khajeh for their contribution and development of new ideas on multil-level converters, grid-connected inverters, harmonics, and electromagnetic interferences.

It was a great opportunity to work in two large R&D centers in Denmark. Many thanks to my colleagues at the Danfoss and Grundfos companies where we worked on different challenging electromagnetic compatibility (EMC) and harmonic mitigations for low- and high-power converters. My special thanks to Dr Dinesh Kumar at Danfoss: we have had interesting technical discussions on many projects, product developments, and proof of concepts.

I would like to thank my professional colleagues and technical experts on IEC, Danish, and Australian standardization committees. We have been working on the development and maintenance of several standards and compatibility levels since 2013.

Finally, I would like to thank Professor Arindam Ghosh, my mentor, colleague, and friend whom I have known since 1999 when I was a PhD student at QUT. He has inspired and helped me with generous support and advice at several stages of my career with unforgettable memories on many joint projects, and professional and social activities. It has been an honor working with him and contributing to the preparation of this book.

Firuz Zare

1

Introduction

Power electronic converters are used in myriad applications. Some of these are adjustable speed motor drive systems, high-voltage direct current (HVDC) power transmission, flexible alternating current transmission systems (FACTS), power conditioning custom power devices, and microgrids. Several power electronic installations use traditional thyristor-based naturally commutated power converters, which have been in use for over half a century. However, with the advent of high-power insulated-gate bipolar transistors (IGBTs), voltage source converters (VSCs) have become increasingly popular in almost all the applications mentioned above.

With the present-day concerns about climate change and its effects on the well-being of all living creatures of our planet, an increased amount of renewable energy sources has been integrated with modern power systems. Traditionally, power is generated through large turbo alternators that are rotated at a fixed speed. Note that the system frequency is directly related to the generator speed ($n = 120\ f/P$, n is the generator speed in rpm, f is the frequency in Hz, and P is the number of poles). Usually, these turbogenerators have large inertia that help in maintaining synchronism during faults or transient disturbances. Renewable generators, on the other hand, provide low inertia and are often integrated through power electronic converters and therefore cannot maintain system frequency. Special control strategies are therefore adopted for the integration of renewable generators.

Renewable energy, as the name signifies, is a form of energy that is replenished constantly. For example, our sun is an abundant source of energy, and it shines throughout the year in all parts of the world. Similarly, wind blows all the time, while its speed depends on the time of day and the terrain. These two are the most prominent types of renewable energy that are used for electricity generation. An excellent resource for renewable energy is the book by Masters [1].

Control of Power Electronic Converters with Microgrid Applications, First Edition.
Arindam Ghosh and Firuz Zare.

Published 2023 by John Wiley & Sons, Inc.

The other forms of renewable energy sources are water (e.g. hydro, wave, and tidal), geothermal, etc. Out of these, hydro and geothermal plants are location dependent. Hydropower is the production of electrical power using the gravitational force of falling or flowing water, where electricity is produced by placing a turbine generator in the path of the flowing water. For this, catchment areas, water heights, and a continuous flow of water are required. Usually, hydro plants are placed in mountainous terrains. Hydropower is the most common form of renewable energy, which accounts for about 16% of the world's electricity generation. The total installed capacity of hydropower in 2020 is 1330 GW [2].

A powerful form of natural energy is generated by the gravitation of the moon and the sun, which causes low and high tides almost twice per day. The movement of the rising and falling sea level alters the potential energy of water that can be converted into electricity by the operation of a power plant. To use this energy, a dam wall is created to enclose a certain amount of seawater in an artificial bay serving the purpose of a reservoir, just like a hydropower plant. When the tide rises, the water enters the reservoir through a turbine which produces electric energy until the seawater inside the reservoir is almost as high as the outside water level. At low tide, the reverse process occurs and the water inside the reservoir exits into the sea through the turbine. Note that these two separate processes are not continuous as there is a pause of about two hours between these two. The tidal power has tremendous potential; however, it is still in the experimental stage of development due to the excessive cost involved. Other forms of waterpower that are also in the experimental stage are ocean current and wave power plants.

Geothermal energy comes from the core of our earth. The center of the earth is 6400 km below the surface. Since the temperature there is about 4200 °C, it is hot enough to melt rock into magma. The molten rock forms the outer core. The heat from the core rises to the earth's mantle, which is the layer that surrounds the core. It is this energy that powers volcanoes, geysers, and hot springs. In a geothermal plant, water is pumped into the earth's mantle and the resultant steam that rises is used for electricity generation using steam turbines. Geothermal plants, however, have a finite lifetime. The energy production ceases when the mantle at the location of the plant cools down due to the continuous extraction of heat energy.

There are two possible ways of generating solar power: through photovoltaic (PV) array and through concentrated solar power (CSP), which is also known as solar thermal power. In CSP, power is generated using mirrors and lenses to concentrate sunlight over a large area onto a receiver. The concentrated light then produces heat energy, which drives steam turbines to produce electricity using thermal generators. It is to be noted that water is not the only source that can be used for heat extraction from CSP: molten nitrite salt and hydrides are also considered for their higher heat retention properties. Spain is the leading country in CSP installation, followed by the United States.

Most of the technologies mentioned above use rotary generators to produce energy without any requirement of power electronic converters. This, however, is not the case for solar PV and wind generators as they require power electronic converters. A PV array produces power at DC voltage, which is then boosted through a DC-DC converter. The DC-DC converter is also often used for maximum power point tracking. The DC-DC converter output is converted into AC through a VSC for grid connection.

There are several types of wind turbines. These are [3]:

- Type 1: Fixed speed in which a squirrel-cage, self-excited induction generator is directly connected to the grid through a transformer. The turbine speed is synchronized with the grid frequency and is therefore (nearly) fixed.
- Type 2: Limited variable speed in which a wound rotor induction generator is connected directly to the grid through a transformer. The generator contains a variable resistor in the rotor circuit, which can control the rotor current quickly to keep the power constant, even during grid or wind disturbance.
- Type 3: Variable speed with partial power electronic conversion using doubly fed induction generator (DFIG). In this, there are a pair of VSCs that are connected back-to-back on the DC side through a capacitor. The grid side converter exchanges power with the grid and holds the DC bus voltage, while the rotor side converter can almost instantaneously control the magnitude and angle of rotor current. The major advantage of the DFIG is that it can bring about a large control of power in the stator circuit while using converters that have a much smaller rating than the machine.
- Type 4: Variable speed with full power electronic conversion in which a permanent magnet synchronous generator is connected to the grid through full-rated back-to-back converters. The turbine, in this case, is allowed to rotate at its optimal aerodynamic speed harnessing maximum power. Also, the need of a bulky gearbox is eliminated since the machine speed is separated from the grid frequency. The turbine side converter converts the generator voltage into DC and the grid side inverter injects power to the grid at rated or prevailing grid frequency.

Recently, several offshore windfarms have been installed. The power from these plants is supplied to the mainland through either submarine DC cables at high voltage or through multiterminal HVDC systems. All of these employ VSCs for power conversion.

There are several smaller generators that are deployed in power distribution systems, though not all of them necessarily use renewable energy. The most prevalent among these are the rooftop solar PV systems, which generate power with an output DC voltage level. These are then converted into AC through DC-AC power converters. There are others such as wind, fuel cells, and microturbines that use power electronic converters. Collectively these generators are called distributed generators (or DGs) because they are distributed throughout power distributions systems and are placed close to where the energy is consumed. Many of the renewable sources (e.g. solar and wind), however, are intermittent in nature. Therefore, storage systems are required to maintain continuity of the power flow. The DGs, together with the energy storage systems, are usually called distributed energy resources (or DERs). The most common energy storage such as battery energy storage systems (BESS) require power converters for converting DC voltage into AC.

From the above discussion, it is evident that power electronic converters play a very crucial role in the modern-day operation of power systems. Therefore, the control of these converters is also very crucial for the smooth and stable operation of power systems. Section 1.1 presents a brief introduction to power electronics.

1.1 Introduction to Power Electronics

Power electronics essentially is power processing. It is the application of electronics, control, and signal processing to adjust, regulate, or control electrical energy. Power electronics consists of power and electronic circuitry. In the power circuitry, DC or AC energy sources are converted to regulate or adjust voltage or current waveforms in the form of DC or AC with specific amplitude or frequency suitable for different applications. Figure 1.1 shows a schematic diagram of a power electronics system, which consists of an input source, a power converter, a load, and a controller.

The input source can be DC (e.g. BESS, solar PV, fuel cell, etc.) or an AC (e.g. grid, wind turbine, etc.). In some applications, the DC source can be in the form of

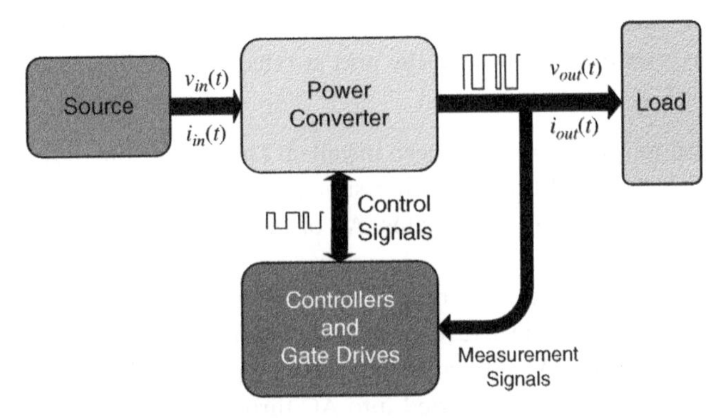

Figure 1.1 Schematic diagram of a power electronic circuit.

capacitors that can store energy. Furthermore, in an AC system, the input can be single- or three-phase. The loads can be either AC or DC. They can operate at either high or low voltage, where the frequency can be variable in the case of AC applications. For example, in home applications, power electronics is used in battery chargers (cell phones, laptops/desktops), electric motors, and induction cooking devices among others.

The power converter consists of semiconductor switching devices and passive elements, such as magnetic devices and capacitors. The semiconductor switching devices, such as MOSFETs (metal oxide silicon field effect transistor) or IGBTs, can operate at high voltage and current ratings that are suitable for different applications.

The controller unit consists of (i) measuring devices including input and output voltage and current signals to monitor and protect the system, (ii) a micro-controller with signal processing capability, and (iii) digital and analog electronic circuits. The controller synthesizes signals in the form of pulses suitable for the power converter to convert the input energy suitable for a load. The interface between the controller and the power converter is through gate drives, which take the control signals based on a pulse pattern and turn the semiconductor switches on and off at high voltage and current amplitudes.

Overall, the main aim of modern power electronic systems is to convert and deliver input power with maximum efficiency, high quality, minimum cost, and weight, in an integrated and high-power density circuit.

The main components used in the controller and the gate drive units are shown in Figure 1.2. As the voltage and current ratings of the controller and gate drives

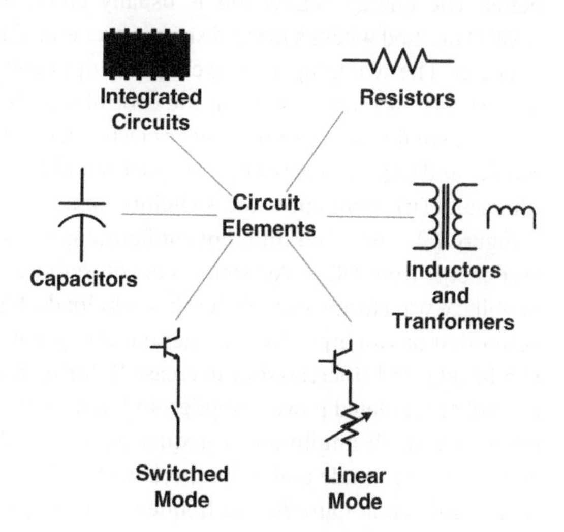

Figure 1.2 The main active and passive components used in controllers and gate drive units.

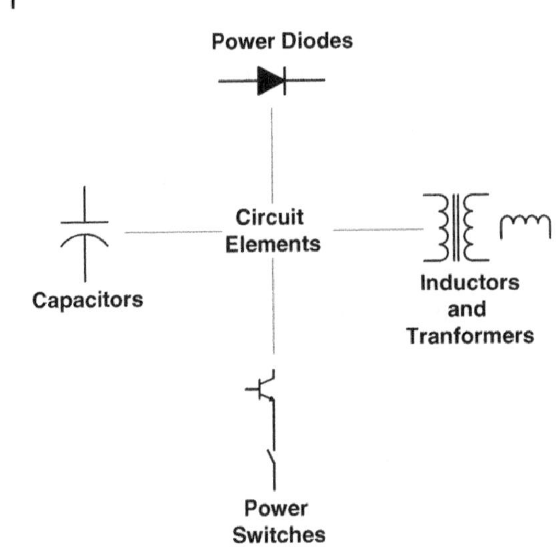

Figure 1.3 The main active and passive components used in power converters.

are very low compared to the power converters, resistors, and operational amplifiers (OPAMPS), linear mode switches are used in these units without any loss due to their high efficiency. On the other hand, their circuitry and design are very complex as the total power electronics system needs to be monitored and controlled through these units.

The power converter consists of four main components, as shown in Figure 1.3. Resistors and power switches in linear mode are not used in the power converters, because they incur significant losses when currents pass through these components. The energy conversion is usually based on a pulse width modulation (PWM) method where a desired signal is generated by pulse patterns at higher frequencies. The switching devices chop the input voltage or current (at high voltage and/or current rating) based on the control signals synthesized by the controller. Thus, the major issues of the power electronics system are (i) the generation harmonics and high-frequency noises which should be controlled and mitigated using filters and (ii) conduction and switching losses.

Figure 1.4 shows four different configurations of power converters that can convert energy from DC or AC sources to adjustable and regulated DC or AC current or voltage waveforms suitable for different loads. Figure 1.4a shows a DC-DC converter that has an input DC voltage source (e.g. battery or PV). The output voltage can be adjusted (increased or decreased) during the operation, as in a DC motor control or regulated power supplies. In Figure 1.4b, the energy from an AC source (fixed or variable amplitude or frequency), can be changed into an AC signal with adjustable amplitude and frequency, as in variable speed motor drive systems, or with regulated amplitude and frequency, as in grid-connected renewable energy

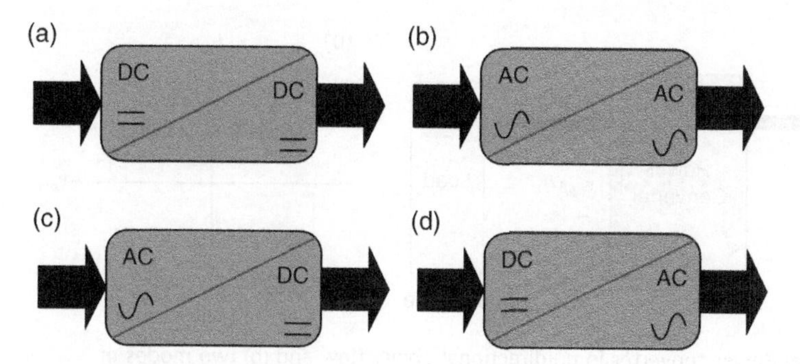

Figure 1.4 Four different configurations of power converters: (a) DC-DC, (b) AC-AC, (c) AC-DC, and (d) DC-AC.

systems. Figure 1.4c shows a power converter can transfer the energy from an AC source (e.g. grid or wind generator) to an adjustable or regulated DC signal (e.g. DC grids, power supply). The input source can be either a single-phase or a three-phase for low- or high-power applications. In the last configuration, shown in Figure 1.4d, a DC source is connected to a power converter and the output AC signal amplitude and frequency can be adjustable (e.g. induction heating and welding) or regulated (e.g. uninterruptable power supply or controllable AC sources).

1.2 Power Converter Modes of Operation

While designing a power converter, its modes of operation should be determined according to the system operation and the load characteristics. The instantaneous values of the load current $i_{out}(t)$ and voltage $v_{out}(t)$ can either be positive or negative in amplitude. These values represent four modes of operation for the power converter, as shown in Figure 1.5. The converter topology will be different when it

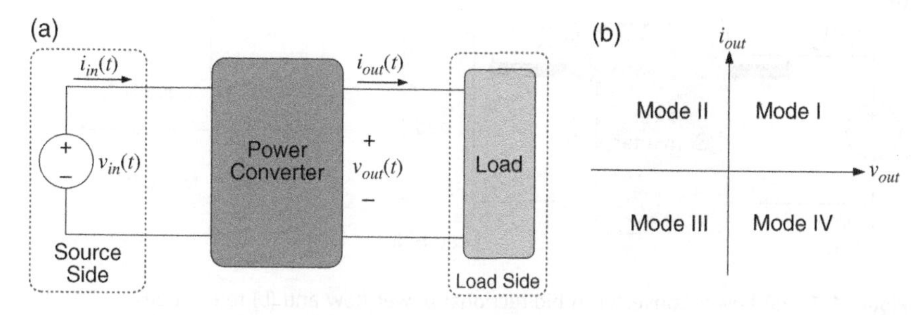

Figure 1.5 (a) Power converter supplying a load and (b) four quadrants of operation.

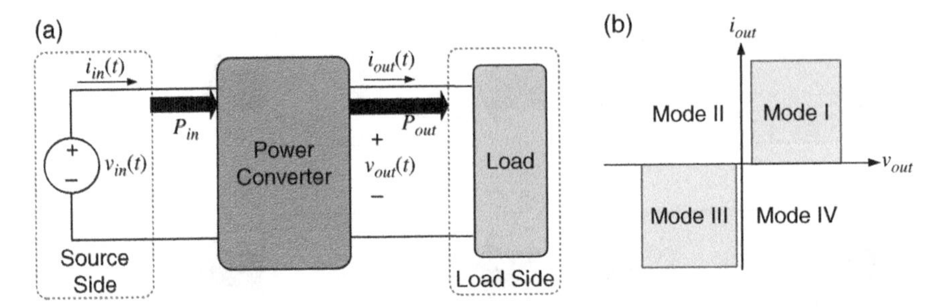

Figure 1.6 (a) Power converter in unidirectional power flow and (b) two modes of operation.

operates in one, two, or four quadrants. These conditions and operating modes are explained in this section.

Figure 1.6 shows a power converter with a unidirectional power flow in which the power is controlled and processed from the input side and transferred to the output side. The converter may operate either in quadrant I (when both voltage and current values are positive) or in quadrant III (when both voltage and current values are negative), or both these quadrants.

A power converter with a bidirectional power flow can operate in four different quadrants and the power can be transferred from the source to the load (consumption) or from the load to the source (regeneration). The converter may operate in any quadrant, based on the instantaneous voltage and current values and the load operating modes, as shown in Figure 1.7. The operating modes of a power converter with different topologies and semiconductor switches (type and configuration) will be explained in the following section.

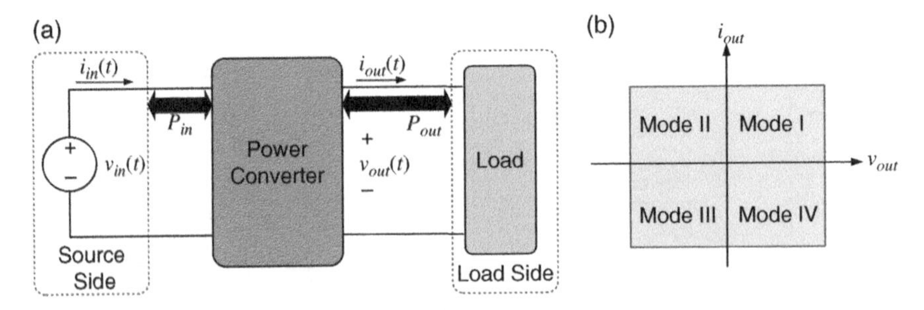

Figure 1.7 (a) Power converter in bidirectional power flow and (b) four modes of operation.

1.3 Power Converter Topologies

Several different topologies are utilized in energy conversion systems. The most common systems are shown in Figure 1.8 and are classified as:

- Low-frequency (at grid frequency 50 or 60 Hz) power converters such as diode rectifiers or controlled rectifiers with slow power switches, such as diodes or silicon-controlled rectifiers (SCR). These converters rectify AC signals (single-phase or a three-phase) to a DC form. These are shown in Figure 1.8a–d.
- High-frequency (at a switching frequency in kHz range) power converters are based on fast semiconductor switching devices such as MOSFETs or IGBTs. These converters are controlled based on PWM signals (modulated signals) and are used in different DC-DC or DC-AC energy conversion systems. The size of passive components utilized in these converters can be reduced if the switching frequency of the PWM signal is increased. This is shown in Figure 1.8e.
- A cascaded topology is based on a combination of a few low- and high-frequency power converters. For example, in Figure 1.8f, two power converters are in

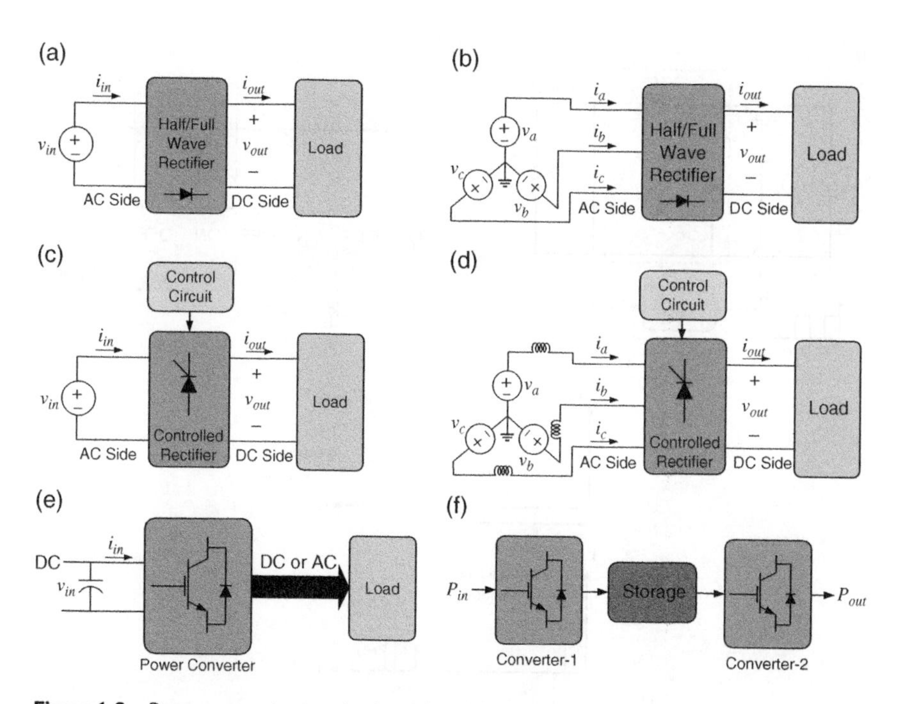

Figure 1.8 Power converter topologies: (a) and (b) low-frequency diode rectifiers, (c) and (d) lower frequency-controlled rectifier, (e) high frequency power converters, and (f) cascade power converter.

cascade with a storage element to convert an AC signal into a DC form and then the resulting DC signal back into an AC signal with adjusted or regulated amplitude and frequency. This is also called back-to-back (B2B) connection.

1.4 Harmonics and Filters

Harmonics and high-frequency noises are the two main aspects of power converters which have a negative impact on the quality and efficiency of the overall system. These phenomena are caused due to low- or high-frequency switching transients of the semiconductor switches in power converters. For example, to change a DC voltage to a desired level that is suitable for a load, a pulse train is applied to a DC-DC converter with a controlled duty cycle in such a way that the average voltage over each switching cycle can be controlled, as shown in Figure 1.9. The DC-DC converter is shown in Figure 1.9a. For the converter, the duty cycle is generated by comparing a reference signal (v_{ref}) with a sawtooth signal (v_{st}) and a gate signal is generated, as shown in Figure 1.9b. It is to be noted

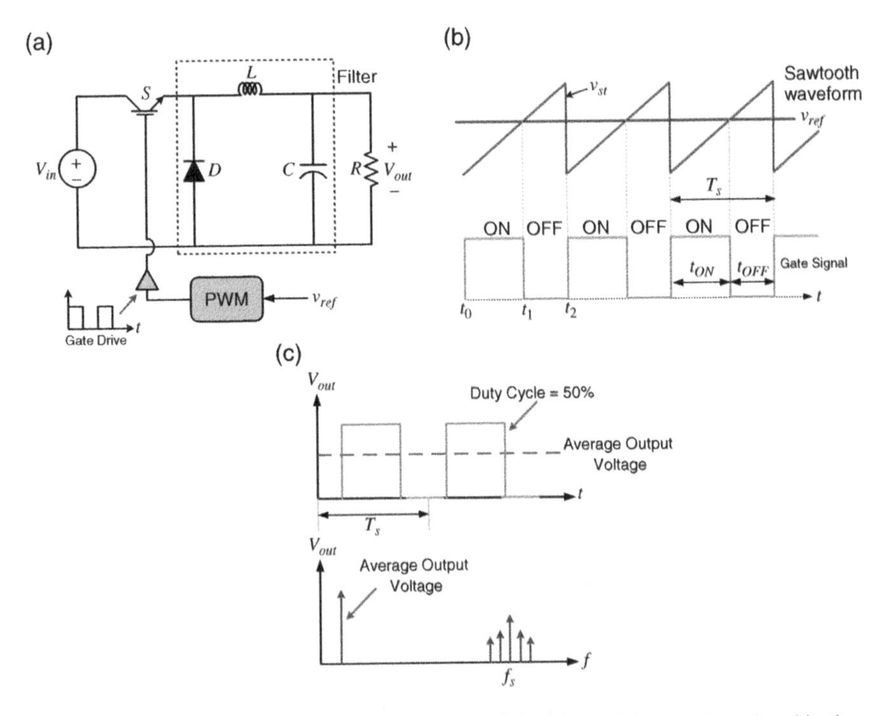

Figure 1.9 (a) A DC-DC converter, (b) its modulated signal, and (c) the output signal in time and frequency domain.

that the cycle time is $T_s = t_{ON} + t_{OFF}$. The switching frequency is $f_s = 1/T_s$, and the percentage duty cycle is defined as $d = (t_{ON}/T_s) \times 100\%$.

Let us assume that the duty cycle is controlled at 50% and the converter is designed such that the average value of the output voltage (V_{out}) for this duty ratio is 50% of the DC value (V_{in}). Although the pulse train is synthesized to control the average value of the output voltage, the proposed pulse waveform has harmonics, as shown in Figure 1.9c. This signal in time domain is not suitable for interfacing with electronic systems and the high-frequency harmonics should be filtered using an LC filter, as shown in Figure 1.9a.

Figure 1.10a shows a single-phase AC-DC converter where the input voltage is supplied from a low-voltage grid. The line current is not sinusoidal, and it is distorted due to the diode rectifier operation and its DC link filter (capacitor). The current harmonic amplitudes must be reduced according to international standardizations. There are several active or passive methods to mitigate current harmonics at the grid side.

Based on the above discussion, a general block diagram of a power electronics system is shown in Figure 1.11, where two filters – one at the grid side and the

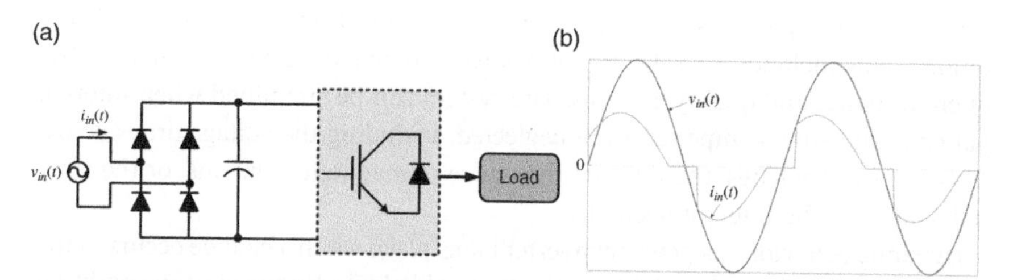

Figure 1.10 (a) A diode rectifier connected to a DC-AC converter and (b) voltage and current waveforms of the diode rectifier.

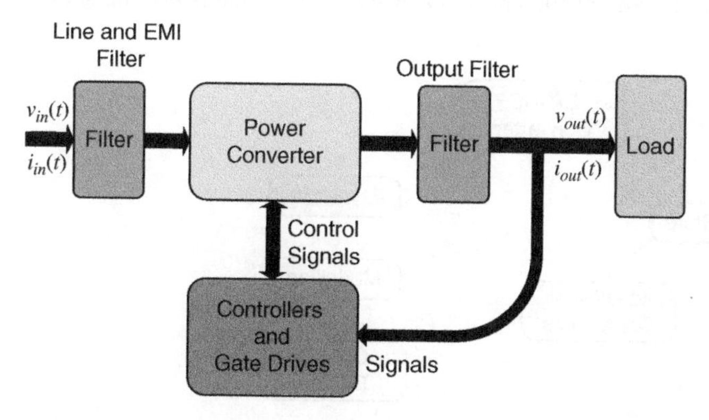

Figure 1.11 A general block diagram of a power electronic system with different filters.

other one at the load side – are utilized to mitigate low- and high-frequency harmonics. The grid side filter consists of two different types of filters: harmonics and electromagnetic interference (EMI) filters. A harmonic filter is designed to mitigate low-order harmonics below the order of 40th or 50th harmonics depending on standardization limits. The EMI filter is designed for high-frequency harmonics, mainly above 150 kHz to suppress conducted emission noise.

1.5 Power Converter Operating Conditions, Modelling, and Control

Power electronics systems are nonlinear as there are several semiconductor switching devices which are turned on and off thereby splitting a power converter circuitry into sub-circuitries. The system might have more subsystems when the inductor current is not continuous during the operation. Figure 1.12 shows all operating conditions of a DC-DC converter, which can operate in either a continuous conduction mode (CCM) or a discontinuous conduction mode (DCM). The steady state analysis is used to design a power converter under different load conditions. This includes the selection of passive and active elements, switching frequency, losses, and quality analysis. The system can be simplified when internal parasitic and stray components are neglected, including the voltage drops across the diodes or switches, the internal resistance of magnetic elements, or the stray inductance of the interconnections.

Dynamic behavior of a power converter takes place when a change occurs in the reference signal or input voltage or the load. This includes the startup condition when a power converter is turned on, as shown in Figure 1.13. In this case, the instantaneous value of the inductor current is increased from zero. The inductor current at the beginning and at the end of each switching cycle is not the same.

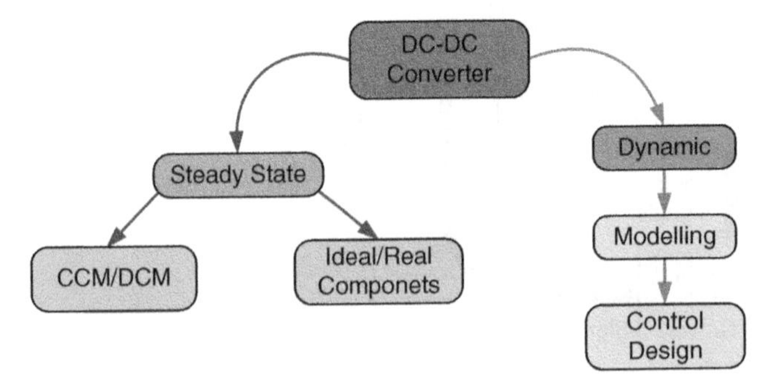

Figure 1.12 Operating conditions of DC-DC converters.

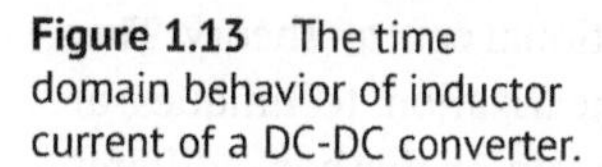
Figure 1.13 The time domain behavior of inductor current of a DC-DC converter.

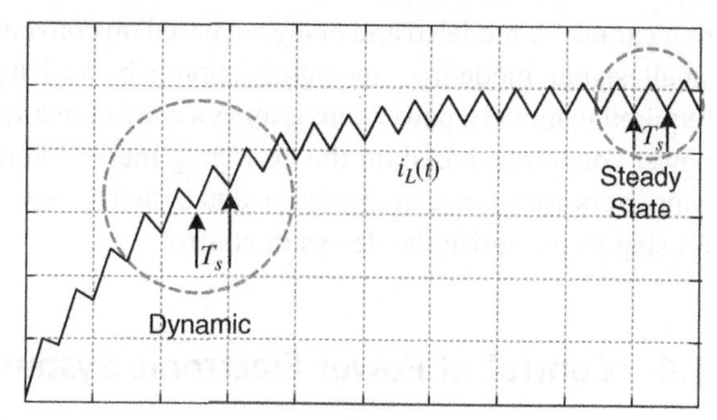

However, when it reaches a steady state after several switching cycles, the inductor current at the beginning and the end of each switching cycle is the same. Thus, the dynamic behavior of a power converter, i.e. reaching the steady state value with minimum transient time, error, and overshoot, can be improved using a proper control system.

The general approach of designing a controller is to find the transfer function of a system. Most power electronics systems are nonlinear with discrete operating modes. For example, Figure 1.14a shows a buck or step-down DC-DC converter operating in CCM where the current through the inductor is always continuous. When the switch is turned "ON" or "OFF," the converter circuitry is changed into two different equivalent circuits, as shown in Figure 1.14b,c. As the power converter is switched on in the frequency range of kHz, it has different subsystems

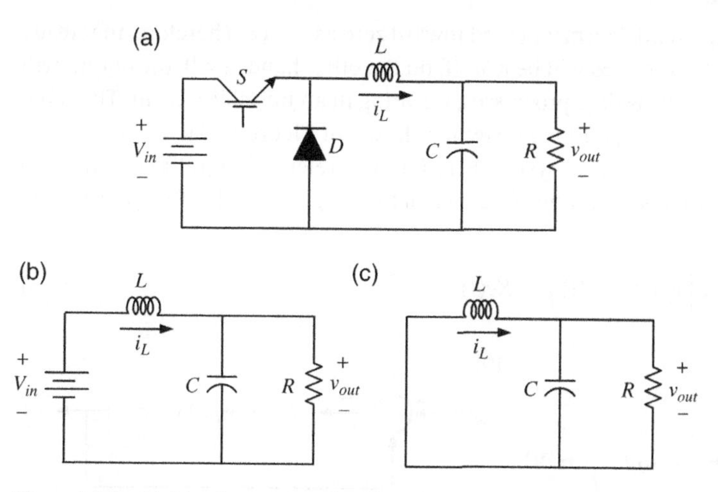

Figure 1.14 (a) A buck converter, (b) when the switch is turned ON, and (c) when the switch is turned OFF.

that cannot be modeled and analyzed based on conventional control theory. Thus, small signal modeling, averaging approach, and linearization techniques are required to model a power converter as a continuous system. In this book, different power converters based on the averaging method and stability analysis of power converters are studied at the device and system levels. Discrete modeling is a helpful step to recognize the delays in control.

1.6 Control of Power Electronic Systems

In this section, the concept of feedback control is briefly discussed and the application of control on a simple power electronic circuit introduced. Our discussion starts with the advantage of feedback control.

1.6.1 Open-loop Versus Closed-loop Control

Consider a first-order system given by the differential equation

$$\dot{y}(t) + \alpha y(t) = u(t) \tag{1.1}$$

where $y(t)$ is the output, $u(t)$ is the input, and α is a scalar. Assume that the system is at rest, i.e. $y(t)|_{t\le 0} = 0$ and a control input $u(t) = K \times u_s$ is applied at time $t = 0$, where u_s is a unit step and K is a scalar constant. Then the system response will be given by

$$y(t) = \frac{K}{\alpha}(1 - e^{-\alpha t}), \quad t \ge 0 \tag{1.2}$$

If $\alpha > 0$, the exponential term will tend toward zero as $t \to \infty$. Therefore, the steady state value of $y(t)$ as $t \to \infty$ will be K/α. If, on the other hand, $\alpha < 0$, the output will tend toward infinity as time progresses, resulting in an unstable system. The schematic diagram of the open-loop system is shown in Figure 1.15a.

The main aim of a control system is to follow a reference input $y_r(t)$ asymptotically. To achieve this, a negative feedback of the output is used to form the control law as

$$u(t) = K\left\{y_{ref}(t) - y(t)\right\} = Ke(t) \tag{1.3}$$

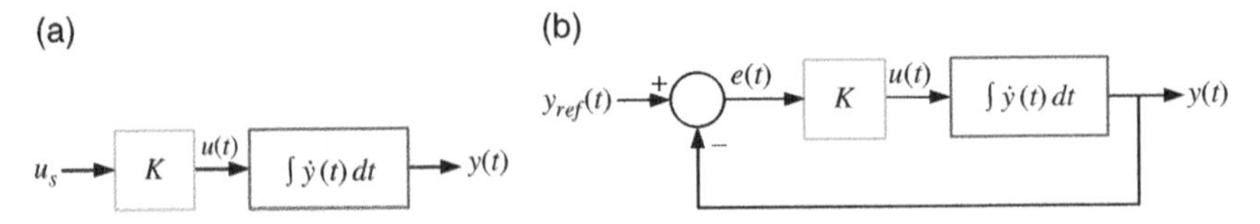

Figure 1.15 (a) Open-loop control system and (b) feedback control system.

where $e(t)$ is defined as the tracking error. Substituting (1.3) in (1.1) and assuming $u(t) = Ke(t)$, we have

$$\dot{y}(t) + (K + \alpha)y(t) = Ky_{ref}(t) \tag{1.4}$$

The closed-loop system is shown in Figure 1.15b. Let us assume that the reference input is a unit step. Then, the output is given by

$$y(t) = \frac{K}{K + \alpha}\left(1 - e^{-(K+\alpha)t}\right), \quad t \geq 0 \tag{1.5}$$

The closed-loop system will remain stable (bounded) so long as $K + \alpha > 0$. If $\alpha > 0$, then the system will be stable for positive values of K. On the other hand, if α is negative, K should be greater than $|\alpha|$. This is one of the advantages of the feedback control. Another important aspect of the feedback is reference tracking, where the output $y(t)$ needs to be close to the reference input $y_{ref}(t)$ in the steady state, which can only be achieved if the system is stable. In that case, the steady state tracking error is defined from (1.5) as

$$e_{ss}(t) = \{y_r(t) - y_r(t)\}|_{t\to\infty} = \frac{K}{K + \alpha} \tag{1.6}$$

The steady state error can be minimized by choosing a large value of K.

Figure 1.16 shows the behavior of the open- and closed-loop systems for $|\alpha| = 0.5$. For the open-loop system, it is assumed that $\alpha > 0$ and $K = 1$. This is shown in

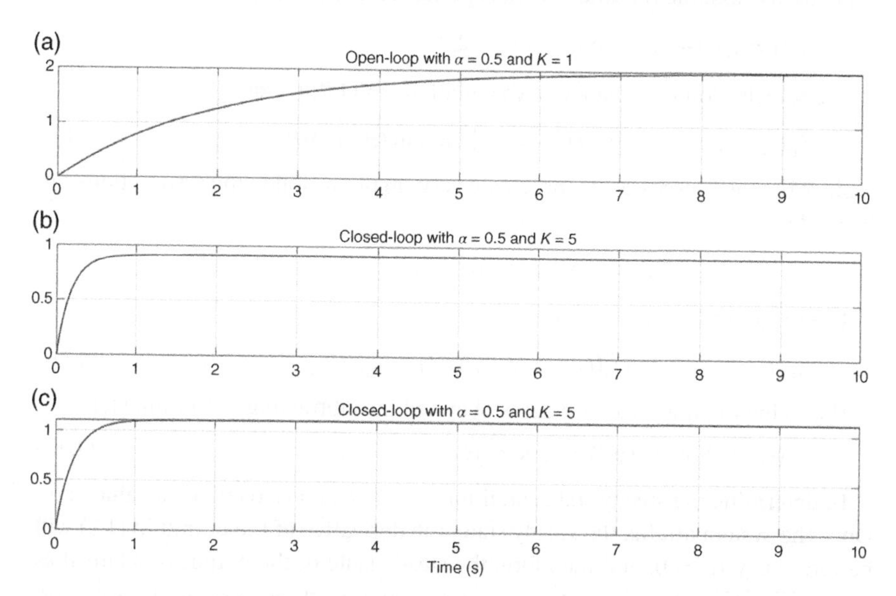

Figure 1.16 (a) Open-loop control system, feedback control system (b) with $\alpha > 0$ and (c) with $\alpha < 0$.

Figure 1.16a, where the output reaches its steady state value of 2. The closed-loop response for $\alpha > 0$ is shown in Figure 1.16b, while Figure 1.16c shows the closed-loop response for $\alpha < 0$. The value of the gain for both these cases is chosen as $K = 5$. Even though the open-loop system is unstable for $\alpha < 0$, the closed-loop system is stable in Figure 1.16c since $K + \alpha > 0$.

1.6.2 Nonlinear Systems

Consider the following system

$$\dot{y}(t) + \alpha y^2(t) = \sin(\theta) \tag{1.7}$$

This is obviously a nonlinear system. Even though there is a vast amount of literature dealing with the stability and control of nonlinear systems, usually linear controllers are designed by linearizing the system around an operating point. For linearization, Taylor series expansion is performed around an operating point, where the second- and the higher-order terms are neglected. This aspect is discussed later in the book. However, we present a simple method here.

Let us assume that the system operates under a steady state operating point of y_0 and θ_0 such that (1.7) can be written as

$$\dot{y}_0(t) + \alpha y_0^2(t) = \sin(\theta_0) \tag{1.8}$$

Let us also assume that the system is perturbed with small increments such that

$$y(t) = y_0(t) + \Delta y(t) \text{ and } \theta = \theta_0 + \Delta\theta$$

The substitution of the above two equations in (1.7) yields

$$\dot{y}_0(t) + \Delta\dot{y}(t) + \alpha\{y_0(t) + \Delta y(t)\}^2 = \sin(\theta_0 + \Delta\theta) \tag{1.9}$$

Since the increments $\Delta y(t)$ and $\Delta\theta$ are very small, the following assumptions can be made

$$\Delta y^2(t) \approx 0, \sin(\Delta\theta) = \Delta\theta, \text{ and } \cos(\Delta\theta) = 1$$

Substituting these in (1.9), we have

$$\dot{y}_0(t) + \Delta\dot{y}(t) + \alpha y_0^2(t) + 2\alpha y_0(t)\Delta y(t) = \sin(\theta_0) + \cos(\theta_0)\Delta\theta \tag{1.10}$$

The following linearized model is obtained by subtracting (1.8) from (1.10)

$$\Delta\dot{y}(t) + 2\alpha y_0(t)\Delta y(t) = \cos(\theta_0)\Delta\theta \tag{1.11}$$

To determine the steady state condition, the first step is to choose a value of θ_0. Once the system attains the steady state, the derivative of the output in (1.8) will be zero, i.e. $\dot{y}_0(t) = 0$, and therefore, the steady state of the output is obtained as $y_0 = \sqrt{\sin(\theta_0)/\alpha}$. For example, if $\alpha = 2$, then the steady state values for $\theta_0 = 10°$, $\theta_0 = 20°$, and $\theta_0 = 30°$ are 0.2947, 0.4135, and 0.5 respectively. Starting from

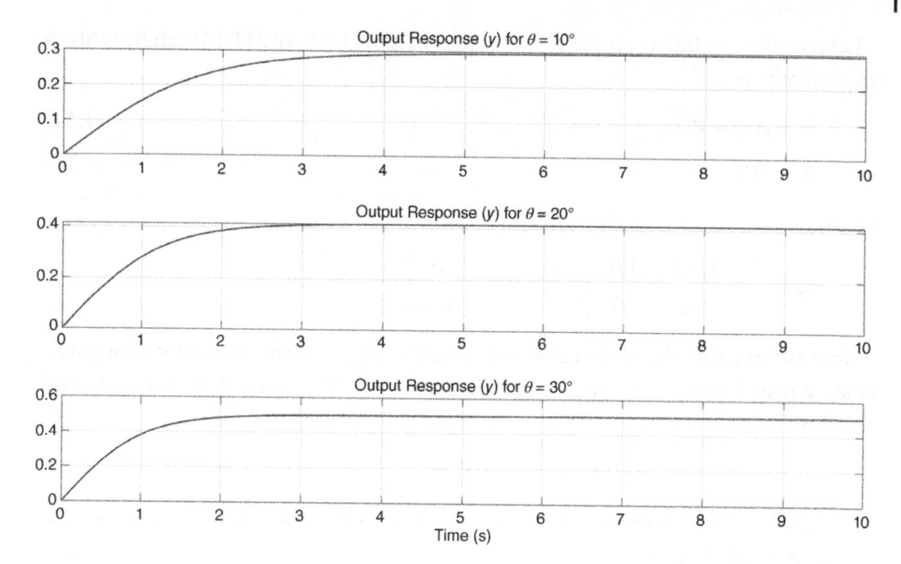

Figure 1.17 The response of the nonlinear system with three different values of θ_0.

$y(t)|_{t=0} = 0$, the response of the system with these values of θ_0 are shown in Figure 1.17. It can be seen that the output attains these values in the steady state.

1.6.3 Piecewise Linear Systems

Consider the buck converter model shown in Figure 1.14a. From Figure 1.14b, the following equations are obtained when the switch is ON.

$$\begin{aligned} \frac{dv_{out}}{dt} &= -\frac{1}{RC}v_{out} + \frac{1}{C}i_L \\ \frac{di_L}{dt} &= -\frac{1}{L}v_{out} + \frac{1}{L}V_{in} \end{aligned} \tag{1.12}$$

On the other hand, when the switch is OFF, the following equations describe the system

$$\begin{aligned} \frac{dv_{out}}{dt} &= -\frac{1}{RC}v_{out} + \frac{1}{C}i_L \\ \frac{di_L}{dt} &= -\frac{1}{L}v_{out} \end{aligned} \tag{1.13}$$

Both these sets of equations are linear. However, the behavior of the circuit is controlled by the duty ratio or duty cycle shown in Figure 1.9c. This is best described in terms of the state space description of the system, which is explained in Chapter 5.

Let us define a state vector as $\mathbf{x} = [v_{out}\ \ i_L]^T$. Then (1.12) and (1.13) can be written respectively as

$$\dot{\mathbf{x}} = \mathbf{A}\mathbf{x} + \mathbf{B}V_{in} \tag{1.14}$$

$$\dot{\mathbf{x}} = \mathbf{A}\mathbf{x} \tag{1.15}$$

where

$$\mathbf{A} = \begin{bmatrix} -1/RC & 1/C \\ -1/L & 0 \end{bmatrix}, \qquad \mathbf{B} = \begin{bmatrix} 0 \\ 1/L \end{bmatrix}$$

Now assume that the switch closes at t_0, opens at t_1, and subsequently closes at t_2, as shown in Figure 1.9b. In the steady state, the duty ratio D is constant, and therefore

$$t_1 - t_0 = DT_s, \quad t_2 - t_1 = (1 - D)T_s \tag{1.16}$$

where D is the duty ratio and T_s is the cycle time. Then the solutions of (1.14) and (1.15) respectively are

$$\mathbf{x}(t_1) = \int_{t_0}^{t_1} (\mathbf{A}\mathbf{x} + \mathbf{B}V_{in})dt + \mathbf{x}(t_0) = \int_{0}^{DT_s} (\mathbf{A}\mathbf{x} + \mathbf{B}V_{in})dt + \mathbf{x}(t_0) \tag{1.17}$$

$$\mathbf{x}(t_2) = \int_{t_1}^{t_2} \mathbf{A}\mathbf{x}dt + \mathbf{x}(t_1) = \int_{0}^{(1-D)T_s} \mathbf{A}\mathbf{x}dt + \mathbf{x}(t_1) \tag{1.18}$$

In the steady state, we have $\mathbf{x}(t_2) = \mathbf{x}(t_0)$. Solutions of (1.17) and (1.18) will yield the description of the system between t_0 and t_2, which is dependent on the duty ratio, which appears in the exponential terms of the solutions of (1.17) and (1.18). Thus, even if the circuit is piecewise linear, the overall behavior of the circuit is nonlinear. Furthermore, the DC-DC converter is controlled by its duty ratio. Therefore, the system will have to be linearized for control design, as is discussed in Chapter 4 (Section 4.1.5) and Chapter 5 (Section 5.11.2).

1.7 Power Distribution Systems

Power systems' voltages and currents can be represented either though their instantaneous components or through their phasor components. The instantaneous voltage of the form $v(t) = V_m \sin(\omega t + \delta)$, where V_m is the voltage magnitude, ω is the angular frequency in rad/s, and δ is its phase angle. Phasor components represent the sinusoidal steady state, i.e. the voltage (or current) magnitude and its angle when the transients have died down and all the quantities in the system are in a pure sinusoidal state. For the instantaneous voltage given above, the phasor component is represented in polar of cartesian form as $V = (V_m/\sqrt{2})e^{j\delta} = (V_m/\sqrt{2})[\cos\delta + j\sin\delta]$. A diagram representing the phasors in a circuit is called the phasor diagram.

One of the main applications of power converters is in power distribution systems. These applications are discussed in Chapters 9 and 10 of this book. Consider, for example, the radial power system shown in Figure 1.18a. It contains a source V_S that supplies an RL load with the impedance of Z_L. Let us assume that the switch S is open. The phasor diagram of this system is shown in Figure 1.18b. The lagging load current (I) has two components: the real component (I_R) and the reactive component (I_Q). It is the real component that is doing any practical work, while the reactive component is present due to the load power factor. However, due to the reactive component, the current magnitude becomes larger. This causes more line voltage drop and larger $R|I|^2$ drop in the line that can lead to excessive heating in the conductors.

When switch S is closed, the capacitor, with a reactance of $-jX_C$, is connected in parallel with the load bus. This will draw a leading current I_C from the system, as

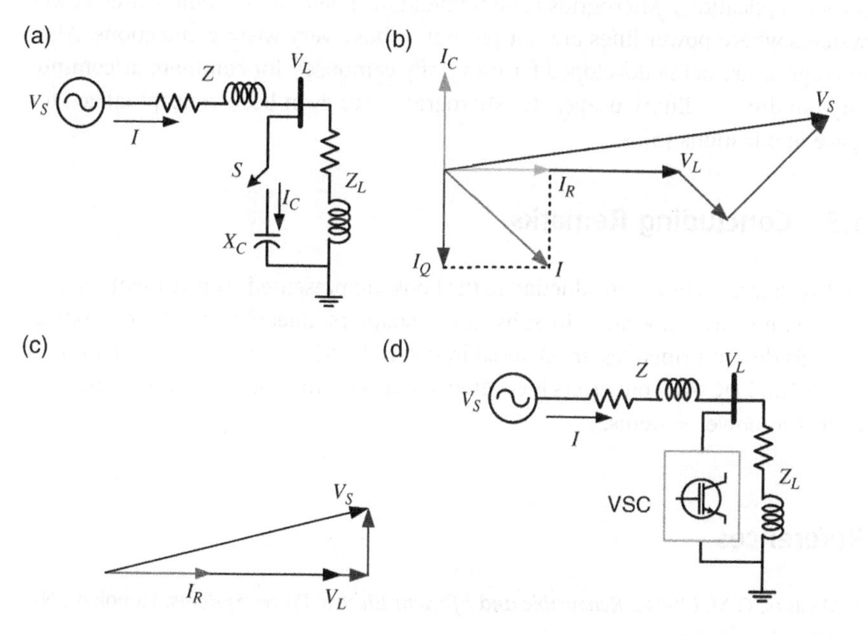

Figure 1.18 (a) A radial distribution system, (b) phasor diagram when the capacitor is not connected, (c) phasor diagram when the capacitor is connected, and (d) power factor correction through a VSC.

shown in Figure 1.18b. If this current is such that $I_C = I_Q$, then the source will only supply the current I_R. Then, the load and the capacitor will draw power from the source at unity power factor, as shown in Figure 11.18c.

The main problem with the above proposition is that the load may change, and therefore fixing the value of the capacitance with all load changes is not feasible. A better approach is to connect a VSC in shunt with the load bus. This VSC, through proper control, can not only correct the power factor but also provide harmonic compensation, balance the load bus voltage, and regulate the bus voltage [4].

A microgrid is a small, localized grid with its embedded control capability. It can operate along with the main utility grid or can also disconnect from the grid and work autonomously while supplying power to its local loads. Microgrids are supplied by their local generators with most of them harnessing power from renewable energy sources. Additionally, a microgrid may even contain battery storage systems. These power supply sources are collectively called DERs. Most of the DERs are connected to microgrids through VSCs and are required to supply power in the autonomous mode, while regulating its bus voltage and frequency. Therefore, converter control plays a significant role in the operation of microgrids. Since microgrids have local generators, they are very suitable for combined heat and power applications. Microgrids have tremendous potential for remote area power systems where power lines are not present or have very weak connections. Also, microgrids are being developed for university campuses, for commercial/community buildings, military usage, etc. Microgrids have even been conceptualized for space applications [5].

1.8 Concluding Remarks

In this chapter, a brief introduction to the book are presented. Topics mentioned in this chapter are elaborated in subsequent chapters. Specifically, several control analysis design principles are covered in detail. Furthermore, the control of both DC-DC and DC-AC converters is covered, along with the applications of these converters to power systems.

References

1 Masters, G.M. (2004). *Renewable and Efficient Electric Power Systems*. Hoboken, NJ: Wiley-InterScience.

2 International Hydropower Association (2021). Hydropower Status Report. https://www.hydropower.org/status-report (accessed 15 May 2022).

3 IEEE PES Wind Plant Collector System Design Working Group (2009). Characteristics of wind turbine generators for wind power plants. IEEE PES General Meeting, Calgary.

4 Ghosh, A. and Ledwich, G. (2002). *Power Quality Improvement Using Custom Power Devices*. New York: Springer Science+Business Media.

5 Ciurans, C., Bazmohammadi, N., Vasquez, J.C. et al. (2021). Hierarchical control of space closed ecosystems: expanding microgrid concepts to bioastronautics. *IEEE Ind. Electron. Mag.* 15 (2): 16–27.

2

Analysis of AC Signals

From a circuit theory point of view, a DC circuit is much simpler to analyze than an AC circuit. In the steady state of a DC circuit, the voltage across a capacitor is open circuited and the current through an inductor is short circuited. The transient response of the capacitor voltages and inductor currents can be solved through differential equations.

AC circuits can be single- or three-phase. Instantaneous active power in a single-phase circuit oscillates at twice the fundamental frequency around a DC value. The instantaneous real power in a three-phase circuit, on the other hand, has a DC value in a balanced system, while it behaves like a single-phase circuit in an unbalanced system. Reactive power in an AC circuit occurs due to the presence of inductors and capacitors. It does not contribute to any useful work but can cause voltage drops and increase line losses. A three-phase circuit can have voltage and/or current unbalance due to unbalance in loads. Also, AC signals can have harmonics, which are nonfundamental frequency waveforms that are superimposed on the fundamental frequency waveforms. Both unbalance and harmonics have detrimental effects on appliances and power apparatus connected to the system.

Control design or analysis in a DC circuit is simpler as all the signals have DC values in the steady state, while in an AC circuit they have a constant magnitude and phase once the transients die down. This allows us to perform a sinusoidal steady state analysis. However, from the control design perspective or for stability analysis, it may be desirable to represent these circuits as equivalent DC circuits. Also, the AC circuits need to remain synchronized always such that all the different interconnected components can function stably.

In this chapter, we discuss the various aspects of AC signals: analysis of unbalance, instantaneous real and reactive power, harmonics, frequency estimation, dq transformation, and phase locking that are important for the integration of power electronic circuits with AC power systems. We start our discussion with the analysis of unbalance.

Control of Power Electronic Converters with Microgrid Applications, First Edition.
Arindam Ghosh and Firuz Zare.

Published 2023 by John Wiley & Sons, Inc.

2.1 Symmetrical Components

Usually, symmetrical components are used to analyze unbalance in a three-phase system in the phasor domain. A three-phase system is termed as balanced when its voltage (or current) waveforms have the same magnitude and are displaced from each other by 120°. Failing these two conditions, the voltages (or currents) are said to be unbalanced. Let us consider a set of three-phase voltages defined by

$$V_a = |V_{ma}|\angle\varphi_a, \quad V_b = |V_{mb}|\angle\varphi_b, \quad V_c = |V_{mc}|\angle\varphi_c \tag{2.1}$$

where $|V_{ma}| \neq |V_{mb}| \neq |V_{mc}|$ and φ_a, φ_b, and φ_c are not phase displaced by 120° from each other. These unbalanced vectors are resolved into three balanced vectors using the following transform

$$\begin{bmatrix} V_{a0} \\ V_{a1} \\ V_{a2} \end{bmatrix} = \frac{1}{3}\begin{bmatrix} 1 & 1 & 1 \\ 1 & a & a^2 \\ 1 & a^2 & a \end{bmatrix}\begin{bmatrix} V_a \\ V_b \\ V_c \end{bmatrix} = \mathbf{T}\begin{bmatrix} V_a \\ V_b \\ V_c \end{bmatrix} \tag{2.2}$$

$$\Rightarrow \mathbf{V}_{a012} = \mathbf{T}\mathbf{V}_{abc}$$

where V_{a0}, V_{a1}, and V_{a2} respectively are the zero-, positive-, and negative-sequence voltage vectors and $a = e^{j120^\circ}$, such that $a = -0.5 + j\sqrt{3}/2$, $a^2 = a^*$ (complex conjugate of a), and $1 + a + a^2 = 0$. Note that the zero- and negative-sequence components will be zero for balanced voltages, when $|V_{ma}| = |V_{mb}| = |V_{mc}|$, $\varphi_b = \varphi_a - 120^\circ$, and $\varphi_c = \varphi_a + 120^\circ$. The phase voltages can be recovered from the symmetrical component voltages using the inverse transform

$$\mathbf{V}_{abc} = \mathbf{T}^{-1}\mathbf{V}_{a012} \tag{2.3}$$

where

$$\mathbf{T}^{-1} = \begin{bmatrix} 1 & 1 & 1 \\ 1 & a^2 & a \\ 1 & a & a^2 \end{bmatrix}$$

Note that we can use the transformations of (2.2) and (2.3) for three-phase currents as well.

Example 2.1 Consider the following set of voltages

$$V_a = 100\angle 10^\circ \text{ V}, \quad V_b = 110\angle -120^\circ \text{ V}, \quad V_c = 90\angle 120^\circ \text{ V}$$

These constitute an unbalanced set. The sequence components are then

$$V_{a0} = -0.5064 + j0.0148 \text{ V}$$
$$V_{a1} = 99.4936 + j5.7883 \text{ V}$$
$$V_{a2} = -0.5064 + j11.5618 \text{ V}$$

These are written in polar form as

$$V_{a0} = 0.51\angle 178.33^\circ \text{ V}, \quad V_{a1} = 99.66\angle 3.33^\circ \text{ V}, \quad V_{a2} = 11.57\angle 92.51^\circ \text{ V}$$

Note that the zero-sequence components for the three phases are the same and are given by

$$V_{a0} = V_{b0} = V_{c0} = 0.51\angle 178.33^\circ \text{ V}$$

The positive sequence phasors are balanced and have the same phase sequence as the original vectors, i.e.

$$V_{a1} = 99.66\angle 3.33^\circ \text{ V}, \quad V_{b1} = 99.66\angle -116.67.51^\circ \text{ V}, \quad V_{c1} = 99.66\angle 123.33^\circ \text{ V}$$

The negative sequence vectors are also balanced, but the phase sequence is reverse of the original phasors, i.e.

$$V_{a2} = 11.57\angle 92.51^\circ \text{ V}, \quad V_{b2} = 11.57\angle 212.51^\circ \text{ V}, \quad V_{c2} = 11.57\angle -27.49^\circ \text{ V}$$

The original unbalanced signals can be written in terms of the symmetrical components as

$$\begin{bmatrix} V_a \\ V_b \\ V_c \end{bmatrix} = \begin{bmatrix} V_{a0} \\ V_{b0} \\ V_{c0} \end{bmatrix} + \begin{bmatrix} V_{a1} \\ V_{b1} \\ V_{c1} \end{bmatrix} + \begin{bmatrix} V_{a2} \\ V_{b2} \\ V_{c2} \end{bmatrix} \tag{2.4}$$

It can be verified that we can express the unbalanced phase voltages using (2.4).

2.1.1 Voltage Unbalanced Factor (VUF)

Voltage unbalance occurs in power distribution systems primarily due to single-phase loads. Usually, domestic power is supplied from one of the three phases and neutral. Even though the utilities try to balance the loads between the phases, the installments of rooftop photovoltaics (PVs) worsens the situation as most of the PV inverters are single-phase. There might be some influence of unbalances in transmission lines or supply transformers on voltage unbalance.

The voltage unbalance of a circuit is defined based on different references such as IEEE and IEC [1, 2]. This is given terms of the ratio of negative sequence to the positive sequence voltage magnitudes as

$$\text{VUF} = \left|\frac{V_{a2}}{V_{a1}}\right| \times 100\% \tag{2.5}$$

The detrimental impacts of voltage unbalance on induction motors are analyzed in [1, 3]. It is desirable to maintain the VUF below 3% for the health of induction motors, which are the workhorse of industry. In Example 2.1, the VUF is 11.61%.

2.1.2 Real and Reactive Power

The complex power (S) in an AC circuit is given by

$$S = P + jQ = VI^*$$

Therefore, for a three-phase circuit, we can write this as

$$S = V_a I_a^* + V_b I_b^* + V_c I_c^* = \mathbf{V}_{abc}^T \mathbf{I}_{abc}^*$$

Using (2.3), the above equation can be written as

$$S = \mathbf{V}_{a012}^T \mathbf{T}^{-T} \left(\mathbf{T}^{-1} \mathbf{I}_{a012}\right)^* \tag{2.6}$$

Now

$$\mathbf{T}^{-T}\left(\mathbf{T}^*\right)^{-1} = \begin{bmatrix} 3 & 0 & 0 \\ 0 & 3 & 0 \\ 0 & 0 & 3 \end{bmatrix}$$

Therefore

$$S = 3\left(V_{a0} I_{a0}^* + V_{a1} I_{a1}^* + V_{a2} I_{a2}^*\right) \tag{2.7}$$

Example 2.2 Consider the voltages of Example 2.1. Assume that these voltages supply an unbalanced load such that the currents are given by

$$I_a = 10\angle -10^\circ \text{ A}, \quad I_b = 11\angle -100^\circ \text{ A}, \quad I_c = 12\angle 110^\circ \text{ A}$$

The sequence components of the currents are

$$\begin{aligned} I_{a0} &= 1.28 - j0.43 = 1.34\angle -18.63^\circ \text{ A} \\ I_{a1} &= 10.67 - j0.02 = 10.67\angle -0.1^\circ \text{ A} \\ I_{a2} &= -2.10 - j1.29 = 2.46\angle -148.48^\circ \text{ A} \end{aligned}$$

Then computing the complex power, we get

$$S = V_a I_a^* + V_b I_b^* + V_c I_c^* = 3\left(V_{a0} I_{a0}^* + V_{a1} I_{a1}^* + V_{a2} I_{a2}^*\right) = (3.14 + j0.116) \times 10^3$$

It is to be noted that, for balanced voltages or currents, the negative- and zero-sequence components will be zero and the symmetrical component transformation will result only in the positive sequence components.

2.2 Instantaneous Symmetrical Components

In Section 2.1, we demonstrate how to represent a set of three voltages or currents in terms of their symmetrical components. The symmetrical component transformation can also be applied to instantaneous voltages and currents [4]. Through this transformation, we can convert three-phase instantaneous quantities into a zero-sequence component and two other phasor components. Let the instantaneous three-phase voltages be given by v_a, v_b, and v_c. We can then define the instantaneous symmetrical components as

$$\begin{bmatrix} v_{a0} \\ v_{a1} \\ v_{a2} \end{bmatrix} = \frac{1}{3}\begin{bmatrix} 1 & 1 & 1 \\ 1 & a & a^2 \\ 1 & a^2 & a \end{bmatrix}\begin{bmatrix} v_a \\ v_b \\ v_c \end{bmatrix} = \mathbf{T}\begin{bmatrix} v_a \\ v_b \\ v_c \end{bmatrix} \tag{2.8}$$

where v_{a0}, v_{a1}, and v_{a2} respectively are the instantaneous zero-, positive-, and negative-sequence components. It is interesting to note that zero sequence is a time-varying real number, while the positive and negative sequences are vectors.

Consider the following set of balanced instantaneous voltages

$$v_a = V_m \sin(\omega t), \quad v_b = V_m \sin(\omega t - 120°), \quad v_c = V_m \sin(\omega t + 120°)$$

Since $v_a + v_b + v_c = 0$, the zero-sequence vector for the balanced signals will be zero. The positive sequence vector is

$$v_{a1} = \frac{1}{3}\{v_a + av_b + a^2 v_c\} = \frac{1}{3}\left\{v_a - \frac{1}{2}(v_b + v_c) + j\frac{\sqrt{3}}{2}(v_b - v_c)\right\} \tag{2.9}$$

Again since $v_a + v_b + v_c = 0$, we have

$$v_a - \frac{1}{2}(v_b + v_c) = \frac{3}{2}v_a = \frac{3V_m}{2}\sin\omega t$$

$$\frac{\sqrt{3}}{2}(v_b - v_c) = \frac{\sqrt{3}}{2}V_m\{\sin(\omega t - 120°) - \sin(\omega t + 120°)\}$$

$$= -\frac{\sqrt{3}}{2}V_m\left(-2\cos\omega t \times \frac{\sqrt{3}}{2}\right) = -\frac{3V_m}{2}\cos\omega t$$

Substitution of the above two equations in (2.9) yields

$$v_{a1} = \frac{V_m}{2}(\sin\omega t - j\cos\omega t) \tag{2.10}$$

The negative sequence vector is

$$v_{a2} = \frac{V_m}{3}\{v_a + a^2 v_b + av_c\} = \frac{V_m}{3}\left\{v_a - \frac{1}{2}(v_b + v_c) + j\frac{\sqrt{3}}{2}(v_c - v_b)\right\}$$

This can be simplified as

$$v_{a2} = \frac{V_m}{2}(\sin \omega t + j \cos \omega t) \tag{2.11}$$

Comparing (2.10) with (2.11), it can be observed that v_{a2} is the complex conjugate of v_{a1} and both these vectors have a magnitude of $V_m/2$. Table 2.1 lists the values of these vectors for one fundamental cycle, where the frequency is chosen as 50 Hz (it can be 60 Hz as well). The rotation of these two vectors over one fundamental frequency is shown in Figure 2.1 for $V_m = 1$. The positive sequence vector starts from $0.5\angle{-90°}$ at time $t = 0$ and moves in the counterclockwise direction, while the negative sequence vector starts its rotation in the clockwise direction from $0.5\angle 90°$.

Table 2.1 Values of the sequence vectors over a fundamental cycle.

ωt (in deg)	v_{a1}	v_{a2}
0	$0.5V_m\angle{-90}^{\circ}$	$0.5V_m\angle 90^{\circ}$
90	$0.5V_m\angle 0^{\circ}$	$0.5V_m\angle 0^{\circ}$
180	$0.5V_m\angle 90^{\circ}$	$0.5V_m\angle{-90}^{\circ}$
270	$0.5V_m\angle 180^{\circ}$	$0.5V_m\angle 180^{\circ}$

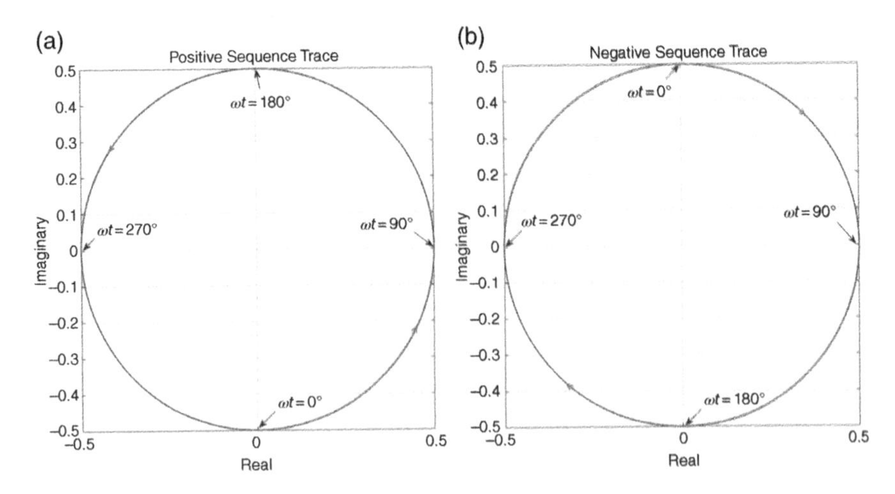

Figure 2.1 Trace of (a) a positive sequence and (b) a negative sequence over a cycle for a balanced system.

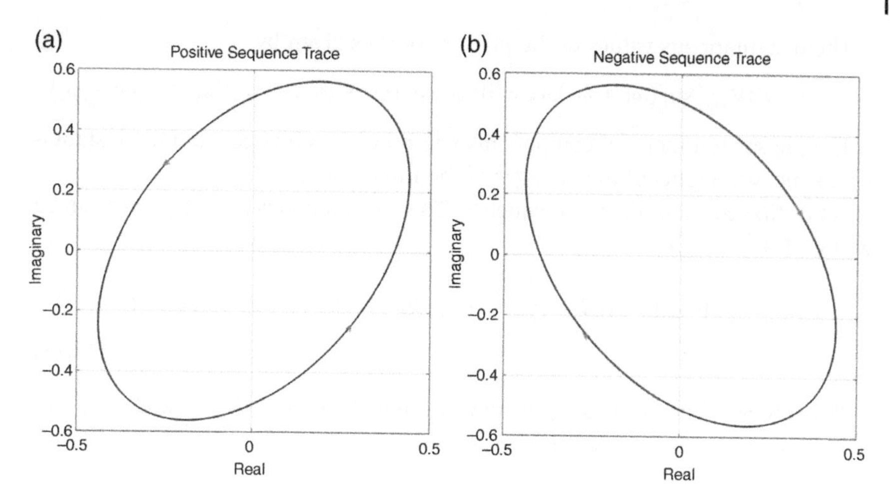

Figure 2.2 Trace of (a) a positive sequence and (b) a negative sequence over a cycle for an unbalanced system.

Example 2.3 Let us now consider what happens when the system is unbalanced. For this, the following instantaneous voltages are chosen

$$v_a = \sin(\omega t), \quad v_b = 0.5\sin(\omega t - 110^\circ), \quad v_c = 1.5\sin(\omega t + 100^\circ)$$

The traces of the two vectors are shown in Figure 2.2. They are not circles, but ellipses. However, these vectors are still complex conjugate of each other, and their directions of rotation are the same as those of the vectors in the balanced system. Obviously, the zero sequence will not be zero in this case since $v_a + v_b + v_c \neq 0$.

2.2.1 Estimating Symmetrical Components from Instantaneous Measurements

Let us define a set of unbalanced phasor voltages as

$$V_a = \frac{|V_{ma}|}{\sqrt{2}} e^{j\varphi_a}, V_b = \frac{|V_{mb}|}{\sqrt{2}} e^{j\varphi_b}, V_c = \frac{|V_{mc}|}{\sqrt{2}} e^{j\varphi_c}$$

The phasor symmetrical components of these quantities are then written from (2.2) as

$$\begin{bmatrix} V_{a0} \\ V_{a1} \\ V_{a2} \end{bmatrix} = \frac{1}{3\sqrt{2}} \begin{bmatrix} |V_{ma}|e^{j\varphi_a} + |V_{mb}|e^{j\varphi_b} + |V_{mc}|e^{j\varphi_c} \\ |V_{ma}|e^{j\varphi_a} + a|V_{mb}|e^{j\varphi_b} + a^2|V_{mc}|e^{j\varphi_c} \\ |V_{ma}|e^{j\varphi_a} + a^2|V_{mb}|e^{j\varphi_b} + a|V_{mc}|e^{j\varphi_c} \end{bmatrix} \tag{2.12}$$

The instantaneous values of the phasor voltages given by

$$v_a = |V_{ma}| \sin(\omega t + \varphi_a), v_b = |V_{mb}| \sin(\omega t + \varphi_b), v_c = |V_{mc}| \sin(\omega t + \varphi_c)$$

The phasor symmetrical components will now be estimated from the instantaneous measurements of the voltages of the three phases.

Let us first consider the zero sequence. The instantaneous zero sequence can be written from (2.8) as

$$v_{a0} = \frac{1}{3}\{|V_{ma}| \sin(\omega t + \varphi_a) + |V_{mb}| \sin(\omega t + \varphi_b) + |V_{mc}| \sin(\omega t + \varphi_c)\} \tag{2.13}$$

The zero-sequence voltage v_{a0} is now multiplied with $e^{-j(\omega t - 90^\circ)}$, and the product is averaged over a time period T_0 to form

$$x_{a0} = \frac{1}{T_0}\int_0^{T_0} v_{a0} e^{-j(\omega t - 90^\circ)} dt = \frac{1}{T_0}\int_0^{T_0} v_{a0}(\sin \omega t + j \cos \omega t) dt \tag{2.14}$$

Substituting (2.13) in the above equation, we get

$$\begin{aligned} x_{a0} = \frac{1}{3T_0}\int_0^{T_0} &\{|V_{ma}| \sin(\omega t + \varphi_a) + |V_{mb}| \sin(\omega t + \varphi_b) \\ &+ |V_{mc}| \sin(\omega t + \varphi_c)\}(\sin \omega t + j \cos \omega t) dt \end{aligned} \tag{2.15}$$

Now consider the following terms

$$|V_{ma}| \sin(\omega t + \varphi_a) \sin \omega t = \frac{|V_{ma}|}{2}\{\cos(\varphi_a) - \cos(2\omega t + \varphi_a)\}$$

$$j|V_{ma}| \sin(\omega t + \varphi_a) \cos \omega t = j\frac{|V_{ma}|}{2}\{\sin(\varphi_a) + \sin(2\omega t + \varphi_a)\}$$

If these two terms are integrated over half a fundamental cycle (i.e. $T_0 = 0.01$ seconds for fundamental frequency of 50 Hz), the double frequency components will be zero. Therefore, adding these two terms and integrating over half a fundamental cycle, the following expression is obtained

$$\begin{aligned} |V_{ma}| \int_0^{T_0} \sin(\omega t + \varphi_a)(\sin \omega t + j \cos \omega t) dt &= T_0 \frac{|V_{ma}|}{2}\{\cos(\varphi_a) + j \sin(\varphi_a)\} \\ &= T_0 \frac{|V_{ma}|}{2} e^{j\varphi_a} \end{aligned} \tag{2.16}$$

In a similar way, the following expressions can be written for the other two phases

$$|V_{mb}|\int_0^{T_0} \sin(\omega t + \varphi_b)(\sin\omega t + j\cos\omega t)dt = T_0\frac{|V_{mb}|}{2}e^{j\varphi_b}$$
$$|V_{mc}|\int_0^{T_0} \sin(\omega t + \varphi_c)(\sin\omega t + j\cos\omega t)dt = T_0\frac{|V_{mc}|}{2}e^{j\varphi_c} \tag{2.17}$$

Therefore substituting (2.16) and (2.17) in (2.15), we have

$$x_{a0} = \frac{1}{3}\left(|V_{ma}|e^{j\varphi_a} + |V_{mb}|e^{j\varphi_b} + |V_{mc}|e^{j\varphi_c}\right) \tag{2.18}$$

Comparing (2.18) with the first row of (2.12), we get

$$V_{a0} = \sqrt{2}x_{a0} \tag{2.19}$$

The instantaneous positive sequence is given from (2.8) as

$$v_{a1} = \frac{1}{3}\{|V_{ma}|\sin(\omega t + \varphi_a) + a|V_{mb}|\sin(\omega t + \varphi_b) + a^2|V_{mc}|\sin(\omega t + \varphi_c)\} \tag{2.20}$$

In the same fashion as (2.14), x_{a1} is defined as

$$x_{a1} = \frac{1}{T_0}\int_0^{T_0} v_{a1}e^{-j(\omega t - 90^\circ)}dt = \frac{1}{T_0}\int_0^{T_0} v_{a1}(\sin\omega t + j\cos\omega t)dt \tag{2.21}$$

Now since both a and a^2 are complex numbers and are independent of t, the same procedure as before is followed to write

$$x_{a1} = \frac{1}{3}\left(|V_{ma}|e^{j\varphi_a} + a|V_{mb}|e^{j\varphi_b} + a^2|V_{mc}|e^{j\varphi_c}\right) \tag{2.22}$$

Comparing (2.22) with the second row of (2.12), we get

$$V_{a1} = \sqrt{2}x_{a1} \tag{2.23}$$

For the negative sequence, x_{a2} is given by

$$x_{a2} = \frac{1}{T_0}\int_0^{T_0} v_{a2}e^{-j(\omega t - 90^\circ)}dt = \frac{1}{T_0}\int_0^{T_0} v_{a2}(\sin\omega t + j\cos\omega t)dt \tag{2.24}$$

This can then be expressed as

$$V_{a2} = \sqrt{2}x_{a2} \tag{2.25}$$

The numerical computations for the sequence component estimation are performed in a digital computer using a fixed sampling frequency. First, from the measurements of the instantaneous voltages (or currents), the instantaneous sequence components are computed at each sampling instant from (2.8). Then the integrals (2.14), (2.21), and (2.24) are computed. One easy way of computing them is to use a moving average filter (MAF). The time window for the MAF is chosen as half a cycle, i.e. $T_0 = 0.01$ seconds. This is then divided into N number of equally spaced samples. The integrals are then the average of the N samples. In an MAF, the samples are stored in an array. Once a new sample arrives, it is placed as the first element of the array, while the other elements are shifted down and the Nth element is discarded. Thus, at any given time, only the latest N elements are stored in the array. An MAF has a fast settling time of 0.01 seconds. Alternatively, a lowpass filter (LPF) can also be used, which will have a slower settling time.

Example 2.4 Let us consider the set of voltages of Example 2.1. The instantaneous voltages are

$$v_a = 100\sqrt{2}\sin(\omega t + 10°), \quad v_b = 110\sqrt{2}\sin(\omega t - 120°),$$
$$v_c = 90\sqrt{2}\sin(\omega t + 120°)$$

where the fundamental frequency is chosen as 50 Hz, i.e. $\omega = 100\pi$ rad/s. The phasor sequence components of these are computed in Example 2.1. Figure 2.3 shows the estimated magnitudes and phases of the sequence components. It can be seen that they are the same as those given in Example 2.1. These values are then used in inverse transformation to obtain the phasor values of the three phases. These are

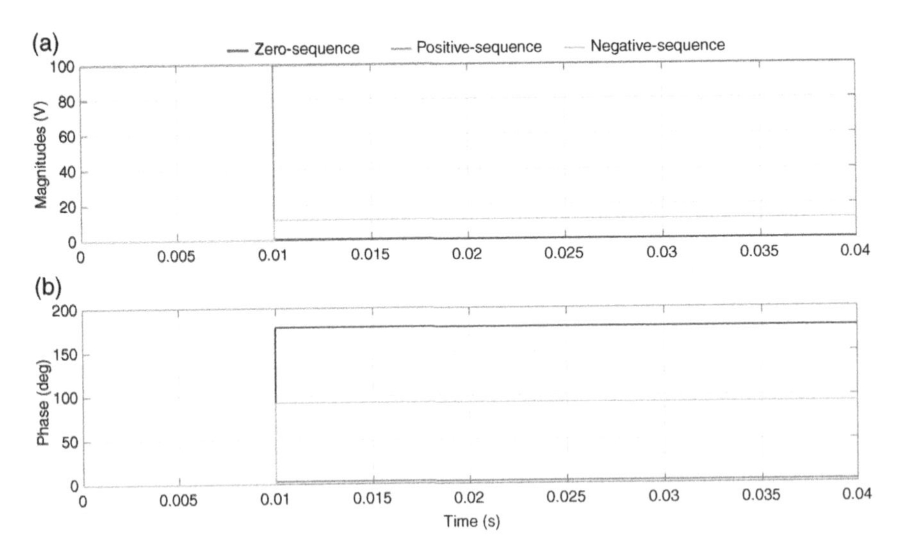

Figure 2.3 Estimated (a) magnitude and (b) phase of the phasor sequence components.

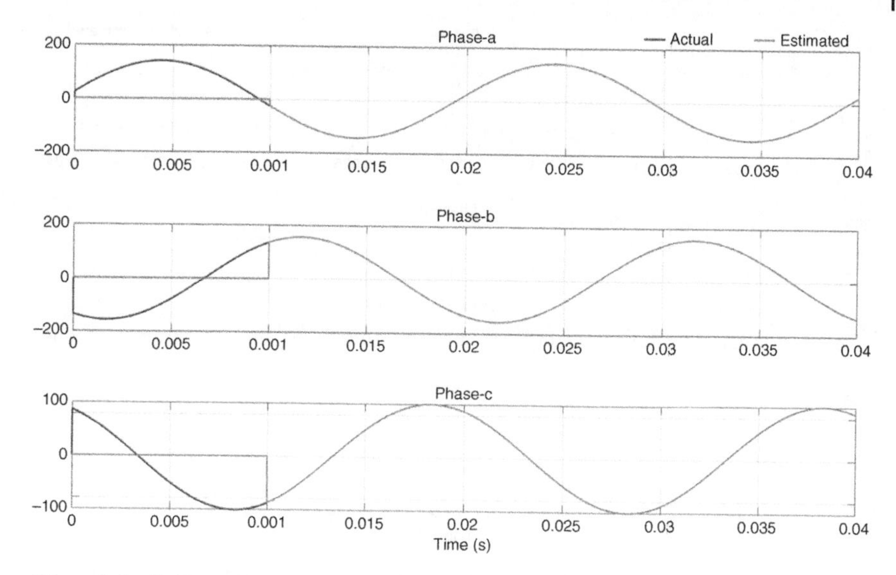

Figure 2.4 Estimated and actual waveforms for the three phases.

then used to recreate the instantaneous waveforms using inverse symmetrical component transform. Figure 2.4 shows the estimated and actual waveforms. The output of the MAF is zero till the buffer is full until half a cycle is complete. Thereafter, the estimated waveforms coincide with the actual waveforms.

Example 2.5 Let us consider the instantaneous voltages of Example 2.4. It has been assumed that these waveforms are corrupted by 3rd, 5th, 7th, 9th, and 11th harmonics with their magnitude being inversely proportional to their harmonic numbers. The symmetrical components are estimated with a time window (T_0) of a half a cycle, i.e. 10 ms. Figure 2.5 shows the instantaneous distorted waveforms and the corresponding fundamental waveforms that are recreated from the estimated sequence components. Comparing these with those in Figure 2.4, it is obvious that the fundamental waveforms are the same for both the cases.

Note that, in Example 2.5, only odd harmonics are considered. Since they have a quarter-wave symmetry, their average over half a cycle is zero. Therefore, T_0 can be chosen as 10 ms. However, if the signals contained both even and odd harmonics or just even harmonics, the MAF window will have to be a full cycle ($T_0 = 20$ ms) to eliminate both even and odd harmonics.

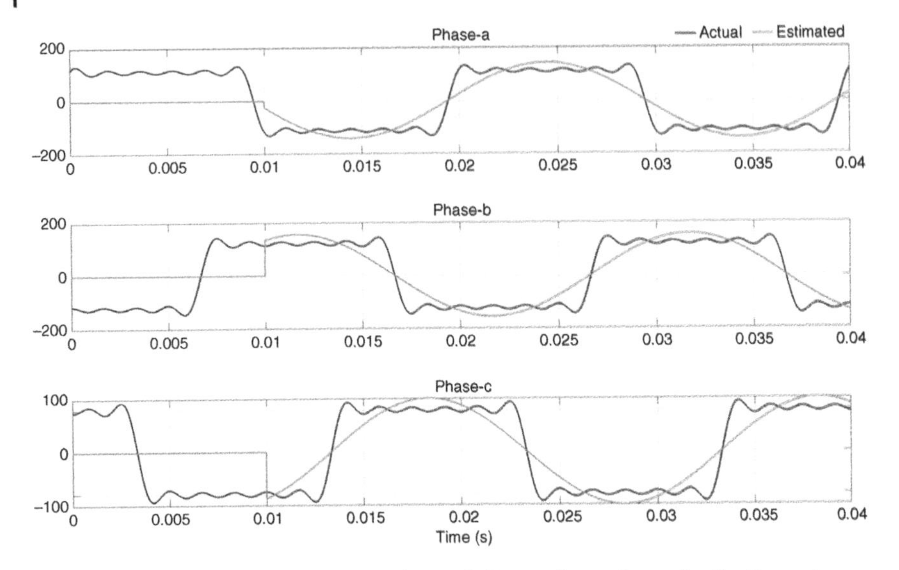

Figure 2.5 Estimated fundamental and actual distorted waveforms for the three phases.

2.2.2 Instantaneous Real and Reactive Power

Let us define a set of three-phase voltages and currents as

$$\mathbf{v}_{abc} = \begin{bmatrix} v_a \\ v_b \\ v_c \end{bmatrix}, \quad \mathbf{i}_{abc} = \begin{bmatrix} i_a \\ i_b \\ i_c \end{bmatrix}$$

Then the instantaneous real power is given by the dot product of these two vectors

$$p = \mathbf{v}_{abc} \cdot \mathbf{i}_{abc} = \mathbf{v}_{abc}^T \mathbf{i}_{abc} = v_a i_a + v_b i_b + v_c i_c \tag{2.26}$$

The instantaneous reactive power is defined as the cross product of these two vectors [5]

$$\mathbf{q}_{abc} = \mathbf{v}_{abc} \times \mathbf{i}_{abc} = \begin{bmatrix} q_a \\ q_b \\ q_c \end{bmatrix} = \begin{bmatrix} \begin{vmatrix} v_b & v_c \\ i_b & i_c \end{vmatrix} \\ \begin{vmatrix} v_c & v_a \\ i_c & i_a \end{vmatrix} \\ \begin{vmatrix} v_a & v_b \\ i_a & i_b \end{vmatrix} \end{bmatrix} \tag{2.27}$$

The instantaneous reactive power can then be given as the algebraic sum of q_a, q_b and, q_c as

$$q = -\frac{q_a + q_b + q_c}{\sqrt{3}} \tag{2.28}$$

The expression of (2.28) retains the polarity of the reactive power. It is also possible to express the real and reactive power in terms of the instantaneous symmetrical components [4]. These are not discussed here.

Let us suppose a voltage $V\angle 0°$ supplies a current $I\angle\delta$. Then the complex power is defined by

$$S = P + jQ = V(I\angle\delta)^* = VI\angle-\delta = VI(\cos\delta - j\sin\delta)$$

This implies that the reactive power is negative if the current leads the voltage ($\delta > 0$), while it is positive when the voltage leads the current ($\delta < 0$). Equation (2.28) preserves the sign notation.

Example 2.6 Let us consider a set of balanced voltages and currents, given by

$$v_a = V\sqrt{2}\sin(\omega t),\ \ v_b = V\sqrt{2}\sin(\omega t - 120°),\ \ V_c = V\sqrt{2}\sin(\omega t + 120°)$$
$$i_a = I\sqrt{2}\sin(\omega t - \phi),\ \ i_b = I\sqrt{2}\sin(\omega t - 120° - \phi),\ \ i_c = I\sqrt{2}\sin(\omega t + 120° - \phi)$$

Then, from (2.26), the instantaneous power is given by

$$\begin{aligned} p = 2VI[&\sin(\omega t)\sin(\omega t - \phi) + \sin(\omega t - 120°)\sin(\omega t - 120° - \phi) \\ &+ \sin(\omega t + 120°)\sin(\omega t + 120° - \phi)] \end{aligned}$$

Noting that

$$\sin A \sin B = \frac{\cos(A - B)}{2} - \frac{\cos(A + B)}{2}$$

we can write

$$\begin{aligned} p &= VI[3\cos(-\phi) - \cos(2\omega t - \phi) - \cos(2\omega t - 120° - \phi) - \cos(2\omega t + 120° - \phi)] \\ &= 3VI\cos\phi = P_{av} \end{aligned}$$

This implies that the instantaneous power is equal to the average power in a balanced three-phase circuit.

Now, q_a in (2.27) is expressed in terms of the voltages and currents as

$$\begin{aligned} q_a &= 2VI\begin{vmatrix} \sin(\omega t - 120°) & \sin(\omega t + 120°) \\ \sin(\omega t - 120° - \phi) & \sin(\omega t + 120° - \phi) \end{vmatrix} \\ &= 2VI[\sin(\omega t + 120° - \phi)\sin(\omega t - 120°) - \sin(\omega t - 120° - \phi)\sin(\omega t + 120°)] \end{aligned}$$

This can be simplified as

$$\begin{aligned} q_a &= VI[\cos(240° - \phi) - \cos(2\omega t - \phi) - \cos(-240° - \phi) + \cos(2\omega t - \phi)] \\ &= VI[\cos(240° - \phi) - \cos(240° + \phi)] = -\sqrt{3}VI\sin\phi \end{aligned}$$

Following the same procedure, we can compute the reactive power for the other two phases and can verify that $q_a = q_b = q_c$ for a balanced circuit. Then the reactive power is computed from (2.28) as

$$q = 3VI\sin\phi = Q_{av}$$

Therefore, the instantaneous reactive power is also the average reactive power.

To see what happens in the case of voltage unbalance, let us define

$$\begin{aligned} &V_a = 1\,\text{kV}, V_b = 1\angle -120°\,\text{kV}, V_c = 1\angle 120°\,\text{kV} \\ &I_a = 15\angle\theta\,\text{A}, I_b = 10\angle(-120° + \theta)\text{A}, I_c = 20\angle(120° + \theta)\,\text{A} \end{aligned}$$

where θ is chosen as $\pm 30°$. Then the average complex power $V_a I_a^* + V_b I_b^* + V_c I_c^*$ is given by

$$\begin{aligned} &(38.97 + j22.5) \times 10^3 \quad \text{for} \quad \theta = -30° \\ &(38.97 - j22.5) \times 10^3 \quad \text{for} \quad \theta = +30° \end{aligned}$$

This means p_{av} = 38.97 kW for both power factor angle and q_{av} = 22.5 kVAr when $\theta = -30°$ (lagging current) and $q_{av} = -22.5$ kVAr when $\theta = +30°$ (leading current). Figure 2.6 shows the instantaneous real and reactive power, where their averages are also shown. The instantaneous quantities oscillate at double the

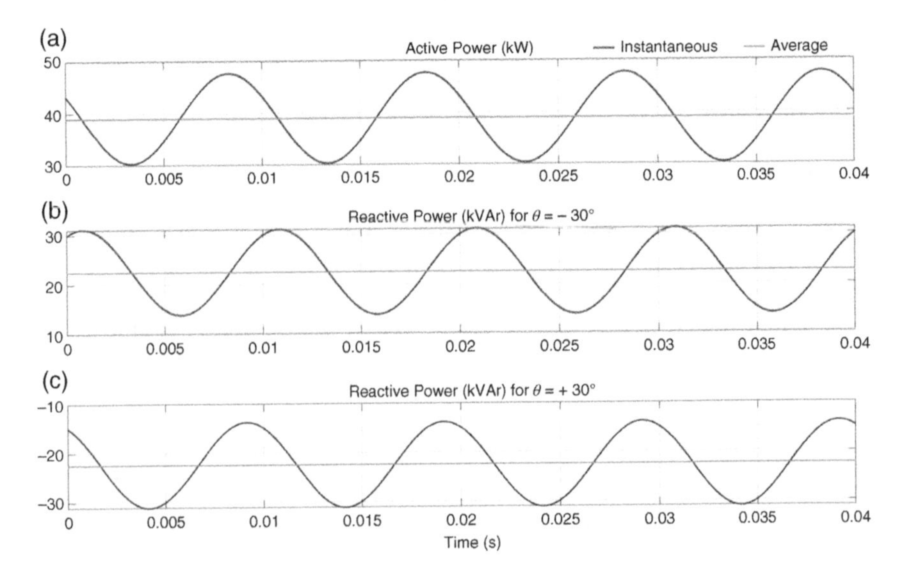

Figure 2.6 Instantaneous real and reactive power. (a) Active power, (b) reactive power for $\theta = -30°$ and (c) reactive power for $\theta = +30°$.

fundamental frequency (100 Hz) around the average values calculated above. In fact, these quantities can be written as

$$p = p_{av} + p_{osc}$$
$$q = q_{av} + q_{osc}$$

where the subscript "*osc*" denotes the double frequency oscillating components.

2.3 Harmonics

Harmonics cause distortion in voltages and currents. These are waveforms that are integer multiples of the fundamental frequency that are impressed on the fundamental frequency waveform. In addition, there may be interharmonics, which are noninteger multiples of the fundamental frequency. Interharmonics that have frequency components below the fundamental frequency are termed subharmonics. Our subsequent discussion is restricted to integer harmonics. Chapter 11 covers the effects of harmonics and their standards in detail.

The rise in the use of power electronic converters and the increasing use of power factor correction capacitors has caused a general rise in the level of harmonics. Harmonics in a power system can increase losses, reduce equipment life, interfere with protection and communication circuits, and reduce the lifetime of appliances. There are several other effects of harmonics that are discussed in [4].

Harmonics in a circuit is usually defined in terms of total harmonic distortion (THD). This is defined as

$$\text{THD} = \frac{\sqrt{\sum_{n=2}^{\infty} V_n^2}}{V_1} \tag{2.29}$$

where V_n is the magnitude of the nth harmonic component and V_1 is magnitude of the fundamental voltage. Based on IEEE or IEC standards, THD is measured up to the 40th or the 50th harmonics.

Example 2.7 Consider a distorted voltage waveform, given in per unit (pu) as

$$v = \sin(\omega t) + \frac{1}{3}\sin(3\omega t) + \frac{1}{5}\sin(5\omega t) + \frac{1}{7}\sin(7\omega t) + \frac{1}{9}\sin(9\omega t) + \frac{1}{11}\sin(11\omega t)$$

The voltage waveforms (distorted and fundamental) and the harmonic spectrum of the distorted waveform are shown in Figure 2.7. The THD of the waveform is given by

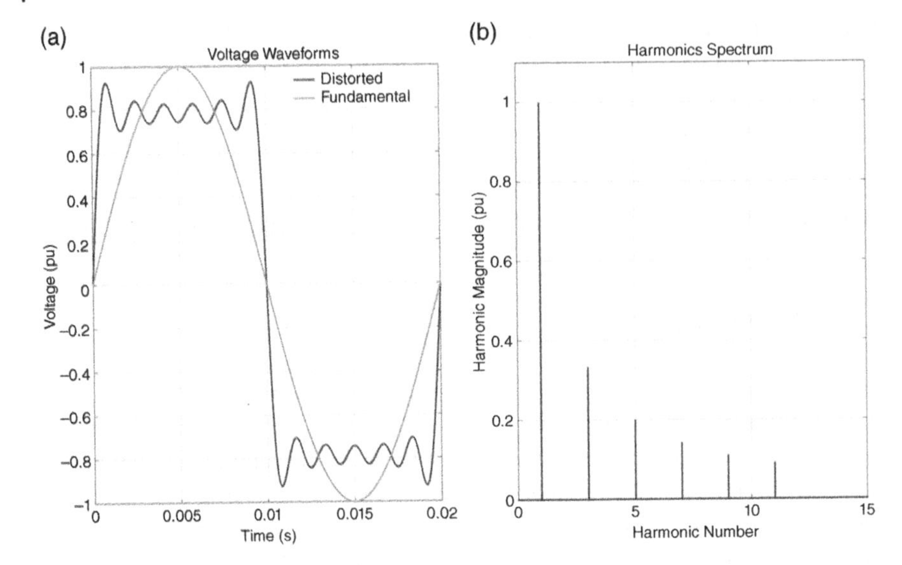

Figure 2.7 (a) Distorted voltage waveform and (b) its harmonic spectrum.

$$\text{THD} = \frac{\sqrt{\left(\frac{1}{3}\right)^2 + \left(\frac{1}{5}\right)^2 + \left(\frac{1}{7}\right)^2 + \left(\frac{1}{9}\right)^2 + \left(\frac{1}{11}\right)^2}}{1} = 0.4383$$

In other words, the THD is 43.83%.

The problem with the definition of THD given in (2.29) is that it becomes infinity if no fundamental waveform is present. To avoid this problem, an index called the distortion index (DIN) (or distortion factor) is used, which is defined as

$$\text{DIN} = \frac{\sqrt{\sum_{n=2}^{\infty} V_n^2}}{\sqrt{\sum_{n=1}^{\infty} V_n^2}} \tag{2.30}$$

Comparing (2.29) with (2.30), we can write

$$\begin{aligned} \text{DIN} &= \frac{THD}{\sqrt{1 + THD^2}} \\ \text{THD} &= \frac{DIN}{\sqrt{1 - DIN^2}} \end{aligned} \tag{2.31}$$

Continuing with Example 2.7, the DIN is computed as

$$\mathrm{DIN} = \frac{\sqrt{\left(\frac{1}{3}\right)^2 + \left(\frac{1}{5}\right)^2 + \left(\frac{1}{7}\right)^2 + \left(\frac{1}{9}\right)^2 + \left(\frac{1}{11}\right)^2}}{\sqrt{1 + \left(\frac{1}{3}\right)^2 + \left(\frac{1}{5}\right)^2 + \left(\frac{1}{7}\right)^2 + \left(\frac{1}{9}\right)^2 + \left(\frac{1}{11}\right)^2}} = 0.4015$$

It can be easily verified that that THD and DIN obey the relations of (2.31).

2.4 Clarke and Park Transforms

In this section, we discuss two transform methods that are used in power systems extensively: Clarke and Park transforms. Their interdependence is also discussed.

2.4.1 Clarke Transform

Let us consider a set of three-phase voltages, v_a, v_b, and v_c. These voltages can be converted from abc- to αβγ-frame by Clarke transform as

$$\begin{bmatrix} v_\alpha \\ v_\beta \\ v_\gamma \end{bmatrix} = \frac{2}{3} \begin{bmatrix} 1 & -\frac{1}{2} & -\frac{1}{2} \\ 0 & \frac{\sqrt{3}}{2} & -\frac{\sqrt{3}}{2} \\ \frac{1}{2} & \frac{1}{2} & \frac{1}{2} \end{bmatrix} \begin{bmatrix} v_a \\ v_b \\ v_c \end{bmatrix} \tag{2.32}$$

where v_γ is the zero-sequence component, which is zero for balanced voltages. Let us consider a set of balanced voltages, given by

$$v_a = V_m \sin(\omega t), \quad v_b = V_m \sin(\omega t - 120°), \quad v_c = V_m \sin(\omega t + 120°)$$

Substituting these in (2.32), we get

$$\begin{aligned} v_\alpha &= \frac{2V_m}{3}[\sin(\omega t) - 0.5\sin(\omega t - 120°) - 0.5\sin(\omega t + 120°)] \\ &= \frac{2V_m}{3}[\sin(\omega t) + 0.5\sin(\omega t)] = V_m \sin(\omega t) \end{aligned} \tag{2.33}$$

$$\begin{aligned} v_\beta &= \frac{V_m}{\sqrt{3}}[\sin(\omega t - 120°) - \sin(\omega t + 120°)] = \frac{V_m}{\sqrt{3}}[-2\cos(\omega t)\sin(120°)] \\ &= -V_m \cos(\omega t) = V_m \sin(\omega t - 90°) \end{aligned} \tag{2.34}$$

$$v_\gamma = \frac{V_m}{3}[\sin\omega t + \sin(\omega t - 120°) + \sin(\omega t + 120°)] = 0 \tag{2.35}$$

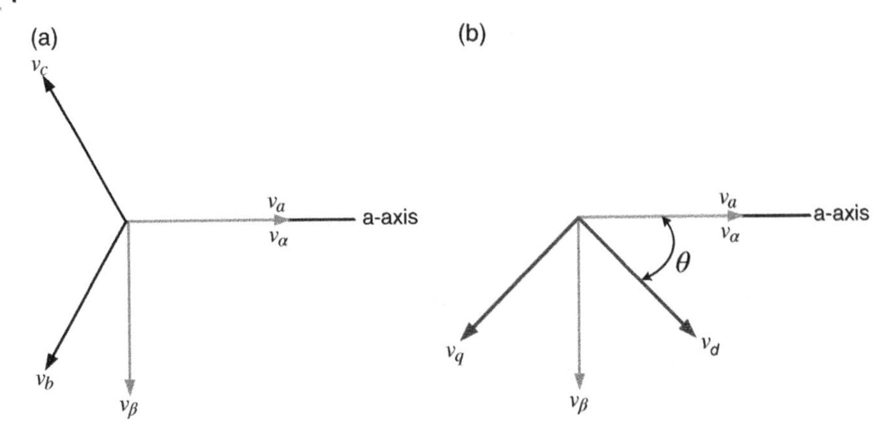

Figure 2.8 (a) Clarke and (b) Park transform of a balanced three-phase voltage.

The graphical representation of the α-β transform is shown in Figure 2.8a, in which the α-axis is aligned with the a-axis and the β-axis lags the a-axis by 90°. The inverse Clarke transform is given by

$$\begin{bmatrix} v_a \\ v_b \\ v_c \end{bmatrix} = \begin{bmatrix} 1 & 0 & 1 \\ -\frac{1}{2} & \frac{\sqrt{3}}{2} & 1 \\ -\frac{1}{2} & -\frac{\sqrt{3}}{2} & 1 \end{bmatrix} \begin{bmatrix} v_\alpha \\ v_\beta \\ v_\gamma \end{bmatrix} \tag{2.36}$$

Example 2.8 Consider the following set of voltages

$$v_a = V_a \sin(\omega t + \varphi_a), \quad v_b = V_b \sin(\omega t + \varphi_b), \quad v_c = V_c \sin(\omega t + \varphi_c)$$

The voltage magnitudes are chosen as $V_a = 100$ V, $V_b = 90$ V, and $V_c = 110$ V, while the angles are selected as $\varphi_a = 0°$, $\varphi_b = -120°$, and $\varphi_c = 120°$. The Clarke transformed voltages are shown in Figure 2.9a. It can be seen that the zero-sequence component is not zero here, and oscillates at 50 Hz with a maximum voltage of 5.77 V. From these voltages, the original waveforms are recovered using the inverse Clarke transform, which are shown in Figure 2.9b.

2.4.2 Park Transform

A set of voltages in the a-b-c plane can be converted into dq components using the transformation matrix

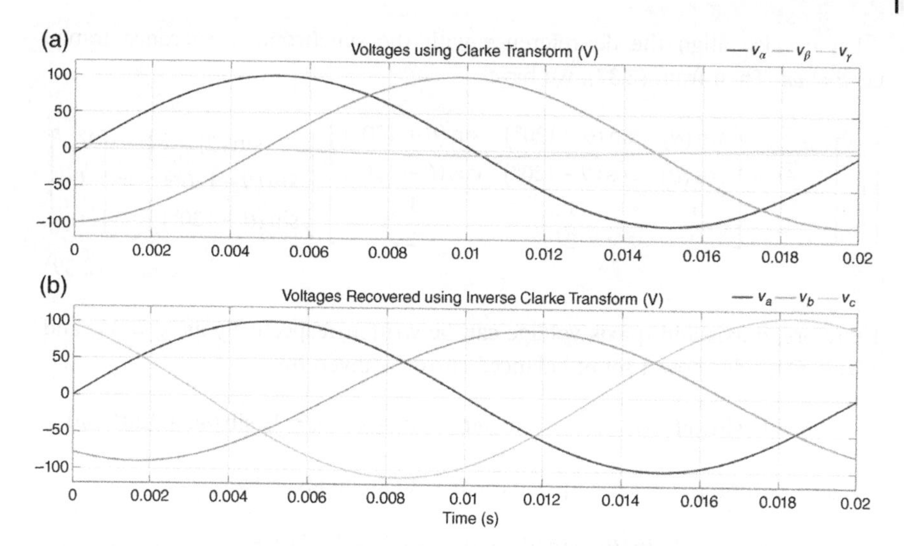

Figure 2.9 (a) Clarke and (b) inverse Clarke transform of unbalanced voltages.

$$\begin{bmatrix} v_d \\ v_q \\ v_0 \end{bmatrix} = \frac{2}{3}\begin{bmatrix} \sin(\theta) & \sin(\theta - 120°) & \sin(\theta + 120°) \\ \cos(\theta) & \cos(\theta - 120°) & \cos(\theta + 120°) \\ \frac{1}{2} & \frac{1}{2} & \frac{1}{2} \end{bmatrix}\begin{bmatrix} v_a \\ v_b \\ v_c \end{bmatrix} \tag{2.37}$$

$$\Rightarrow \mathbf{v}_{dq0} = \mathbf{T}\mathbf{v}_{abc}$$

where $\theta = \omega t + \delta$ is the angle between the rotating and fixed frames at time t and δ is the initial phase shift of the voltage. This is graphically shown in Figure 2.8b. The inverse transform is

$$\begin{bmatrix} v_a \\ v_b \\ v_c \end{bmatrix} = \begin{bmatrix} \sin(\theta) & \cos(\theta) & 1 \\ \sin(\theta - 120°) & \cos(\theta - 120°) & 1 \\ \sin(\theta + 120°) & \cos(\theta + 120°) & 1 \end{bmatrix}\begin{bmatrix} v_d \\ v_q \\ v_0 \end{bmatrix} \tag{2.38}$$

$$\Rightarrow \mathbf{v}_{abc} = \mathbf{T}^{-1}\mathbf{v}_{dq0}$$

2.4.3 Real and Reactive Power

Consider a set of balanced voltages, given by

$$v_a = V_m \sin(\omega t), \quad v_b = V_m \sin(\omega t - 120°), \quad v_c = V_m \sin(\omega t + 120°)$$

Let us also align the dq reference with the synchronous reference frame, i.e. $\theta = \omega t$. Then from (2.37), we have

$$\begin{bmatrix} v_d \\ v_q \\ v_0 \end{bmatrix} = \frac{2V_m}{3}\begin{bmatrix} \sin(\theta) & \sin(\theta-120^\circ) & \sin(\theta+120^\circ) \\ \cos(\theta) & \cos(\theta-120^\circ) & \cos(\theta+120^\circ) \\ \frac{1}{2} & \frac{1}{2} & \frac{1}{2} \end{bmatrix}\begin{bmatrix} \sin(\theta) \\ \sin(\theta-120^\circ) \\ \sin(\theta+120^\circ) \end{bmatrix} = \begin{bmatrix} V_m \\ 0 \\ 0 \end{bmatrix} \tag{2.39}$$

Therefore, d-axis and q-axis voltage can be written respectively as $v_d = V_m$ and $v_q = 0$. Consider now a set of balanced currents, given by

$$i_a = I_m \sin(\omega t - \varphi),\quad i_b = I_m \sin(\omega t - 120^\circ - \varphi),\quad i_c = I_m \sin(\omega t + 120^\circ - \varphi)$$

Again assuming $\theta = \omega t$, d–q–0 axis current can be given by

$$\begin{bmatrix} i_d \\ i_q \\ i_0 \end{bmatrix} = \frac{2I_m}{3}\begin{bmatrix} \sin(\theta) & \sin(\theta-120^\circ) & \sin(\theta+120^\circ) \\ \cos(\theta) & \cos(\theta-120^\circ) & \cos(\theta+120^\circ) \\ \frac{1}{2} & \frac{1}{2} & \frac{1}{2} \end{bmatrix}\begin{bmatrix} \sin(\theta-\varphi) \\ \sin(\theta-120^\circ-\varphi) \\ \sin(\theta+120^\circ-\varphi) \end{bmatrix} = \begin{bmatrix} I_m\cos(\varphi) \\ -I_m\sin(\varphi) \\ 0 \end{bmatrix} \tag{2.40}$$

We thus have $i_d = I_m \cos(\varphi)$ and $i_q = -I_m \sin(\varphi)$.

In the three-phase balanced circuit with the currents and voltages defined above, the rms voltage and current are $V_m/\sqrt{2}$ and $I_m/\sqrt{2}$ respectively. Then the real and reactive power are given respectively by

$$P = \frac{3}{2}V_m I_m \cos(\varphi)$$
$$Q = \frac{3}{2}V_m I_m \sin(\varphi)$$

With the d-axis- and q-axis voltages and currents obtained from (2.39) and (2.40), the real and reactive power can be written as

$$P = \frac{3}{2}v_d i_d$$
$$Q = -\frac{3}{2}v_d i_q \tag{2.41}$$

In general, however, when $\delta \neq 0$, we have

$$v_d = V_m \cos(\delta), \quad v_q = -V_m \sin(\delta)$$
$$i_d = I_m \cos(\delta - \varphi), \quad i_q = -I_m \sin(\delta - \varphi)$$

Then the real and reactive power are

$$\begin{aligned} P &= \frac{3}{2}\left(v_d i_d + v_q i_q\right) \\ Q &= -\frac{3}{2}\left(v_d i_q - v_q i_d\right) \end{aligned} \tag{2.42}$$

2.4.4 Analyzing a Three-phase Circuit

Consider the simple radial system shown in Figure 2.10. The differential equations governing the circuit are

$$\begin{aligned} \frac{di_a}{dt} &= -\frac{R}{L}i_a + \frac{1}{L}(v_{Sa} - v_{Ra}) \\ \frac{di_b}{dt} &= -\frac{R}{L}i_b + \frac{1}{L}(v_{Sb} - v_{Rb}) \\ \frac{di_c}{dt} &= -\frac{R}{L}i_c + \frac{1}{L}(v_{Sc} - v_{Rc}) \end{aligned} \tag{2.43}$$

Taking inverse dq0 transform of (2.38), the three equations given in (2.43) can be combined as

$$\begin{aligned} &\frac{d}{dt}\left(\mathbf{T}^{-1}\mathbf{i}_{dq0}\right) = -\frac{R}{L}\mathbf{T}^{-1}\mathbf{i}_{dq0} + \frac{1}{L}\mathbf{T}^{-1}\left(\mathbf{v}_{Sdq0} - \mathbf{v}_{Rdq0}\right) \\ &\Rightarrow \frac{d}{dt}\left(\mathbf{T}^{-1}\right)\mathbf{i}_{dq0} + \mathbf{T}^{-1}\frac{d}{dt}\left(\mathbf{i}_{dq0}\right) = -\frac{R}{L}\mathbf{T}^{-1}\mathbf{i}_{dq0} + \frac{1}{L}\mathbf{T}^{-1}\left(\mathbf{v}_{Sdq0} - \mathbf{v}_{Rdq0}\right) \\ &\Rightarrow \frac{d}{dt}\left(\mathbf{i}_{dq0}\right) = -\mathbf{T}\frac{d}{dt}\left(\mathbf{T}^{-1}\right)\mathbf{i}_{dq0} - \frac{R}{L}\mathbf{i}_{dq0} + \frac{1}{L}\left(\mathbf{v}_{Sdq0} - \mathbf{v}_{Rdq0}\right) \end{aligned} \tag{2.44}$$

Let us assume $\theta = \omega t$. Then, taking time derivative of $\mathbf{T}^{-1}$ in (2.38), we have

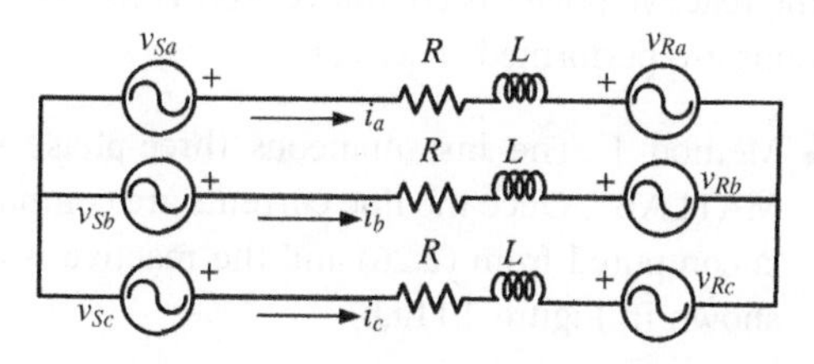

Figure 2.10 A simple radial system.

$$\dot{\mathbf{T}}^{-1} = \omega \begin{bmatrix} \cos(\omega t) & -\sin(\omega t) & 0 \\ \cos(\omega t - 120°) & -\sin(\omega t - 120°) & 0 \\ \cos(\omega t + 120°) & -\sin(\omega t + 120°) & 0 \end{bmatrix}$$

Therefore

$$\mathbf{T}\dot{\mathbf{T}}^{-1} = \frac{2\omega}{3} \begin{bmatrix} \sin(\omega t) & \sin(\omega t - 120°) & \sin(\omega t + 120°) \\ \cos(\omega t) & \cos(\omega t - 120°) & \cos(\omega t + 120°) \\ \frac{1}{2} & \frac{1}{2} & \frac{1}{2} \end{bmatrix} \begin{bmatrix} \cos(\omega t) & -\sin(\omega t) & 0 \\ \cos(\omega t - 120°) & -\sin(\omega t - 120°) & 0 \\ \cos(\omega t + 120°) & -\sin(\omega t + 120°) & 0 \end{bmatrix}$$

The solution of the above equation is

$$\mathbf{T}\dot{\mathbf{T}}^{-1} = \begin{bmatrix} 0 & -\omega & 0 \\ \omega & 0 & 0 \\ 0 & 0 & 0 \end{bmatrix} \tag{2.45}$$

Hence (2.44) is rewritten using (2.45) as

$$\frac{d}{dt}\left(\mathbf{i}_{dq0}\right) = \begin{bmatrix} -\frac{R}{L} & \omega & 0 \\ -\omega & -\frac{R}{L} & 0 \\ 0 & 0 & 0 \end{bmatrix} \mathbf{i}_{dq0} + \frac{1}{L}\left(\mathbf{v}_{Sdq0} - \mathbf{v}_{Rdq0}\right) \tag{2.46}$$

Example 2.9 Consider the circuit of Figure 2.10, where the system is assumed to be balanced. The parameters are

$$V_{Sa} = \frac{11}{\sqrt{3}}\angle 30°\ \text{kV}, V_{Ra} = \frac{11}{\sqrt{3}}\angle 0°\ \text{kV}, R = 2.42\ \Omega, L = 77\ \text{mH}$$

Using phasor analysis, the three-phase real power is computed as 2.54 MW and the reactive power is computed as 0.41 MVAr. Then two different sets of simulations are performed. These are:

- Method 1: The instantaneous three-phase system of (2.43) is simulated in MATLAB®. Once the line currents are computed, the instantaneous real power is computed from (2.26) and the reactive is computed using (2.28). These are shown in Figure 2.11a,b.

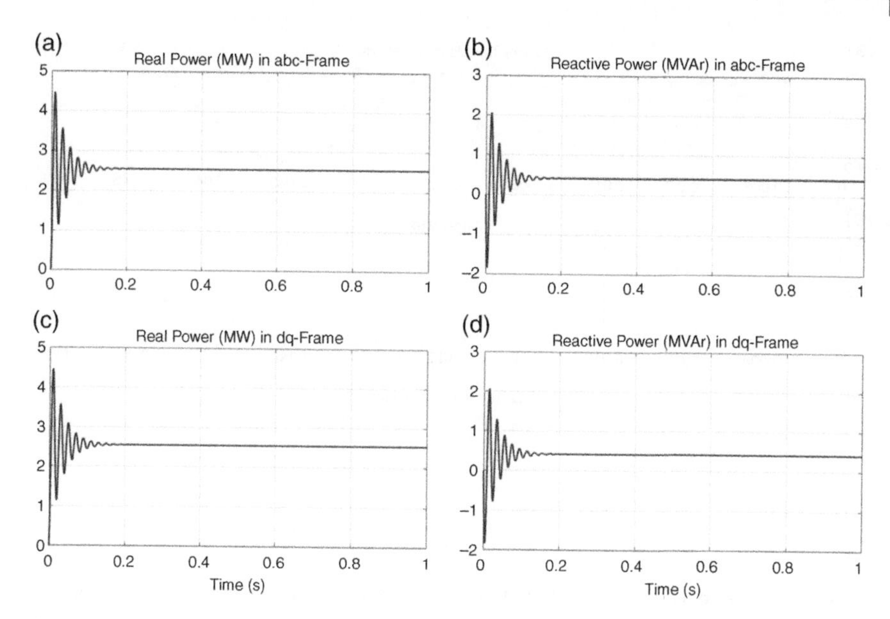

Figure 2.11 Simulation results of the circuit of Figure 2.10 using instantaneous and dq-domains. (a) Real power in abc-frame, (b) reactive power in abc-frame, (c) Real power in dq-frame, (d) reactive power in dq-frame.

- Method 2: The instantaneous sending and receiving end voltages are then transformed into dq-domain. The system is then simulated using (2.46). The instantaneous real and reactive power are then obtained using (2.41). The results are shown in Figure 2.11c,d. Comparing Figure 2.11a with Figure 2.11c, and Figure 2.11b with Figure 2.11d, it can be stated that both these methods yield the same results. Specifically, the steady state the real power is identical for both the methods. Similarly, the reactive power is also identical.

2.4.5 Relation Between Clarke and Park Transforms

To determine a relationship between Clarke and Park transforms, we define

$$\theta = \tan^{-1}\left(\frac{v_\beta}{v_\alpha}\right) \tag{2.47}$$

Then from Figure 2.8b, we can write

$$\begin{bmatrix} v_d \\ v_q \\ v_0 \end{bmatrix} = \begin{bmatrix} \cos\theta & \sin\theta & 0 \\ -\sin\theta & \cos\theta & 0 \\ 0 & 0 & 1 \end{bmatrix} \begin{bmatrix} v_\alpha \\ v_\beta \\ v_\gamma \end{bmatrix} \tag{2.48}$$

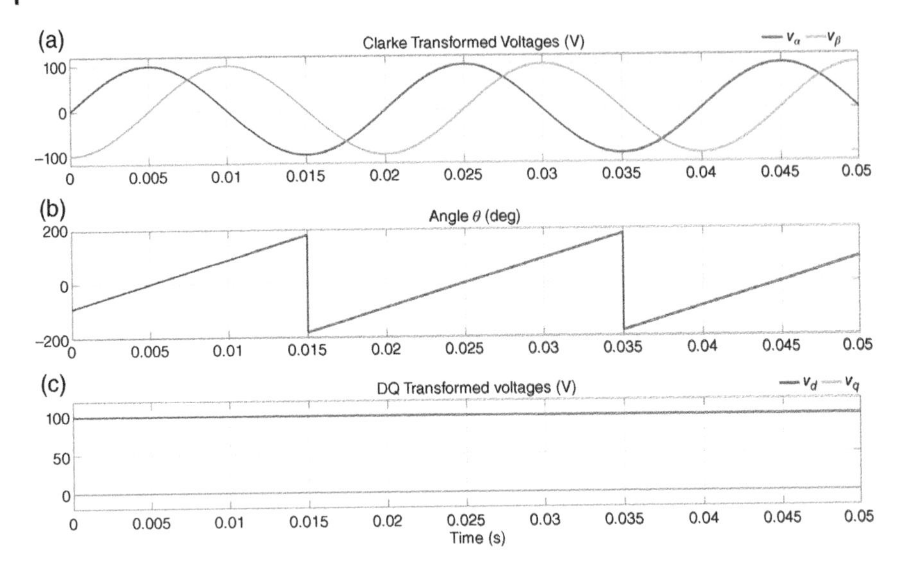

Figure 2.12 (a) αβ components, (b) the angle θ, and (c) dq components of a balanced system.

Example 2.10 Let us consider a set of balanced voltages with a peak voltage $V_m = 100$ V. The system frequency is assumed to be 50 Hz. The αβ components are plotted in Figure 2.12a. These components vary sinusoidally, where v_α is aligned with v_a. From these values, the angle θ is computed using the "atan2" function in MATLAB® (Version 2021a). This is shown in Figure 2.12b. Note that it varies between −180° and 180° in 20 ms. Alternatively, if the "atan" function is used θ will vary between −90° and 90° in 10 ms. This would have given an erroneous result. The dq components are shown in Figure 2.12c. It can be seen that v_d is a constant and equal to V_m, while v_q is 0, as expected. When the signals are sinewave, then the q-axis component is zero, whereas for cosine wave signals the d-axis component is zero. However, in general, both the d- and the q-axis components are DC. This is an advantage from the point of control system design.

2.5 Phase Locked Loop (PLL)

The basic aim of a phase locked loop (PLL) is to make a signal to track another. A typical schematic diagram of a PLL is shown in Figure 2.13. Here the aim is to keep the output signal x_0 synchronized with the input signal x_i both in terms of frequency and phase. The PLL, shown in Figure 2.13, contains a phase detector (PD), an LPF, and a voltage-controlled oscillator (VCO).

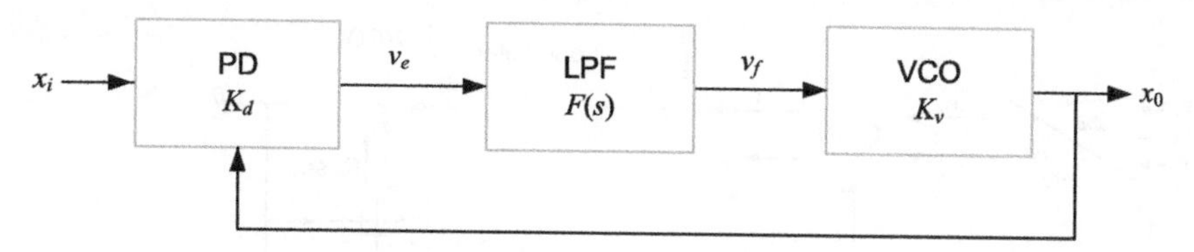

Figure 2.13 Schematic diagram of a PLL.

Let the input reference (x_i) and output (x_0) signals be given by

$$\begin{aligned} x_i(t) &= A \sin(\omega_i t + \delta_i) \\ x_0(t) &= B \sin(\omega_0 t + \varphi_0) \end{aligned} \tag{2.49}$$

These two signals have a phase difference. The accuracy of PLL is defined by its ability to detect the phase difference between two signals. The information about the error in the phase difference between the two signals is then used to control the frequency of the loop. If the two sinewaves of (2.49) are normalized to pu values and are plotted on the same graph, the following three different behaviors can be observed:

- If the zero-crossings of the two signals coincide all the time, then they have the same frequency and phase.
- If the zero-crossings occur at fixed time differences in every cycle, then the signals are not in phase, but their frequencies are the same.
- If the zero-crossings' time differences between the signals vary continuously, then the signals do not have the same frequency and/or phase.

The reference signal and the output of the VCO are fed to the phase detector. The error signal v_e from the PD passes through the LPF to remove any high-frequency elements from the error. The output of the LPF is fed to the VCO, which tries to reduce the phase difference between the signals and hence between the frequencies. Initially, the loop will be out of the lock, and the error voltage will pull the frequency of the VCO toward that of the reference, until it cannot reduce the error any further and the loop is locked. When the phase is locked, a steady state error voltage is produced. By using an amplifier between the PD and the VCO, the actual error between the signals can be reduced to very small levels. However, some voltage must always be present at the control terminal of the VCO as this is what produces the correct frequency.

2.5.1 Three-phase PLL System

The block diagram of the three-phase PLL system is shown in Figure 2.14, where the voltages v_a, v_b, and v_c are measured from a utility bus. The estimated angle $\hat{\theta}$

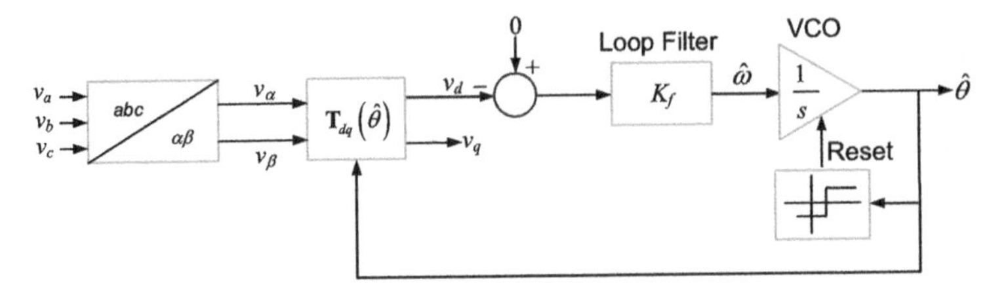

Figure 2.14 Block diagram of a three-phase PLL.

needs to be synchronized with the utility voltage angle θ. Let us assume that the input (utility) voltages are balanced and are given by

$$\mathbf{v}_{abc} = \begin{bmatrix} v_a \\ v_b \\ v_c \end{bmatrix} = V_m \begin{bmatrix} \sin(\theta) \\ \sin(\theta - 120°) \\ \sin(\theta + 120°) \end{bmatrix} \tag{2.50}$$

These voltages are then converted to an αβ-frame to obtain the vector $\mathbf{v}_{\alpha\beta} = [v_\alpha \; v_\beta]^T$ using (2.32). These are then converted into dq-axis quantities by the estimated angle $\hat{\theta}$ using [6]

$$\begin{bmatrix} v_d \\ v_q \end{bmatrix} = \begin{bmatrix} \cos\hat{\theta} & \sin\hat{\theta} \\ -\sin\hat{\theta} & \cos\hat{\theta} \end{bmatrix} \begin{bmatrix} v_\alpha \\ v_\beta \end{bmatrix} = \mathbf{T}_{dq}(\hat{\theta})\mathbf{v}_{\alpha\beta} \tag{2.51}$$

Combining (2.32) and (2.50) with (2.51) we have

$$\begin{aligned} \mathbf{v}_{dq} &= \frac{2V_m}{3}\mathbf{T}_{dq}(\hat{\theta}) \begin{bmatrix} 1 & -\frac{1}{2} & -\frac{1}{2} \\ 0 & \frac{\sqrt{3}}{2} & -\frac{\sqrt{3}}{2} \end{bmatrix} \begin{bmatrix} \sin(\theta) \\ \sin(\theta - 120°) \\ \sin(\theta + 120°) \end{bmatrix} \\ &= V_m \begin{bmatrix} \cos\hat{\theta} & \sin\hat{\theta} \\ -\sin\hat{\theta} & \cos\hat{\theta} \end{bmatrix} \begin{bmatrix} \sin(\theta) \\ -\cos(\theta) \end{bmatrix} \end{aligned}$$

Of particular interest is the d-axis voltage, which can be written from the above equation as

$$v_d = V_m\left[\cos\hat{\theta}\sin\theta - \sin\hat{\theta}\cos(\theta)\right] = V_m \sin\left(\theta - \hat{\theta}\right) = V_m \sin(\varphi) \tag{2.52}$$

where $\varphi = \theta - \hat{\theta}$.

The main aim is to force φ to zero such that the difference between $\hat{\theta}$ and θ asymptotically tends to zero. This can be achieved when v_d is equal to zero. Thus, it is compared with 0 and then the difference is passed through a loop filter with a transfer function of $K_f(s)$. The output of the loop filter is fed into the VCO, which is an integrator. The output of the VCO is the estimated angle $\hat{\theta}$. It is to be noted that

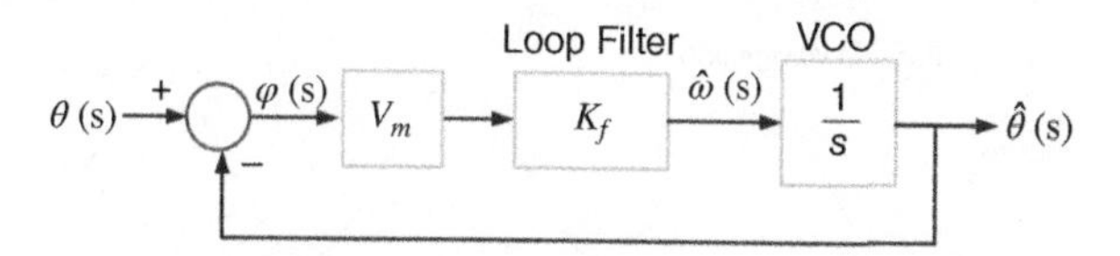

Figure 2.15 Linearized block diagram of the three-phase PLL.

this angle is to be restricted between 0 and 2π (i.e. between 0° and 360°). Thus, the integrator is reset every time it reaches 2π, as is indicated in Figure 2.14.

Assuming $\varphi = \theta - \hat{\theta} \approx 0$, (2.52) is linearized as

$$\Delta v_d = V_m \varphi \tag{2.53}$$

Also, from Figure 2.14, the estimated angular frequency is given by

$$\hat{\omega} = \frac{d\hat{\theta}}{dt} = K_f \Delta v_d \tag{2.54}$$

The linearized block diagram of the PLL system is shown in Figure 2.15. The loop filter is a proportional plus integral controller of the form

$$K_f(s) = \frac{sK_P + K_I}{s} \tag{2.55}$$

where K_P and K_I respectively are the proportional and integral gains. The closed-loop transfer function is then given by

$$\frac{\hat{\theta}(s)}{\theta(s)} = \frac{V_m(sK_P + K_I)}{s^2 + V_m K_P s + V_m K_I} \tag{2.56}$$

Example 2.11 Consider a set of balanced voltages, given by

$$v_a = V_m \sin(\omega t), \quad v_b = V_m \sin(\omega t - 120°), \quad v_c = V_m \sin(\omega t + 120°)$$

where $V_m = 9$ kV. The frequency at the beginning is 49 Hz and is changed to 51 Hz at 0.5 seconds. From the measurement of these instantaneous voltages, we need to produce current references that are given by

$$i_a = I_m \sin(\omega t + \phi), \quad i_b = I_m \sin(\omega t + \phi - 120°), \quad i_c = I_m \sin(\omega t + \phi + 120°)$$

where the power factor angle is chosen as $\phi = -90°$ and I_m is the magnitude of the current, which is taken as 3 kA. Note that the PLL output angle $\theta = \omega t$. Therefore, the currents for the three phases are obtained as

$$i_a = I_m \sin(\theta - 90°), \quad i_b = I_m \sin(\theta - 210°), \quad i_c = I_m \sin(\theta + 30°)$$

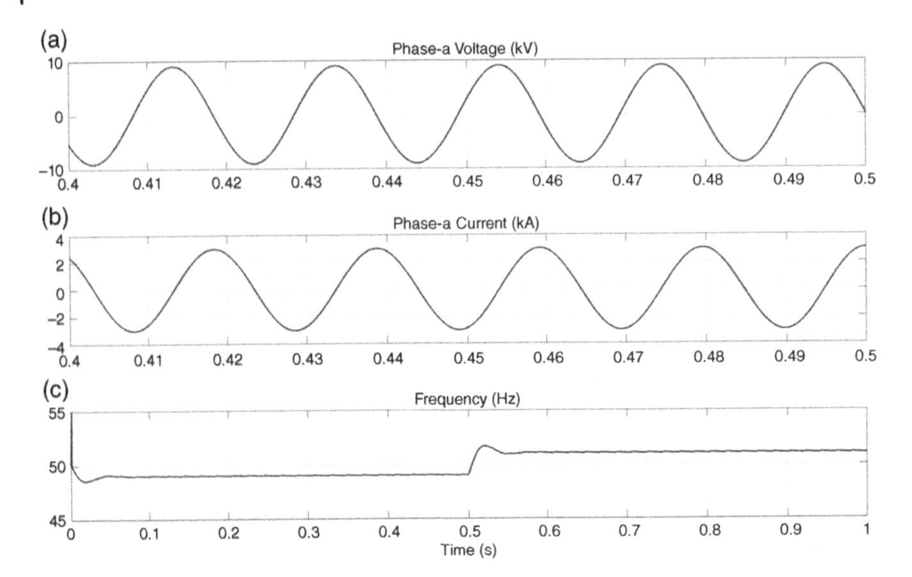

Figure 2.16 PLL performance: (a) reference voltage, (b) synthesized current, and (c) output frequency.

The proportional gain K_P is chosen as 8, while the integral gain K_I is 1000. The results are shown in Figure 2.16, where the phase-a voltage, current, and the estimated frequency can be seen. It is obvious that the current waveform lags the voltage waveform by 90°, while the frequency change is tracked accurately. It can also be seen that the output frequency reflects the change in the input frequency.

2.5.2 PLL for Unbalanced System

The three-phase PLL system discussed in Section 2.5.1 is valid for balanced systems. Consider now a set of unbalanced voltages, given by

$$\mathbf{v}_{abc} = \begin{bmatrix} v_a \\ v_b \\ v_c \end{bmatrix} = \begin{bmatrix} V_{ma} \sin(\omega t + \varphi_a) \\ V_{mb} \sin(\omega t + \varphi_b - 120°) \\ V_{mc} \sin(\omega t + \varphi_c + 120°) \end{bmatrix} \tag{2.57}$$

where $V_{ma} \neq V_{mb} \neq V_{mc}$, $\varphi_a \neq \varphi_b \neq \varphi_c$, and ω is an unknown frequency. Obviously, the derivations presented in Section 2.5.1 will not be valid. However, even if the signals are unbalanced, they have the same frequency. Therefore, we will synthesize a balanced three-phase waveform from the signal of one of the three phases.

From (2.57), phase-a voltage is given by

$$\hat{v}_a = V_{ma} \sin(\omega t + \varphi_a)$$

From this voltage, the voltages of the other two phases that have the same magnitude and are phase displaced by 120° from each other will be estimated. These balanced estimated voltages can then be used in the algorithm given in Section 2.5.1. The derivative of the voltage $\hat{v}_a$ is given as

$$\lambda = \frac{d\hat{v}_a}{dt} = \frac{V_{ma}}{\omega} \cos(\omega t + \varphi_a) \tag{2.58}$$

Then the estimates of the other two phases are

$$\begin{aligned} \hat{v}_b &= V_{ma} \sin(\omega t + \varphi_a - 120\text{o}) \\ &= V_{ma} \sin(\omega t + \varphi_a) \cos(120°) - V_{ma} \cos(\omega t + \varphi_a) \sin(120°) \\ &= -\frac{1}{2}\hat{v}_a - \frac{\sqrt{3}}{2}\omega\lambda \end{aligned} \tag{2.59}$$

$$\hat{v}_c = V_{ma} \sin(\omega t + \varphi_a + 120°) = -\frac{1}{2}\hat{v}_a + \frac{\sqrt{3}}{2}\omega\lambda \tag{2.60}$$

The block diagram of the PLL is shown in Figure 2.17. Note that the term ω in (2.58) can be chosen as the fundamental frequency. This will result in a negligible error in the computation of λ provided that system frequency does not have a large deviation from the fundamental frequency.

Example 2.12 Consider a set of unbalanced voltages, given by

$$v_a = 9 \sin(\omega t) \text{ kV}, \quad v_b = 8 \sin(\omega t - 120°) \text{ kV}, \quad v_c = 10 \sin(\omega t + 120°) \text{ kV}$$

The frequency at the beginning is 49 Hz and is changed to 51 Hz at 0.5 seconds. We must produce a set of balanced current references of peak magnitude of 3 kA, while the phase-a of the current is in phase with phase-a of the measured voltage. The results are shown in Figure 2.18, where the synthesized current is in phase with the phase-a voltage. The frequency, however, has more ripple than that

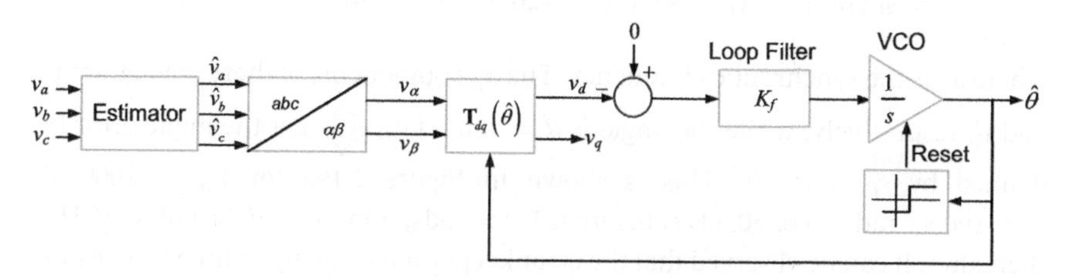

Figure 2.17 Block diagram of a PLL for unbalanced voltages.

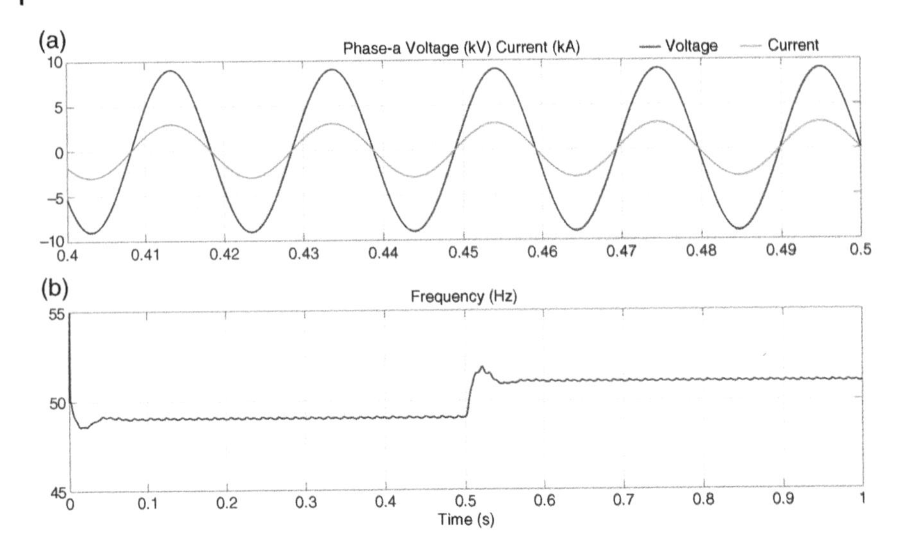

Figure 2.18 Performance of PLL for unbalanced signals: (a) reference voltage and synthesized current and (b) output frequency.

observed in Figure 2.16. The assumption that ω in (2.59) and (2.60) is the fundamental frequency is the cause of these ripples.

2.5.3 Frequency Estimation of Balanced Signal Using αβ Components

Consider a set of balanced supply voltages, given by

$$v_a = V_m \sin(\omega t), \quad v_b = V_m \sin(\omega t - 120°), \quad v_c = V_m \sin(\omega t + 120°)$$

where both the magnitude V_m and frequency ω are unknown. The αβ components of these voltage are denoted by v_α and v_β respectively, while their angle is given from (2.47) as $\theta = \tan^{-1}(v_\beta/v_\alpha)$.

Let us now define a set of balanced voltages with a known frequency as

$$v'_a = \sin(\omega_s t), \quad v'_b = \sin(\omega_s t - 120°), \quad v'_c = \sin(\omega_s t + 120°)$$

where ω_s is the synchronous frequency. The αβ components of these voltage are v'_α and v'_β respectively, while the angle is $\theta' = \tan^{-1}\left(v'_\beta/v'_\alpha\right)$. Let the angle error be defined by $\varepsilon_\theta = \theta - \theta'$. This is shown in Figure 2.19a for $V_m = 1000$ V, $\omega = 100.8\pi$ rad/s (i.e. 50.4 Hz) before 0.12 seconds, and $\omega_s = 100\pi$ rad/s (50 Hz) thereafter. It can be observed that the error keeps on increasing as time progresses. We now design a proportional plus integral controller of the form

$$\rho = K_P \varepsilon_\theta + K_I \int \varepsilon_\theta \, dt \tag{2.61}$$

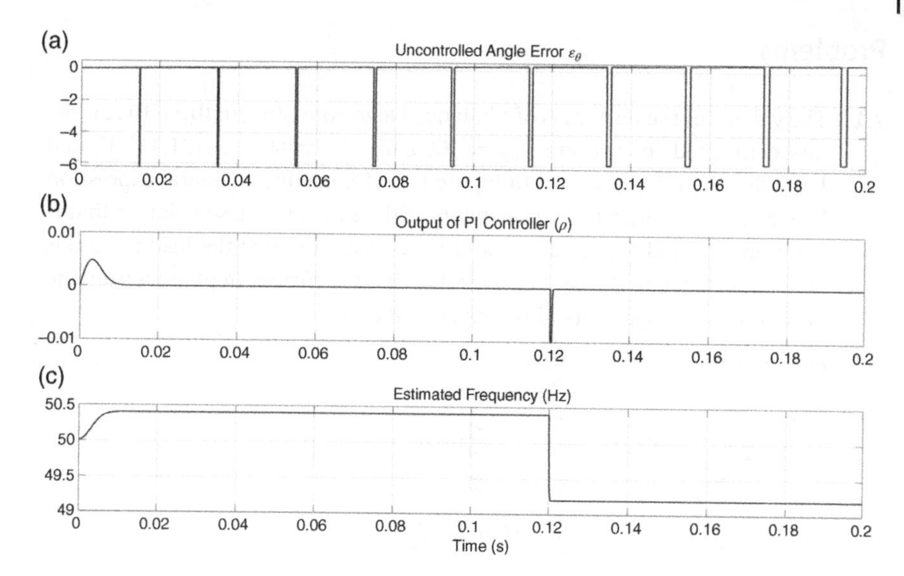

Figure 2.19 Frequency estimation in αβ components: (a) angle error, (b) PI controller output, and (c) estimated frequency.

The frequency estimate is then given by

$$\hat{\omega} = \omega_s + \rho \text{ rad/s} \tag{2.62}$$

The frequency of the supply voltage is changed from 50.4 to 49.2 Hz at 0.12 seconds. The controller gains are chosen as $K_P = K_I = 1$. Figure 2.19b shows the output ρ of the proportional plus integral (PI) controller. It can be seen to have a jump discontinuity when the frequency changes. Nevertheless, it has a fast settling time. The estimated frequency that is shown in Figure 2.19c also has a small settling time following the change in frequency. In this example it has been assumed that the supply voltage and the synthesized voltage have the same phase. The settling time of the frequency estimator can be considerably larger when this is not true. The frequency detection scheme under unbalanced conditions using Clarke's transform is discussed in [7].

2.6 Concluding Remarks

In this chapter, various aspects of sinusoidal signals are discussed. We discuss how system unbalance can be analyzed, how to estimate the symmetrical components, Clarke and Park transforms, and PLL. In general, DC/AC converters may have to work under unbalanced and/or distorted voltage conditions. Therefore, their working principle under different conditions must be systematically analyzed. The concepts discussed here can be used for the feedback control design of DC/AC converters.

Problems

2.1 The voltage at the terminals of a balanced load consisting of three 10 Ω resistors connected in wye are $V_{ab} = 100\angle 0°$, $V_{bc} = 80.8\angle{-121.44°}$ V, and $V_{ca} = 90\angle 130°$ V, as shown in Figure P2.1. Determine a general expression between the symmetrical components of the line and phase voltages that is between V_{ab1} and V_{an1} and V_{ab2} and V_{an2}. Assuming that the load neutral is not connected with the source neutral n, find the current in phase-a from the symmetrical components of the given voltages.

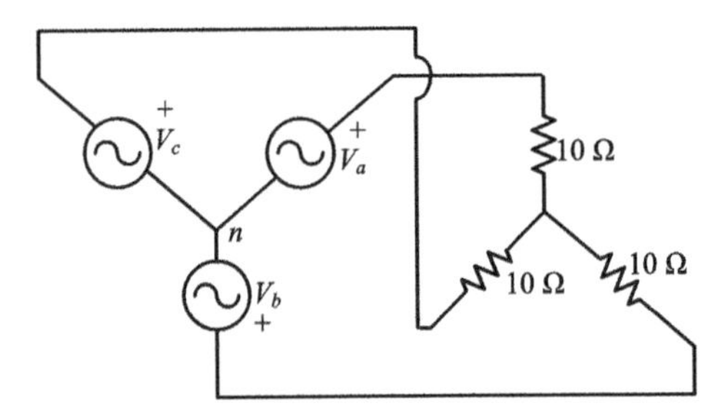

Figure P2.1 Circuit of Problem 2.1.

2.2 In the circuit shown in Figure P2.2, the open delta load is supplied by a balanced three-phase supply. The value of the load impedance is $Z = 18 + j10\ \Omega$. Assume that the line-to-line supply voltage is $V_{ab} = 400\angle 0°$ V.

(a) Find the currents I_{ab} and I_{bc}.
(b) Using the values calculated in (a), find I_{ab0}, I_{ab1}, and I_{ab2}.
(c) Using the values of I_{ab0}, I_{ab1}, and I_{ab2} obtained in (b), find the sequence components of the line current I_{a0}, I_{a1}, and I_{a2}.

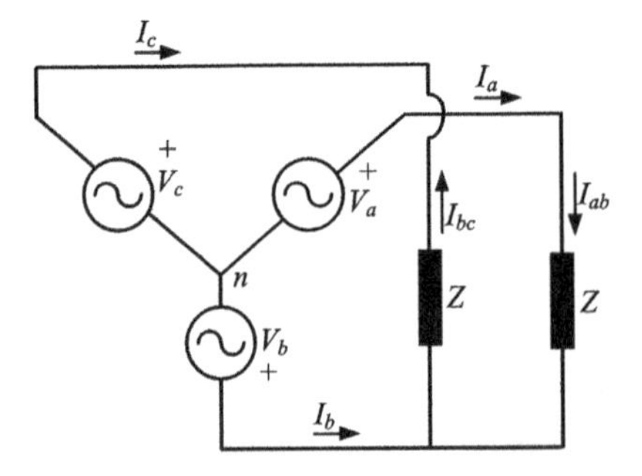

Figure P2.2 Open delta circuit of Problem 2.2.

2.3 Consider the wye-connected distribution system, the single line diagram of which is shown in Figure P2.3. In this, a balanced load voltage supplies an unbalanced load. The system parameters are:
Source voltage: $V_s = 400$ (L-L)
Feeder impedance: $Z_f = 0.08 + j0.04\ \Omega$
Unbalanced wye-connected load: $Z_{La} = 10 + j4.51\ \Omega$, $Z_{Lb} = 8.5 + j3.83\ \Omega$, and $Z_{Lc} = 2 + j9\ \Omega$
Determine the percentage VU of the load voltage V_L.

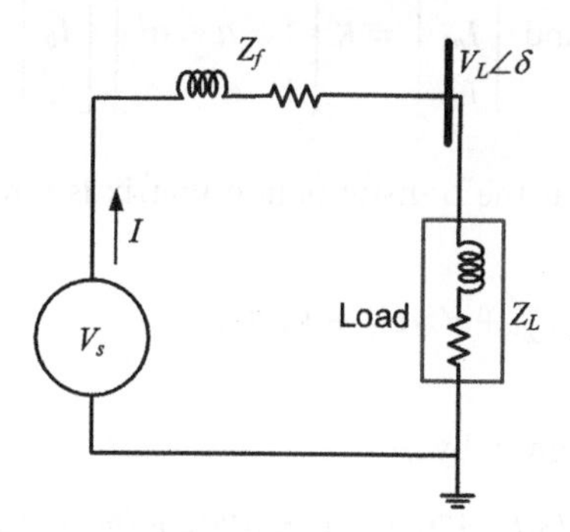

Figure P2.3 Distribution system supplying unbalanced load of Problem 2.3.

2.4 Consider the circuit shown in Figure P2.4, in which a harmonically distorted voltage source supplies a load. The system fundamental frequency is 50 Hz. The instantaneous source voltage is given by

$$V_S = 325.269\left[\sin(\omega t) + \frac{\sin(5\omega t)}{5} + \frac{\sin(7\omega t)}{7}\right] \text{V}$$

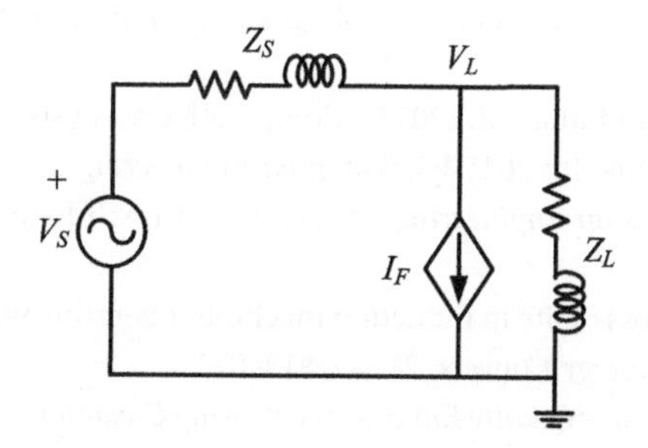

Figure P2.4 Distribution system for VU calculation of Problem 2.4.

The line and load impedances respectively are

$$Z_S \Rightarrow R_S = 2\ \Omega,\ \ L_S = 0.01\ \text{H},\quad Z_L \Rightarrow R_L = 10\ \Omega,\ \ L_L = 0.5\ \text{H}$$

The current I_F draws a fundamental frequency current of 1.5 A with phase angle of 0°. Find the THD of the load bus voltage.

2.5 The symmetrical component transform of (2.2) is rewritten as

$$\begin{bmatrix} V_{a0} \\ V_{a1} \\ V_{a2} \end{bmatrix} = K \begin{bmatrix} 1 & 1 & 1 \\ 1 & a & a^2 \\ 1 & a^2 & a \end{bmatrix} \begin{bmatrix} V_a \\ V_b \\ V_c \end{bmatrix} \text{ and } \begin{bmatrix} I_{a0} \\ I_{a1} \\ I_{a2} \end{bmatrix} = K \begin{bmatrix} 1 & 1 & 1 \\ 1 & a & a^2 \\ 1 & a^2 & a \end{bmatrix} \begin{bmatrix} I_a \\ I_b \\ I_c \end{bmatrix}$$

Determine the value of K such that the transformation matrix is power invariant, i.e.

$$S = V_a I_a^* + V_b I_b^* + V_c I_c^* = V_{a0} I_{a0}^* + V_{a1} I_{a1}^* + V_{a2} I_{a2}^*$$

2.6 Consider a set of balanced voltages, given by

$$v_a = 100\cos(\omega t),\quad v_b = 100\cos(\omega t - 120^\circ),\quad v_c = 100\cos(\omega t + 120^\circ)$$

Determine the d-axis and the q-axis component of the voltages.

Notes and References

Some of the signal analysis techniques are presented in [4]. PLL is a thoroughly researched area. There are several techniques that discuss PLL implementations [8–11].

1 Gnacinski, P. (2008). Windings temperature and loss of life of an induction machine under voltage unbalance combined with over- or undervoltages. *IEEE Trans. Energy Convers.* 23 (2): 363–371.

2 Rodriguez, A.D., Fuentes, F.M., and Matta, A.J. (2015). Comparative analysis between voltage unbalance definitions. In: *2015 Workshop on Engineering Applications – International Congress on Engineering (WEA)*, 1–7. https://doi.org/10.1109/WEA.2015.7370122.

3 Pillay, P. and Manyage, M. (2006). Loss of life in induction machines operating with unbalanced supplies. *IEEE Trans. Energy Convers.* 21 (4): 813–822.

4 Ghosh, A. and Ledwich, G. (2002). *Power Quality Enhancement Using Custom Power Devices*. New York: Springer Science+Business Media.

5 Peng, F.Z. and Lai, J.S. (1996). Generalized instantaneous reactive power theory for three-phase power systems. *IEEE Trans. Instrum. Meas.* 45 (1): 293–297.

6 Chung, S.K. (2000). A phase tracking system for three phase utility interface inverter. *IEEE Trans. Power Electron.* 15 (3): 431–438.

7 Canteli, M.M., Fernandez, A.O., Eguíluz, L.I., and Estébanez, C.R. (2006). Three-phase adaptive frequency measurement based on Clarke's transformation. *IEEE Trans. Power Delivery* 21 (3): 1101–1105.

8 Hsich, G.C. and Hung, J.C. (1996). Phase-locked loop techniques: a survey. *IEEE Trans. Ind. Electron.* 43 (6): 609–615.

9 Kaura, V. and Blasko, V. (1997). Operation of phase locked loop system under distorted utility conditions. *IEEE Trans. Ind. Appl.* 33 (1): 58–63.

10 Liccardo, F., Marino, P., and Raimondo, G. (2011). Robust and fast three-phase PLL tracking system. *IEEE Trans. Ind. Electron.* 58 (1): 221–231.

11 Thacker, T., Boroyevich, D., Burgos, R., and Wang, F. (2011). Phase-locked loop noise reduction via phase detector implementation for single-phase systems. *IEEE Trans. Ind. Electron.* 58 (6): 2482–2490.

3

Review of SISO Control Systems

Many industrial processes can be controlled through a single input, which can be used for controlling one specific output. Consider, for example a DC motor. The speed of the motor can be controlled by controlling the armature voltage. Even though there are other parameters in the motor circuit (e.g. mechanical torque), the input–output relationship between armature voltage and output speed can be written by a single linear ordinary differential equation (ODE). Systems that can be adequately described by a single ODE are usually referred to as single-input, single-output (SISO) systems in continuous-time domain. In this chapter, we discuss the control of SISO systems. For control analysis and design, a SISO system is represented by its transfer function, which is the ratio of the Laplace transform of the output over the Laplace transform of the input, assuming the initial conditions of the ODEs remain zero. Once a system is described by its transfer function, we can analyze the system and synthesize its control using several tools that have been formulated over the years. In the following, we present a review of these techniques – starting with system pole-zero description, followed by time response, Routh–Hurwitz's stability analysis, root locus, frequency response methods, the Nyquist stability criterion, and system gain and phase margins (PMs). Basic control actions such as PID (proportional plus integral plus derivative) and lag–lead compensators are also discussed.

A discrete-time control or digital control system is the basic backbone of industrial control as most controllers are realized using microprocessors these days. In this chapter, we introduce the basic concepts of digital control. A discrete-time system is represented by a z-transform and difference equation. It is to be noted that the control analysis techniques for both continuous time and discrete time are essentially the same. We therefore do not discuss the digital control in detail here; control design using the difference equations is covered in detail in Chapter 6.

Control of Power Electronic Converters with Microgrid Applications, First Edition.
Arindam Ghosh and Firuz Zare.

Published 2023 by John Wiley & Sons, Inc.

3.1 Transfer Function and Time Response

Consider a linear time system that is represented by the block diagram of Figure 3.1, where Y_r is the reference input and Y is the output. The open-loop transfer function of the system is $G(s)H(s)$. The transfer function of the closed-loop system is given by

$$\frac{Y(s)}{Y_r(s)} = \frac{G(s)}{1 + G(s)H(s)} \tag{3.1}$$

Note that, if $H(s) = 1$, then the system will be called a unity feedback system.

The characteristic equation of the system is given by

$$1 + G(s)H(s) = 0 \tag{3.2}$$

It is essentially a polynomial in s. The stability of a system is governed by the roots of the characteristic equation, which are also called the poles of the system. A system is stable if all the poles are on the left half of the s-plane. If any of the roots have a positive real part (i.e. any pole is on the open right-half s-plane), the closed-loop system will be unstable. The system can be marginally stable if simple poles are on the $j\omega$-axis, except at the origin. However, the system will be unstable if there are multiple poles on this axis.

3.1.1 Steady State Error and DC Gain

Let the open-loop transfer function of the system be defined by

$$G(s)H(s) = \frac{P(s)}{s^N Q(s)}$$

where $P(s)$ and $Q(s)$ are polynomials in s and the system has N poles at the origin. The system order is then equal to the total number of roots of the polynomial $Q(s)$ plus N, while the system type is defined by N. For example, the system will be called Type-0 if $N = 0$ or Type-1 if $N = 1$, and so on.

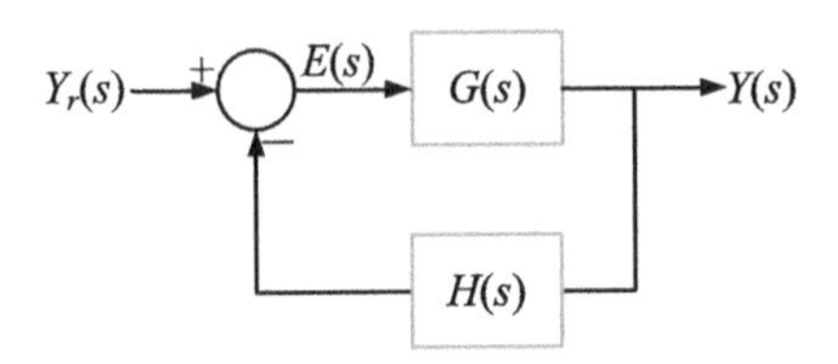

Figure 3.1 Block diagram of a closed-loop system.

From Figure 3.1, the error equation is given by

$$E(s) = Y_r(s) - H(s)Y(s) = Y_r(s) - G(s)H(s)E(s)$$
$$= \frac{Y_r(s)}{1 + G(s)H(s)} = Y_r(s)\frac{s^N Q(s)}{P(s) + s^N Q(s)} \tag{3.3}$$

The steady error is obtained from (3.3), where a system with a fixed reference input will have a fixed error as time t tends to infinity, provided that the system is stable. A system cannot have any steady state error if any of the poles are either on the right-half s-plane or on the imaginary axis, except for the poles located at the origin. For a stable system, the steady state error can be determined using the final value theorem of Laplace transform as

$$e_{ss} = \lim_{s\to 0} s \times \frac{s^N Y_r(s)Q(s)}{P(s) + s^N Q(s)} = \lim_{s\to 0} \left[s^{N+1} Y_r(s)\right] \times \frac{Q(0)}{P(0) + \left[\lim_{s\to 0} s^N\right] \times Q(0)} \tag{3.4}$$

Usually, the reference inputs are step, ramp, and parabolic. The steady state error depends on the types of input and the system types, as listed in Table 3.1.

The DC gain (K_{dc}) of a system is defined as the steady state output of the system, when the input $Y_r(s)$ in (3.1) is a unit step, i.e.

$$y_{ss} = \lim_{t\to\infty} y(t) = \lim_{s\to 0} s\frac{1}{s}\frac{G(s)}{1 + G(s)H(s)} = \frac{G(0)}{1 + G(0)H(0)} = K_{dc} \tag{3.5}$$

In a unity feedback system, the steady state error is zero when the DC gain is 1.

Table 3.1 The steady state error depending on the system type for different inputs.

	Input $Y_r(s)$ type		
System type	**Step (1/s)**	**Ramp ($1/s^2$)**	**Parabolic ($1/s^3$)**
Type-0 ($N = 0$)	Finite	Infinite	Infinite
Type-1 ($N = 1$)	Zero	Finite	Infinite
Type-2 ($N = 2$)	Zero	Zero	Finite
Type-3 or above ($N \geq 3$)	Zero	Zero	Zero

3.1.2 System Damping and Stability

Let us consider a second-order system that is traditionally written in the form of

$$\frac{Y(s)}{Y_r(s)} = \frac{\omega_n^2}{s^2 + 2\xi\omega_n s + \omega_n^2} \tag{3.6}$$

where ω_n is the undamped natural frequency of the system and ξ is the damping ratio. Note that the system will have a DC gain of 1 when $\xi > 0$, i.e. the poles are on the left half of the s-plane. In general, the system response will be governed by the damping ratio ξ. The roots of the characteristic equation of the system are given by

$$s_{1,2} = -\xi\omega_n \pm j\omega_n\sqrt{1-\xi^2} \tag{3.7}$$

In (3.7), the roots will have positive real parts when ξ is negative. This means a negatively damped system is unstable. For non-negative values of the damping ratio ξ, the system can be classified into the following four different categories:

- Undamped (or critically damped): when $\xi = 0$; the roots are on the imaginary axis of the s-plane at $\pm j\omega_n$. Since there is no damping in the system, a sustained oscillation occurs at the frequency of ω_n, and therefore, it is called the undamped natural frequency.
- Underdamped: when $0 < \xi < 1$: the roots are at the locations given by (3.7). The unit step response of the system is then given by

$$y(t) = 1 - \frac{e^{-\xi\omega_n t}}{\sqrt{1-\xi^2}} \sin\left(\omega_d t + \tan^{-1}\frac{\sqrt{1-\xi^2}}{\xi}\right) \tag{3.8}$$

where ω_d is the damped natural frequency of the system, given by

$$\omega_d = \omega_n\sqrt{1-\xi^2}$$

From (3.8), it can be surmised that the system oscillates with a frequency of ω_d and the transient dies out faster as the damping ratio increases.

- Critically damped: when $\xi = 1$; both the roots are at the same location on the real axis at $-\omega_n$. This means that the system will exponentially reach the steady state of 1 for a unit step input.
- Overdamped: when $\xi > 1$; it can be seen from (3.7) that the system will have two positive roots, both on the real axis. The system response will be sluggish.

Example 3.1 Consider a system of the form (3.6) with $\omega_n = 9$ rad/s. The input is assumed to be a unit step. The system response is shown in Figure 3.2 for different values of ξ. It is evident from Figure 3.2b that the system damping increases as the damping ratio increases. Moreover, the system takes longer to attain the steady state for the overdamped case.

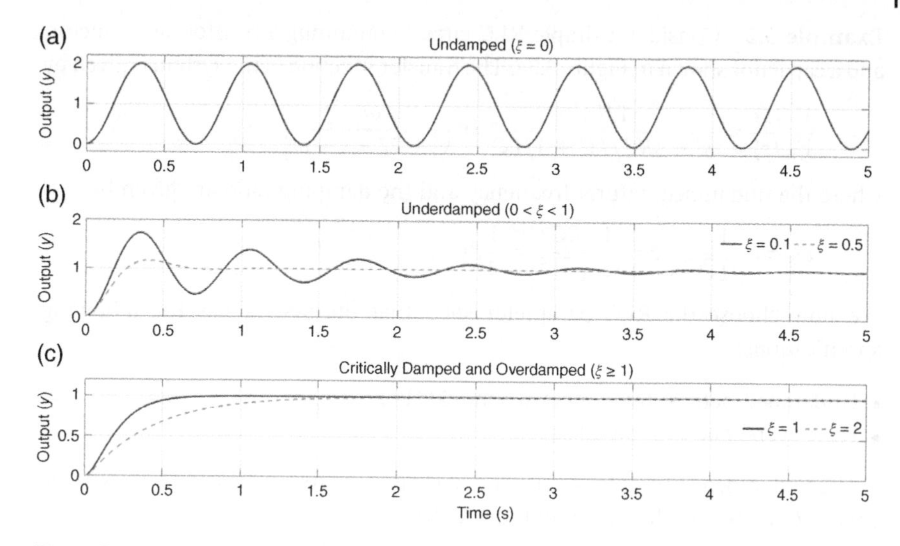

Figure 3.2 Step response of a second-order system: (a) undamped, (b) underdamped, and (c) critically damped and overdamped.

3.1.3 Shaping a Second-order Response

For a second-order system, the following terms are defined to quantify the system response:

- Rise time: the time required for the system response to rise from 10 to 90% of its final value.
- Settling time: the time required for the system response to attain the steady state. Usually, the steady state here means that the response reaches and stays within 2% of its final value.
- Maximum (or peak) overshoot (M_P): the maximum value the response curve reaches when measured from unity. This is usually defined in terms of percentage by the formula

$$M_P = e^{-\left(\xi/\sqrt{1-\xi^2}\right)\pi} \times 100\% \tag{3.9}$$

- Peak time (t_P): the time required for the response to reach its first peak (which is usually the maximum overshoot). It is given by

$$t_P = \frac{\pi}{\omega_d} \tag{3.10}$$

Example 3.2 Consider a simple RLC circuit containing a resistor, an inductor, and a capacitor shown in Figure 3.3a. The transfer function of the system is given by

$$\frac{V_C(s)}{V_{in}(s)} = \frac{1/LC}{s^2 + sRC/LC + 1/LC} = \frac{\omega_n^2}{s^2 + 2\xi\omega_n s + \omega_n^2}$$

where the undamped natural frequency and the damping ratio are given by

$$\omega_n = \frac{1}{\sqrt{LC}}, \quad \xi = \frac{1}{2}\frac{RC}{\sqrt{LC}} = \frac{1}{2}RC\omega_n$$

We now choose the RLC parameter such that the system has the following specifications:

- Undamped natural frequency: around 50 Hz
- Peak overshoot: around 20%.

The undamped natural frequency for 50 Hz is $\omega_n = 100\pi = 314.1593$ rad/s. If we choose $C = 500\ \mu F$, then L should ideally be

$$L = \frac{1}{C\omega_n^2} = 20.26 \text{ mH}$$

Instead, we choose $L = 20$ mH. This will slightly change ω_n to 316.23 rad/s or 50.33 Hz.

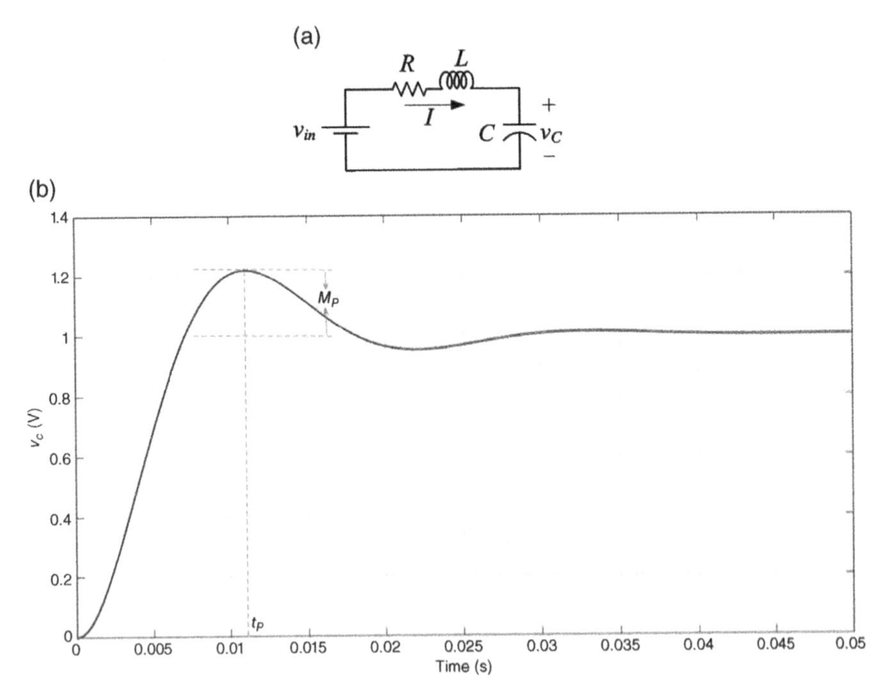

Figure 3.3 (a) An RLC circuit and (b) its step response.

From (3.9), the peak overshoot is given as

$$0.2 = e^{-\left(\xi/\sqrt{1-\xi^2}\right)\pi} \Rightarrow -1.6094 = -\frac{\xi}{\sqrt{1-\xi^2}}\pi$$

Solving the above equation, we get $\xi = 0.456$. Therefore

$$R = \frac{2\xi}{C\omega_n} = 5.77\,\Omega$$

Instead, we choose $R = 5.5\ \Omega$ to have the following transfer function

$$\frac{V_C(s)}{V_{in}(s)} = \frac{10^5}{s^2 + 275s + 10^5}$$

This gives us the following:

- Damping ratio (ξ) = 0.435
- Peak overshoot (M_P) = 21.94%
- Damped natural frequency (ω_d) = 284.77 rad/s
- Peak time (t_P) = 0.011 seconds.

The unit step response of the system is shown in Figure 3.3b. The peak time and peak overshoot are the same as those calculated above.

3.1.4 Step Response of First- and Higher-order Systems

Consider the following first-order system

$$\frac{Y(s)}{Y_r(s)} = \frac{1}{s\tau + 1} \tag{3.11}$$

where τ is called the time constant of the system. When the reference input Y_r is a unit step, (3.11) is rewritten as

$$Y(s) = \frac{1}{s}\frac{1}{s\tau + 1} = \frac{1}{s} - \frac{1}{s + 1/\tau} \tag{3.12}$$

Taking the inverse Laplace transform of (3.12), we get

$$y(t) = 1 - e^{-t/\tau} \tag{3.13}$$

Note that when $t = \tau$, $y(t)$ is equal to 0.6321. Therefore, the time constant (τ) of the system is defined as the time required by the system to reach 63.21% of its final value. The smaller the time constant, the faster the time response. Figure 3.4 shows the step response of the system of (3.11) with two different values of the time constant. It can be seen that the system reaches the steady state faster when $\tau = 0.5$ s.

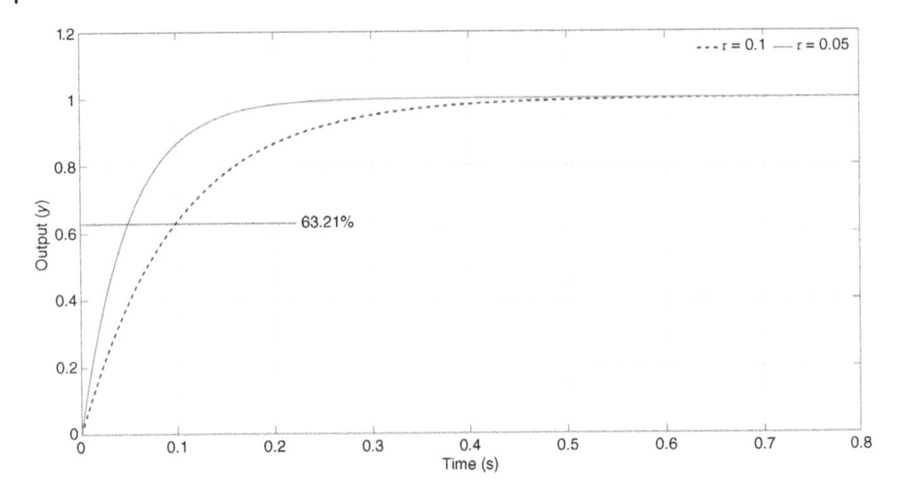

Figure 3.4 Step response of a first-order system with two different values of time constants.

A third-order system can have three real roots or a real root and a pair of complex conjugate roots, e.g.

$$\frac{Y(s)}{Y_r(s)} = \frac{1}{s\tau + 1} \frac{\omega_n^2}{s^2 + 2\xi\omega_n s + \omega_n^2} \tag{3.14}$$

The transient response of such a system will depend on the relative position of these roots. Consider, for example, a third-order system where the second-order term has $\omega_n = 9$ rad/s and $\xi = 0.1$. The roots of the second-order polynomial are located at $-0.9 \pm j8.955$. The system response for two different values of τ is shown in Figure 3.5. For $\tau = 1$ s, the root of the first-order polynomial is at -1. Hence it is close to the decaying term of the second-order poles, and thus the system almost exhibits a critically damped response. On the other hand, when $\tau = 0.2$ s, the root of the first-order polynomial is at -5. Therefore, the system response is dictated by the complex conjugate poles, resulting in an oscillatory response, as in a second-order system. Since the system response is governed by the closed-loop poles that are closest to the $j\omega$-axis, they are called the dominant poles.

3.2 Routh–Hurwitz's Stability Test

As mentioned in Section 3.1, the stability of a closed-loop system will depend on the roots of the characteristic equation. Consider the characteristic equation

$$C(s) = s^n + a_{n-1}s^{n-1} + a_{n-2}s^{n-2} + \cdots + a_1 s + a_0 = 0 \tag{3.15}$$

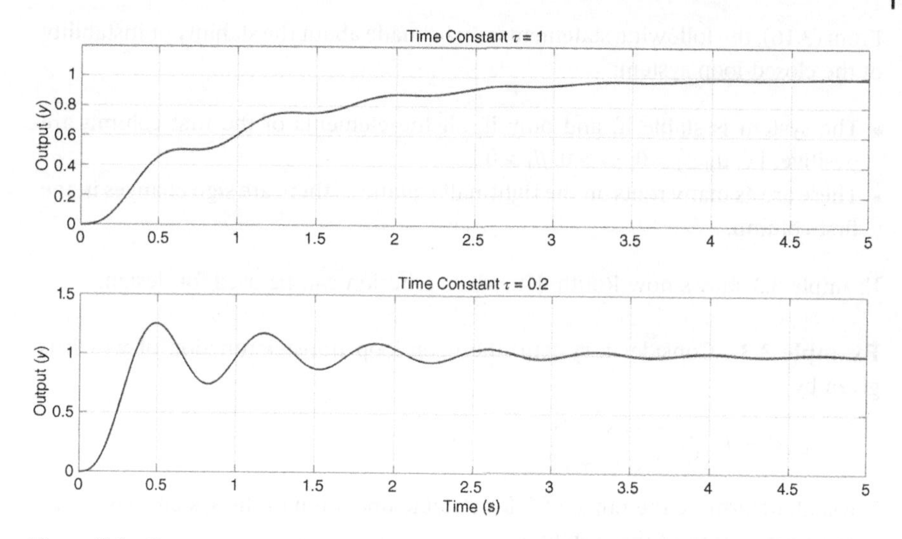

Figure 3.5 Step response of a third-order system with two different values of τ.

It is easy to evaluate the roots of the system using MATLAB® with the "roots" command. From the perspective of checking the overall stability, Routh–Hurwitz's criterion has been losing prominence these days. However, from the design perspective, it is still very useful. We briefly discuss this method in this section.

The first step is to determine if the polynomial $C(s)$ is Hurwitz or not. The necessary, but not sufficient, condition for the polynomial $C(s)$ to be Hurwitz is that all its coefficients should be positive and there must not be any zero coefficients except for a_0. If this condition is satisfied, the Routh's table is constructed, which, for (3.15), is of the form

$$\begin{array}{lllll} s^n & 1 & a_{n-2} & a_{n-4} & \cdots \\ s^{n-1} & a_{n-1} & a_{n-3} & a_{n-5} & \cdots \\ s^{n-2} & \alpha_1 & \alpha_2 & \alpha_3 & \cdots \\ s^{n-3} & \beta_1 & \beta_2 & \cdots & \cdots \\ \vdots & & & & \\ s^0 & a_0 & & & \end{array} \tag{3.16}$$

where

$$\begin{aligned} \alpha_1 &= \frac{a_{n-1}a_{n-2} - a_{n-3}}{a_{n-1}}, \quad \alpha_2 = \frac{a_{n-1}a_{n-4} - a_{n-5}}{a_{n-1}} \\ \beta_1 &= \frac{\alpha_1 a_{n-3} - \alpha_2 a_{n-1}}{\alpha_1}, \quad \beta_2 = \frac{\alpha_1 a_{n-5} - \alpha_3 a_{n-1}}{\alpha_1} \end{aligned} \tag{3.17}$$

From (3.16), the following statements can be made about the stability or instability of the closed-loop system:

- The system is stable if, and only if, all the elements of the first column are positive, i.e. $a_{n-1} > 0$, $\alpha_1 > 0$, $\beta_1 > 0 \ldots$
- There are as many roots on the right-half s-plane as there are sign changes in the first column.

Example 3.3 shows how Routh–Hurwitz's criterion can be used for design.

Example 3.3 Consider a system, the open-loop transfer function of which is given by

$$G(s) = K\frac{(s+1)(s+2)}{s^3(s+5)(s+25)}$$

We shall determine the range of K for a stable operation of the system. The characteristic equation of the system is

$$C(s) = s^5 + 30s^4 + 125s^3 + Ks^2 + 3Ks + 2K = 0$$

Then Routh's table will be

$$\begin{array}{llll} s^5 & 1 & 125 & 3K \\ s^4 & 30 & K & 2K \\ s^3 & \dfrac{3750-K}{30} & \dfrac{90K-2K}{30} & \\ s^2 & \alpha & 2K & \\ s^1 & \beta & & \\ s^0 & 2K & & \end{array}$$

where

$$\alpha = \frac{\dfrac{3750K-K^2}{30} - \dfrac{30 \times 88K}{30}}{\dfrac{3750-K}{30}} = \frac{1110K-K^2}{3750-K}$$

$$\beta = \frac{\alpha\dfrac{88K}{30} - 2K\left(\dfrac{3750-K}{30}\right)}{\alpha}$$

To ensure that all the elements of the first column are positive, the following conditions must be satisfied.From s^0 row, $K > 0$.

From s^3 row

$$\frac{3750-K}{30} > 0 \Rightarrow K < 3750$$

1) If the above condition is satisfied, then α will be positive when

$$(1110 - K)K > 0$$

Since $K > 0$, we have $K < 1110$.

Now $\beta > 0$ implies

$$\beta = \frac{88K}{30} - \frac{2K}{\alpha}\left(\frac{3750 - K}{30}\right) > 0$$
$$= \frac{1}{30}\left[88K - 2K\frac{(3750 - K)^2}{1110K - K^2}\right] > 0$$

The above equation can be simplified as

$$-90K^2 + 112680K - 28125000 > 0$$
$$\Rightarrow 90K^2 - 112680K + 28125000 > 0$$

The solution of the above quadratic equation is

$$(K + 344.2625)(K + 907.7375) < 0$$

The above equation is valid in the range $344.2625 < K < 907.7375$.

Note that conditions (1–3) are also satisfied by condition (4). Now for $K = 344.2625$, there is a complex conjugate pair at $\pm j2.98$, indicating a sustained oscillation at a frequency of 2.98 rad/s. Similarly, for $K = 907.7375$, there is another set of complex conjugate pairs at $\pm j5.3$, indicating a sustained oscillation at this frequency as well. The behavior of this system is investigated when we discuss root locus in Section 3.3.

3.3 Root Locus

The transient response of a closed-loop system is dictated by its closed-loop poles. These poles are the roots of the characteristic equation. Sometimes the closed-loop poles can be placed at certain desired locations by using a simple gain (proportional control). Root locus is a plot of the system poles when this gain changes from zero to infinity. Consider the characteristic equation

$$1 + KG(s)H(s) = 0$$
$$\Rightarrow KG(s)H(s) = -1 \tag{3.18}$$

where K is the system gain. Eq. (3.18) gives the angle condition of

$$\angle G(s)H(s) = \pm 180°(2k + 1), k = 0, 1, 2, \cdots$$
$$\Rightarrow \angle G(s)H(s) = \pm 180°, \pm 540°, \pm 900°, \cdots \tag{3.19}$$

and the magnitude condition of

$$|KG(s)H(s)| = 1 \tag{3.20}$$

The root locus is a plot in the complex plane (s-plane) in which the values of s fulfill the angle condition. The roots of the characteristic equation for a given value of gain can be determined from the magnitude condition.

There are several rules for plotting a root locus. These rules were important when they were plotted manually. With the current day availability of powerful software tools like MATLAB®, most of these rules have become superfluous. Therefore, only the most critical ones are discussed here. For more information, readers should consult a textbook on the subject.

3.3.1 Number of Branches and Terminal Points

Consider an open-loop transfer function of the form

$$KG(s)H(s) = K\frac{(s+z_1)(s+z_2)\cdots(s+z_m)}{(s+p_1)(s+p_2)\cdots(s+p_n)}, n \geq m \tag{3.21}$$

The characteristic equation is then given by

$$(s+p_1)(s+p_2)\cdots(s+p_n) + K(s+z_1)(s+z_2)\cdots(s+z_m) = 0 \tag{3.22}$$

From (3.22) it can be surmised that when K is equal to zero the closed-loop poles are equal to the open-loop poles. Since the root locus is a plot in which K changes from 0 to ∞, the starting point of any root locus is at the open-loop poles. Again, as $K \to \infty$, m number of poles will terminate in the open-loop zeros. The rest $n - m$ number of poles will move to infinity. This is illustrated with the help of Example 3.4.

Example 3.4 Let us consider an open-loop system with three poles and a zero, given by

$$KG(s)H(s) = K\frac{s+5}{s^3+3s^2+5s+6}$$

The poles are at -2, $-0.5 \pm j1.66$ and the zero is at -5. Now we choose a large value of gain, say $K = 10^{12}$. The characteristic equation is given by

$$s^3 + 3s^2 + 10^{12}s + 5 \times 10^{12} = 0$$

The roots of the characteristic equation are at -5, $-1 \pm j10^6$. From this it is evident that one of the poles terminates at the system zero, while the other two poles move toward infinity. The root locus is shown in Figure 3.6.

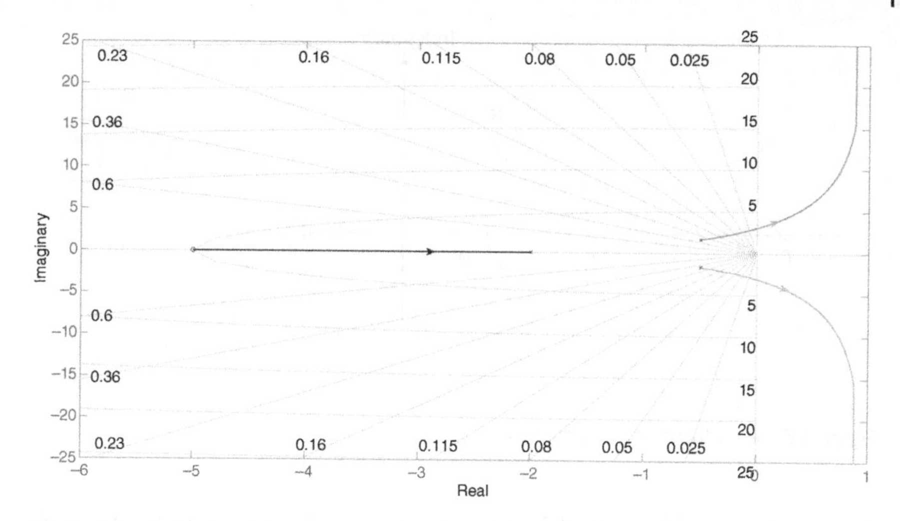

Figure 3.6 Root locus plot of the system of Example 3.4.

From the above example, following can be concluded:

- There are as may branches of the root locus as there are open-loop poles.
- The root locus always starts at the open-loop poles.
- Given that the open-loop system has m - zeros and n - poles, a total number of m - loci terminates at the open-loop zeros.
- The rest $m - n$ number of loci asymptotically approaches infinity.
- The complex conjugate poles have a constant real component when they approach infinity.

3.3.2 Real Axis Locus

The root locus on the real axis depends only on the poles and zeros on this axis. The complex conjugate pairs of poles and zeros have no influence on the real axis locus. Consider, for example, the partial root locus shown in Figure 3.7, in which three points (A, B, and C) are identified on the real axis. The sum total of the angle contributions of the complex conjugate poles all along the real axis is 0° (i.e. $\phi - \phi$) or 360° (i.e. $\phi - 360° - \phi$). These poles cannot have any contribution on the real axis locus. Now consider the segment between points A and B. The angular contribution of the pole in point A is −180°, whereas the angular contribution of the pole at C and the zero at B cancel out. Therefore, this segment will be on the real axis locus. For the segment between points B and C, the angular contribution of the

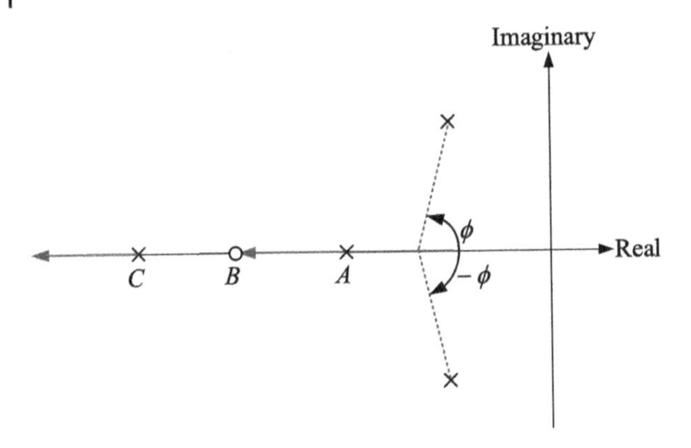

Figure 3.7 Root locus on the real axis.

pole in point A and the zero at B cancel out, whereas the angular contribution of the pole at C is 0°, and hence this segment cannot be on the real axis locus. Finally, for the segment to the left C, the angular contribution of the pole in point A and the zero at B cancel out, whereas the angular contribution of the pole at C is −180°. Thus, this segment will be on the real axis locus. In the real axis locus shown in Figure 3.7, the pole in point A terminates in the zero at B, and the pole at point C progresses to infinity along the real axis.

Following the logic mentioned above, the portions of the real axis that fall on the real axis can be determined using the following simple logic:

Choose a test point on the real axis. If the total number of real axis poles and zeros to the right of this point is odd, then this point is on the real axis locus.

Consider Example 3.5.

Example 3.5 Let us consider the following open-loop system

$$KG(s)H(s) = K\frac{s + z_1}{(s + p_1)(s + p_2)}$$

where z_1, p_1, and p_2 are strictly real and positive. We shall draw the root locus for the following three conditions.

a) $p_1 > z_1 > p_2$
b) $p_1 > p_2 > z_1$
c) $z_1 > p_1 > p_2$.

Case (a) is like the plot shown in Figure 3.7. This is shown in Figure 3.8a in which one pole terminates at the zero and the other pole moves to infinity. Case (b) is also like Case (a), except that the direction of the pole movement reverses for

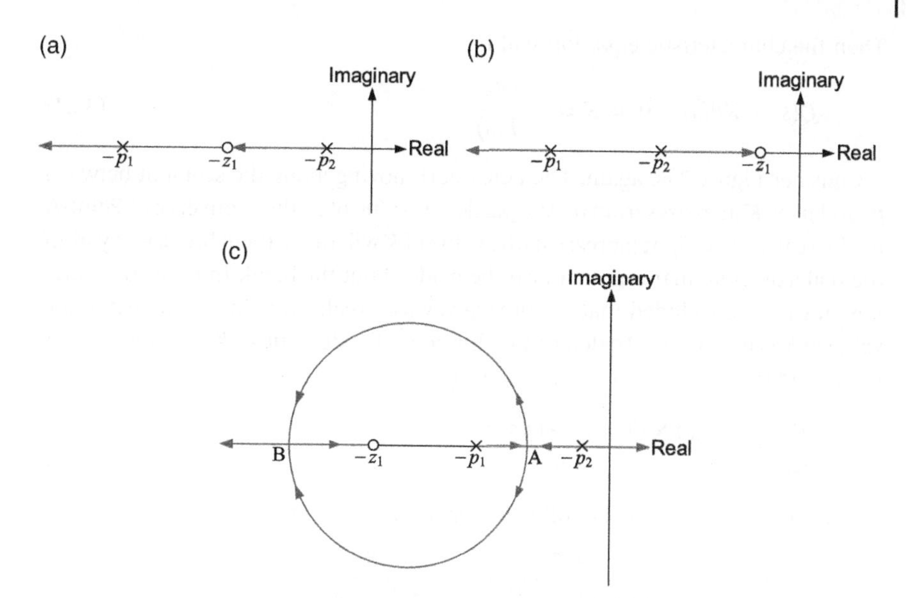

Figure 3.8 (a, b and c) Root locus plots for the three cases of Example 3.5.

one of the two poles. This is shown in Figure 3.8b. Case (c), however, is completely different. From our discussion above, we know that the real axis locus will be in the segment between p_1 and p_2, and the segment to the left of z_1. We also know that one of the two poles will terminate at the zero. However, there is no way to restrict the movement of the poles on the real axis only. Therefore, the poles must break away from the real axis somewhere in the segment between p_1 and p_2 (at point A) and break in somewhere in the segment on the left of z_1, say at Point-B. Thereafter, one pole will terminate at the zero, while the other pole will move toward infinity. This is shown in Figure 3.8c.

3.3.3 Breakaway and Break-in Points

It is obvious from Example 3.5 that sometimes a root locus breaks away from or breaks into the real axis. Just by inspection, it is easy to determine from which segment it will break away and to which segment it will converge (break in). However, we cannot determine the exact value at which the locus will break in or break away. To determine this, consider an open-loop transfer function of the form

$$KG(s)H(s) = K\frac{P(s)}{Q(s)} \tag{3.23}$$

Then the characteristic equation will be

$$Q(s) + KP(s) = 0 \Rightarrow K = -\frac{Q(s)}{P(s)} \tag{3.24}$$

Consider Figure 3.8c again. The poles start moving along the segment between p_1 and p_2 as K increases from 0. At a particular value of K, they converge at Point-A on the real axis. A slight increase in the value of K will make them break away from the real axis. A similar argument can be made about the break-in Point-B. Therefore, it can be concluded that the breakaway (or break-in) point occurs when the value of K is maximum. To determine this point, the derivate of K with respect to s is equated to zero. Thus from (3.24), we get

$$\frac{dK}{ds} = -\frac{P(s) \times {}^{dQ(s)}/_{ds} - Q(s) \times {}^{dP(s)}/_{ds}}{P^2(s)} = 0 \tag{3.25}$$

Example 3.6 Consider the following open-loop system

$$KG(s)H(s) = K\frac{s+5}{(s+1)(s+3)} = K\frac{s+5}{s^2+4s+3}$$

From (3.25), we have

$$\left(s^2 + 4s + 3\right) \times 1 - (s+5) \times (2s+4) = 0$$
$$\Rightarrow s^2 + 10s + 17 = 0$$

The solution of the quadratic gives $s_{1,2} = -3.17, -7.83$. The root locus plot for the system is shown in Figure 3.9.

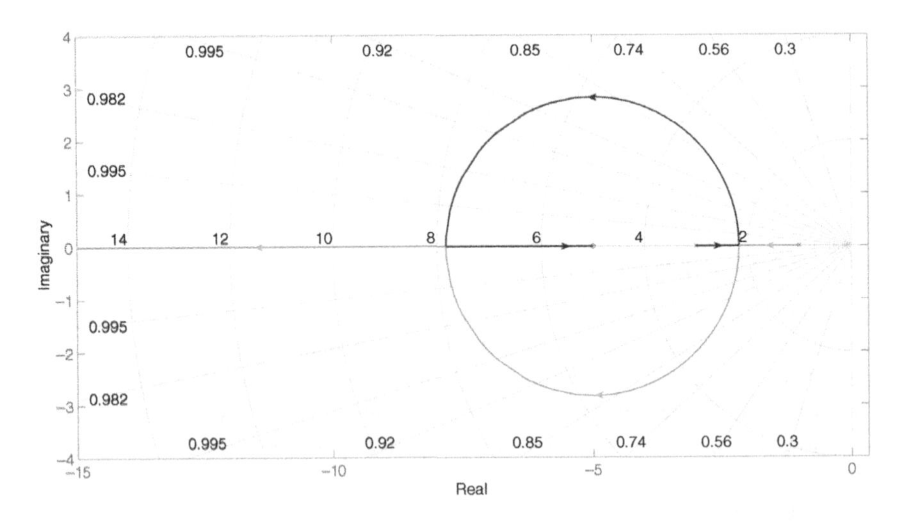

Figure 3.9 Root locus plot for the system of Example 3.6.

Example 3.7 Let us now revisit Example 3.3, where the open-loop transfer function is given by

$$G(s) = K\frac{(s+1)(s+2)}{s^3(s+5)(s+25)}$$

It was shown that the system remains stable for $344.2625 < K < 907.7375$. The root locus plot is shown in Figure 3.10a, while the zoomed portion around the imaginary axis is shown in Figure 3.10b. It is obvious from the figures that, of the three

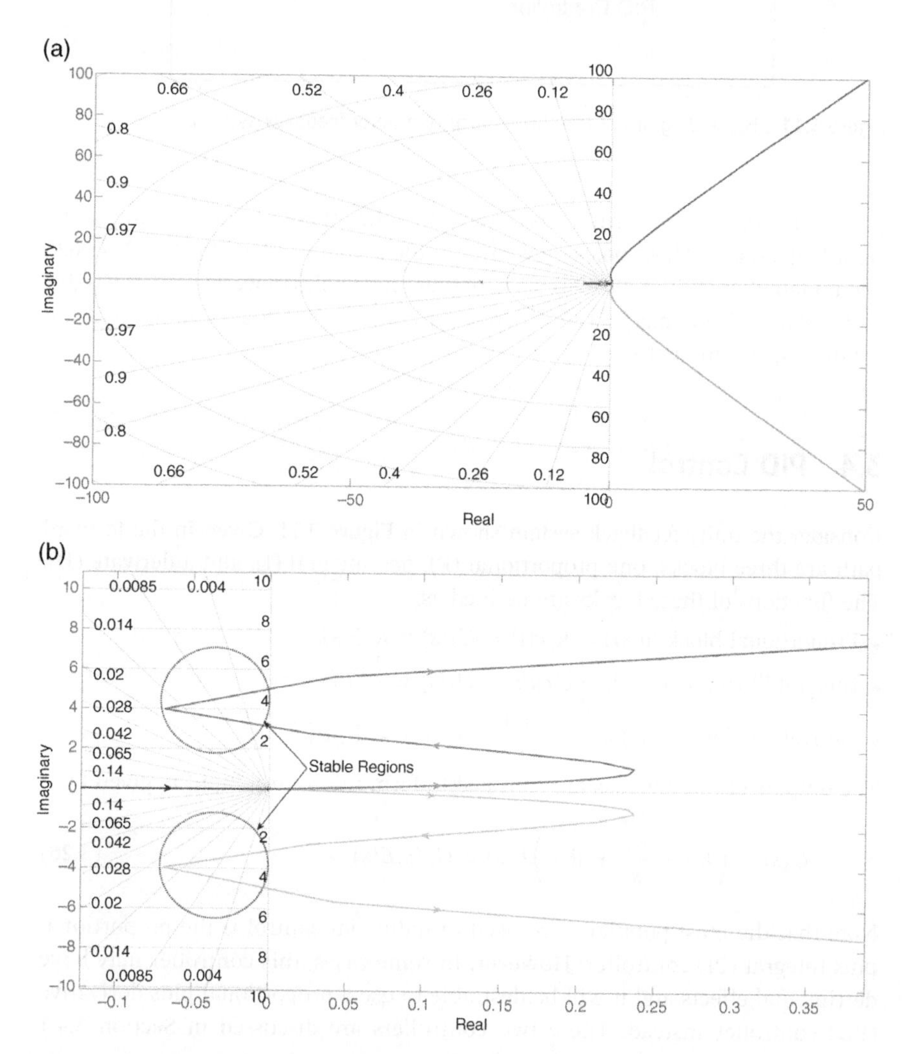

Figure 3.10 (a) Root locus plot for the system of Example 3.7 and (b) zoomed portion around the imaginary axis.

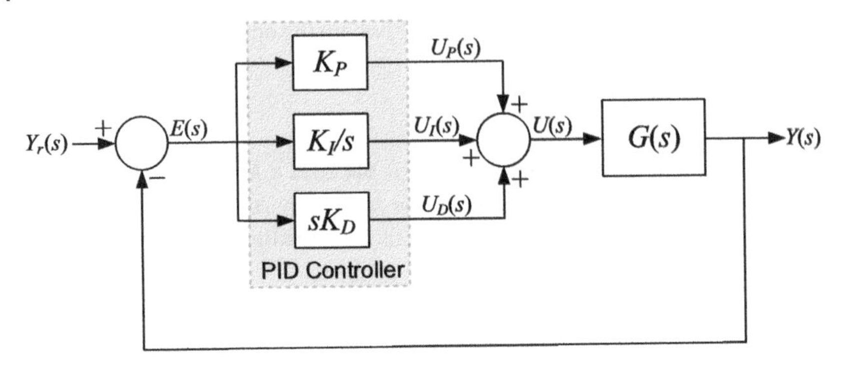

Figure 3.11 Block diagram of the PID control of a unity feedback system.

poles at the origin, one terminates at −1, while the other two start moving in the right-half s-plane. These poles move over to the left-half s-plane for $K = 344.2625$ and remain there till $K = 907.7375$. Thereafter, these poles move again to the right-half s-plane. Their imaginary axis crossing points are at the locations that are obtained in Example 3.3.

3.4 PID Control

Consider the unity feedback system shown in Figure 3.11. Given in the forward path are three blocks: one proportional (P), one integral (I), and a derivate (D). The functions of these blocks are defined as:

- Proportional block: $u_P(t) = K_P e(t) \Rightarrow U_P(s) = K_P E(s)$
- Integral block: $u_I(t) = K_I \int e(t)dt \Rightarrow U_I(s) = \frac{K_I}{s} E(s)$
- Derivative block: $u_D(t) = K_D \frac{de(t)}{dt} \Rightarrow U_D(s) = sK_D E(s)$

The outputs of these blocks are summed to form the control input u, given by

$$U(s) = \left(K_P + \frac{K_I}{s} + sK_D\right)E(s) = G_C(s)E(s) \tag{3.26}$$

Note that the most popular form used in industrial control is the proportional plus integral (PI) controller. However, in some cases, this controller may have detrimental effects and it will be desirable to use a proportional plus derivative (PD) controller instead. These two controllers are discussed in Section 3.4.1 and 3.4.2.

3.4.1 PI Controller

Consider an open-loop transfer function of the form

$$G(s) = \frac{(s + z_1)(s + z_2)\cdots(s + z_m)}{(s + p_1)(s + p_2)\cdots(s + p_n)}, n \geq m \tag{3.27}$$

The closed-loop transfer function of the system with a P-type controller is

$$\frac{Y(s)}{Y_r(s)} = \frac{K_P(s + z_1)(s + z_2)\cdots(s + z_m)}{(s + p_1)(s + p_2)\cdots(s + p_n) + K_P(s + z_1)(s + z_2)\cdots(s + z_m)} \tag{3.28}$$

The DC gain of the system then is

$$K_{dc} = \frac{K_P \times z_1 \times z_2 \times \cdots \times z_m}{p_1 \times p_2 \times \cdots \times p_n + K_P \times z_1 \times z_2 \times \cdots \times z_m} \tag{3.29}$$

The DC gain can be brought closer to 1, if a large value of K_P is used. However, as we have observed before, a large gain can also move the poles to the right-half s-plane causing system instability.

Instead, a PI controller is used which is of the form

$$G_C(s) = K_P + \frac{K_I}{s} = \frac{sK_P + K_I}{s}$$

The closed-loop transfer function of the system then is

$$\frac{Y(s)}{Y_r(s)} = \frac{(sK_P + K_I)(s + z_1)(s + z_2)\cdots(s + z_m)}{s(s + p_1)(s + p_2)\cdots(s + p_n) + (sK_P + K_I)(s + z_1)(s + z_2)\cdots(s + z_m)} \tag{3.30}$$

The steady state output with a unit step response will then be

$$y_{ss} = \lim_{s \to 0} \frac{(sK_P + K_I)(s + z_1)(s + z_2)\cdots(s + z_m)}{(sK_P + K_I)(s + z_1)(s + z_2)\cdots(s + z_m)} = 1 \tag{3.31}$$

This implies that the DC gain will be 1 and the system will not have any steady state error. We can therefore conclude that a PI controller is used to eliminate any steady state error. Also note that the system defined in (3.27) is that of a Type-0 system. The inclusion of the PI controller in the forward loop changes the system Type to 1, and hence, from Table 3.1, it is obvious that the steady state error will be zero for a step input.

Example 3.8 This example demonstrates a procedure for the choice of PI controller gains. Consider a unity feedback control system with

$$G(s) = \frac{1}{s(s+4)}$$

It will be controlled by a PI controller of the form

$$G_C(s) = \frac{10s + K_I}{s}$$

We want to determine the limiting value of K_I. The open-loop transfer function of the system is

$$G(s)G_C(s) = \frac{10s + K_I}{s^2(s+4)}$$

Therefore, the characteristic equation of the system is given by

$$s^2(s+4) + 10s + K_I = s(s^2 + 4s + 10) + K_I = 0$$

The Routh's table then is

$$\begin{array}{lll} s^3 & 1 & 10 \\ s^2 & 4 & K_I \\ s^1 & \dfrac{40 - K_I}{40} & \\ s^0 & K_I & \end{array}$$

From s^1 row, the limiting value of the integral gain is obtained as $K_I < 40$.

The characteristic equation can be rewritten as

$$1 + K_I G_1(s) = 0, \quad G_1(s) = \frac{1}{s(s^2 + 4s + 10)}$$

This is in the form of (3.18) and hence is suitable for the root locus plot.

The root locus plot is shown in Figure 3.12. It can be seen from the figure that the roots cross the imaginary axis over the right-half s-plane. For $K_I = 40$, the complex conjugate poles are located at $\pm j3.16$ and the real pole will be located at -4. The system step response with this value of the integral gain is shown in Figure 3.13, where a sustained oscillation can be observed. The step system response with a different value of $K_I = 20$ is also shown in Figure 3.13. The system is stable in this case with a steady state error of zero.

3.4.2 PD Controller

Consider a unity feedback system with an open-loop transfer function of

$$G(s) = \frac{1}{s^2}$$

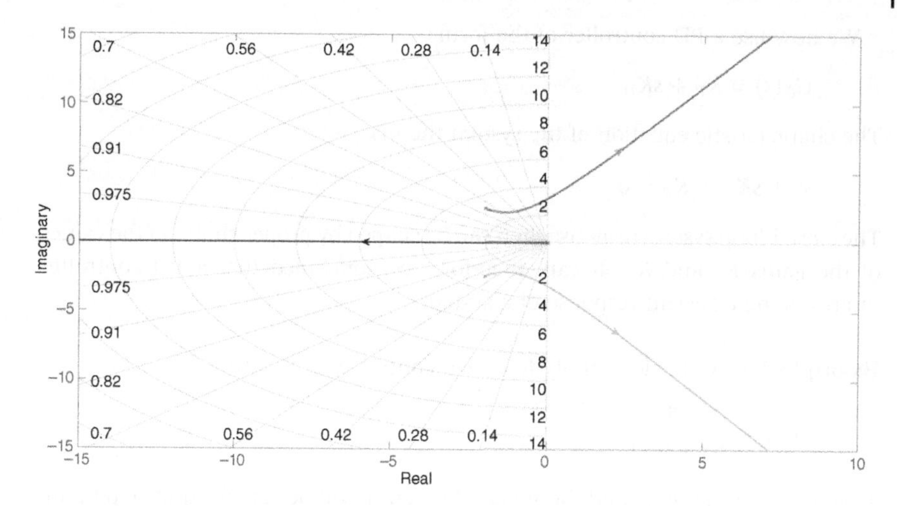

Figure 3.12 Root locus plot for the system of Example 3.8.

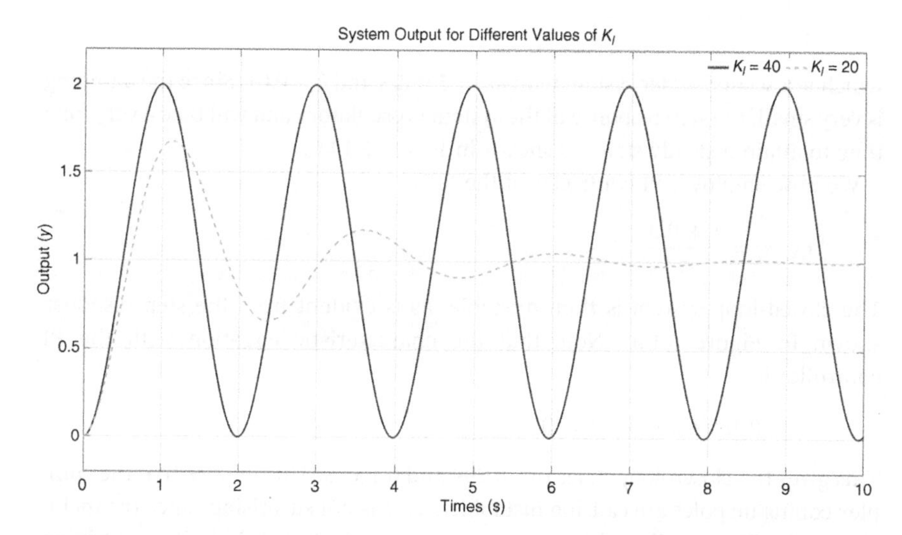

Figure 3.13 Step response with two different values of integral gain in Example 3.8.

With only a proportional controller, the characteristic equation of the system will be

$$s^2 + K_P = 0$$

The system will have two closed-loop poles at $\pm\sqrt{K_P}$ and hence the system will have a sustained oscillation when subjected to a step input.

We now use a PD controller of the form

$$G_C(s) = K_P + sK_D$$

The characteristic equation of the system then is

$$s^2 + sK_D + K_P = 0$$

The closed-loop system response then can be shaped by proper choice of the values of the gains K_P and K_D. It can, therefore, be concluded that a PD controller improves the transient response of a system.

Example 3.9 Consider a unity feedback control system with

$$G(s) = \frac{25}{s(s+0.1)}$$

This is a Type-1 system, and the steady state error will be zero for unit step input. However, without any controller, the closed-loop transfer function is given by

$$\frac{Y(s)}{Y_r(s)} = \frac{25}{s^2 + 0.1s + 25}$$

which is a second-order system with $\omega_n = 5$ rad/s and $\xi = 0.01$. Since the damping is very small, the step response of the system is oscillatory and will take a very long time to attain a steady state, as shown in Figure 3.14a.

We now employ a PI controller of the form

$$G_C(s) = \frac{s + 0.3}{s}$$

The closed-loop system is then unstable, as is evident from the step response shown in Figure 3.14b. Note that the characteristic equation with the PI controller is

$$s^3 + 0.1s^2 + 25s + 7.5 = 0$$

This gives the closed-loop poles of -0.299 and $0.1 \pm j5$. This implies that the complex conjugate poles are causing instability. This is not surprising, since the inclusion of the PI controller, changes the system type to 2, and this will result in an unstable response when the input is a unit step.

Alternatively, we employ the following PD controller instead of the PI controller

$$G_C(s) = 25 + 5s$$

The inclusion of the PD controller does not change the system type, and hence the steady state error to a step input will be zero. The closed-loop transfer function then is

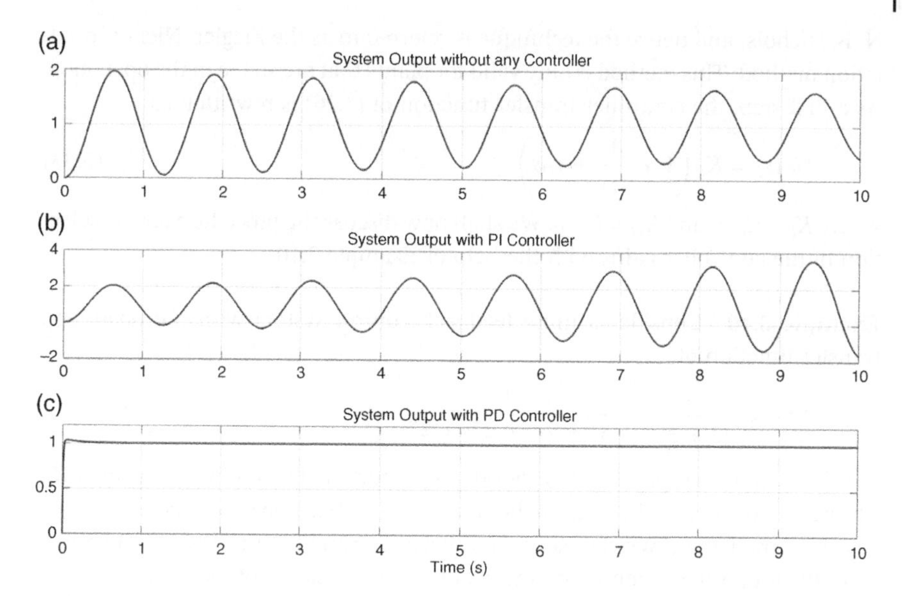

Figure 3.14 Step response of the system of Example 3.9 with (a) no controller, (b) PI controller, and (c) PD controller.

$$\frac{Y(s)}{Y_r(s)} = \frac{125s + 625}{s^2 + 125.1s + 625}$$

This implies that the system has a DC gain of 1 and the closed-loop poles are stable. The step response of the system is shown in Figure 3.14c.

Often the derivative part of the PID (or PD) controller is not implemented in the form shown in Figure 3.11. An ideal derivative action can generate spikes every time the set point changes. The PID controller, with the modified derivative action, is given by

$$G_C(s) = K_P + \frac{K_I}{s} + K_D \frac{Ns}{s + N} \tag{3.32}$$

where $0 < N < 200$. Note that the derivative of (3.32) contains a lowpass filter with the transfer function of $N/(s + N)$ that will eliminate the high-frequency terms.

3.4.3 Tuning of PID Controllers

Often, PID controllers are tuned by trial-and-error methods. Even though this usually produces satisfactory results, a more systematic approach is needed for tuning the controllers. We present a method here that was developed by J. G. Ziegler and

N. B. Nichols, and hence the technique is referred to as the Ziegler–Nichols oscillation method. This method is only valid for plants that are stable in the open loop. As a first step, the controller transfer function of (3.26) is rewritten as

$$G_C(s) = K_P\left(1 + \frac{1}{\tau_I s} + \tau_D s\right) \tag{3.33}$$

where $K_I = K_P/\tau_I$ and $K_D = K_P\tau_D$. We shall now discuss the procedure for the selection of the controller gains with the help of Example 3.10.

Example 3.10 Consider a unity feedback control system with the open-loop transfer function of

$$G(s) = \frac{1}{(s+3)^3} = \frac{1}{s^3 + 9s^2 + 27s + 27}$$

The first step in this process is to consider a proportional control with a gain of K_c. The next step is to find the value of the gain K_c at which the open-loop poles exhibit a sustained oscillation. We can easily find this gain using Routh–Hurwitz's method. With the proportional controller, the characteristic equation of the system is given by $s^3 + 9s^2 + 27s + 27 + K_c = 0$. The partial Routh–Hurwitz's table is then formed as

$$\begin{array}{lll} s^3 & 1 & 27 \\ s^2 & 9 & 27 + K_c \\ s^1 & \dfrac{27 \times 9 - (27 + K_c)}{9} & \end{array}$$

Note that the system will have a sustained oscillation when the element of s^1 row is zero, i.e.

$$27 \times 9 - (27 + K_c) = 0 \Rightarrow K_c = 27 \times 8 = 216$$

We now form an auxiliary polynomial from row s^2 with this value of K_c

$$9s^2 + 27 + 216 = 0 \Rightarrow s = \pm j\sqrt{\frac{243}{9}} = \sqrt{27}$$

This means that the undamped frequency is $\omega_n = \sqrt{27} = 5.196$ rad/s, which is equivalent to $f_n = \omega_n/2\pi = 0.827$ Hz. Therefore, the undamped oscillation period is $t_n = 1/f_n = 1.21$ s. This oscillation period is shown in Figure 3.15a. The empirical formulas for the selection of the time constants based on K_c and t_n are given in Table 3.2. From this table, we get

$$K_P = 0.6 \times 216 = 129.6$$
$$K_I = \frac{K_P}{\tau_I} = \frac{129.6}{0.5 \times 1.21} = 214.357$$
$$K_D = K_P\tau_D = 129.6 \times 0.125 \times 1.21 = 19.589$$

The system response with a unit step input is shown in Figure 3.15b.

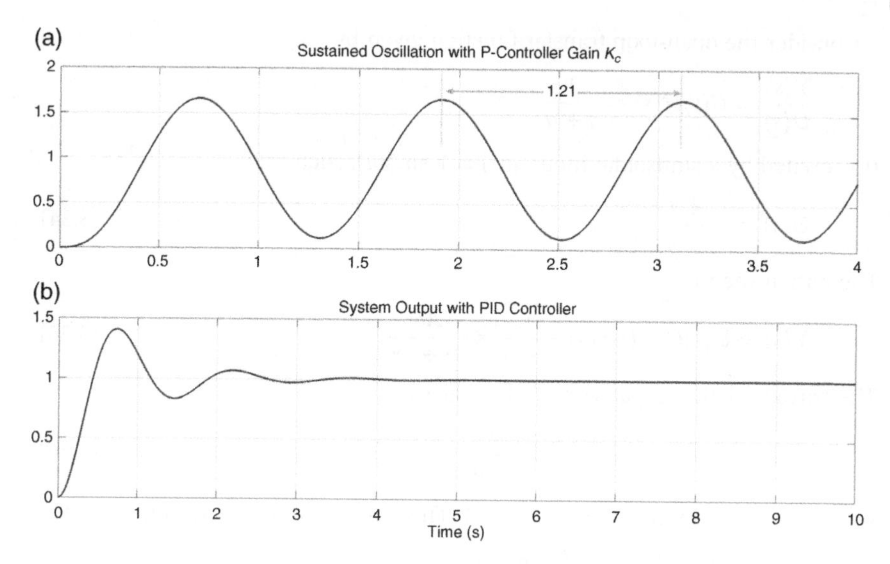

Figure 3.15 (a) Sustained oscillation time period and (b) step response of the system of Example 3.10 with PID controller.

Table 3.2 The selection of PID parameters.

Type	K_P	τ_I	τ_D
P	$0.5K_c$		
PI	$0.45K_c$	$t_n/1.2$	
PID	$0.6K_c$	$0.5t_n$	$0.125t_n$

3.5 Frequency Response Methods

When a physical system is suddenly excited by a sinusoidal waveform, it goes through a transient before attaining steady state, which is called the sinusoidal steady state. In an electrical system, we often perform phasor analysis, assuming that system is in the sinusoidal steady state. In this state, the voltages and currents are sinusoids with a constant frequency and amplitude. Similarly, the frequency response analysis in a control system is the steady state response of a system when it is excited by a sinusoidal input. In general, however, the frequency of the sinusoidal input is varied over a large range to study the resulting output.

Consider the open-loop transfer function given by

$$\frac{Y(s)}{U(s)} = G(s)H(s) = \frac{1}{s+\alpha}$$

It is excited by a sinusoidal input, $u(t) = V\sin(\omega t)$, such that

$$U(s) = \frac{A\omega}{s^2+\omega^2} \tag{3.34}$$

The output then is

$$Y(s) = U(s)G(s)H(s) = \frac{1}{s+\alpha} \times \frac{A\omega}{s^2+\omega^2} \tag{3.35}$$

The partial fraction expansion of (3.35) gives

$$Y(s) = \frac{A}{s+j\omega} + \frac{A^*}{s-j\omega} + \frac{B}{s+\alpha} \tag{3.36}$$

where A^* is the complex conjugate of A. The time response of the output is then

$$y_{ss}(t) = Ae^{-j\omega t} + A^*e^{j\omega t} + Be^{-\alpha t}$$

If $\alpha > 0$, i.e. the system is stable, then the exponential term $e^{-\alpha t}$ will go to zero. Therefore, we have the following steady state output

$$y_{ss}(t) = Ae^{-j\omega t} + A^*e^{j\omega t} \tag{3.37}$$

Now from (3.35) and (3.36), the following values are obtained.

$$A = G(-j\omega)H(-j\omega)\frac{V}{-j2},\ A^* = G(j\omega)H(j\omega)\frac{V}{j2}$$

Now define

$$G(j\omega)H(j\omega) = |GH|e^{j\varphi} \tag{3.38}$$

Therefore

$$A = -\frac{V}{j2}|GH|e^{-j\varphi},\ A^* = \frac{V}{j2}|GH|e^{j\varphi}$$

Substituting the expressions in (3.37), we have

$$y_{ss}(t) = \frac{V}{j2}|GH| \times \left[-e^{-j(\omega t+\varphi)} + e^{j(\omega t+\varphi)}\right] = V|GH|\sin(\omega t+\varphi) \tag{3.39}$$

Equation (3.39) indicates that the output of a stable system, when excited by a sinusoidal signal of a particular frequency, is also a sinusoid of the same frequency at the sinusoidal steady state. The output can be written from (3.39) as

$$Y(j\omega) = V|GH|\angle\varphi \tag{3.40}$$

From (3.40), it can be summarized that the amplitude of the output gets multiplied by the gain of the transfer function and its angle gets phase shifted by the phase of the transfer function. We can then write

$$|GH| = \left|\frac{Y(j\omega)}{U(j\omega)}\right|, \quad \varphi = \angle\frac{Y(j\omega)}{U(j\omega)}$$

3.5.1 Bode Plot

Consider an open-loop transfer function of the form

$$G(s)H(s) = \frac{\omega_n^2}{s(s+\alpha)\left(s^2 + 2\xi\omega_n s + \omega_n^2\right)} \tag{3.41}$$

Replacing s by $j\omega$, the transfer function is rewritten as

$$G(j\omega)H(j\omega) = \frac{\omega_n^2}{j\omega(j\omega+\alpha)\left(j2\xi\omega_n\omega + \omega_n^2 - \omega^2\right)} \tag{3.42}$$

A. Magnitude Plot: From (3.42), the magnitude condition is given as

$$|G(j\omega)H(j\omega)| = \frac{\omega_n^2}{\omega \times \sqrt{\omega^2+\alpha^2} \times \sqrt{4(\xi\omega_n\omega)^2 + \left(\omega_n^2 - \omega^2\right)^2}} \tag{3.43}$$

In Bode plots, the magnitude is often calculated in decibels (dB). A dB is defined as 20 times the log of the gain. Therefore, the gain $|G(j\omega)H(j\omega)|$ in (3.43) is given as

$$|G(j\omega)H(j\omega)| = 20\left[\log\frac{1}{\omega} + \log\frac{1}{\sqrt{\omega^2+\alpha^2}} + \log\frac{\omega_n^2}{\sqrt{4(\xi\omega_n\omega)^2 + \left(\omega_n^2-\omega^2\right)^2}}\right] \text{dB} \tag{3.44}$$

Equation (3.44) indicates that each individual term can be plotted separately and then can be added together to form the composite magnitude plot. Example 3.11 illustrates this.

Example 3.11 Consider the following open-loop transfer function

$$G(s)H(s) = \frac{9}{s(s+3)(s^2+3s+9)} = \frac{1}{3}\frac{1}{s\left(\frac{s}{3}+1\right)(s^2+3s+9)}$$

The transfer function has a pole at the origin (integral term), a first-order pole at −3, and a second-order term with $\omega_n = 3$ rad/s and $\xi = 0.5$. Replacing s by $j\omega$, the above transfer function is rewritten as

$$G(j\omega)H(j\omega) = \frac{1}{3} \times \frac{1}{j\omega} \times \frac{1}{\left(\frac{j\omega}{3} + 1\right)} \times \frac{1}{\left(1 + \frac{j\omega}{3} - \frac{\omega^2}{9}\right)}$$

A semilog scale is usually used for Bode plots. Note that $20 \log(1/3) = -9.54$ dB and is constant. For an integral term with a gain of 1, we have the following

$$20 \log \left|\frac{1}{j\omega}\right|\Bigg|_{\omega = 0.1} = 20 \text{ and } 20 \log \left|\frac{1}{j\omega}\right|\Bigg|_{\omega = 1} = 0$$

Therefore, the integral term drops 20 dB per decade, which is usually written as having a slope of −20 dB/decade. Using a similar argument, the slope of a derivative term is +20 dB/decade. The magnitude plots of the gain and the integral terms are shown in Figure 3.16.

For the first-order term, we have

$$20 \log \left|\frac{1}{\frac{j\omega}{3} + 1}\right| = \begin{cases} 0 \text{ for } \omega \ll 3 \\ -20 \log(\omega/3) \text{ for } \omega \gg 3 \end{cases}$$

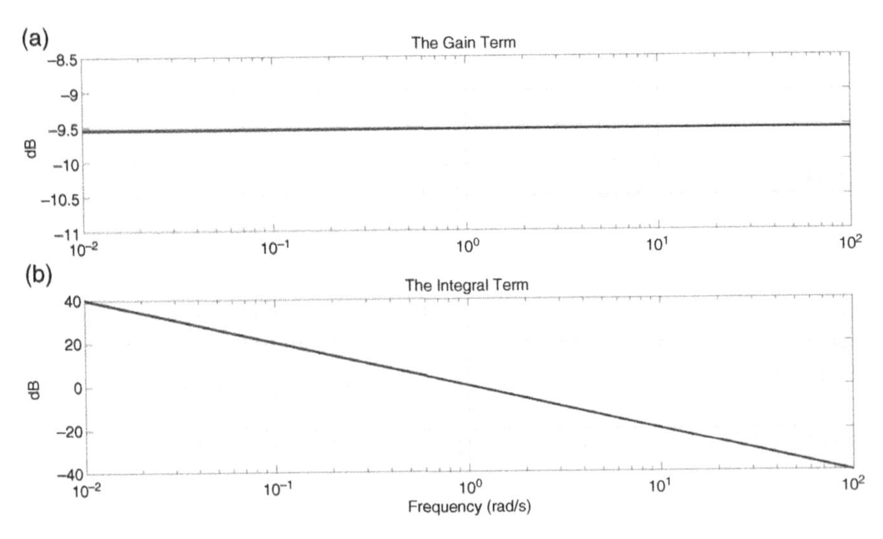

Figure 3.16 The magnitude plots of (a) the gain and (b) the integral terms.

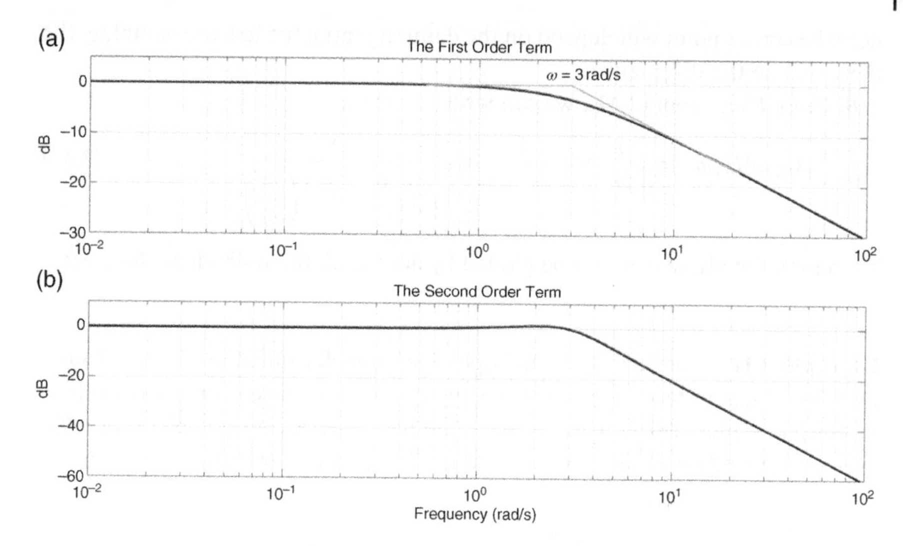

Figure 3.17 The magnitude plots (a) the first- and (b) the second-order terms.

The Bode plot of the first-order term is a combination of two asymptotes that intersect at $\omega = 3$ rad/s, as shown in Figure 3.17a. One asymptote is a 0-dB straight line, while the other has a slope of −20 dB/decade. The frequency $\omega = 3$ rad/s is called the corner frequency. The Bode plot, however, does not have a discontinuity and is a smooth curve, as shown in Figure 3.17a. Note that a first-order term in the numerator like $(j\omega/\omega_c + 1)$ will have a rise at the rate of 20 dB/decade from the corner frequency of ω_c.

In a similar way, the Bode magnitude plot of the second-order term can also be obtained. This term is given by

$$\left| \frac{1}{\left(1 + \frac{j\omega}{3} - \frac{\omega^2}{9}\right)} \right| = -20\log\sqrt{\frac{\omega^2}{9} + \left(1 - \frac{\omega^2}{9}\right)^2}$$

Therefore

$$\left| \frac{1}{\left(1 + \frac{j\omega}{3} - \frac{\omega^2}{9}\right)} \right| = \begin{cases} 0 \text{ for } \omega \ll 3\,(=\omega_n) \\ -40\log\left(\frac{\omega}{\omega_n}\right) \quad \omega \gg 3\,(=\omega_n) \end{cases}$$

The gain plot of the second-order term also has two asymptotes, where the slope of one of the asymptotes will be −40 dB/decade. The magnitude plot of the second-order term is shown in Figure 3.17b. The bump divergence from the asymptotes near the corner point will depend on the damping ratio: the less the damping, the more will be the divergence.

B. Phase Plot: From (3.43), we can write

$$\angle G(j\omega)H(j\omega) = \angle\frac{1}{j\omega} + \angle\frac{1}{j\omega + a} + \angle\frac{1}{j2\xi\frac{\omega}{\omega_n} + 1 - \left(\frac{\omega}{\omega_n}\right)^2} \tag{3.45}$$

Therefore, the phase can also be plotted by adding all the individual elements.

Example 3.12 Consider again the open-loop transfer function given in Example 3.11. The gain term does not have any phase and hence does not contribute to the phase plot. The integral term ($1/j\omega$) will have a constant phase of $-90°$ as it is independent of the frequency value. The phase of the first-order term is

$$\angle\frac{1}{\frac{j\omega}{3} + 1} = -\tan^{-1}\left(\frac{\omega}{3}\right)$$

Therefore, we can write

$$\angle\frac{1}{\frac{j\omega}{3} + 1} = \begin{cases} 0° \text{ for } \omega \approx 0 \\ -45° \text{ for } \omega = 3 \\ -90° \text{ as } \omega \to \infty \end{cases}$$

The phase plot of the first-order term is shown in Figure 3.18a. The phase of the second-order term is

$$\angle\left|\frac{1}{\left(1 + \frac{j\omega}{3} - \frac{\omega^2}{9}\right)}\right| = -\tan^{-1}\left(\frac{\frac{\omega}{3}}{1 - \frac{\omega^2}{9}}\right)$$

We can therefore surmise

$$\angle\left|\frac{1}{\left(1 + \frac{j\omega}{3} - \frac{\omega^2}{9}\right)}\right| = \begin{cases} 0° \text{ for } \omega \approx 0 \\ -90° \text{ for } \omega = \omega_n \\ -180° \text{ as } \omega \to \infty \end{cases}$$

The phase plot of the second-order term is shown in Figure 3.18b.

The composite Bode plot of the system of Examples 3.11 and 3.12 is shown in Figure 3.19. Note that from this plot, we can determine the gain and phase of the system for a particular value of frequency. For example, for $\omega = 10$ rad/s, the gain is -61 dB and the phase is $-325°$. This can also be calculated analytically by substituting $s = j10$ in the open-loop transfer function.

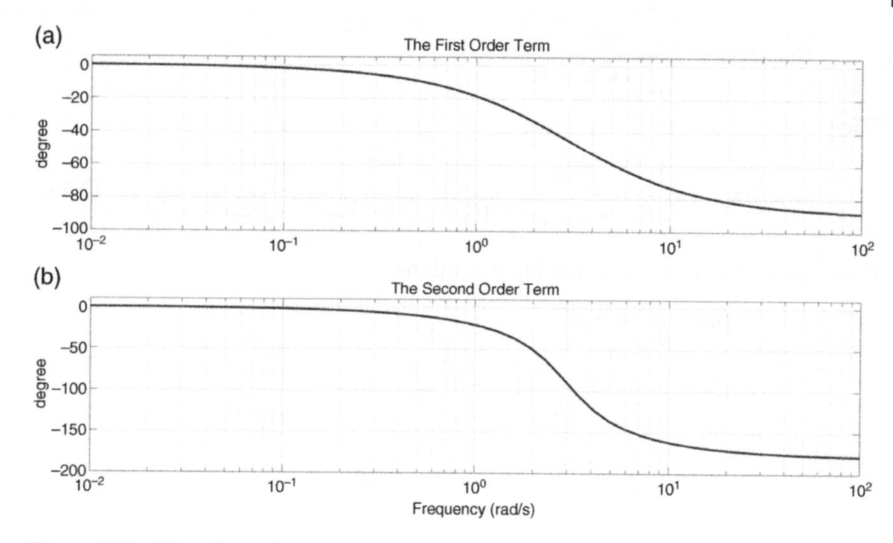

Figure 3.18 The phase plots of (a) the first- and (b) second-order terms.

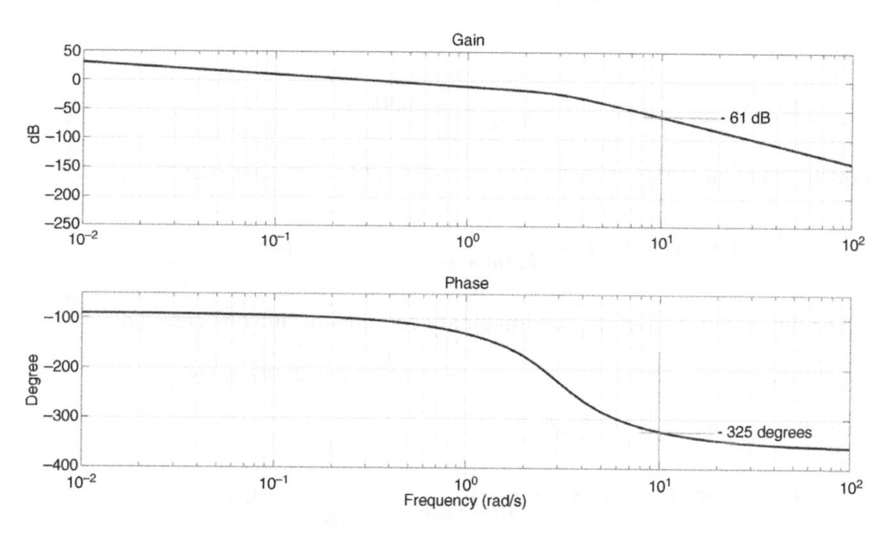

Figure 3.19 The composite Bode plot of Examples 3.11 and 3.12.

3.5.2 Nyquist (Polar) Plot

A Nyquist plot (or polar plot) is the graph of the magnitude of the transfer function $G(j\omega)H(j\omega)$ versus its phase angle in polar coordinates as the frequency changes from $-\infty$ to ∞. Consider, for example, the following transfer function

$$G(s)H(s) = \frac{50}{(s+1)^2(s+10)}$$

Then

$$G(j\omega)H(j\omega) = \frac{50}{(j\omega+1)^2(j\omega+10)} = \frac{50}{(10-12\omega^2)+j(21\omega-\omega^3)} \tag{3.46}$$

From (3.46), we have the following conditions

$$G(j\omega)H(j\omega)|_{\omega=0} = 5\angle 0^\circ$$

$$G(j\omega)H(j\omega)|_{\omega=-\infty} = \left.\frac{50}{(j\omega)^3}\right|_{\omega\to-\infty} = 0\angle 270^\circ$$

$$G(j\omega)H(j\omega)|_{\omega=\infty} = \left.\frac{50}{(j\omega)^3}\right|_{\omega\to\infty} = 0\angle -270^\circ$$

To determine the direction of rotation, we first determine the imaginary axis crossing point, where the real part of $G(j\omega)H(j\omega)$ is equal to zero. From (3.46), this is calculated as

$$10-12\omega^2, \quad \Rightarrow \omega = \pm\sqrt{\frac{10}{12}} = \pm\ 0.91\ \text{rad/s}$$

For the imaginary axis crossing at $\omega = -j0.91$, (3.46) is rewritten as

$$G(-j0.91)H(-j0.91) = \left.\frac{50}{j(21\omega+\omega^3)}\right|_{\omega=-j0.91} = j2.72$$

For the real axis crossing, the imaginary component must be zero, i.e.

$$21\omega-\omega^3 = 0 \quad \Rightarrow \omega = 0 \text{ and } \omega = \pm\sqrt{21} = \pm 4.583\ \text{rad/s}$$

Therefore, from (3.46), we have

$$G(\pm j4.483)H(\pm j4.483) = \left.\frac{50}{10-12\omega^2}\right|_{\omega=\pm j4.483} = -0.21$$

The Nyquist plot is shown in Figure 3.20. The plot starts at $\omega = -\infty$, for which the magnitude of $G(j\omega)H(j\omega)$ is zero and has a phase angle of 270°. Then at $\omega = -j4.483$ rad/s, it crosses the real axis at Point-B. As the frequency decreases, it crosses the imaginary axis at Point-A when $\omega = -j0.91$ rad/s. As the frequency reduces further to $\omega = 0$, the magnitude becomes +5 on the real axis. The rest of the plot is a mirror image of the above. Thus, the plot travels in a clockwise direction from $\omega = 0$ and enters the origin with an angle of 90° as ω tends to infinity.

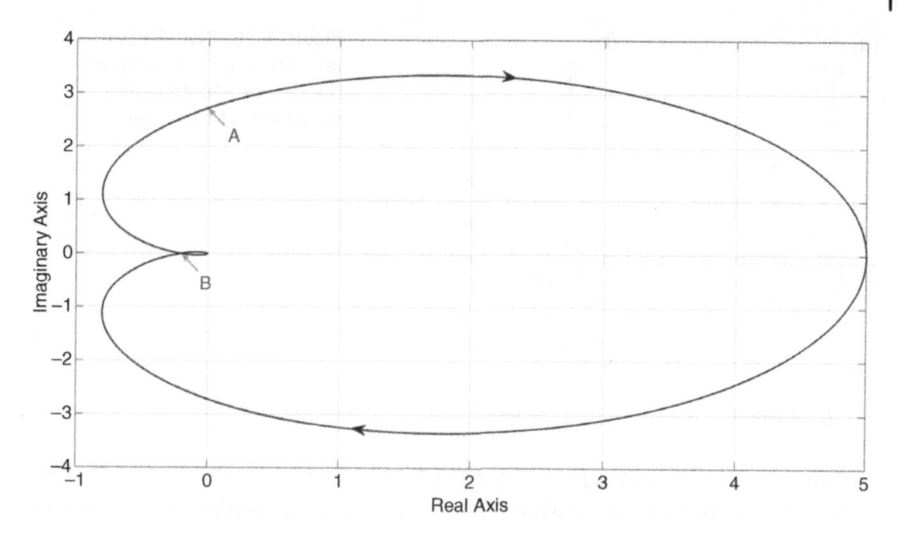

Figure 3.20 Nyquist plot of the transfer function given in (3.45).

An open-loop transfer function can be written in the frequency domain as $G(j\omega)H(j\omega) = |GH(\omega)| \angle \delta(\omega)$, where both the magnitude and the phase angle vary as ω varies. These two quantities are plotted separately against frequency in a Bode plot. In the polar plot, however, as the name suggests, the locus of the tip of the vector $G(j\omega)H(j\omega)$ is plotted in the complex plane as the frequency changes. Both these plots convey certain information that are critical for control system performance evaluation. However, these can be explained better through the Nyquist stability criterion, which is discussed in Section 3.5.3.

3.5.3 Nyquist Stability Criterion

In this section, the Nyquist stability criterion is discussed, without presenting the formal proof of the theorem. An open-loop transfer function can be written as

$$G(s)H(s) = \frac{q(s)}{p(s)}$$

Let us define the following rational function

$$F(s) = 1 + G(s)H(s) = \frac{p(s) + q(s)}{p(s)} \tag{3.47}$$

Note that the zeros of $F(s)$ are the roots of the characteristic equation, while its poles are also the poles of the open-loop system $G(s)H(s)$. The first step is to form a closed contour in the s-plane that encloses the entire right half of the plane.

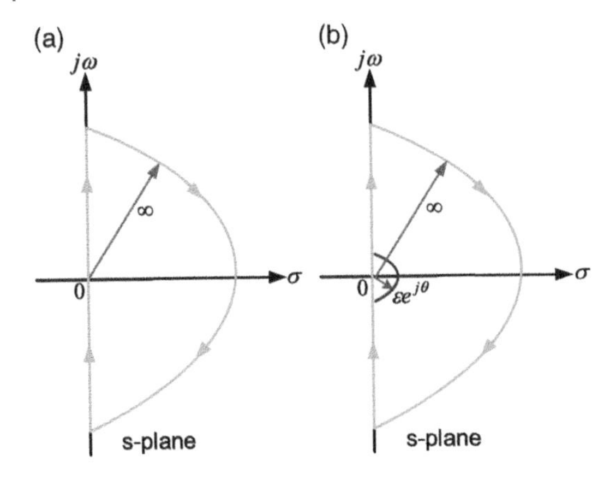

Figure 3.21 Nyquist path (a) without a singularity on the $j\omega$ axis and (b) with a singularity at the origin.

The contour consists of the $j\omega$ axis as ω varies from $-\infty$ to ∞ and a semicircular path with a radius of ∞, as shown in Figure 3.21a. Clearly, the direction of travel is clockwise. This path is usually called the Nyquist path and it encloses all the poles and zeros $F(s)$ that have positive real parts. It is also necessary that the Nyquist path does not pass through any pole or zero of $F(s)$. Therefore, if $F(s)$ has any pole or zero along the $j\omega$ axis, then a detour is taken to avoid this point, as shown in Figure 3.21b. Here ε is a very small number and θ varies from $-180°$ to $180°$.

The Nyquist stability criterion is then expressed as

$$Z' = N' + P' \tag{3.48}$$

where

Z' = number of zeros of $F(s)$ that are in the right-half s-plane.

P' = number of poles of $F(s)$ that are in the right-half s-plane.

N' = number of clockwise encirclements of the origin in the $F(j\omega)$ - plane.

Consider the contours shown in Figure 3.22a. Notice that $F(j\omega)$is the sum of a unit vector and $G(j\omega)H(j\omega)$. Thus $1 + G(j\omega)H(j\omega)$ is identical to the vector drawn from $-1+j0$ point to a point on the path formed by the vector $G(j\omega)H(j\omega)$, as shown in Figure 3.22b. Therefore, the encirclement of the origin in the $F(j\omega)$-plane is equivalent to the encirclement of the $-1+j0$ point in the $G(j\omega)H(j\omega)$-plane. Therefore, we redefine the Nyquist stability criterion as

$$Z = N + P \tag{3.49}$$

where

Z = number of zeros of the characteristic equation $1 + G(s)H(s) = 0$ (or closed-loop poles) that are in the right-half s-plane.

P = number of poles of $G(s)H(s)$ (i.e. open-loop poles) that are in the right-half s-plane.

N = number of clockwise encirclements of the $-1+j0$ point in the $G(j\omega)H(j\omega)$ - plane.

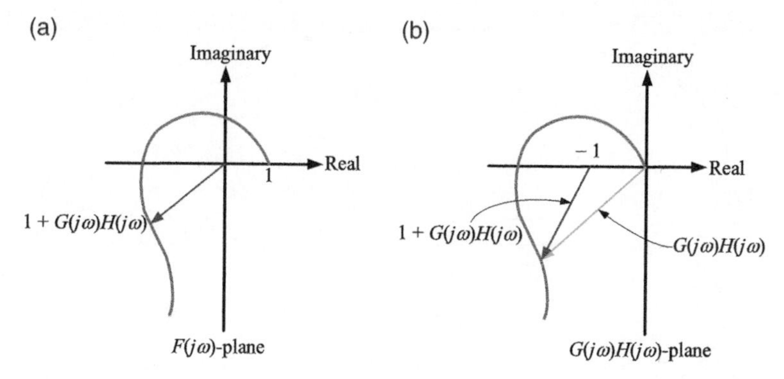

Figure 3.22 Polar plots in two different planes: (a) $F(j\omega) = 1 + G(j\omega)H(j\omega)$ plane and (b) $G(j\omega)H(j\omega)$ plane.

For stability, we must have $Z = 0$. Now suppose an open-loop system has m number of poles on the right-half s-plane, i.e. $P = m$. Then for the system to remain stable, we must have $N = -m$. This means that the Nyquist path must have m number of counterclockwise encirclements of the $-1 + j0$ point.

Example 3.13 Consider the open-loop transfer function, given by

$$G(s)H(s) = \frac{s^2 - 1}{s^3 - 5s^2 + 1.53s + 10.764}$$

It has got poles at 3.9, 2.3, and −1.2. Since there are two open-loop poles in the right-half s-plane, we have $P = 2$. Now

$$G(j\omega)H(j\omega) = \frac{-\omega^2 - 1}{(5\omega^2 + 10.764) + j(1.53\omega - \omega^3)}$$

The terminal values of the plot are

$$G(j\omega)H(j\omega)|_{\omega=0} = \frac{-1}{10.764} = 0.093\angle -180^\circ$$

$$G(j\omega)H(j\omega)|_{\omega\to\infty} = \left.\frac{1}{j\omega^3}\right|_{\omega\to\infty} = 0\angle -90^\circ$$

The Nyquist plot is shown in Figure 3.23. To find the real axis intercept, equate the imaginary part to zero, i.e.

$$(1.53\omega - \omega^3) = 0$$

$$\Rightarrow \omega = 0 \text{ and } \omega = \pm\sqrt{1.53} = \pm 1.2369$$

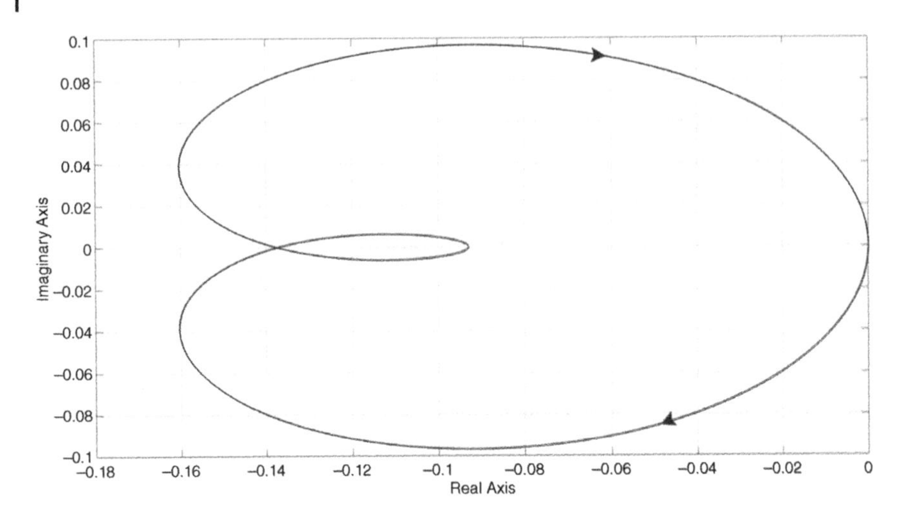

Figure 3.23 Nyquist plot of the system of Example 3.13.

Now

$$G(j\omega)H(j\omega)|_{\omega\,=\,1.2369} = -0.1374 = 0.1374\angle-180^{\circ}$$

From the Nyquist plot of Figure 3.23, where there is no encirclement of the $-1+j0$ point and thus $N = 0$. Therefore, from (3.49), we get $Z = 2$, i.e. there are two closed-loop poles on the right-half s-plane. It can be verified from the characteristic equation of the system that the closed-loop poles are located at $-1.23,\ 2.61 \pm j1.06$.

Example 3.14 This is a design example, where we shall determine the range of K_P for a stable operation of the system given by the following characteristic equation

$$1 + G(s)H(s) = s^3 + 4K_Ps^2 + (K_P + 3)s + 10 = 0$$

From the characteristic equation, the following open-loop transfer function is obtained

$$G(s)H(s) = K_P\frac{4s^2 + s}{s^3 + 3s + 10}$$

The open-loop poles are at -1.7, $0.85 \pm j2.273$, and hence $P = 2$. Thus, to have a stable operation ($Z = 0$), we must have two counterclockwise encirclements of the $-1+j0$ point ($N = -2$). Now

$$\begin{aligned}G(j\omega)H(j\omega) &= K_P\frac{j\omega - 4\omega^2}{10 + j\omega(3-\omega^2)}\\ &= K_P\frac{\omega(\omega - 4\omega^2) - 40\omega^2 + j[10\omega + 4\omega^3(3-\omega^2)]}{100 + \omega(3-\omega^2)^2}\end{aligned}$$

The terminal points are then

$$G(j\omega)H(j\omega)|_{\omega=0} = 0\angle 0^\circ, \quad G(j\omega)H(j\omega)|_{\omega\to\infty} = 0\angle -90^\circ$$

The real axis crossing will occur when

$$10\omega + 4\omega^3(3-\omega^2) = 0$$

The above equation has roots at $\omega = 0$ and $\omega = \pm 1.92$ rad/s. Hence

$$G(j\omega)H(j\omega)|_{\omega=1.92} = -1.47K_P$$

For the frequency of $\omega = 1.92$ rad/s, we can write

$$1 + G(j\omega)H(j\omega) = 1 - 1.47K = 0$$

$$\Rightarrow K = \frac{1}{1.47} = 0.68$$

The Nyquist plot for $K = 0.68$ is shown in Figure 3.24a. It passes through $-1 + j0$ point indicating that the system is critically stable, i.e. undamped with two complex conjugate poles on the imaginary axis. We know that the Nyquist plot needs two counterclockwise encirclements of the $-1 + j0$ point for stability. Therefore, from Figure 3.24a, we can surmise that, for stability, K must be greater than 0.68. The Nyquist plot for $K = 0.9$ is shown in Figure 3.24b. Since this plot encircles the $-1 + j0$ point twice in the counterclockwise direction, the system is stable for this value of K. This result can also be verified using Routh–Hurwitz's criterion.

3.6 Relative Stability

Consider the three Nyquist plots of the unity feedback system shown in Figure 3.25. These plots are for three different values of the open-loop gain $|G(j\omega)|$. Assume that the open-loop system is stable. Therefore, for stability, there shall not be any encirclement of the $-1 + j0$ point. When this gain is small, the Nyquist path does not encircle this point (Curve-1). As the gain increases, the Nyquist path changes – it passes through $-1 + j0$ point (Curve-2) or encircles this point (Curve-3). This implies that the system is critically stable when it is on Curve-2 and is unstable when it is on Curve-3. Therefore, it can be surmised that, with the gain $|G(j\omega)|$, the system stability changes. In this section, we evaluate how far a system is from instability using both Nyquist and Bode plots.

3.6.1 Phase and Gain Margins

Consider the Nyquist plot of the open-loop transfer function of the unity feedback system shown in Figure 3.26a. Also plotted in this figure is a circle with unit radius. The Nyquist plot intersects the unit circle at point ρ_1 and the real axis at point ρ_2. These two intersections points are defined as

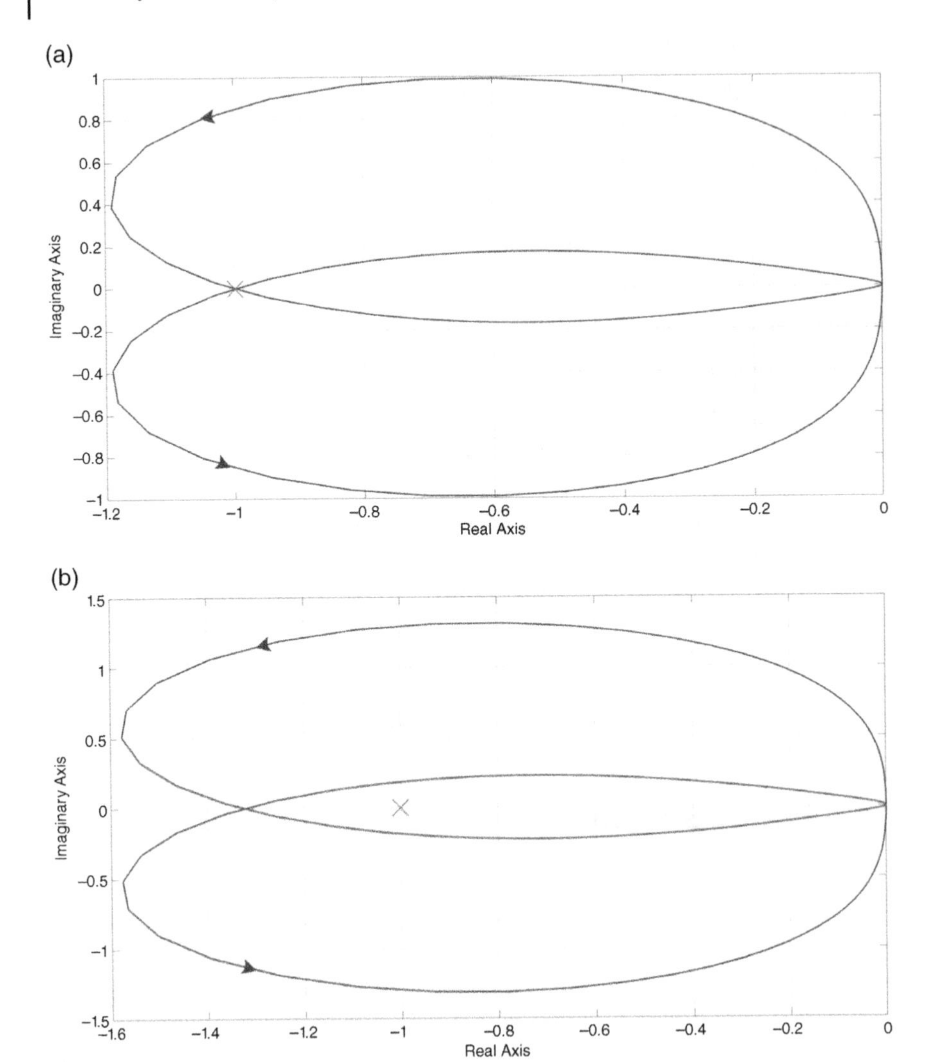

Figure 3.24 Nyquist plot for the (a) undamped system and (b) stable system of Example 3.14.

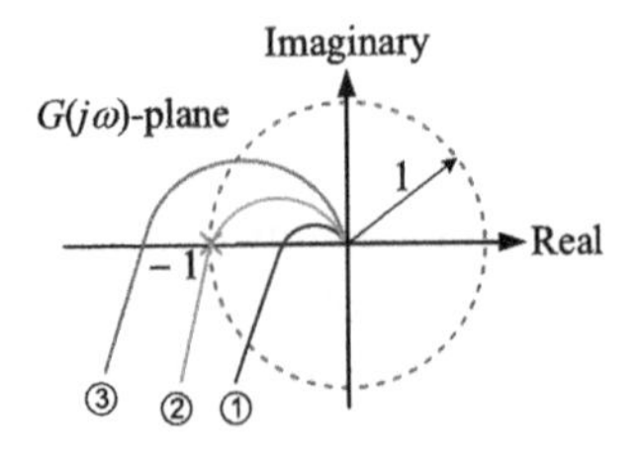

Figure 3.25 Nyquist plots as the open-loop gain $|GH|$ of a unity feedback system changes.

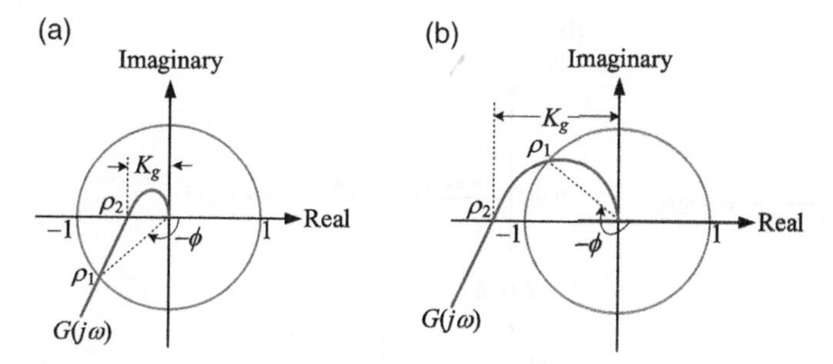

Figure 3.26 Nyquist plot of a system with (a) positive phase and GMs and (b) negative phase and GMs.

- ρ_1: gain crossover point. At this point, the plot is on the unit circle and hence the gain $|G(j\omega)|$ is equal to 1. The frequency at which this occurs is called the gain crossover frequency ω_g.
- ρ_2: phase crossover point. At this point, the plot intersects the negative real axis and hence the phase $\angle G(j\omega)$ is equal to $-180°$. The frequency at which this occurs is called the phase crossover frequency ω_p.

Now note that in Curve-2 of Figure 3.25 the gain and phase crossover points of the critically stable curve are the same, and it occurs when the curve intersects the unit circle. Therefore, for critical stability we can write $G(j\omega) = -1 + j0 = 1\angle -180°$, i.e. the gain should be 1 and the phase should be $-180°$.

The PMs are defined as the additional phase lag required at the gain crossover point to bring the system in the verge of instability. In Figure 3.26a, the angle that $G(j\omega)$ makes at gain crossover point ρ_1 is $(360° - \phi)$. Noting that at the critical point the phase must be $-180°$, the PM is defined by how many additional degrees the system will require to reach this critical point, i.e.

$$\text{PM} = \gamma = 360° - \phi - 180° = 180° - \phi \tag{3.50}$$

In Figure 3.26a, since $\phi < 180°$, the PMs $\gamma > 0$, and the system is said to have positive PM. On the other hand, since $\phi > 180°$ in Figure 3.26b, the system is said to have a negative PM, i.e. $\gamma < 0$.

From Figure 3.26a, we find that $|G(j\omega)| < 1$ at phase crossover frequency. Let us denote the gain at phase crossover frequency as $|G(j\omega_p)| = K_g$. The gain margin (GM) defines how much additional gain is required to push the system on the verge of instability. It is given in terms of decibels by

$$\text{GM} = 0 - 20\log K_g = 20\log\left(\frac{1}{K_g}\right)\ \text{dB} \tag{3.51}$$

From (3.51), it is obvious that the GM will be negative if $K_g > 1$. The system with the Nyquist plot of Figure 3.26b has both negative GM and negative PM. However, the system's stability is affected by either of them being negative.

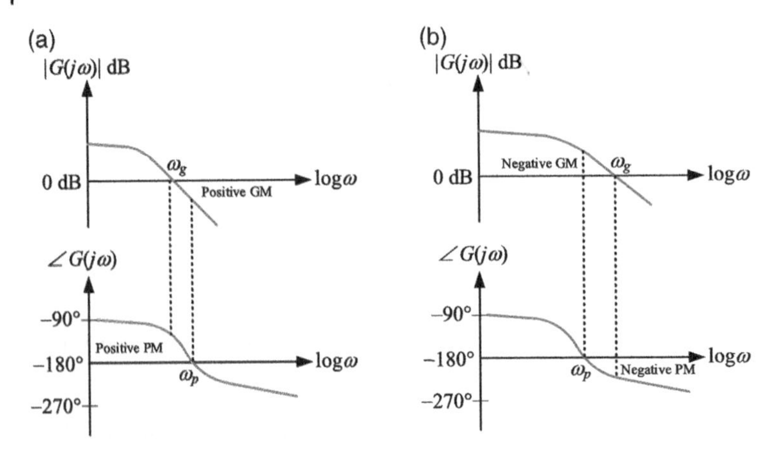

Figure 3.27 Bode plot of a system with (a) positive phase and GMs and (b) negative phase and GMs.

The Bode plot of a system with positive GM and negative PM is shown in Figure 3.27a, in which, the phase is greater than $-180°$ at the gain crossover frequency ω_g, while the gain is below 0 dB at the phase crossover frequency ω_p. In Figure 3.27b, a system with both negative GM and negative PM is shown. At ω_g, the phase is below $-180°$, while the gain is above 0 dB at ω_p. A system is stable when both gain and PMs are positive.

Example 3.15 The open-loop transfer function of a unity feedback control system is

$$G(s)H(s) = \frac{K_P}{s(s^2 + s + 2)}$$

We shall determine the value of K_P such that the PM of the system is 50°. The transfer function can be rewritten as

$$G(j\omega)H(j\omega) = \frac{K_P}{j\omega(2 - \omega^2 + j\omega)}$$

At the gain crossover frequency ω_g, the phase must be equal to $-180° + 50° = -130°$. Therefore

$$\angle G(j\omega_g)H(j\omega_g) = -90° - \tan^{-1}\left(\frac{\omega_g}{2 - \omega_g^2}\right) = -130°$$

$$\Rightarrow \tan^{-1}\left(\frac{\omega_g}{2 - \omega_g^2}\right) = 40°$$

Solving the above equation, we get the following quadratic equation

$$\omega_g^2 + 1.192\omega_g - 2 = 0$$

Solving the quadratic equation gives us $\omega_g = 0.94, \; -3.13$ rad/s. At the gain crossover frequency, the gain must be equal to 1 and hence

$$|G(j0.94)H(j0.94)| = \left|\frac{K_P}{j0.94(2-0.94^2+j0.94)}\right| = \frac{K_P}{1.371} = 1$$

From the above equation, we get $K_P = 1.371$.

Note that the phase crossover will occur when the imaginary component is zero (see Figure 3.27a). Therefore

$$\omega_p\left(2-\omega_p^2\right) = 0$$

$$\Rightarrow \omega_p = 0 \text{ and } \omega_p = \pm\sqrt{2}\,\text{rad/s} = \pm 1.4142\,\text{rad/s}$$

We then get

$$K_g = |G(j1.4141)H(j1.4142)| = 0.6855$$

The GM is then $20\log\left({}^{1}\!/_{K_g}\right) = 20\log\left({}^{1}\!/_{0.6855}\right) = 3.28$ dB. The Bode plot, where the margins and crossover points are also indicated, is shown in Figure 3.28.

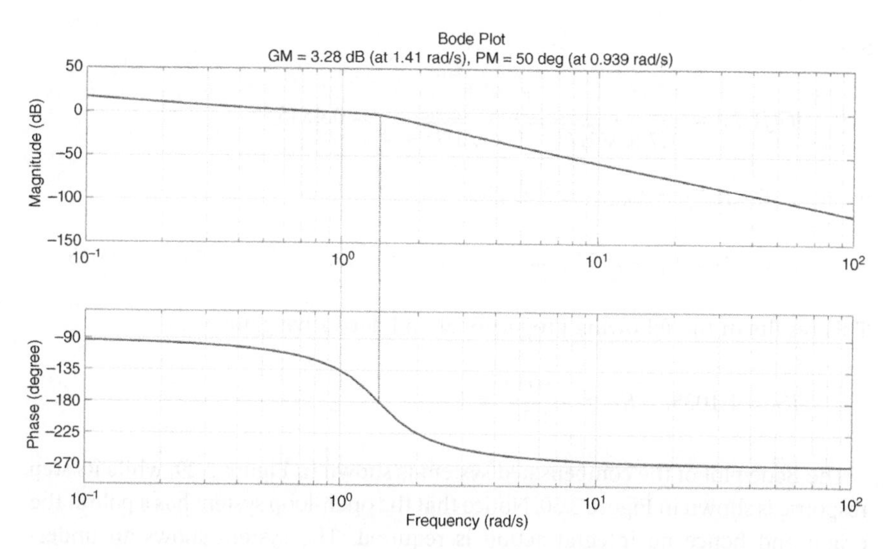

Figure 3.28 Bode plot for the system designed in Example 3.15.

Example 3.16 A unity feedback system with an open-loop transfer function of

$$G(s) = \frac{4}{s(s+1)(s+2)}$$

will be controlled by a PD controller of the form $G_C(s) = K_P + sK_D$. The controller will be designed such that the compensated system has a PM of 50° at the gain crossover frequency (ω_g) of 1.7 rad/s.

At the gain crossover frequency

$$\angle G(j1.7) = -90^\circ - \tan^{-1}(1.7) - \tan^{-1}\left(\frac{1.7}{2}\right) = -189.9^\circ$$

At this frequency, the phase of the compensated system must be equal to $(50-180)^\circ = -130^\circ$. Therefore

$$\angle G_C(j1.7) - 189.9^\circ = -130^\circ$$

$$\Rightarrow \angle G_C(j1.7) = 59.9^\circ$$

Now we can define G_C at the gain crossover as

$$G_C(j1.7) = |G_C(j1.7)|e^{j59.9^\circ} = |G_C(j1.7)| \times (0.5015 + j0.8651)$$

Also note that at the gain crossover

$$|G_C(j1.7)| \times |G(j1.7)| = 1$$

Since

$$|G(j1.7)| = \frac{4}{1.7 \times \sqrt{1.7^2+1} \times \sqrt{1.7^2+4}} = 0.4545$$

we have $|G_C(j1.7)| = 1/0.4545 = 2.2$. Then

$$G_C(j1.7) = K_P + j1.7K_D = 2.2 \times (0.5015 + j0.8651) = 1.1035 + j1.9035$$

This results in the following proportional and derivative gains

$$K_P = 1.1035, \quad K_D = \frac{1.9035}{1.7} = 1.2$$

The Bode plot of the compensated system is shown in Figure 3.29, while its step response is shown in Figure 3.30. Notice that the open-loop system has a pole at the origin and hence no integral action is required. The system shows an underdamped response with a zero steady state error.

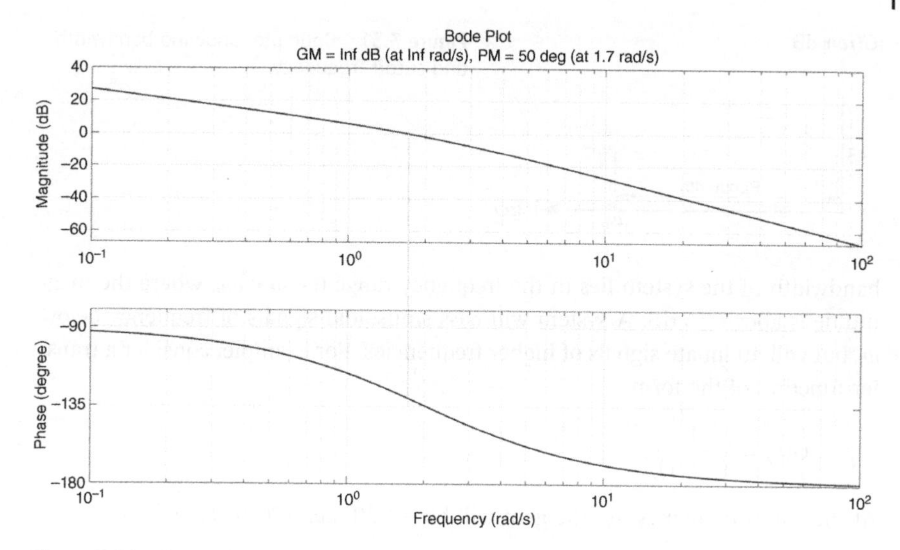

Figure 3.29 Bode plot of the compensated system of Example 3.16.

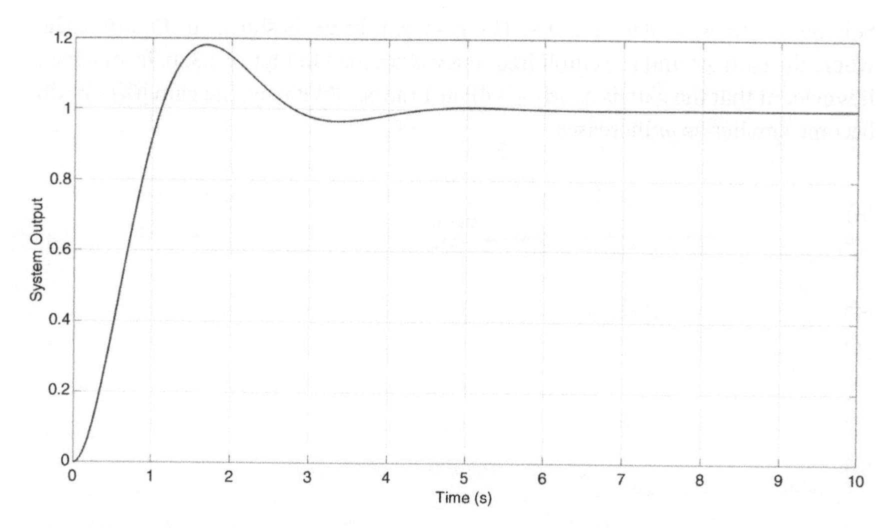

Figure 3.30 Step response of the compensated system of Example 3.16.

3.6.2 Bandwidth

The bandwidth of a system defines how well the system will be able to pass a sinusoidal input. Consider the Bode plot shown in Figure 3.31. In this, the magnitude becomes −3 dB at ω_b, which is called the cutoff frequency of the system. The

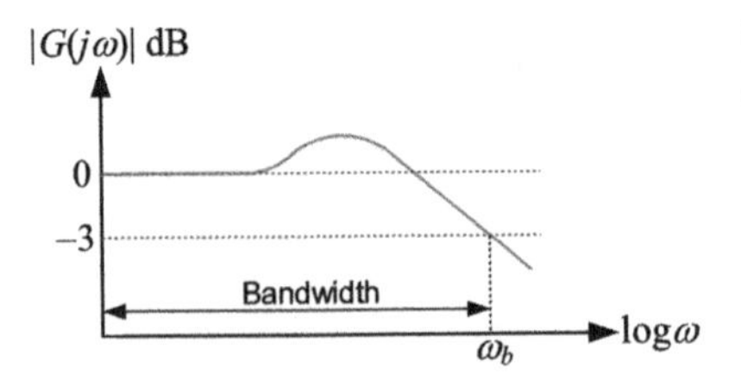

Figure 3.31 Bode plot showing bandwidth and cutoff frequency.

bandwidth of the system lies in the frequency range $0 < \omega < \omega_b$, where the magnitude is above −3 dB. A system will pass sinusoidal signals of frequency below ω_b but will attenuate signals of higher frequencies. For example, consider a transfer function of the form

$$G(s) = \frac{1}{(s+1)}$$

At the cutoff frequency ω_b, the gain will be −3 dB, i.e. 0.7079. Then

$$|G(j\omega_b)| = \frac{1}{\sqrt{\omega_b^2 + 1}} = 0.7079$$

Solving we get $\omega_b = 0.9976$ rad/s. The gain plot in dB is shown in Figure 3.32a, where the gain around the cutoff frequency is zoomed in Figure 3.32b, from which it is evident that the gain is nearly −3 dB at 1 rad/s. Obviously, the gain $|G(j\omega)|$ will become smaller as ω increases.

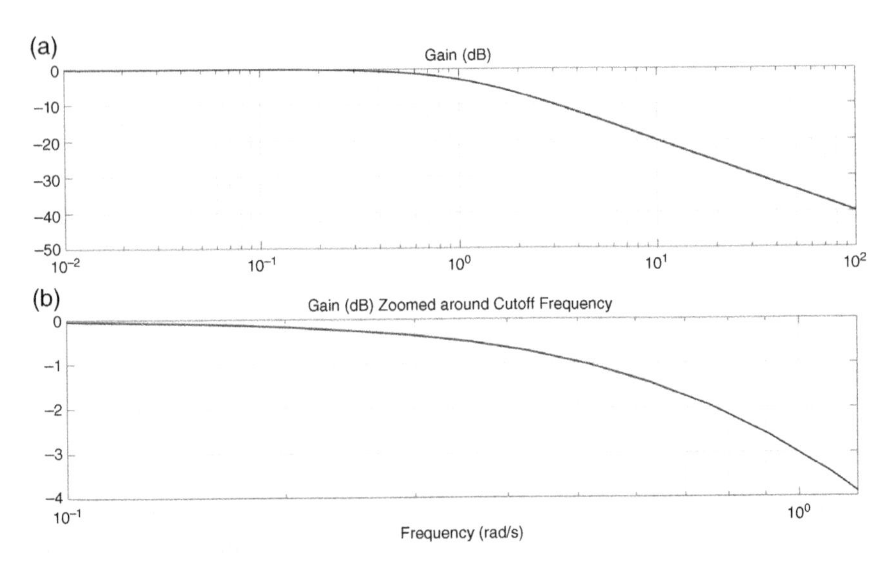

Figure 3.32 Log magnitude of a first-order system. (a) gain plot and (b) gain around the cutoff frequency.

Example 3.17 Consider a unity feedback system with an open-loop transfer function of

$$G(s) = \frac{25}{s^2 + 10\xi s}$$

The closed-loop transfer function is

$$\frac{G(s)}{1 + G(s)} = \frac{25}{s^2 + 10\xi s + 25}$$

This indicates that the closed-loop system has an undamped natural frequency of 5 rad/s. Now we obtain the following cutoff frequencies for different values of the damping ratio

$$\xi = 0.1, \quad \omega_b = 7.71 \text{ rad/s}$$
$$\xi = 0.5, \quad \omega_b = 6.36 \text{ rad/s}$$
$$\xi = 0.9, \quad \omega_b = 3.72 \text{ rad/s}$$

The step response of the system for these three different values of damping ratio is shown in Figure 3.33. From these, the following observations can be made:

- As the damping ratio increases, the cutoff frequency decreases.
- As the damping ratio increases, the peak overshoot decreases.

Therefore, the peak overshoot and cutoff frequency are directly related to each other.

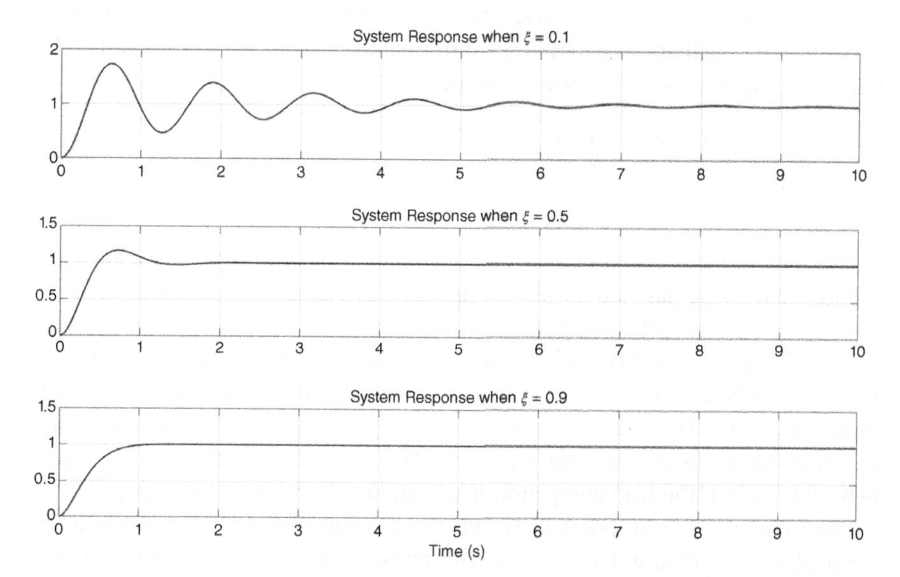

Figure 3.33 Step response of the system of Example 3.17 for three different values of the damping ratio.

3.7 Compensator Design

Compensators are basically controllers. They shape the response of the closed-loop system, just like the PID controller discussed in Section 3.4. These are, however, called compensators as they can compensate the phase of an open-loop system. There are two basic forms of compensators: phase lead and phase lag. They can also be combined to form a lead–lag compensator. Phase lead compensators improve the transient response but have little or no effect on the steady state properties. Therefore, they are similar in their performances to PD controllers. Phase lag compensators, on the other hand, improve the steady state performance, albeit at the expense of transient response. Consequently, they are akin to PI controllers. Lead–lag compensators combine the two and hence can alter both the transient response and the steady state performance, just like a PID controller.

3.7.1 Lead Compensator

A lead (or phase lead) compensator has the following transfer function

$$G_C(s) = \frac{s + 1/\tau}{s + 1/\alpha\tau} = \alpha\frac{\tau s + 1}{\alpha\tau s + 1}, \alpha < 1 \tag{3.52}$$

It can be seen that $G_C(j0) = \alpha$ and $G_C(j\infty) = 1$. The polar plot of the lead compensator is shown in Figure 3.34a.

The maximum value of the phase lead angle is denoted by ϕ_m. This is the angle between the imaginary axis and tangent drawn from the origin to the semicircle shown in Figure 3.34a. Then we can write

$$\sin\varphi_m = \frac{(1-\alpha)/2}{(1+\alpha)/2} = \frac{1-\alpha}{1+\alpha} \tag{3.53}$$

This equation gives us a relation between the maximum phase lead angle and α.

The Bode plot of the system, using asymptotic approximation, is shown in Figure 3.34b. This has two corner frequencies: one at $\omega = 1/\tau$ and the other at $\omega = 1/\alpha\tau$. Since $\alpha < 1$, the numerator term dominates the response, holding it to 0 dB before $\omega = 1/\tau$ and 20 dB/decade rise thereafter till $\omega = 1/\alpha\tau$. The denominator term is 0 dB before $\omega = 1/\alpha\tau$ and a fall of −20 dB/decade thereafter. The overall magnitude curve becomes constant after $\omega = 1/\alpha\tau$, since the numerator and the denominator terms cancel each other out. The phase plot also shows a similar behavior. In fact, the lead compensator will show a highpass filter response.

The frequency of the point at which the tangent touches the semicircle is denoted by ω_m (Figure 3.34a). From the phase plot, we can surmise that the maximum phase lead occurs at the geometric mean of the two corner frequencies, i.e.

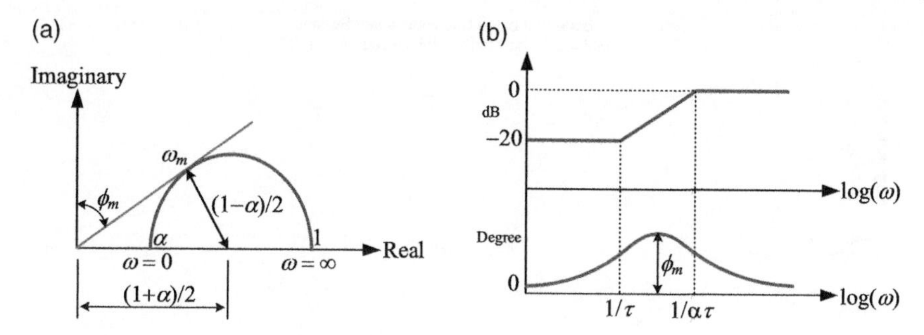

Figure 3.34 (a) Polar plot and (b) Bode plot of a lead compensator.

$$\log(\omega_m) = \frac{1}{2}\left[\log\left(\frac{1}{\tau}\right) + \log\left(\frac{1}{\alpha\tau}\right)\right] = \frac{1}{2}\log\left(\frac{1}{\alpha\tau^2}\right) = \log\left(\frac{1}{\sqrt{\alpha}\tau}\right)$$

Therefore, we can write

$$\omega_m = \frac{1}{\sqrt{\alpha}\tau} \tag{3.54}$$

Example 3.18 Consider a unity feedback system with an open-loop transfer function of

$$G(s) = \frac{1}{s^2}$$

The Bode plot of the uncompensated system is shown in Figure 3.35. We now want to design a phase lead compensator such that the PM of the compensated system is 45°.

The designed lead compensator will have the following form

$$G_C(s) = \alpha K_c \frac{\tau s + 1}{\alpha\tau s + 1}, \quad \alpha K_c = 1$$

The phase lead condition of (3.52) gives

$$\sin 45^\circ = \frac{1-\alpha}{1+\alpha} \Rightarrow \alpha = 0.1716$$

At the maximum phase lead frequency of $\omega_m = 1/\sqrt{\alpha}\tau$, we have

$$|G_C(j\omega_m)| = \frac{\tau \times \dfrac{j}{\sqrt{\alpha}\tau} + 1}{\alpha\tau \times \dfrac{j}{\sqrt{\alpha}\tau} + 1} = \frac{j + \sqrt{\alpha}}{j\alpha + \sqrt{\alpha}} = \sqrt{\frac{1+\alpha}{\alpha^2+\alpha}} = \sqrt{\frac{1}{\alpha}}$$

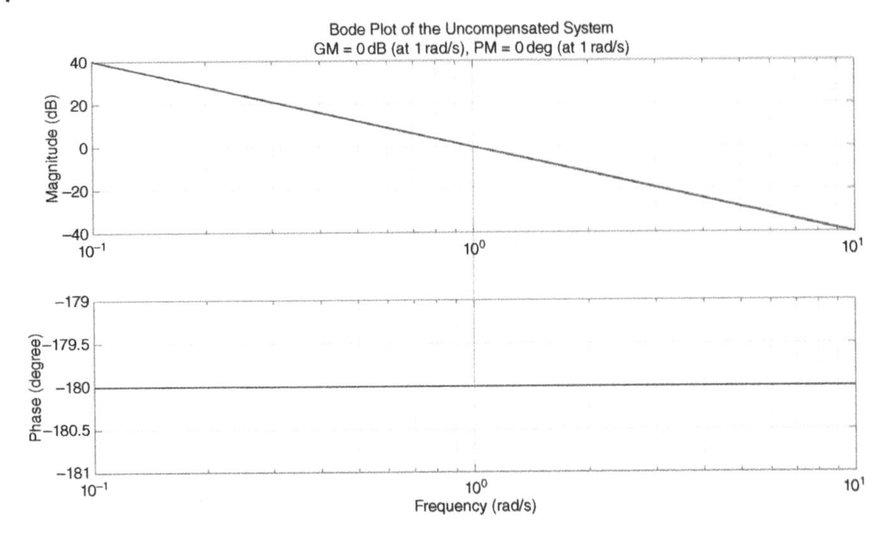

Figure 3.35 Bode plot of the uncompensated system of Example 3.18.

The frequency at which the maximum phase lead occurs to be the gain crossover frequency, i.e.

$$|G_C(j\omega_m)| + |G(j\omega_m)| = 0 \text{ dB}$$

$$\Rightarrow 20\log\left(\frac{1}{\omega_m^2}\right) = -20\log\left(\frac{1}{\sqrt{\alpha}}\right)$$

Therefore

$$\omega_m^2 = \frac{1}{\sqrt{\alpha}} \Rightarrow \omega_m = 1.5537 \text{ rad/s}$$

Again from (3.54), we get

$$\tau = \frac{1}{\omega_m\sqrt{\alpha}} = 1.5538$$

Therefore, the lead compensator is

$$G_C(s) = \frac{1.5538\,s + 1}{0.2666\,s + 1}$$

The Bode plot of the compensated system is shown in Figure 3.36.

Note that a simple unit feedback of the uncompensated system will result in an undamped system with the poles located at $\pm j1$. The step response of the closed-loop system is shown in Figure 3.37, where it is well damped. We can therefore reiterate that a phase-lead compensator improves the transient response of the system, just like a PD controller.

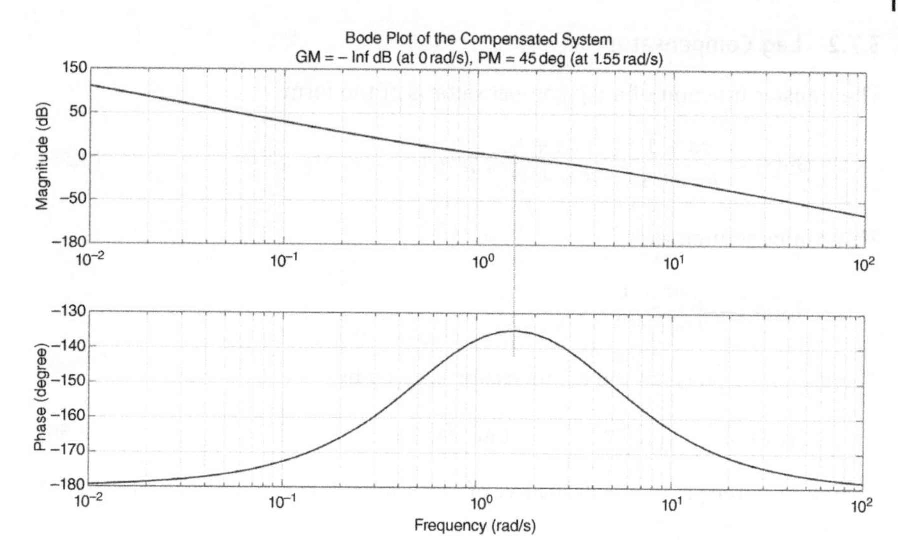

Figure 3.36 Bode plot of the compensated system of Example 3.18.

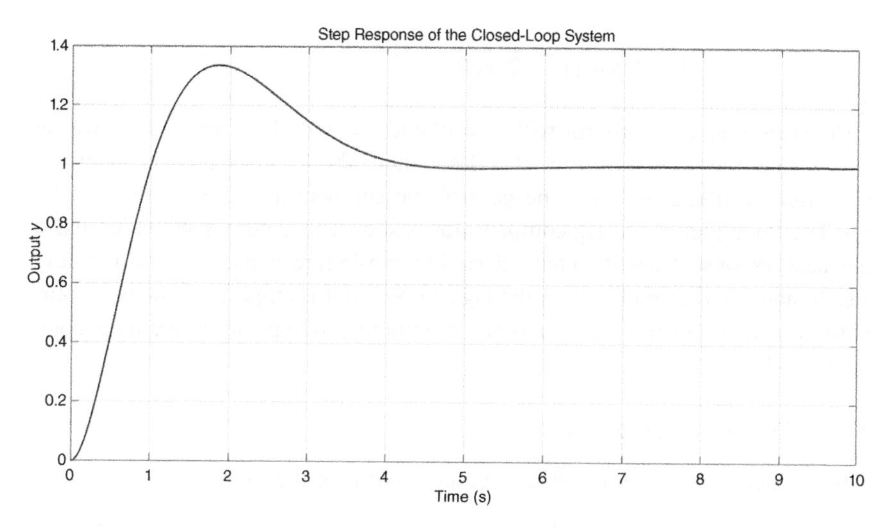

Figure 3.37 The closed-loop step response of the compensated system of Example 3.18.

3.7.2 Lag Compensator

The transfer function of a lag compensator is of the form

$$G_C(s) = \frac{\tau s + 1}{\beta \tau s + 1} = \frac{1}{\beta} \frac{s + 1/\tau}{s + 1/\beta\tau}, \quad \beta > 1 \tag{3.55}$$

This can be written as

$$G_C(j\omega) = \frac{j\omega\tau + 1}{j\omega\beta\tau + 1}$$

Therefore, the phase of the compensator is given by

$$\angle\{G_C(j\omega)\} = \tan^{-1}(\alpha\tau\omega) - \tan^{-1}(\tau\omega) \tag{3.56}$$

To minimize the phase, the derivative of $\angle\{G_C(j\omega)\}$ is taken with respect to ω and is equated to zero to get

$$\begin{aligned} &\frac{\tau}{(\omega\tau)^2 + 1} - \frac{\beta\tau}{(\omega\beta\tau)^2 + 1} = 0 \\ &\Rightarrow \omega = \pm\sqrt{\frac{\beta - 1}{\beta\tau^2(\beta - 1)}} = \pm\frac{1}{\tau\sqrt{\beta}} \end{aligned} \tag{3.57}$$

Consider a lag compensator with $\beta = 10$ and $\tau = 0.01$. Then from (3.57) we get $\omega = 31.62$ rad/s. Substituting this frequency, the compensator is given by $G_C(j31.62) = 0.32\angle - 54.9°$. The gain of the compensator at 31.62 rad/s is 10 dB. The Bode plot of the lag compensator is shown in Figure 3.38, which has a low pass response. If a system has a desirable transient response but unsatisfactory steady state characteristics, then the lag compensator is employed. The basic purpose is to increase the open-loop gains without moving the closed-loop poles appreciably.

3.7.3 Lead–lag Compensator

The transfer function of a lead–lag compensator is of the form

$$G_C(s) = \left(\frac{s + 1/\tau_1}{s + \beta/\tau_1}\right) \times \left(\frac{s + 1/\tau_2}{s + 1/\beta\tau_2}\right), \quad \beta > 1 \tag{3.58}$$

The Bode plot of this compensator is shown in Figure 3.39. It is obvious that it is a combination of a lag network and a lead network. The response of this network can be shaped by the proper choice of the gains and time constants.

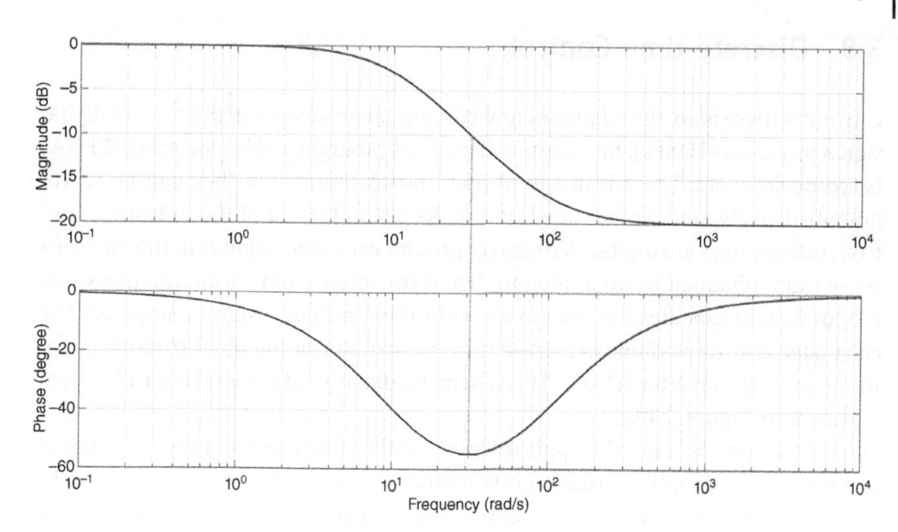

Figure 3.38 Bode plot of the lag compensator with $\beta = 10$ and $\tau = 0.01$.

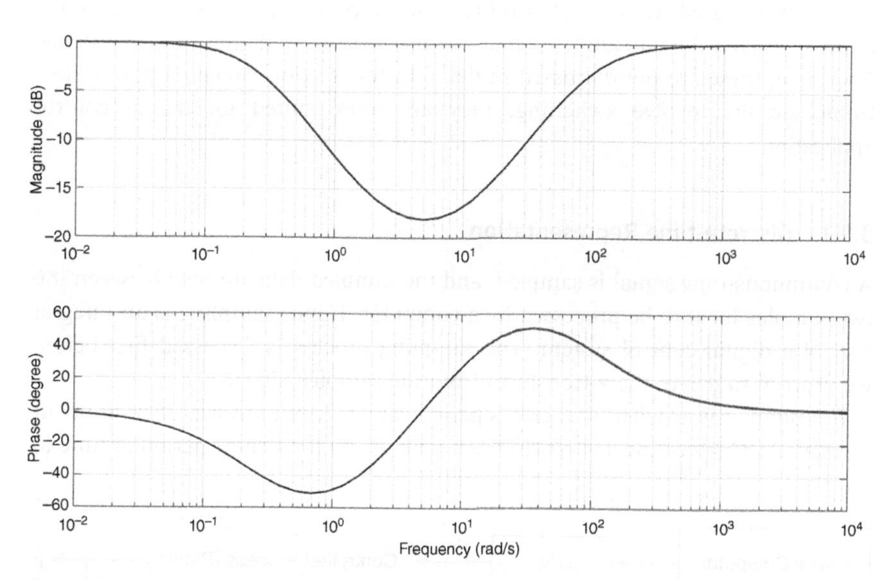

Figure 3.39 Bode plot of a lead–lag compensator with $\beta = 10$, $\tau_1 = 0.1$ and $\tau_2 = 0.4$.

3.8 Discrete-time Control

Currently there is an abundance of low-cost digital processors in the form of digital signal processors (DSPs), microcontrollers, field-programmable gate array (FPGA) based devices, etc. The advantage of these processors is that they can be reprogrammed easily and can be tuned online. As a result, most of the industrial controllers these days are digital. A digital controller uses input signals in the form of a pulse train, obtained by an analog-to-digital converter (ADC) from the measured output data. It executes the control algorithm that resides inside a computer. The calculated control variable is then sent to the actuator of the plant through a digital-to-analog converter (DAC). The schematic diagram of a digital control system is shown in Figure 3.40.

A digital control is not the application of continuous-time controller in a digital computer. The sampling process gives it an altogether different characteristic. In digital control, we use z-transform and difference equations as opposed to Laplace transform and differential equations used in continuous-time systems. However, there are similarities between these two different approaches. Most of the techniques that are used in classical continuous-time control systems (such as roots locus, frequency domain analysis, etc.) can also be used for digital control systems. Therefore, these are not discussed in detail in this section. Since all power electronic circuits involve switching, they are more suited for digital control application.

3.8.1 Discrete-time Representation

A continuous-time signal is sampled, and the sampled data are held between the two samples for it to be processed by a computer. Hence sampling is an integral part of a digital control system. The sampling process is discussed first before we proceed to derive the z-transform from the sampled signal.

The sampler is a switch that closes periodically to input physical measurement through ADC. Once the switch closes, it remains on for a short period of time to

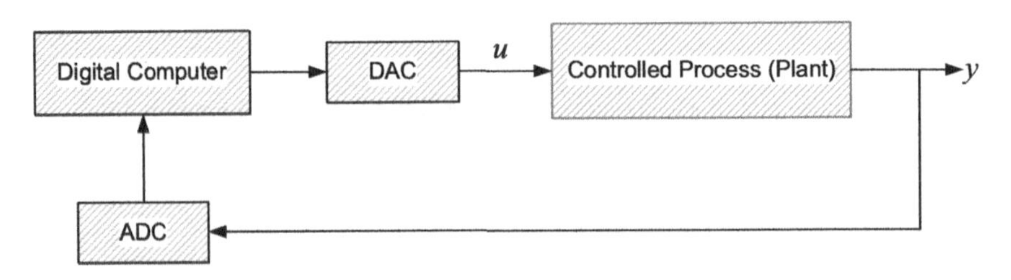

Figure 3.40 Schematic diagram of a digital control system.

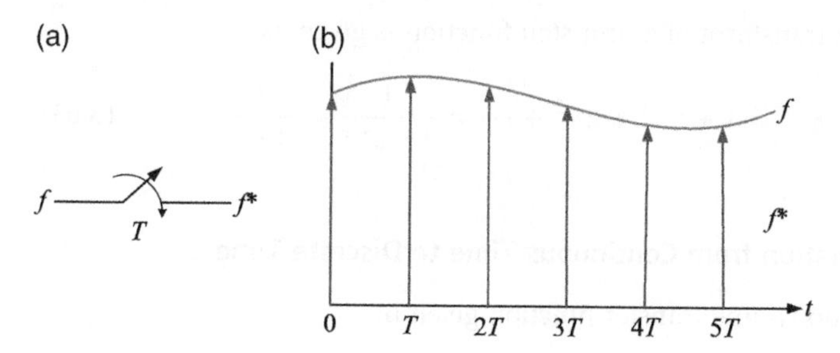

Figure 3.41 The sampling process: (a) ideal sampler and (b) impulse modulated sampling waveform.

capture the signal. In practice, however, this time period is very small compared to the sampling time T and, therefore, is neglected. The sampling process then can be graphically represented, as in Figure 3.41a, where the input to the sampler (S) is the signal $f(t)$, while the output is a train of impulses $f^*(t)$. The sampler in this case is called an ideal sampler. The train of impulses for an arbitrary signal is shown in Figure 3.41b and is expressed mathematically as

$$\begin{aligned} f^*(t) &= \sum_{k=0}^{\infty} f(kT)\delta(t-kT) \\ &= f(0)\delta(t) + f(T)\delta(t-T) + f(2T)\delta(t-2T) + \cdots \end{aligned} \tag{3.59}$$

Taking the Laplace transform on both sides of (3.59), we get

$$F^*(s) = \mathcal{L}\{f^*(t)\} = \int_0^{\infty} \left\{ \sum_{k=0}^{\infty} f(kT)\delta(t-kT) \right\} e^{-st} dt = \sum_{k=0}^{\infty} f(kT)\, e^{-skT} \tag{3.60}$$

3.8.2 The z-transform

The Laplace transform of the train of impulses given by (3.60) is not a rational function, given the presence of the e^{-sT} terms on the right-hand side. Thus, it is desirable to have a transform that converts $F^*(s)$ into a rational function $F(z)$. This is called the z-transform of $f(t)$. To convert $F^*(s)$ into $F(z)$, we substitute

$$z = e^{sT} \tag{3.61}$$

Then the z-transform is given by

$$F(z) = F^*(s)|_{z=e^{sT}} = \sum_{k=0}^{\infty} f(kT)\, z^{-k} \tag{3.62}$$

Using (3.62), the z-transform of a unit step function is given as

$$F(z) = \sum_{k=0}^{\infty} z^{-k} = 1 + z^{-1} + z^{-2} + \cdots = \frac{1}{1-z^{-1}} = \frac{z}{z-1} \tag{3.63}$$

3.8.3 Transformation from Continuous Time to Discrete Time

Consider a continuous-time transfer function given by

$$G(s) = \frac{b_m s^m + b_{m-1}s^{m-1} + \cdots + b_1 s + b_0}{a_n s^n + a_{n-1}s^{n-1} + \cdots + a_1 s + a_0}, \quad n \geq m \tag{3.64}$$

Assuming that the system has n-distinct poles, (3.64) can be expanded in a partial fraction as

$$G(s) = \sum_{i=1}^{n} \frac{K_i}{s + p_i} \tag{3.65}$$

where p_i, $i = 1, 2, \ldots, n$ are the poles and K_is their respective residues. Equation (3.65) can be written in time domain (impulse response) as

$$g(t) = \sum_{i=1}^{n} K_i e^{-p_i t} \tag{3.66}$$

Therefore, we have

$$g(kT) = \sum_{i=1}^{n} K_i e^{-Kp_i T} \tag{3.67}$$

Let us now consider a function $f(t) = e^{-at}$. Using (3.62), its z-transform is given by

$$F(z) = \sum_{k=0}^{\infty} e^{-\alpha kT} z^{-k} = 1 + e^{-\alpha T} z^{-1} + e^{-2\alpha T} z^{-2} + e^{-3\alpha T} z^{-3} + \cdots$$
$$= \frac{1}{1 - e^{-\alpha T} z^{-1}} = \frac{z}{z - e^{-\alpha T}}$$

Using the above result, the discrete-time transfer function $G(z)$ of (3.67) is given as

$$G(z) = \sum_{i=1}^{n} \frac{K_i}{1 - e^{-p_i T} z^{-1}} \tag{3.68}$$

3.8.4 Mapping s-Plane into z-Plane

Note from (3.61) that $z = e^{sT}$. Therefore, as $s = \sigma + j\omega$ is a complex number so is $z = e^{sT}$. We shall now determine the region of stability in the z-plane, given that the region of stability in the s-plane is the closed left half. Assume that a pole is at a

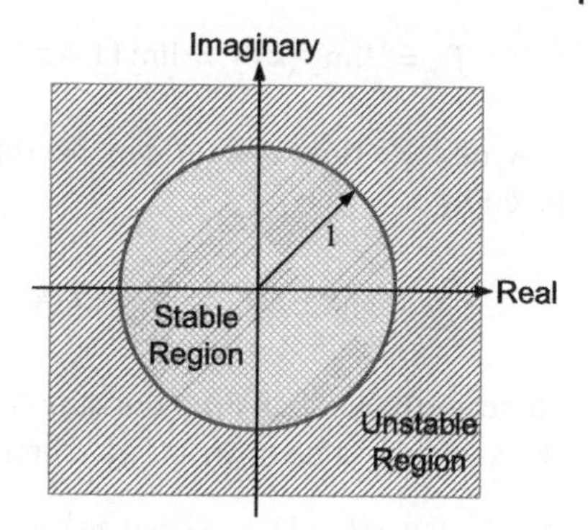

Figure 3.42 Stable and unstable operating regions on the z-plane.

location $s = j\omega$ on the imaginary axis on the s-plane. Then, in the z-domain, it will be at $z = 1 \angle \omega T$. Now if the pole remains on the imaginary axis on the s-plane, then correspondingly it will have a magnitude of 1 on the z-plane and it will turn a full circle every time ωT changes by 360°. Thus, the imaginary axis is mapped into a circle with unit radius, as shown in Figure 3.42. This circle is usually called the unit circle.

Now assume that $s = \sigma + j\omega$. Then

$$z = e^{(\sigma + j\omega)T} = e^{\sigma T} \angle(\omega T)$$
$$\Rightarrow |z| = e^{\sigma T}$$

If $\sigma < 0$, i.e. the pole is on the -half s-plane, then, in the z-domain $0 < |z| < 1$, i.e. the pole in the z-domain will be inside the unit circle. Conversely, if $\sigma > 0$, i.e. the pole is on the right-half s-plane, then $|z| > 1$. This means that the stable region in the s-plane is mapped inside the unit circle in the z-plane, and hence it can be stated that a discrete-time system is stable when its poles are inside the unit circle in the z-plane.

3.8.5 Difference Equation and Transfer Function

There are several properties of z-transform. We shall state only two of them that are important for our discussions. Let us define $F(z) = \mathfrak{z}[f(t)]$. Then

- Real Translation (Shifting Theorem): $\mathfrak{z}[f(t - nT)] = z^{-n}F(z)$

$$\mathfrak{z}[f(t + nT)] = z^n \left[F(z) - \sum_{k=0}^{n-1} f(kT) z^{-k} \right]$$

- Final Value Theorem: If $(1 - z^{-1})F(z)$ does not have a pole on or outside the unit circle (i.e. if the system is stable such that the final value exists), then

$$f_{ss} = \lim_{k \to \infty} f(kT) = \lim_{z \to 1} (1 - z^{-1}) F(z) \tag{3.69}$$

A discrete-time system can be represented by a finite difference equation, given by

$$\sum_{i=0}^{n} a_i y[(k-i)T] = \sum_{i=0}^{m} b_i u[(k-i)T], \quad n \geq m \tag{3.70}$$

Henceforth, we shall drop the sampling time T for brevity, where $y(k)$ will signify the sampled value of $y(t)|_{t=kT}$. Then (3.70) can be expanded as

$$a_0 y(k) + a_1 y(k-1) + \cdots + a_n y(k-n) = b_0 u(k) + b_1 u(k-1) + \cdots + b_m u(k-m) \tag{3.71}$$

Using the shifting property of the z-transform, we get $\mathfrak{z}[f(t-n)] = z^{-n} F(z)$. Therefore, the transfer function of (3.70) can be determined as

$$\frac{Y(z)}{U(z)} = \frac{b_0 + b_1 z^{-1} + \cdots + b_m z^{-m}}{a_0 + a_1 z^{-1} + \cdots + a_n z^{-n}} = \frac{1}{z^{n-m}} \frac{b_0 z^m + b_1 z^{m-1} + \cdots + b_m}{a_0 z^n + a_1 z^{n-1} + \cdots + a_n} \tag{3.72}$$

Assume that $U(z)$ is a unit step function of the form given in (3.63). Then from (3.72), we have

$$Y(z) = \frac{1}{1 - z^{-1}} \frac{b_0 + b_1 z^{-1} + \cdots + b_m z^{-m}}{a_0 + a_1 z^{-1} + \cdots + a_n z^{-n}}$$

Using the final value theorem of (3.69), the DC gain is obtained as

$$y_{ss} = \lim_{z \to 1} (1 - z^{-1}) \left(\frac{1}{1 - z^{-1}} \frac{b_0 + b_1 z^{-1} + \cdots + b_m z^{-m}}{a_0 + a_1 z^{-1} + \cdots + a_n z^{-n}} \right) = \frac{b_0 + b_1 + \cdots + b_m}{a_0 + a_1 z + \cdots + a_n} \tag{3.73}$$

Therefore, the DC gain is the sum of numerator coefficients over the denominator coefficients. The major advantage of the difference equation is that it can be used in recursive computation. For example, consider (3.71). This equation can be rewritten for the kth sample as

$$a_0 y(k) = -a_1 y(k-1) - \cdots - a_n y(k-n) + b_0 u(k) + b_1 u(k-1) + \cdots + b_m u(k-m)$$

Then replacing k by $k + 1$, we get

$$\begin{aligned} a_0 y(k+1) = & -a_1 y(k) - \cdots - a_n y(k-n-1) + b_0 u(k+1) + b_1 u(k) \\ & + \cdots + b_m u(k-m-1) \end{aligned}$$

Therefore, we find that $y(k + 1)$ is computed with the new input $u(k + 1)$ and shifting the previous values of inputs and outputs by one.

If $n = m - 1$ in (3.70), the transfer function is written as

$$\frac{Y(z)}{U(z)} = \frac{1}{z^{-1}} \frac{b_0 z^m + b_1 z^{m-1} + \cdots + b_m}{a_0 z^n + a_1 z^{n-1} + \cdots + a_n} = \frac{b_0 + b_1 z^{-1} + \cdots + b_m z^{-m}}{a_0 z^{-1} + a_1 z^{-2} + \cdots + a_n z^{-n-1}}$$

This results in the following difference equation

$$a_0 y(k-1) + a_1 y(k-2) + \cdots + a_n y(k-n-1) = b_0 u(k) + b_1 u(k-1) + \cdots + b_m u(k-m)$$

We can then see that the system becomes noncausal, i.e. $y(k-1)$ depends on $u(k)$. This implies that the output at $k-1$ depends on the future input at k, which is not possible from a control design perspective. Thus, to ensure causality, it is stipulated that $n \geq m$.

3.8.6 Digital PID Control

Consider the PID controller given in (3.26). There are several methods for representing s in discrete form, like forward Euler, backward Euler, or trapezoidal form. The backward Euler method is given by

$$s = \frac{1 - z^{-1}}{T} \tag{3.74}$$

where T is the sampling time. Then we can write (3.26) as

$$\begin{aligned} \frac{U(z)}{E(z)} &= K_P + \frac{K_I T}{1 - z^{-1}} + \frac{K_D}{T}(1 - z^{-1}) \\ &= \frac{K_P(1 - z^{-1}) + K_I T + \frac{K_D}{T}(1 - z^{-1})^2}{1 - z^{-1}} \end{aligned} \tag{3.75}$$

In difference equation form, (3.75) is written as

$$u(k) - u(k-1) = \alpha_0 e(k) + \alpha_1 e(k-1) + \alpha_2 e(k-2) \tag{3.76}$$

where

$$\alpha_0 = K_P + K_I T + \frac{K_D}{T}, \quad \alpha_1 = -K_P - 2\frac{K_D}{T}, \quad \alpha_2 = \frac{K_D}{T}$$

In Chapter 6, we discuss control design in the discrete domain, where several control design methods are discussed.

3.9 Concluding Remarks

In this chapter, a review of the classical control system is presented, along with various techniques that are used for the control design of power electronic converters. Continuous-time control systems are discussed in some detail. We discuss the control of power converters using classical techniques in Chapter 4. An introduction to discrete-time control is also covered. Most of the techniques that are used in continuous-time systems are also valid for discrete-time systems and, therefore, are not discussed in detail. One of the most important aspects of discrete-time systems is the representation of SISO system in difference equation form. This has been extensively used in power converter control systems. Therefore, this is introduced in the chapter; their use in the design of controllers in the difference equation domain is discussed in Chapter 6.

Problems

3.1 Find the transfer function $E_2(s)/E_1(s)$ of the operational amplifier (OPAMP) circuit shown in Figure P3.1.

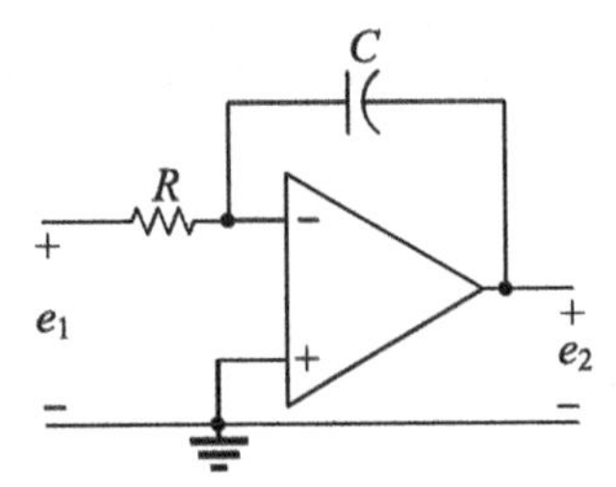

Figure P3.1 OPAMP circuit of Problem 3.1.

3.2 Consider the unity feedback system of where the plant transfer function is given by (Figure P3.2)

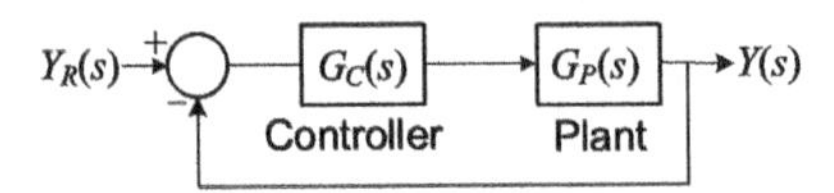

Figure P3.2 Unity feedback control system of Problem 3.2.

$$G_P(s) = \frac{1}{s^2 + 6s + 9}$$

(a) Find the steady state error to a unit step input (i.e. $Y_r(s) = 1/s$) when the controller is a proportional controller with $G_C(s) = K_P$.

(b) We now replace the proportional controller with a PI controller of the form$G_C(s) = K_P + K_I/s$. Determine the values K_P and $K_I$of such that the damping ratio and the undamped natural frequency of the dominant complex conjugate poles are 0.5 and 3 rad/s respectively. The remain closed-loop pole is assumed to be placed at −3.

(c) For the controller designed in (b), find the steady state error to a unit ramp input.

3.3 For the system, the root locus for which is shown in Figure P3.3, find the characteristic equation and the values of σ_1 and ω_0.

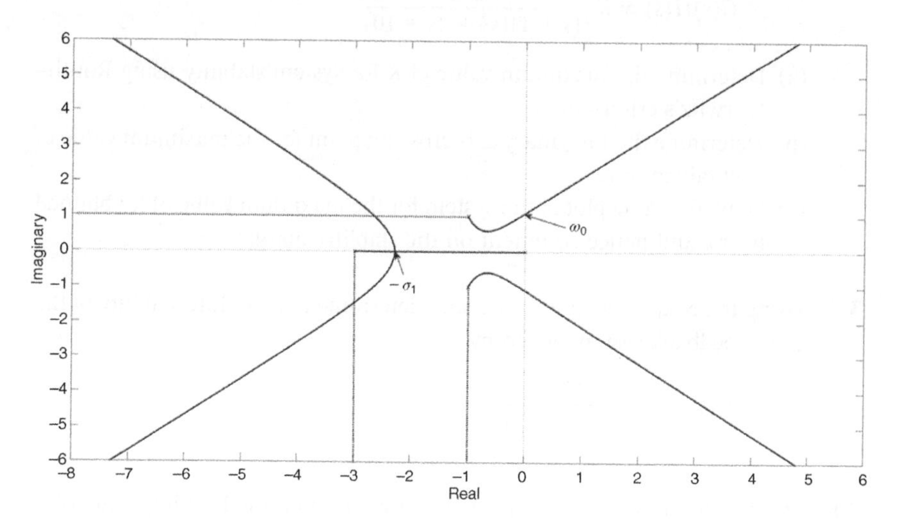

Figure P3.3 Root locus for Problem 3.3.

3.4 Consider again the unity feedback system shown in Figure P3.2, where

$$G_P(s) = \frac{400}{s^2 + 40s}$$

$$G_C(s) = K_P + \frac{K_I}{s}, \quad K_P > 1$$

(a) Draw the root locus as K_I changes from zero to infinity.
(b) Find the relation between K_P and K_I for the absolute stability of the system.

3.5 The open-loop transfer function of a unity feedback control system is given by

$$G(s) = K\frac{s + \beta}{s^2(s + 2)}$$

Draw the root locus and the value (or range) of the parameter β when

(a) There is only one breakaway point at $s = 0$.
(b) When there are two breakaway points including the one at $s = 0$.

3.6 Consider the open-loop transfer function of

$$G(s)H(s) = K\frac{s+2}{s(s+1)(s^2+4s+10)}$$

(a) Determine the maximum value of K for system stability using Routh–Hurwitz's criterion.
(b) Determine the imaginary axis crossing point for the maximum value of K obtained in (a).
(c) Draw the Bode plot of the system for the maximum value of K obtained in (a), and hence comment on the stability margins.

3.7 Using the Nyquist stability criterion, determine the absolute stability of the unity feedback system, given by

$$G(s) = \frac{50}{(s+1)^2(s+10)}$$

3.8 Determine through the Nyquist stability criterion, the limiting value of K for the absolute stability of the system shown in Figure P3.8, where $K > 0$.

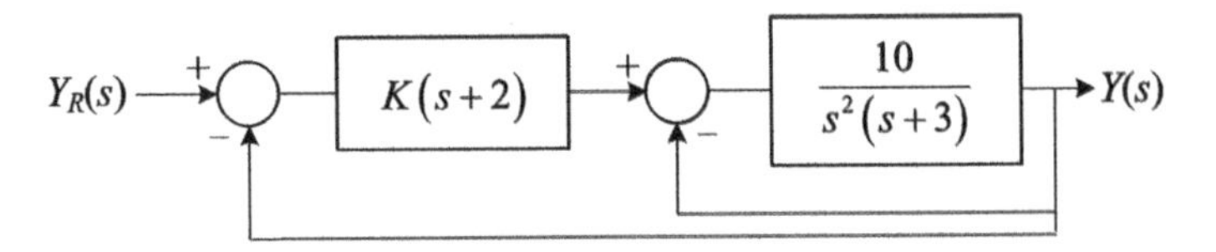

Figure P3.8 Feedback system of Problem 3.8.

3.9 A unity feedback system has the open-loop transfer function of

$$G(s) = \frac{K}{(s+1)^n}$$

Using the Nyquist stability criterion, determine the range of K for stable operation, when (a) $n = 2$ and when (b) $n = 3$.

3.10 The open-loop transfer function of a unity feedback control system is given by

$$G(s) = \frac{K}{s(s^2+s+4)}$$

Determine the value of K such that the PM of the system is 30°. What is the GM of the system for this value of K?

3.11 The open-loop transfer function of a unity feedback control system is given by

$$G(s) = \frac{1 + \tau s}{s(s+1)(1+0.01s)}$$

Determine the smallest possible value of τ such that the system has an infinite GM.

3.12 Consider a second-order system of the form

$$\frac{Y(s)}{Y_r(s)} = \frac{\omega_n^2}{s^2 + 2\xi\omega_n s + \omega_n^2}$$

Show that the bandwidth ω_b is given by

$$\omega_b = \omega_n \times \sqrt{1 - 2\xi^2 + \sqrt{4\xi^4 - 2\xi^2 + 2}}$$

3.13 Consider the feedback control system of Figure P3.2, where

$$G_C(s) = K_P + \frac{K_I}{s} + sK_D$$

Let ω_b be the gain crossover frequency at which the compensated system has a PM of ϕ_m. Then prove that

$$K_P = \frac{\cos\theta}{\left|G_P(j\omega_g)\right|} \text{ and } \omega_g K_D - \frac{K_I}{\omega_g} = \frac{\sin\theta}{\left|G_P(j\omega_g)\right|}$$

where $\theta = \angle G_C(j\omega_g)$.

3.14 In the feedback control system of Figure P3.2, the plant and the controller are given by

$$G_P(s) = \frac{4}{s(s+1)(s+2)} \text{ and } G_C(s) = K_P + sK_D$$

Using the results of Problem 3.13, find K_P and K_D such that the compensated system has a PM of 50° at the crossover frequency of 1.7 rad/s.

3.15 In the feedback control system of Figure P3.2, the plant transfer function is given by

$$G_P(s) = \frac{1}{s^2}$$

Design a phase lead compensator such that the PM of the compensated system is 45°.

3.16 In the feedback control system of Figure P3.2, the plant transfer function is given by

$$G_P(s) = \frac{80}{s(s+4)}$$

Design a phase lag compensator such that the PM of the compensated system is at least 50°.

3.17 Find the z-transform of the following sequences:

$$f(kT) = \begin{cases} 0 & \text{for } k = 0 \text{ and even integers} \\ 1 & \text{for } k = \text{odd integers} \end{cases}$$

$$f(kT) = \begin{cases} 1 & \text{for } k = 0 \text{ and even integers} \\ -1 & \text{for } k = \text{odd integers} \end{cases}$$

$$f(kT) = kTa^k$$

Notes and References

Since this chapter presents a review, an in-depth coverage of different topics is not presented. A prior knowledge of classical control technique and Laplace transform has been assumed. There are several good textbooks that cover the materials presented in this chapter in detail. We have tried to sum up the most relevant topics in one chapter. However, the studies can be supplemented by going through the books listed in [1–5]. All these books are easy to comprehend and can therefore supplement the materials presented in this chapter. There are several textbooks that cover different aspects of digital control. A couple of them are listed below [6, 7]. The Ziegler–Nichols method of PID tuning has been around for a long time. Reference [8] provides a good introduction to this method.

1 Kuo, B.C. (1987). *Automatic Control System*, 5e. Englewood Cliffs, NJ: Prentice-Hall.
2 Ogata, K. (2010). *Modern Control Engineering*, 5e. Englewood Cliffs, NJ: Prentice-Hall.
3 Dorf, R.C. and Bishop, R.H. (2005). *Modern Control Systems*, 10e. Upper Saddle River, NJ: Pearson Education.
4 Nagrath, I.J. and Gopal, M. (1975). *Control Systems Engineering*. New York: Wiley.
5 Franklin, G.F., Powell, J.D., and Emami-Naeini, A. (2019). *Feedback Control of Dynamic Systems*, 8e. Upper Saddle River, NJ: Pearson Education.

6 Kuo, B.C. (1980). *Digital Control System*. Tokyo: Holt-Saunders Japan.
7 Franklin, G.F. and Powell, J.D. (1998). *Digital Control of Dynamic Systems*, 3e. Englewood Cliffs, NJ: Prentice Hall.
8 Goodwin, G.C., Graebe, S.F., and Salgado, M.E. (2001). *Control System Design*. Englewood Cliffs, NJ: Prentice-Hall.

4

Power Electronic Control Design Challenges

In many power electronics systems, the voltage or current waveform across a load should be regulated or controlled regardless of internal or external disturbances and variations in the power electronics parameters. To elaborate on these issues, we discuss DC-DC converters in this chapter, where both step-up (boost) and step-down (buck) converters are considered. A DC-DC converter, connected to two different DC sources and a resistive load, is shown in Figure 4.1. In this chapter, not only is the control of DC-DC converters discussed, but the characteristics of these converter with respect to the sources is also studied.

4.1 Analysis of Buck Converter

Assume that the buck converter is connected to the battery source (v_{in}) with no voltage fluctuation and the aim of the controller is to regulate the output voltage (v_o). The circuit diagram of the converter is shown in Figure 4.2. To find the transfer function of the system, we need to analyze the system when the switch is on and when it is off.

When the switch is turned on, the diode is turned off as the positive input voltage (positive DC voltage v_{in}) appears across the diode. The circuit diagram is shown in Figure 4.3a. The voltage across the inductor is extracted during this time interval (on-time) as

$$v_L(t) = v_{in}(t) - v_o(t) \tag{4.1}$$

When the switch is turned off, the diode is turned on as the current through the inductor is positive in the circuit diagram, shown in Figure 4.3b. Therefore, the voltage across the inductor is extracted during this time interval (off-time) as

$$v_L(t) = -v_0(t) \tag{4.2}$$

Control of Power Electronic Converters with Microgrid Applications, First Edition.
Arindam Ghosh and Firuz Zare.

Published 2023 by John Wiley & Sons, Inc.

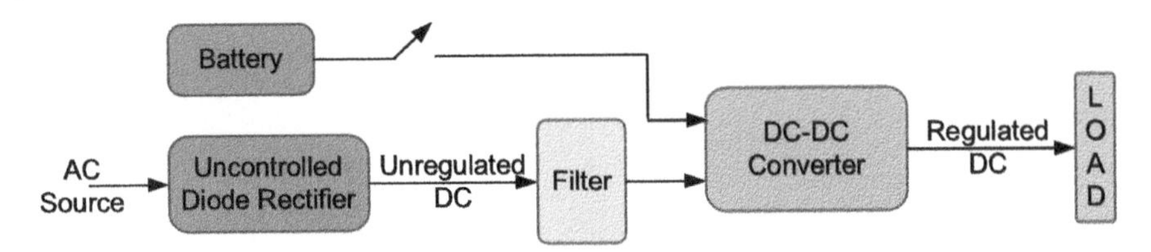

Figure 4.1 The block diagram of a DC-DC converter connected to two different DC sources.

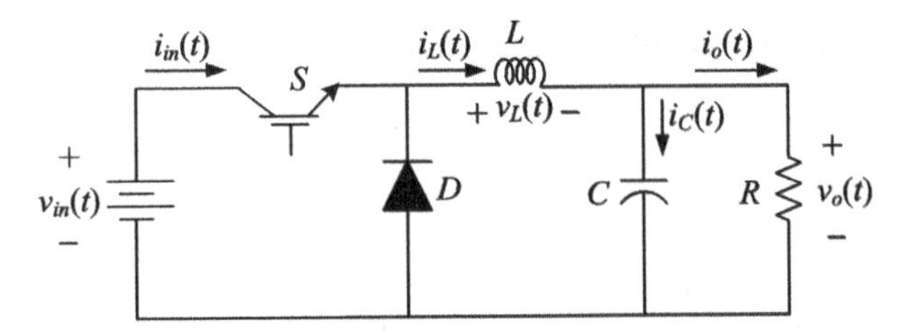

Figure 4.2 The circuit diagram of a buck converter connected to a battery (V_{in}) and a resistive load (R).

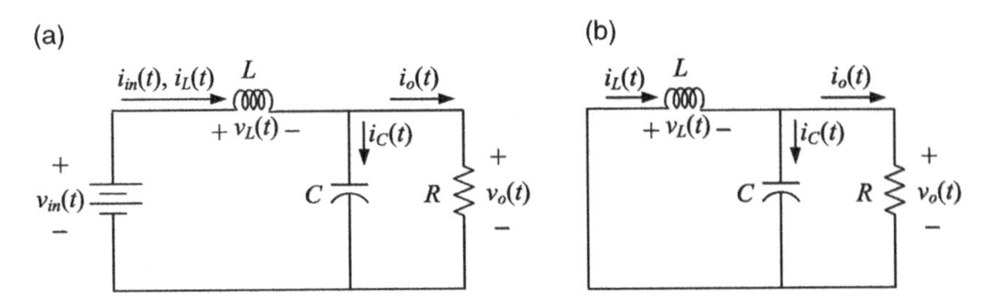

Figure 4.3 The circuit diagram of a buck converter when the switch is (a) on and (b) off.

Assuming the input and output voltage ripples are negligible, and the system is in the steady state, the inductor voltage equations are rewritten as

$$v_L(t) = \begin{cases} V_{in} - V_0 & \text{when the switch is turned on} \\ -V_0 & \text{when the switch is turned off} \end{cases} \tag{4.3}$$

Note that the steady state quantities are denoted by uppercase letters, while instantaneous quantities are denoted by lowercase letters.

The duty cycle (or ratio) $d(t)$ of the converter is defined as the ratio of the on-time, t_{ON} to the switching cycle time, T_s. When the input and output voltages have no variation, the duty cycle is constant (denoted by D) and is expressed as

$$d(t) = D = \frac{t_{ON}}{T_s}$$
$$d'(t) = 1 - d(t) \tag{4.4}$$

Based on the above analysis and operating modes, the voltage and current waveforms of the buck converter over one switching cycle are shown in Figure 4.4. It is assumed that the switch and the diode are ideal (with no voltage drop when they are turned on), the current through the inductor is continuous, and the output voltage ripple is negligible.

When the converter has not attained a steady state, the inductor current and the output voltage are changed at the end of each switching cycle. Once the converter is in the steady state, the average voltage across the inductor and the current through the capacitor should be zero. The average voltage across the inductor over one switching cycle can then be calculated as

$$\overline{v_L(t)} = \frac{1}{T_s}\int_0^{T_s} v_L(t)\ dt = 0 \tag{4.5}$$

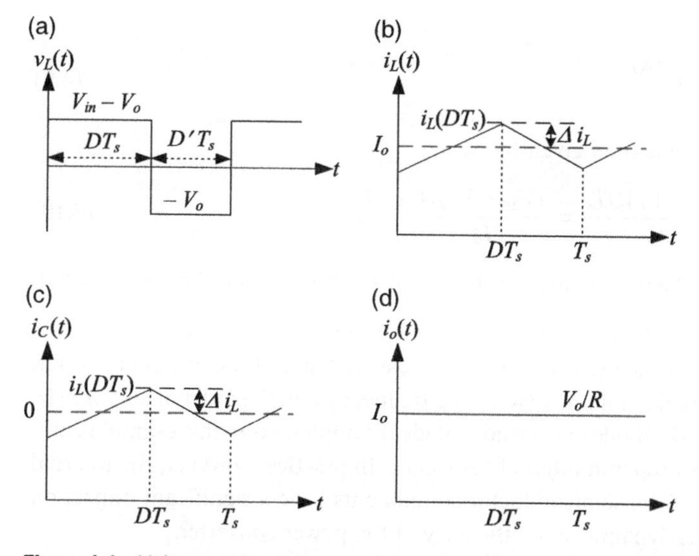

Figure 4.4 Voltage and current waveforms of a buck converter over a switching cycle in the steady stare: (a) voltage across the inductor, (b) current through the inductor, (c) current through the capacitor, and (d) load current.

Based on (4.3) and (4.4), the average voltage of (4.5) can be expressed as

$$\overline{v_L(t)} = \frac{1}{T_s}\left[\int_0^{DT_s} (V_{in} - V_0)\ dt + \int_{DT_s}^{T_s} (-V_0)\ dt\right] = 0 \tag{4.6}$$

The solution of (4.6) yields

$$D = \frac{V_0}{V_{in}} \tag{4.7}$$

4.1.1 Designing a Buck Converter

To design a buck converter, the inductance and capacitance values must be determined. The value of the inductor can be selected based on the system operating mode and the expected voltage and current ripples. For example, during the time period when the switch is turned on, the inductor voltage can be expressed in terms of its current as

$$v_L(t) = L\frac{di_L(t)}{dt} \tag{4.8}$$

From (4.3), the voltage across the inductor is obtained. Therefore, when the switch is on, we have

$$V_{in} - V_o = L\frac{2\Delta i_L}{DT_s} \tag{4.9}$$

The inductor ripple is then given by

$$\Delta i_L = \frac{(V_{in} - V_o)DT_s}{2L} = \frac{(V_{in} - V_o)D}{2Lf_s} \tag{4.10}$$

For the design of a buck converter, the following aspects must be kept in mind:

- One of the main design parameters is the switching frequency, f_s, which is selected based on the quality, efficiency, cost, and size of the power converter. In low power applications, the switching frequency is in the order of tens of kHz.
- The switch and the diode are considered ideal components in this example without any voltage drop and internal resistance. In practice, however, the internal characteristics of the semiconductor components have a significant impact on the steady state, dynamic, and efficiency of the power converter.
- The inductor is considered to be without any internal resistance (copper loss) and any parasitic capacitor between the turns of the winding. These parameters can affect the voltage drop, ripple, transfer function, and high-frequency noise emission.

Example 4.1 Consider an ideal buck converter, the parameters of which are given in Table 4.1. We first have to choose the duty ratio. From (4.7), we find that for an input voltage of 12 V the duty ratio should be kept at $D = 5/12 = 0.417$ to keep the output as 5 V across the load. For a 5 V output and 1 Ω resistance, the output current is

$$I_{Load} = I_o = \frac{5}{1} = 5\ \text{A}$$

For 10% current ripple, we have $\Delta i_L = 0.5$ A. Therefore, from (4.10), the inductance is calculated as

$$L = \frac{(V_{in} - V_o)D}{2\Delta i_L f_s} = \frac{7 \times 0.417}{2 \times 0.5 \times 20 \times 10^3} = 0.146 \times 10^{-3} = 0.146\ \text{mH}$$

To determine the value of the capacitance, the average current through the capacitor can be calculated as shown in Figure 4.5. During the time interval t_1

Table 4.1 Buck converter design parameters.

Quantities	Values
Switching frequency, $f_s = \dfrac{1}{T_s}$	20 kHz
Input voltage, V_{in}	12 V
Output voltage, V_o	5 V
Load resistance, R	1 Ω
Inductor current ripple, Δi_L	10%
Output voltage ripple, Δv_o	2%

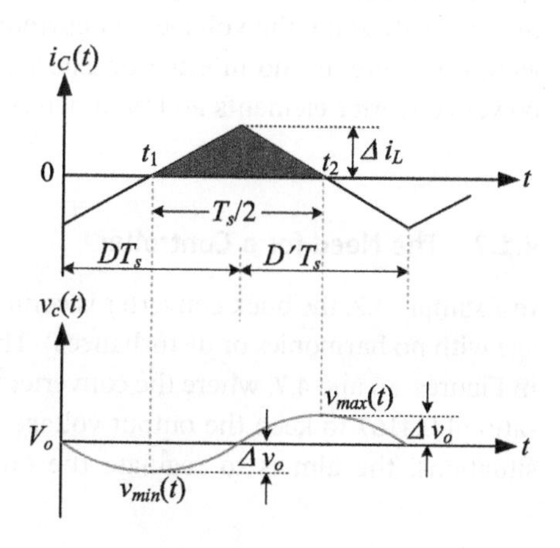

Figure 4.5 Capacitor current and voltage waveforms.

to t_2 (half of the switching cycle), the current through the capacitor is positive. The voltage ripple, Δv_o, is then given by

$$i_C(t) = C\frac{dv_C(t)}{dt} \tag{4.11}$$

The surface area of the capacitor current between t_1 and t_2 is

$$\int_{t_1}^{t_2} v_c(t)dt = \frac{1}{C}\int_{t_1}^{t_2} i_c(t)dt \tag{4.12}$$

The left side of (4.12) is equal to $2\Delta v_o$, i.e. two times the voltage ripple across the capacitor (see Figure 4.5). Thus, the voltage ripple amplitude in terms of the converter parameters is given by

$$2\Delta v_c = 2\Delta v_o = \frac{T_s \Delta i_L}{4C} \tag{4.13}$$

Therefore, the value of the capacitor is computed as

$$C = \frac{\Delta i_L}{8\Delta v_o f_s} \tag{4.14}$$

For the values given in Table 4.1, we get

$$C = \frac{0.5}{8 \times (0.02 \times 5) \times 20 \times 10^3} = 0.03125 \times 10^{-3} = 31.25\ \mu\text{F}$$

Example 4.2 The buck converter of Table 4.1 is simulated with the designed values of the inductance and capacitance. The cold start transients are shown in Figure 4.6, while the voltage and currents in the steady state condition are shown in Figure 4.7. The duty ratio is chosen as $5/12 = 0.4167$. It can be seen that, in the steady state, the voltage and current ripples are the same as the designed values. If there are no internal or external disturbances or any variation in the power converter elements and parameters, the output voltage is exactly 5 V.

4.1.2 The Need for a Controller

In Example 4.2, the buck converter is connected to an ideal battery (constant voltage with no harmonics or disturbances). The open-loop system response is shown in Figures 4.6 and 4.7, where the converter is turned on and off with an exact duty ratio of 0.4167 to keep the output voltage constant at 5 V. However, in practical situations, the aim is to regulate the output voltage at 5 V regardless of the

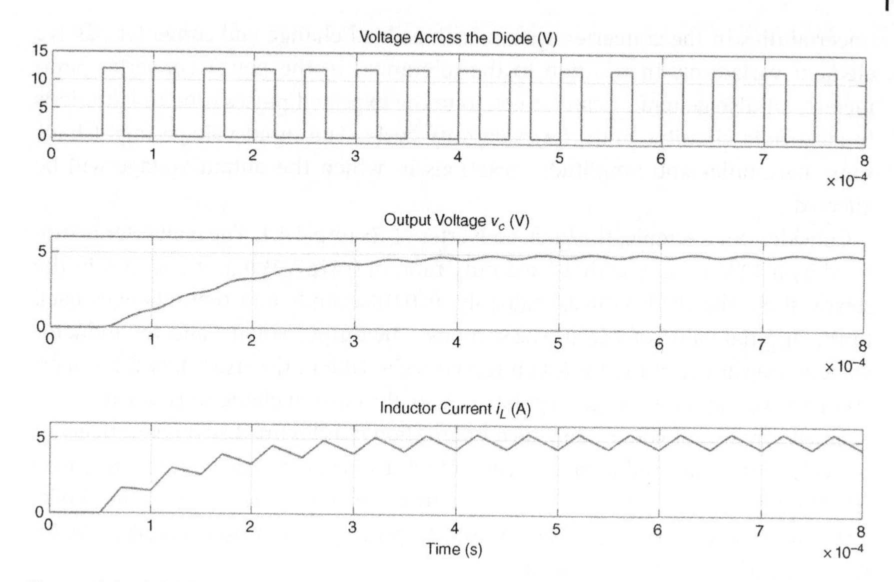

Figure 4.6 Initial starting transients of the buck converter.

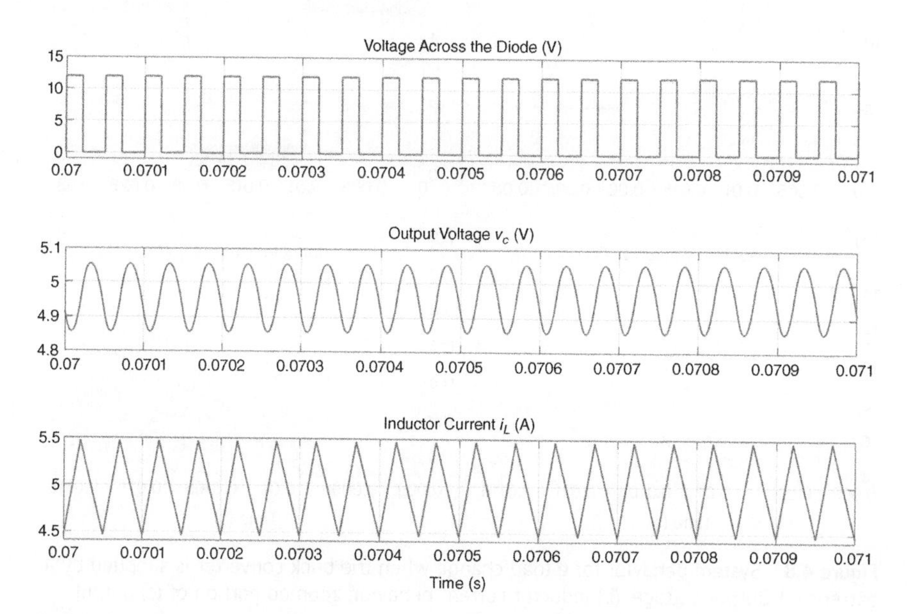

Figure 4.7 Steady state voltage and current ripples with the buck converter.

uncertainties in the converter system such as load change and converter passive element variations. In addition to the tolerances in the power converter parameters, a pulse generator cannot generate the expected pulses (due to variations in electronic circuits), with the exact duty cycle. The input voltage might have some harmonics and amplitude variations in which the output voltage will be affected.

Consider, for example, the buck converter of Example 4.1. The converter is supplied by a 12 V battery with a fixed duty ratio of 0.4167. When it operates in the steady state, the load reduces suddenly at 0.01 seconds and then changes back to the original value of 1 Ω at 0.02 seconds. The output voltage and the inductor current are shown in Figure 4.8a,b respectively. Due to the fixed duty ratio operation, the voltage does not get regulated. Also, the current chatters around 0 A. The zoomed versions of voltage and current are shown in Figure 4.8c,d respectively. It is evident that the inductor current enters a discontinuous conduction mode (DCM), where the current is prevented from becoming negative by the diode. Thus, to regulate the voltage at 5 V and to prevent the current entering DCM, the duty ratio must be controlled.

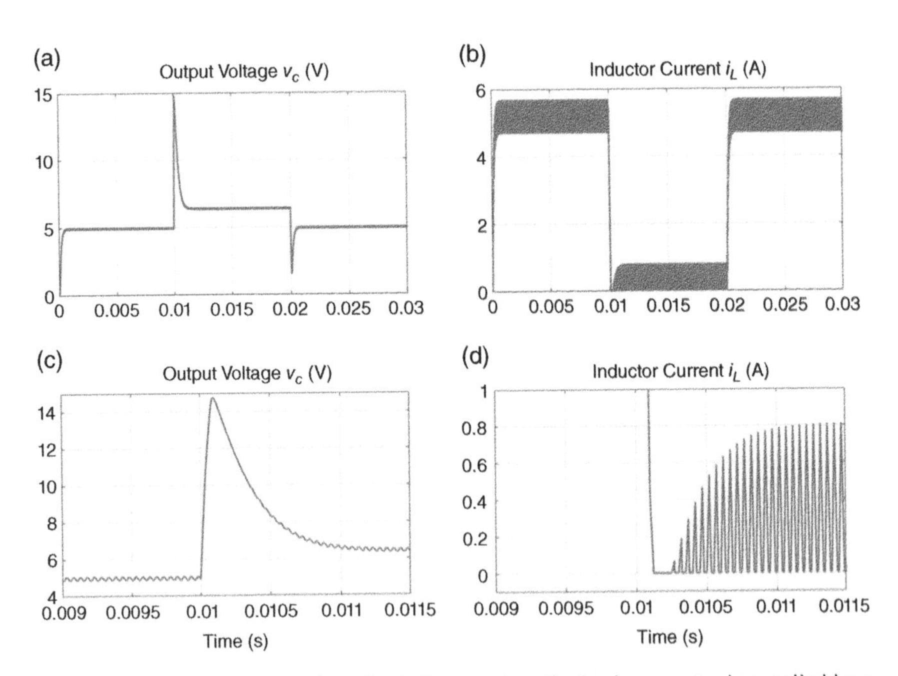

Figure 4.8 System behavior for a load change when the buck converter is supplied by a battery: (a) Output voltage, (b) inductor current behavior, zoomed portion of (c) output voltage and (d) inductor current.

Consider another example where the buck converter is connected to a diode rectifier. The output voltage should be at 5 V. This situation is completely different from the previous case – changing the input DC source from a battery to a diode rectifier that produces unregulated DC voltage. Here the voltage amplitude is not constant and introduces voltage harmonics. These are obvious from the response shown in Figure 4.9.

Due to the input voltage variation and disturbances, the duty cycle should be changed continuously in order to generate a regulated voltage of 5 V across the load. Therefore, there is a need to design and implement a controller based on negative feedback. This is to control the duty cycle automatically based on the error generated by any voltage disturbance and variation.

A general circuit diagram of a buck converter with negative feedback is shown in Figure 4.10, where the control algorithm is assumed to be executed by a microcontroller. An ideal DC source is shown in Figure 4.10a, where the input voltage (battery) has no harmonics, and it is expected to be constant at 12 V. The output voltage amplitude is low (5 V) and it can be directly measured by the controller. The output voltage is compared to a reference value (5 V in this case), while the error should be minimized through a proper controller. The pulse generator (pulse width modulator, PWM) can generate pulses with a variable duty cycle to control

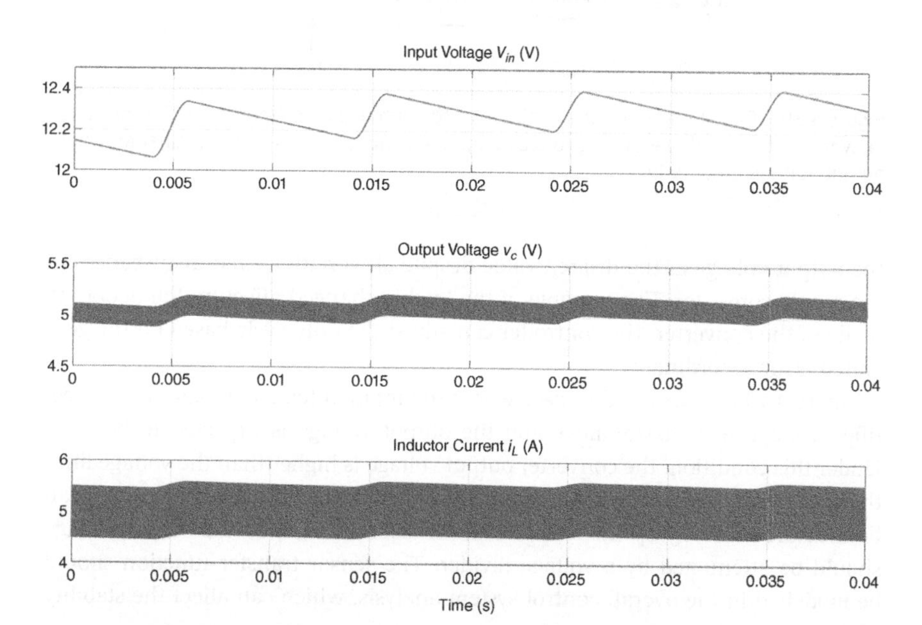

Figure 4.9 Voltage and current behavior when the buck converter is supplied by an uncontrolled bridge rectifier.

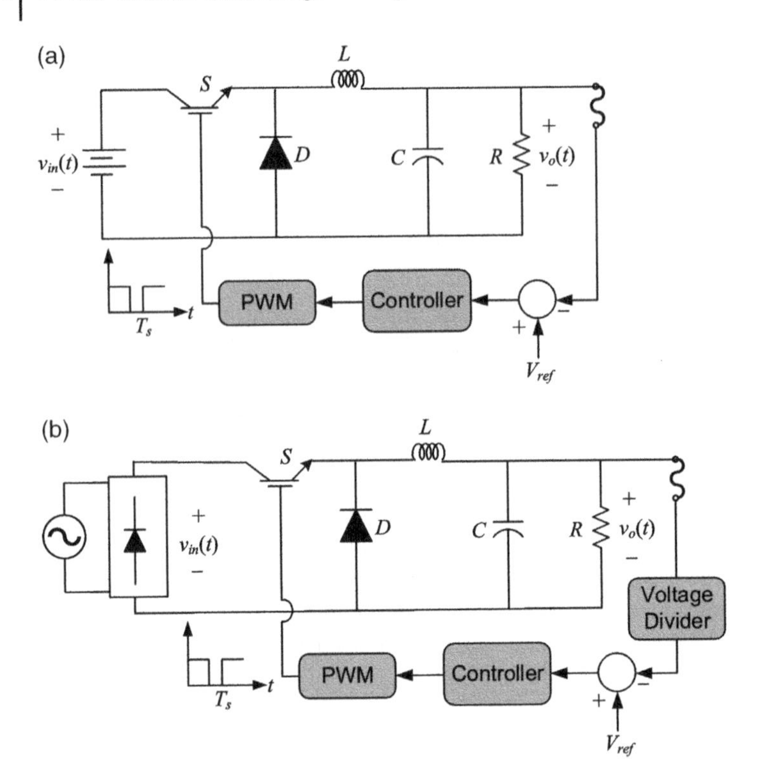

Figure 4.10 Circuit diagrams of a buck converter with negative feedback and a controller: (a) with a battery and low-output voltage amplitude and (b) with a diode rectifier and high-output voltage amplitude.

the output voltage at the desired value despite all variations and disturbances in the power converter. For example, if the load is changed affecting the operating mode of the converter, the controller can adjust the duty cycle based on the generated error accordingly.

Figure 4.10b shows another case where the input voltage fluctuates (e.g. a rectified voltage with harmonics), and the output voltage is expected to be 24 V. Under this condition, the converter output voltage is higher than the voltage limit that can be safely measured by a microcontroller (typically voltages below 5 V can be measured directly by it). Thus, for the measurement, first, the output voltage should be attenuated by a voltage divider. The sensor transfer function should be modelled in the overall control system analysis, which can affect the stability of the system.

To design a controller for both systems, a transfer function and a model of the converter is required. The state variables of inductor current (i_L) and capacitor voltage (v_c) are defined to extract and derive the equations for all subcircuits when semiconductor switches are turned on and off. In Sections 4.1.3–4.1.7, we introduce several stages to find the transfer function of a power converter. The averaging method has been used to find a continuous model of the discrete subcircuits. As the final equations are nonlinear (some of the variables are multiplied together), a linearization technique is utilized to simplify the equations into linear time-invariant form. Finally, the Laplace transform of the linearized equations is used to find the transfer function, based on which a controller can be designed.

4.1.3 Dynamic State of a Power Converter

The dynamic behavior of a power converter is observed when the current through an inductor or the voltage across a capacitor is changed. During this time, the values of the voltage and current at the beginning and at the end of each switching cycle are not the same. This phenomenon happens when a change occurs in the reference signal, the input voltage, or the load. A similar behavior can be observed during the start-up condition when the power converter is turned on with zero initial condition; the current and the voltage in a power converter are increased and changed over several switching cycles until these parameters reach the steady state stage. This is obvious from Figure 4.6c. It can be seen that the inductor current increases at the beginning of every switching cycle till the steady state is reached at around 0.7 ms. Thus, the main aim of designing a proper control system is to achieve the optimum dynamic behavior of a power converter. This means the robust operation of the converter with respect to disturbances, such that it reaches the steady state condition with minimum transient time, steady state error, and over- or undershoot.

4.1.4 Averaging Method

Figure 4.3 shows the buck converter operation modes assuming the current through the inductor is always continuous. The state variables for the converter are the capacitor voltage $v_c(t)$ and the inductor current $i_L(t)$. When the switch is turned on, we get the following equations from Figure 4.3a

$$v_L(t) = L\frac{di_L}{dt} = v_{in}(t) - v_o(t) \tag{4.15}$$

$$i_C(t) = C\frac{dv_o}{dt} = i_L(t) - i_o(t) = i_L(t) - \frac{v_o(t)}{R} \tag{4.16}$$

On the other hand, when the switch is turned off, the converter operation mode changes to the circuit shown in Figure 4.3b. From this circuit, the dynamic equations are written as

$$v_L(t) = -v_o(t) \tag{4.17}$$

$$i_C(t) = i_L(t) - i_o(t) = i_L(t) - \frac{v_o(t)}{R} \tag{4.18}$$

DC-DC converters are developed in such a way that the current through the inductors and the voltage across the capacitors are controlled when semiconductor switches are turned on and off. In fact, there is at least one current loop for each inductor when one of the semiconductor switches is turned on or off; otherwise, a significant overvoltage ($^{di}/_{dt}$) appears across the inductor and can damage the converter. Consider, for example, Figure 4.3. It can be seen that, when the switch is turned on and off, the inductor current branch is not open circuited, and the current is circulated through the capacitor and the load. Similarly, the voltage across the capacitor must not be short circuited when one of the switches is turned on or off. This will damage the capacitor.

The inductor current or the capacitor voltage changes instantaneously based on the subcircuits and the operating modes of the converter. For control and stability analysis, either the high frequency or the instantaneous variation of the current or the voltage waveforms is not essential. However, the average variation of the current or the voltage signal over each switching cycle should be analyzed. These average values of the voltage and current are used in power converters as they represent the low-frequency behavior of the signal during the steady state and transient conditions.

In general, over a time period T_s, the average value of a variable is given by

$$\bar{y}(t) = \frac{1}{T_s}\int_0^{T_s} y(t)dt$$

where $\bar{y}$ denotes the average of the quantity y. This method of averaging will remove the high-frequency switching ripple over one switching cycle, while the average value can be changed from one switching period to the next such that low-frequency components are retained [1]. Consider (4.15), which defines the inductor and capacitor voltage and current when the switch is closed. We replace the input voltage $v_{in}(t)$ and output voltage $v_o(t)$ by their low-frequency average values of $\bar{v}_{in}(t)$ and $\bar{v}_o(t)$ respectively to get

$$v_L(t) = L\frac{di_L}{dt} = \bar{v}_{in}(t) - \bar{v}_o(t) \tag{4.19}$$

Similarly, from (4.17) we have

$$v_L(t) = -\bar{v}_o(t) \tag{4.20}$$

The low-frequency average of the inductor voltage is then given as

$$\bar{v}_L(t) = \frac{1}{T_s}\int_t^{t+T_s} v_L(\tau)d\tau = d(t)[\bar{v}_{in}(t) - \bar{v}_0(t)] - d'(t)\bar{v}_0(t)$$

where $d'(t) = 1 - d(t)$. Then (4.20) can be written as

$$\bar{v}_L(t) = L\frac{d\bar{i}_L(t)}{dt} = d(t)\bar{v}_{in}(t) - \bar{v}_0(t) \quad (4.21)$$

In the steady state, since $d = {}^{v_0}/_{v_{in}}$, the average voltage is zero. However, during transients, the average current varies linearly with the input voltage if the high-frequency ripple is neglected. Similarly, the low-frequency average of the capacitor current is rewritten from (4.16) and (4.18) as

$$\bar{i}_C(t) = C\frac{d\bar{v}_o(t)}{dt} = \bar{i}_L(t) - \frac{\bar{v}_o(t)}{R} \quad (4.22)$$

4.1.5 Small Signal Model of Buck Converter

The averaged model of the inductor voltage in (4.21) is nonlinear since this is defined in terms of a product of $\bar{v}_{in}(t)$ and $d(t)$. Usually, a nonlinear differential equation is linearized using the Taylor series. Consider, for example, the following nonlinear equation

$$\dot{y}(t) = f(y, u) \quad (4.23)$$

where $u(t)$ is the input. Equation (4.23) will be linearized around an operating point, where the steady state values are denoted by uppercase letters and the perturbed values are prefixed by capital delta, i.e.

$$y(t) = Y + \Delta y(t), \quad u(t) = U + \Delta u(t)$$

Then in the Taylor series expansion, the second- and higher-order terms are neglected to obtain a linear form of (4.23), given by

$$\Delta\dot{y}(t) = \frac{\partial f(y,u)}{\partial y}\Delta y(t) + \frac{\partial f(y,u)}{\partial u}\Delta u(t) \quad (4.24)$$

For the buck converter, the following quantities are defined

$$\bar{v}_o(t) = V_0 + \Delta v_o(t)$$
$$\bar{i}_L(t) = I_L + \Delta i_L(t)$$
$$\bar{v}_{in}(t) = V_{in} + \Delta v_{in}(t)$$
$$d(t) = D + \Delta d(t)$$
$$d'(t) = 1 - d(t) = D' - \Delta d(t)$$

Then the linearized model of (4.21) is obtained in the similar fashion of (4.24) as

$$L\frac{d\Delta i_L(t)}{dt} = D\Delta v_{in}(t) + V_{in}\Delta d(t) - \Delta v_0(t) \tag{4.25}$$

The output voltage equation of (4.22) is already linear and is rewritten in small signal form as

$$C\frac{d\Delta v_o(t)}{dt} = \Delta i_L(t) - \frac{\Delta v_o(t)}{R} \tag{4.26}$$

4.1.6 Transfer Function of Buck Converter

Equations (4.25) and (4.26) are now used to determine the transfer function of the buck converter. The Laplace transforms of these two equations are

$$sL\Delta i_L(s) = D\Delta v_{in}(s) + V_{in}\Delta d(s) - \Delta v_0(s) \tag{4.27}$$

$$sC\Delta v_o(s) = \Delta i_L(s) - \frac{\Delta v_o(s)}{R} \tag{4.28}$$

Substituting $\Delta i_L(s)$ from (4.27) in (4.28), we have

$$sL\left(sC + \frac{1}{R}\right)\Delta v_o(s) = D\Delta v_{in}(s) + V_{in}\Delta d(s) - \Delta v_0(s)$$

$$\Rightarrow \left(LCs^2 + \frac{L}{R}s + 1\right)\Delta v_o(s) = D\Delta v_{in}(s) + V_{in}\Delta d(s)$$

The above equation can then be rewritten as

$$\Delta v_o(s) = G_{vin}^{buck}(s)\Delta v_{in}(s) + G_d^{buck}(s)\Delta d(s) \tag{4.29}$$

where

$$G_{vin}^{buck}(s) = \frac{D}{LCs^2 + \frac{L}{R}s + 1} \text{ and } G_d^{buck}(s) = \frac{V_{in}}{LCs^2 + \frac{L}{R}s + 1}$$

Note that, when the buck converter is supplied by a constant DC source such as battery, $\Delta v_{in} = 0$ and the transfer function (4.29) is given simply by $\Delta v_o(s) = G_d^{buck}(s)\Delta d(s)$.

4.1.7 Control of Buck Converter

In this section, we discuss the control of a buck converter, where the main aim is to control the output voltage. Consider the feedback control structure of Figure 4.10a, where the firing signal is generated through the intersection of the control signal $v_{cont}(t)$ with a sawtooth waveform of height V_{ST}, as shown in Figure 4.11a. If the

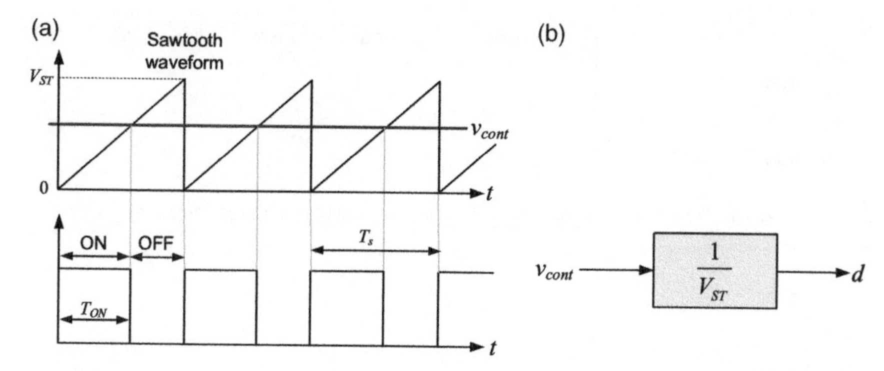

Figure 4.11 (a) Sawtooth and control waveforms and (b) transfer function of the PWM.

control signal is greater than the sawtooth signal, the output of the PWM is high; otherwise, it is low. The duty cycle is then given by

$$d(t) = \frac{T_{ON}}{T_s} = \frac{v_{cont}(t)}{V_{ST}}$$

Then the transfer function of the PWM is

$$H_{PWM}(s) = \frac{d(s)}{v_{cont}(s)} = \frac{1}{V_{ST}} \tag{4.30}$$

This is shown in Figure 4.11b.

Example 4.3 Consider the buck converter of Example 4.2, where the input voltage is supplied by a constant 12 V source. The output voltage is assumed to be 5 V such that the duty ratio is 0.4167. The peak of the sawtooth waveform V_{ST} is chosen as 1 such that $d(t) = v_{cont}(t)$. The buck converter open-loop transfer function then is

$$G_d^{buck}(s) = \frac{V_{in}}{LCs^2 + \frac{L}{R}s + 1} = \frac{12}{4.56 \times 10^{-9}s^2 + 1.46 \times 10^{-4}s + 1}$$

The DC gain of the open-loop system is 12. The root locus of the converter is shown in Figure 4.12, from which it is obvious that the system will always remain stable. Let us now assume that the converter is controlled by a proportional controller with a gain of K_P. The closed loop transfer function is then given by

$$G_{CL}(s) = \frac{K_P G_d^{buck}(s)}{1 + K_P G_d^{buck}(s)} = \frac{12K_P}{4.56 \times 10^{-9}s^2 + 1.46 \times 10^{-4}s + 1 + 12K_P}$$

The DC gain of the closed-loop system is $12K_P/(1 + 12K_P)$. Then for $K_P = 2$, the DC gain will be 0.96, while for $K_P = 20$, the DC gain will be 0.996. However, with a proportional plus integral (PI) controller, there will not be any steady state error. Figure 4.13 shows the response of the small signal model of the buck converter for

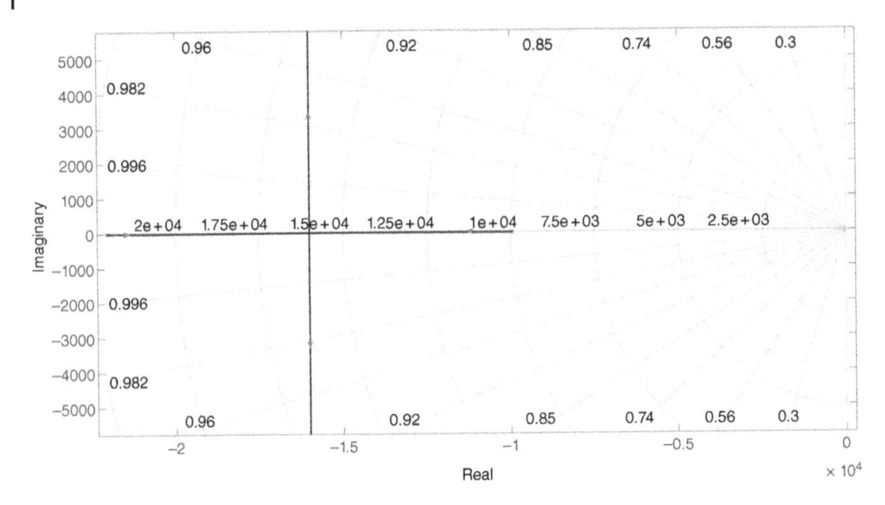

Figure 4.12 Root locus plot of the small signal model of the buck converter.

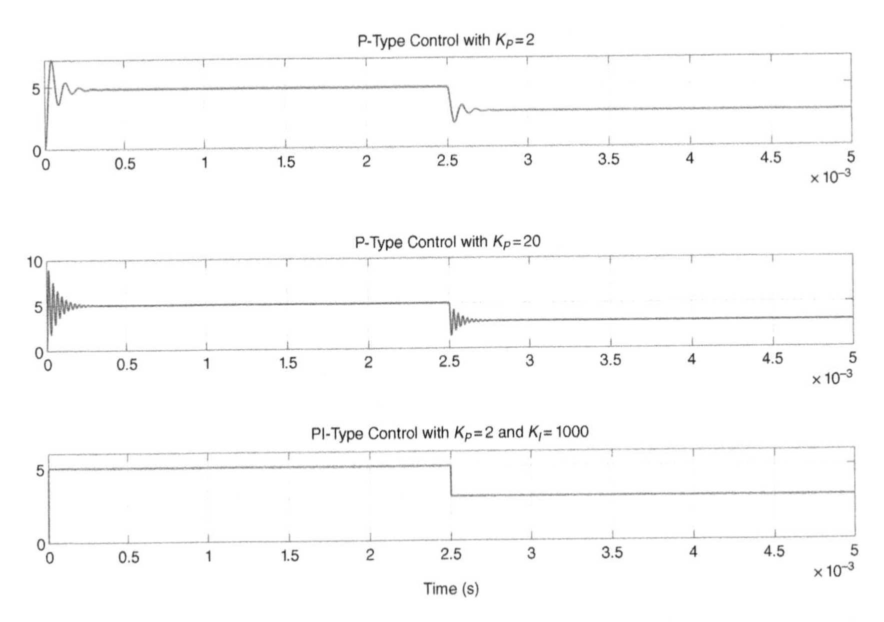

Figure 4.13 Dynamic response of the buck converter small signal model to P-type and PI-type controllers.

these two values of proportional gain, along with a PI controller with $K_P = 2$ and $K_I = 1000$, where the reference voltage changes from 5 V to 3 V at 2.5 ms. It can be seen that, as the value of K_P increases, the steady state error decreases. However, this causes larger transients, as evident by comparing Figure 4.13a with Figure 4.13b. The response of the PI controller is much superior to that of the proportional controller.

The closed-loop control system of a DC-DC converter is shown in Figure 4.14. The response of the buck converter with the PI controller is shown in Figure 4.15, where the reference voltage is chosen as 5 V initially. It is then reduced to 3 V after 0.15 seconds and increased to 7 V at 0.3 seconds. It is evident that the voltage tracking is accurate, and the duty ratio changes a per the requirements.

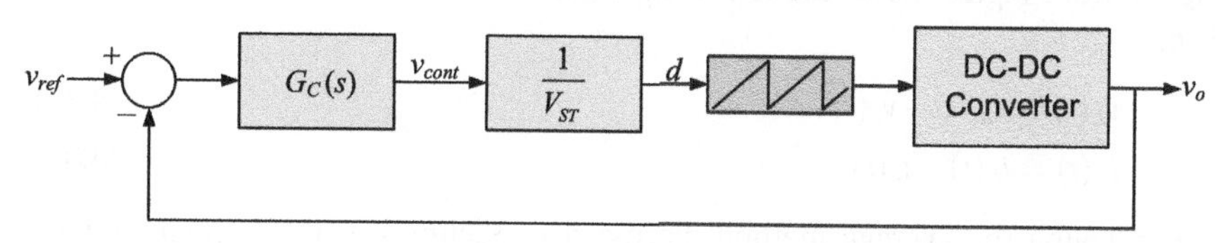

Figure 4.14 Closed-loop control of a DC-DC converter.

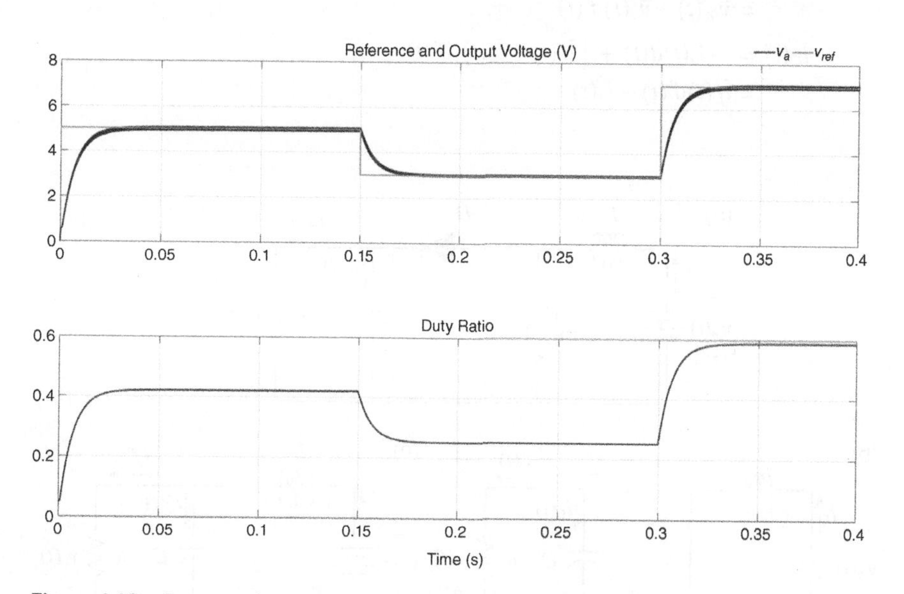

Figure 4.15 Output voltage response and duty ratio of the buck converter.

4.2 Transfer Function of Boost Converter

The circuit diagram of a boost converter is shown in Figure 4.16a. Its transfer function can be derived based on the averaging method and the small signal analysis explained in Section 4.1. There are two subcircuits associated with the operating mode of the converter when the semiconductor switch is turned on and off, as shown in Figure 4.16b,c respectively. From Figure 4.16b, we get the following equations when the switch is on

$$v_L(t) = v_{in}(t) \tag{4.31}$$

$$i_C(t) = -i_o(t) \tag{4.32}$$

Again, from Figure 4.16c, the following equations are obtained when the switch is off

$$v_L(t) = v_{in}(t) - v_o(t) \tag{4.33}$$

$$i_C(t) = i_L(t) - i_o(t) \tag{4.34}$$

Employing the average method discussed in Section 4.1, the average voltage across the inductor and the current through the capacitor are calculated as

$$\begin{aligned}\bar{v}_L(t) &= \bar{v}_{in}(t)d(t) + [\bar{v}_{in}(t) - \bar{v}_o(t)]d'(t) \\ &= \bar{v}_{in}(t) - \bar{v}_o(t)d'(t)\end{aligned} \tag{4.35}$$

$$\begin{aligned}\bar{i}_C(t) &= -\bar{i}_o(t)d(t) + \left[\bar{i}_L(t) - \bar{i}_o(t)\right]d'(t) \\ &= \bar{i}_L(t)d'(t) - \bar{i}_o(t)\end{aligned} \tag{4.36}$$

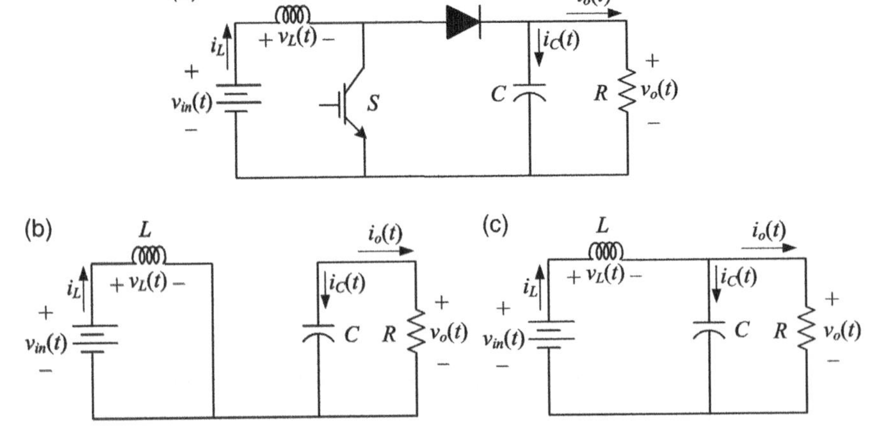

Figure 4.16 (a) A boost converter, its equivalent circuit when the switch is (b) on and (c) off.

Equations (4.35) and (4.36) are linearized as

$$L\frac{d\Delta i_L(t)}{dt} = \Delta v_{in}(t) + V_o\Delta d(t) - D'\Delta v_o(t) \tag{4.37}$$

$$C\frac{dv_o(t)}{dt} = D'\Delta i_L(t) - I_L\Delta d(t) - \Delta i_o(t) \tag{4.38}$$

From Figure 4.16, we find that $i_o(t) = v_o(t)/R$. Also, the steady state relation between the inductor current and the output current is given by $D^{'}I_L = I_0 = V_o/R$. Substituting these in (4.38), we have

$$C\frac{d\Delta v_o(t)}{dt} = D'\Delta i_L(t) - \frac{V_o}{D'R}\Delta d(t) - \frac{\Delta v_o(t)}{R} \tag{4.39}$$

The Laplace transform of (4.37) is

$$Ls\Delta i_L(s) = \Delta v_{in}(s) + V_o\Delta d(s) - D'\Delta v_o(s) \tag{4.40}$$

Now taking the Laplace transform of (4.39) and substitution in (4.40) results in

$$Cs\Delta v_o(s) = \frac{D'}{sL}[\Delta v_{in}(s) + V_o\Delta d(s) - D'\Delta v_o(s)] - \frac{V_o}{D'R}\Delta d(s) - \frac{\Delta v_o(s)}{R}$$

$$\Rightarrow \left[Cs + \frac{D'^2}{sL} + \frac{1}{R}\right]\Delta v_o(s) = \frac{D'}{sL}\Delta v_{in}(s) + \left(\frac{D'V_o}{sL} - \frac{V_o}{D'R}\right)\Delta d(s)$$

$$\Rightarrow \left[\frac{LC}{D'}s^2 + \frac{L}{RD'}s + D'\right]\Delta v_o(s) = \Delta v_{in}(s) + V_o\left(1 - \frac{L}{D'^2R}s\right)\Delta d(s)$$

Rearranging the above equation, we get

$$\Delta v_o(s) = G_{vin}^{boost}(s)\Delta v_{in}(s) + G_d^{boost}(s)\Delta d(s) \tag{4.41}$$

where

$$G_{vin}^{boost}(s) = \frac{1}{\frac{LC}{D'}s^2 + \frac{L}{RD'}s + D'} \quad \text{and} \quad G_d^{boost}(s) = \frac{V_o\left(1 - \frac{L}{D'^2R}s\right)}{\frac{LC}{D'}s^2 + \frac{L}{RD'}s + D'}$$

4.2.1 Control of Boost Converter

We shall now discuss the control of a boost converter, based on the parameters listed in Table 4.2. It is assumed that the desired output voltage is 100 V. Then the duty ratio is 0.5, i.e. $D^{'} = 0.5$. Then the transfer function of the boost converter is

$$G_d^{boost}(s) = \frac{V_o\left(1 - \frac{L}{D'^2R}s\right)}{\frac{LC}{D'}s^2 + \frac{L}{RD'}s + D'} = \frac{-0.04s + 100}{2\times 10^{-8}s^2 + 2\times 10^{-5}s + 0.5}$$

Table 4.2 Boost converter parameters.

Quantities	Values
Switching frequency, f_s	20 kHz
Input voltage, V_{in}	50 V
Load resistance, R	10 Ω
Inductor, L	1 mH
Capacitor, C	100 μF

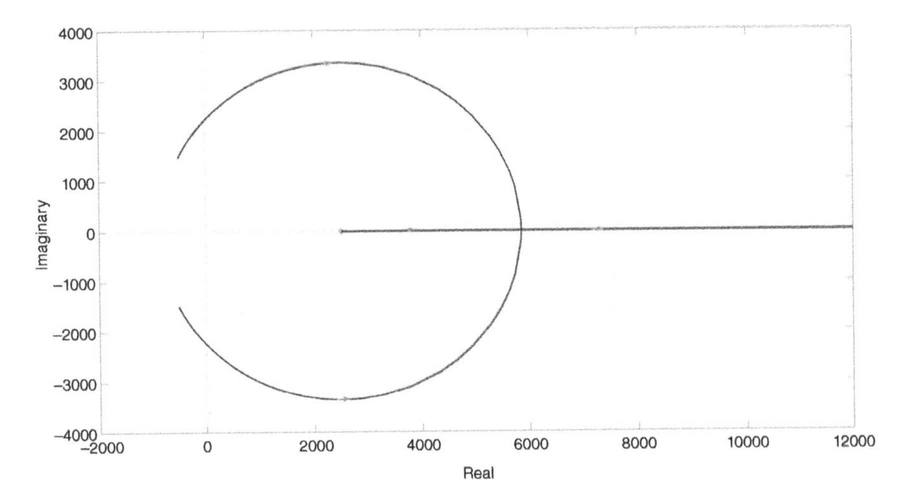

Figure 4.17 Root locus plot of the boost converter.

The root locus plot is shown in Figure 4.17. Note that the system has an open-loop zero on the right-half s-plane. Therefore, the roots move over to the right-half s-plane after starting from stable positions. For a proportional gain of K_P, the characteristic equation of the system is given by

$$2 \times 10^{-7}s^2 + \left(2 \times 10^{-4} - 0.04K_P\right)s + 0.5 + 100K_P = 0$$

Therefore, for the system to remain stable, we must have $0.0002 - 0.04K_P > 0$, i.e. $K_P < 0.005$. The system will become unstable for a very small gain. If, on the other hand, we choose a very small gain, the steady state error will be high.

Let us now consider a PI controller of the form $G_C(s) = (K_Ps + K_I)/s$. Then the open-loop transfer function of the system is

$$G_C(s)G_d^{boost}(s) = \frac{(K_P s + K_I)(-0.04s + 100)}{2 \times 10^{-7}s^3 + 0.0002s^2 + 0.5s}$$

The characteristic equation of the closed-loop system is then given by

$$2 \times 10^{-7}s^3 + (0.0002 - 0.04K_P)s^2 + (0.5 - 0.04K_I + 100K_P)s + 100K_I = 0$$

To find the limits of the control parameters, Routh's table is considered, which is given below

$$\begin{array}{lll} s^3 & 2 \times 10^{-7} & (0.5 - 0.04K_I + 100K_P) \\ s^2 & (0.0002 - 0.04K_P) & 100K_I \\ s^1 & \alpha & \\ s^0 & 100K_I & \end{array}$$

where

$$\alpha = \frac{(0.0002 - 0.04K_P)(0.5 - 0.04K_I + 100K_P) - 2 \times 10^{-5}K_I}{(0.0002 - 0.04K_P)}$$

Since all the coefficients of the first column of each row must be positive, from the s^2 row it can be written as

$0.0002 - 0.04K_P > 0$, i.e. $K_P < 0.005$.

Let us choose $K_P = 0.0045$. With this value of the proportional gain, the characteristic equation can be rewritten as

$$K_1 \frac{-0.04s + 100}{2 \times 10^{-7}s^3 + 2 \times 10^{-5}s^2 + 0.95s} + 1 = 0$$

Figure 4.18 shows the root locus plot as K_I increases from 0 to high values.

It is evident from Figure 4.18 that the system remains stable only for a limited range of values of K_I. To check the range, we again refer to Routh's table. Since α must be greater than 0, we have

$$\alpha = \frac{2 \times 10^{-5}(0.2225 - 0.04K_I) - 2 \times 10^{-5}K_I}{2 \times 10^{-5}} = 0.2225 - 1.04K_I > 0$$

This gives $K_I < 0.2139$. Let us choose $K_I = 0.2$. Then the roots of the closed-loop system are placed at $-39.4 \pm j2169.5$ and -21.2. The closed-loop response of the system is shown in Figure 4.19. Even though the controller can track the reference voltage, the low gains make disturbance rejection difficult. The consequence of this is that the system will become unstable even for small changes in the system, such as reference voltage or load resistance.

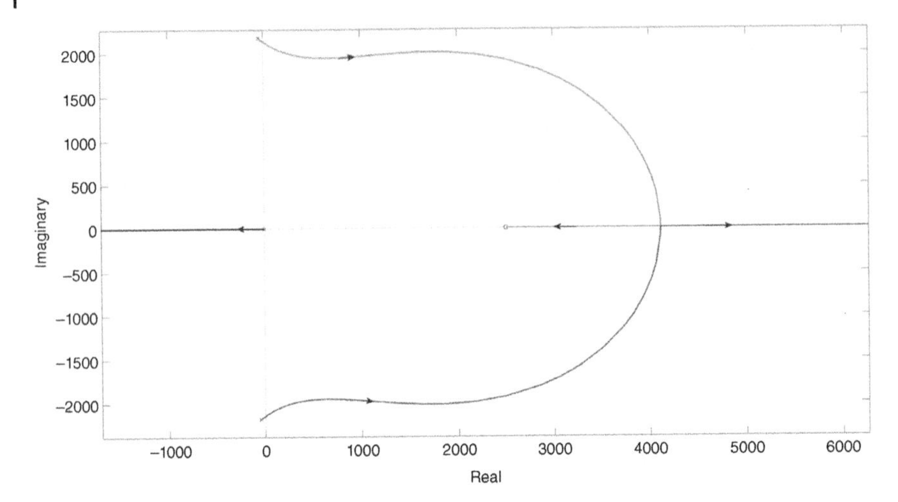

Figure 4.18 Root locus plot of the boost converter when K_I increases from 0 to high values.

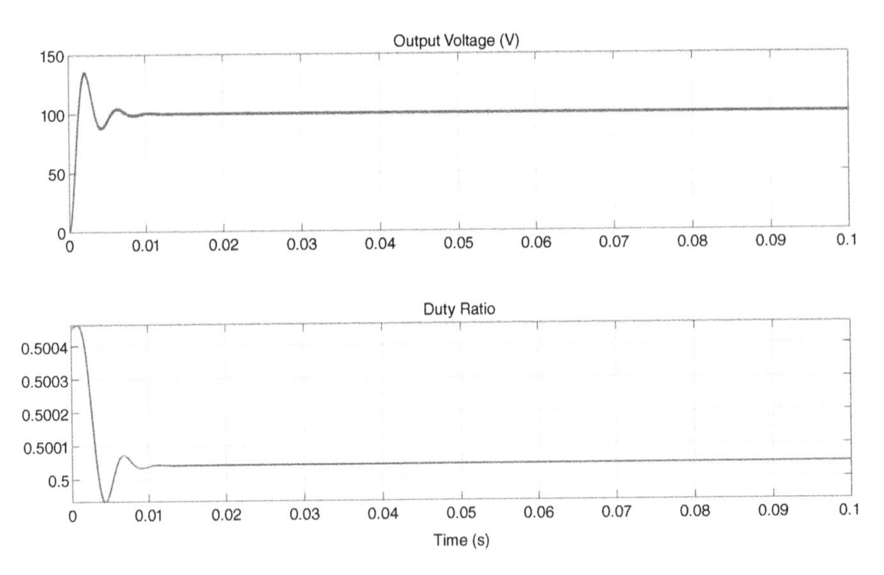

Figure 4.19 Boost converter response with PI controller.

4.2.2 Two-loop Control of Boost Converter

The two-loop control system of the boost converter is shown in Figure 4.20a, in which the inner current is a faster loop, while the outer voltage loop is slower. The inner current loop, shown in Figure 4.20b, has a controller denoted by $G_{CI}(s)$.

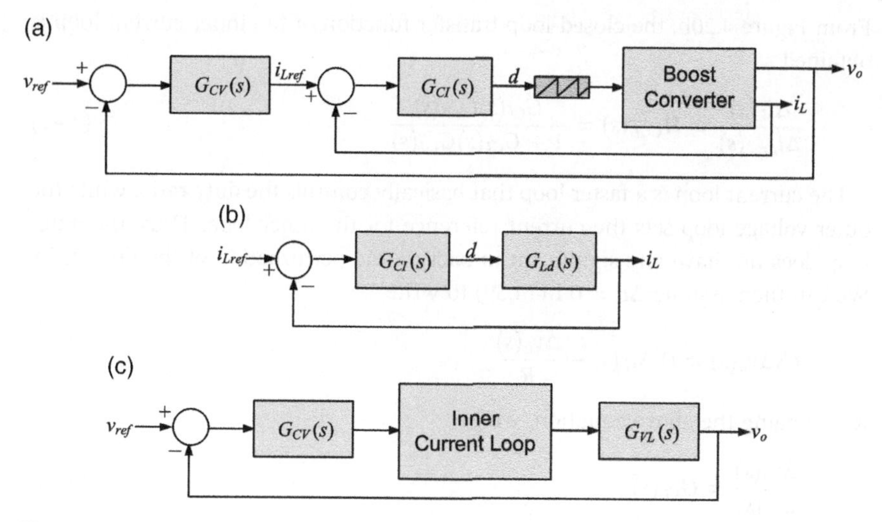

Figure 4.20 (a) Two-loop control of boost converter, (b) inner current loop, and (c) outer voltage loop.

The outer voltage loop has a controller $G_{CV}(s)$ as shown in Figure 4.20c. The transfer functions of the different subsystems are derived below.

Assuming that the input voltage is constant, i.e. $\Delta v_{in} = 0$, we first derive the transfer function of the inner current loop. For this, (4.39) is rewritten in the Laplace domain as

$$\Delta v_o(s) = \frac{R}{RCs+1}\left[D'\Delta i_L(s) - \frac{V_o}{D'R}\Delta d(s)\right]$$

Substituting the above equation in (4.40) and since input voltage is constant, we have

$$\begin{aligned} Ls\Delta i_L(s) &= V_o\Delta d(s) - D'\Delta v_o(s) \\ &= V_o\Delta d(s) - \frac{D'R}{RCs+1}\left[D'\Delta i_L(s) - \frac{V_o}{D'R}\Delta d(s)\right] \end{aligned}$$

Rearranging the above equation, the open-loop transfer function of the current loop transfer is given by

$$\frac{\Delta i_L(s)}{\Delta d(s)} = G_{Ld}(s) \tag{4.42}$$

where

$$G_{Ld}(s) = \frac{V_o(Cs + 2/R)}{LCs^2 + L/R\,s + D'^2}$$

From Figure 4.20b, the closed-loop transfer function of the inner current loop is obtained as

$$\frac{\Delta i_L(s)}{\Delta i_{Lref}(s)} = H_{CCL}(s) = \frac{G_{CI}(s)G_{Ld}(s)}{1 + G_{CI}(s)G_{Ld}(s)} \tag{4.43}$$

The current loop is a faster loop that basically controls the duty ratio, while the outer voltage loop sets the current reference for the inner loop. Thus, the outer loop does not have any significant impact on the perturbation of the duty ratio. We can then assume $\Delta d = 0$ in (4.39) to write

$$Cs\Delta v_o(s) = D'\Delta i_L(s) - \frac{\Delta v_o(s)}{R}$$

Rearranging the above equation, we get

$$\frac{\Delta v_o(s)}{\Delta i_L(s)} = G_{VL}(s) \tag{4.44}$$

where

$$G_{VL}(s) = \frac{D'R}{RCs + 1}$$

From Figure 4.20c, the closed-loop transfer function of the boost converter is then given by

$$\frac{\Delta v_o(s)}{\Delta v_{ref}(s)} = H_{CL}(s) = \frac{G_{CV}(s)H_{CCL}(s)G_{VL}(s)}{1 + G_{CV}(s)H_{CCL}(s)G_{VL}(s)} \tag{4.45}$$

Example 4.4 Consider the boost converter, the parameters of which are given in Table 4.2. First, we consider the inner current controller. For chosen parameters, the open-loop transfer function is given as

$$G_{Ld}(s) = \frac{0.01s + 20}{1 \times 10^{-7}s^2 + 1 \times 10^{-4}s + 0.25}$$

The Bode plot of the system is shown in Figure 4.21. The system has a phase margin of 89.43° at the gain crossover frequency of 15.9 kHz, i.e. at 10^5 rad/s. The aim here is to design a compensator that has a high phase margin (say 60°) at the gain crossover frequency of 1.59 kHz, i.e. 10^4 rad/s.

The open-loop transfer function of the system is rewritten by setting $s = j\omega$ as

$$G_{Ld}(j\omega) = \frac{20 + j0.01\omega}{0.25 - 1 \times 10^{-7}\omega^2 + j1 \times 10^{-4}\omega}$$

At the desired gain crossover of $\omega_g = 10^4$ rad/s 100, $G_{Ld}(j\omega_g)$ is

$$G_{Ld}(j\omega_g) = 10.4\angle -95.45^\circ$$

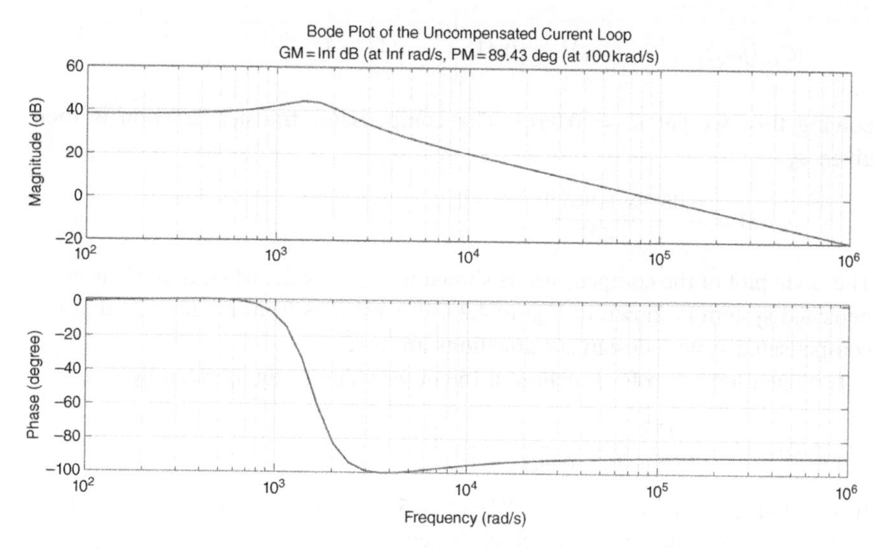

Figure 4.21 Bode plot of the uncompensated current loop.

At the gain crossover frequency ω_g, the phase must be equal to $-180° + 60° = -120°$. We must therefore add a phase lag of $120° - 95.45° = 24.55°$ through a compensator.

For this, a phase lag compensator is employed, which is given in (3.55) as

$$G_{CI}(s) = K_c \frac{s + 1/\tau}{s + 1/\beta\tau}, \quad \beta > 1$$

Note that a lag compensator is really the inverse of a lead compensator. We can thus have the following by modifying (3.53)

$$\sin \phi_m = \frac{\beta - 1}{\beta + 1}$$

Solving this, the value of β is found as

$$\beta = \frac{1 + \sin(24.55°)}{1 - \sin(24.55°)} = 2.42$$

Then replacing α by β in (3.53), we get

$$\tau = \frac{1}{\sqrt{\beta}\omega_g} = 6.43 \times 10^{-5}$$

Note that, at the gain crossover frequency ω_g, the total gain of the compensated system should be 1, i.e.

$$\left|G_{Ld}(j\omega_g)\right| \times \left|G_{CI}(j\omega_g)\right| = 10.4K_c\left|\frac{s + 1/\tau}{s + 1/\beta\tau}\right| = 1$$

Solving this, we get $K_c = 0.0618$. The compensator transfer function is then given by

$$G_{CI}(s) = \frac{0.0618s + 961.1}{s + 6427}$$

The Bode plot of the compensator is shown in Figure 4.22, while that of the compensated system is shown in Figure 4.23. It is obvious from Figure 4.23 that the compensated system design specifications are met.

The open-loop transfer function of the outer voltage loop is given by

$$G_{VL}(s) = \frac{D'R}{RCs + 1} = \frac{5}{0.001s + 1}$$

It is obvious that the system has a DC gain of 5. To eliminate any steady state error, a PI controller is employed, which is chosen as

$$G_{CV}(s) = 5 + \frac{1000}{s} = \frac{5s + 1000}{s}$$

Then the Bode plot of the compensated voltage loop is shown in Figure 4.24.

The response of the designed controller is shown in Figure 4.25. At the beginning, the reference output voltage is set as 100 V, which is subsequently changed

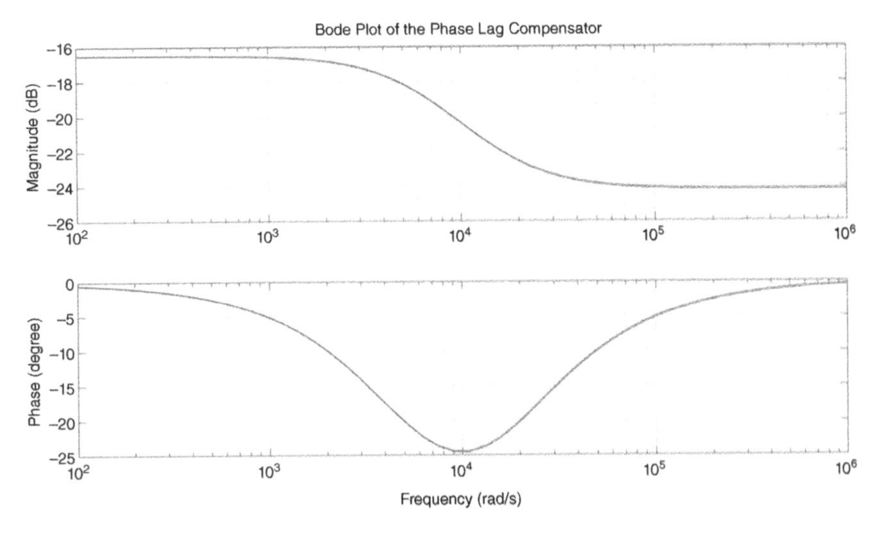

Figure 4.22 Bode plot of the lag compensator.

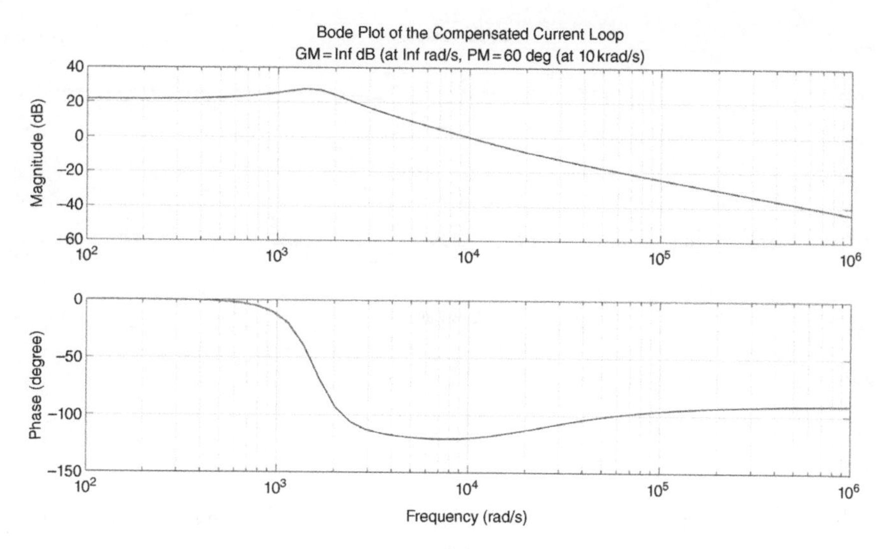

Figure 4.23 Bode plot of the compensated current loop.

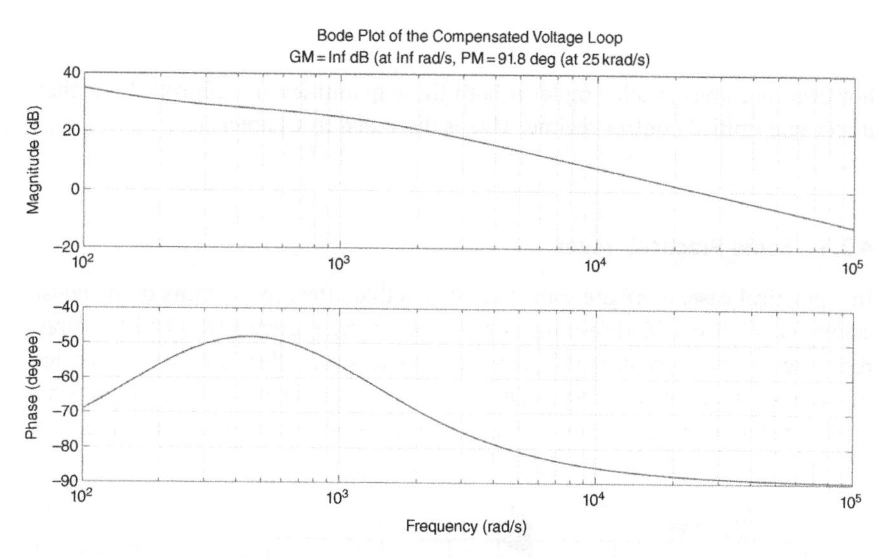

Figure 4.24 Bode plot of the compensated voltage loop.

to 150 V at 0.25 seconds. The controller tracks the change in the reference voltage within 0.05 seconds.

Both capacitor voltage and inductor current are used in this two-loop controller, where each individual controller is designed separately. The control design,

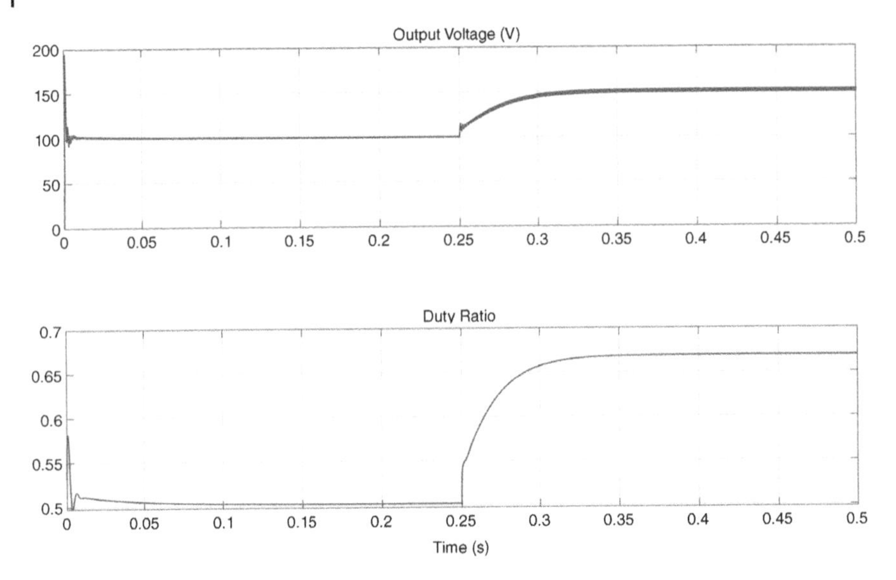

Figure 4.25 Response of the two-loop control system.

however, becomes much simpler if both these quantities are controlled together under one unified control regime. This is discussed in Chapter 5.

4.2.3 Some Practical Issues

In a practical case, there are some parameters that affect the stability of the boost converter. Figure 4.26 shows the copper loss of the inductor (R_L) and the stray resistance of the capacitor (R_C). These parameters can affect the overall transfer function of the boost converter. The inductance and capacitance values depend on the system quality, cost, and size (design parameters), which can change the

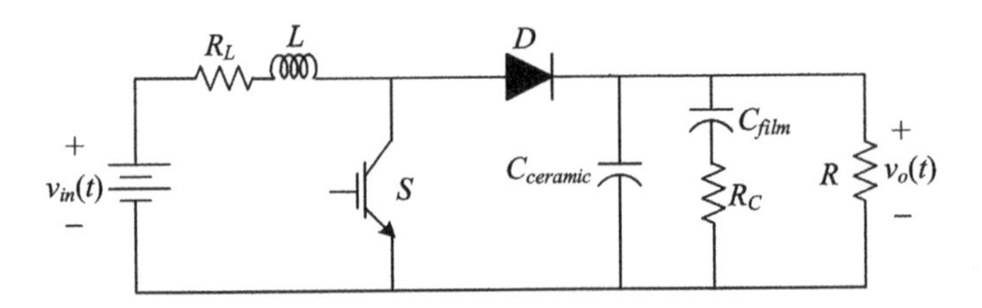

Figure 4.26 Boost converter with stray components.

transfer function of the boost converter. Depending on low- or high-power applications, a combination of different types of DC capacitors can be different – the value of R_C can change depending on the capacitor type – such as a ceramic or a film capacitor. Both resistors can change and affect the location of zero of the converter.

The switching frequency of power converters is reduced in high-power and high-voltage applications due to losses. Furthermore, due to the thermal issues of the semiconductor switches, the efficiency of the system reduces. Thus, to control the current ripple, the inductance value can also be increased based on the switching frequency, and this will change the transfer function of the boost converter.

4.3 Concluding Remarks

This chapter discusses the control of two different types of DC-DC converters, namely the buck and boost converters. Mainly, the classical control aspects are discussed here. This chapter shows that the classical design can be very challenging. The two-loop control design is a fairly arduous process. In Chapter 5, we present state space analysis and design. In this approach, the control design of DC-DC converters becomes much simpler as the state feedback system contains both the inductor current and the output capacitor voltage. The desired response then can be obtained by the choice of the feedback parameters.

Problems

4.1 In a boost converter, what is the main impact of lossy components in the system stability? Analyze and compare the system stability of the following converters:

- an ideal boost converter with $L = 1$ mH and $C = 1$ μF
- a real case boost converter with $L = 1$ mH, $C = 1$ μF, and $R_L = 10 \times 10^{-3}$ Ω. Neglect all other parasitic components and losses.

4.2 Repeat Problem 4.1 for a buck converter and explain the impact of R_L on the system stability compared to the boost converter.

4.3 If a DC-DC converter is connected to a single-phase AC system to supply 48 V DC source, what are the main sources of disturbances? Draw the entire circuit diagram and analyze the stability of the system in order to reduce the disturbance effects.

4.4 A boost converter is connected to a single-phase AC system and the main aim is to control the grid current with unity power factor. Design and analyze the control system with the following parameters:

- $V_{dc} = 300$ V, $V_{in} = 220$ V (rms) at 50 Hz
- $P_0 = 1$ kW, $\Delta V_{dc} = 5\,\%$, and f_{SW} less than 50 kHz.

  Justify the selection of inductor and capacitor values as well.

Notes and References

The control of DC-DC converters is discussed in several books. The book by Erickson and Maksimovic presents the control aspects in detail (see Chapters 7–9) [1]. The book by Mohan, Undeland, and Robbins has an illuminating chapter on DC-DC converters [2]. The control aspects of DC-DC converters are also covered in [3].

1 Erikson, R.W. and Maksimovic, D. (2020). *Fundamental of Power Electronics*, 3e. Springer.
2 Mohan, N., Undeland, T.M., and Robbins, W.P. (2002). *Power Electronics: Converters, Applications and Design*, 3e. New York: Wiley.
3 Krein, P. (2014). *Elements of Power Electronics*, 2e. Oxford University Press.

5

State Space Analysis and Design

In this chapter, we have a thorough look at state space modeling and the control of linear systems. If parameters of linear systems do not vary with time, they are usually called linear time invariant or LTI systems. The discussions in this chapter are limited to LTI systems. Different concepts, such as eigenvalues, eigenvectors (right and left), controllability, observability, controller, and observer design are presented. Both continuous and discrete-time systems are discussed in a generalized framework.

In the classical approaches, the control analysis and synthesis of a linear system depend on transfer functions, which are derived assuming that the system is at rest. The state space domain analysis does not suffer from this limitation. In fact, the solution of a state equation depends on the initial condition. Therefore, starting from any arbitrary initial condition, the system behavior can be specified for any arbitrary time.

The state space approach is very crucial not only for control design but also for analyzing the stability of a system. The nature of power systems containing power electronic converters is nonlinear. Again, there might be systems that contain multiple converters, such as a microgrid. Such systems are linearized around an operating point to obtain a linear system description. The stability evaluation of the system is then performed through eigenvalue analysis, where one or more parameters can be varied to ascertain that the system remains stable under various disturbances. There is a drastic difference between this approach and that of root locus. In the latter, the characteristic equation of a system is written in the form $1 + KG(s)H(s) = 0$. Then the gain K is varied from zero to infinity to trace the movement of the system poles. It is not always easy to decompose a characteristic equation in the above form. However, the use of state space modeling makes this redundant.

The other advantage of the state space representation is that it can easily include multi-input, multi-output (MIMO) systems. The state space analysis technique is based on a dynamic representation of systems in matrix–vector forms, where both

Control of Power Electronic Converters with Microgrid Applications, First Edition.
Arindam Ghosh and Firuz Zare.

Published 2023 by John Wiley & Sons, Inc.

input and output vectors can have more than one element. However, the analysis of the system will remain the same, irrespective of the number of elements in the input or output vectors. Furthermore, the state feedback control design process, in general, is the same for single-input, single-output (SISO) and MIMO systems.

5.1 State Space Representation of Linear Systems

5.1.1 Continuous-time Systems

Consider an nth-order differential equation, given by

$$\overset{(n)}{y} + a_{n-1}\overset{(n-1)}{y} + a_{n-2}\overset{(n-2)}{y} + \cdots + a_1\dot{y} + a_0 y = u \tag{5.1}$$

where y is the output, u is the input, and $\overset{(k)}{x}$ is the kth time derivative of x, i.e. $\overset{(k)}{x} = d^k x/dt^k$. The differential equation of (5.1) contains no derivative of the input signal. In general, however, differential equations containing derivatives of input can also be represented in the state space form, as is discussed in Section 5.4.1. With respect to (5.1), let us define

$$\begin{aligned} x_1 &= y \\ x_2 &= \dot{y} \\ &\vdots \\ x_{n-1} &= \overset{(n-2)}{y} \\ x_n &= \overset{(n-1)}{y} \end{aligned} \tag{5.2}$$

Then comparing (5.1) with (5.2), the following expressions are obtained

$$\begin{aligned} \dot{x}_1 &= x_2 \\ \dot{x}_2 &= x_3 \\ &\vdots \\ \dot{x}_{n-1} &= x_n \\ \dot{x}_n &= \overset{(n)}{y} = u - a_{n-1}x_n - a_{n-2}x_{n-1} - \cdots - a_1 x_2 - a_0 x_1 \end{aligned} \tag{5.3}$$

The state vector of the system is defined as

$$\mathbf{x} = \begin{bmatrix} x_1 \\ x_2 \\ \vdots \\ x_{n-1} \\ x_n \end{bmatrix} \tag{5.4}$$

Then (5.3) can be written in the state space matrix–vector form as

$$\dot{\mathbf{x}} = \mathbf{A}\mathbf{x} + \mathbf{B}u \tag{5.5}$$

where the system matrices are

$$\mathbf{A} = \begin{bmatrix} 0 & 1 & 0 & \cdots & 0 & 0 \\ 0 & 0 & 1 & \cdots & 0 & 0 \\ \vdots & \vdots & \vdots & \ddots & \vdots & \vdots \\ 0 & 0 & 0 & \cdots & 0 & 1 \\ -a_0 & -a_1 & -a_2 & \cdots & -a_{n-2} & -a_{n-1} \end{bmatrix}, \quad \mathbf{B} = \begin{bmatrix} 0 \\ 0 \\ \vdots \\ 0 \\ 1 \end{bmatrix} \tag{5.6}$$

The output equation is then given from (5.2) as

$$y = x_1 = \mathbf{C}\mathbf{x} \tag{5.7}$$

where the output matrix is

$$\mathbf{C} = [1 \; 0 \; \cdots \; 0 \; 0] \tag{5.8}$$

Equations (5.5) and (5.6) represent a special form called the controllable canonical form or phase variable canonical form. This is widely used for controllability tests and state feedback control designs, as are discussed elsewhere in this chapter.

5.1.2 Discrete-time Systems

Consider the following nth-order difference equation

$$y(k+n)+a_{n-1}y(k+n-1)+a_{n-2}y(k+n-2)+\cdots+a_1y(k+1)+a_0y(k)=u(k) \tag{5.9}$$

In a similar fashion to Section 5.1.1, we can define

$$\begin{aligned} &x_1(k)=y(k) \\ &x_1(k+1)=x_2(k)=y(k+1) \\ &x_2(k+1)=x_3(k)=y(k+2) \\ &\vdots \\ &x_{n-1}(k+1)=x_n(k)=y(k+n-1) \\ &x_n(k+1)=y(k+n)=u(k)-a_{n-1}x_n(k)-a_{n-2}x_{n-1}(k)-\cdots-a_1x_2(k)-a_0x_1(k) \end{aligned} \tag{5.10}$$

Then for the state vector of (5.4), the discrete-time state space description is given by

$$\mathbf{x}(k+1) = \mathbf{A}\mathbf{x}(k) + \mathbf{B}u(k) \tag{5.11}$$

$$y(k) = \mathbf{C}\mathbf{x}(k) \tag{5.12}$$

where the matrices **A** and **B** are the same as given by (5.6) and the **C** matrix is given by (5.8).

5.2 Solution of State Equation of a Continuous-time System

The solution of a state equation is similar to solving an nth-order differential equation. We shall start the discussion by defining what is called a state transition matrix.

5.2.1 State Transition Matrix

A state equation is called linear homogeneous when the input signal is zero. Substituting $u = 0$ (5.5), the homogeneous state equation is given by

$$\dot{\mathbf{x}}(t) = \mathbf{A}\mathbf{x}(t) \tag{5.13}$$

Let $\boldsymbol{\varphi} \in \Re^{n\times n}$ be the state transition matrix of the system in (5.13). Then it must satisfy

$$\dot{\boldsymbol{\varphi}}(t) = \mathbf{A}\boldsymbol{\varphi}(t) \tag{5.14}$$

Since the state equation of (5.13) has no forcing function, its solution will depend on the initial condition $\mathbf{x}(0)$ at time $t = 0$, given by

$$\mathbf{x}(t) = e^{\mathbf{A}t}\mathbf{x}(0) \tag{5.15}$$

Let us define

$$\boldsymbol{\varphi}(t) = e^{\mathbf{A}t} \tag{5.16}$$

It can easily be verified that $\boldsymbol{\varphi}(t)$ satisfies (5.14), i.e.

$$\dot{\boldsymbol{\varphi}}(t) = \mathbf{A}e^{\mathbf{A}t} = \mathbf{A}\boldsymbol{\varphi}(t)$$

The solution of the state equation can be written in terms of the state transition matrix by substituting (5.16) in (5.15) as

$$\mathbf{x}(t) = \boldsymbol{\varphi}(t)\mathbf{x}(0) \tag{5.17}$$

Taking the Laplace transform of both sides of (5.13), we get

$$s\mathbf{X}(s) - \mathbf{x}(0) = \mathbf{A}\mathbf{X}(s)$$
$$\Rightarrow \mathbf{X}(s) = (s\mathbf{I} - \mathbf{A})^{-1}\mathbf{x}(0)$$

The inverse Laplace transform of the above equation is

$$\mathbf{x}(t) = \mathcal{L}^{-1}\left[(s\mathbf{I}-\mathbf{A})^{-1}\right]\mathbf{x}(0)$$

Comparing (5.17) with the above equation, the state transition matrix is obtained as

$$\boldsymbol{\varphi}(t) = \mathcal{L}^{-1}\left[(s\mathbf{I}-\mathbf{A})^{-1}\right] \tag{5.18}$$

Example 5.1 Let us consider a homogeneous state equation of the form

$$\dot{\mathbf{x}} = \begin{bmatrix} 0 & 1 \\ -4 & -5 \end{bmatrix}\mathbf{x}(t)$$

Then

$$(s\mathbf{I}-\mathbf{A})^{-1} = \begin{bmatrix} s & -1 \\ 4 & s+5 \end{bmatrix}^{-1} = \frac{1}{s^2+5s+4}\begin{bmatrix} s+5 & 1 \\ -4 & s \end{bmatrix} = \frac{1}{(s+4)(s+1)}\begin{bmatrix} s+5 & 1 \\ -4 & s \end{bmatrix}$$

Through partial fraction expansion of the above equation, we get

$$(s\mathbf{I}-\mathbf{A})^{-1} = \frac{1}{3}\begin{bmatrix} -\dfrac{1}{s+4}+\dfrac{4}{s+1} & -\dfrac{1}{s+4}+\dfrac{1}{s+1} \\ \dfrac{4}{s+4}-\dfrac{4}{s+1} & \dfrac{4}{s+4}-\dfrac{1}{s+1} \end{bmatrix}$$

Therefore, the state transition matrix is

$$\boldsymbol{\varphi}(t) = \mathcal{L}^{-1}\left[(s\mathbf{I}-\mathbf{A})^{-1}\right] = \begin{bmatrix} -\dfrac{1}{3}e^{-4t}+\dfrac{4}{3}e^{-t} & -\dfrac{1}{3}e^{-4t}+\dfrac{1}{3}e^{-t} \\ \dfrac{4}{3}e^{-4t}-\dfrac{4}{3}e^{-t} & \dfrac{4}{3}e^{-4t}-\dfrac{1}{3}e^{-t} \end{bmatrix}$$

Therefore, the solution of the state equation can then be written as

$$\mathbf{x}(t) = \begin{bmatrix} -\dfrac{1}{3}e^{-4t}+\dfrac{4}{3}e^{-t} & -\dfrac{1}{3}e^{-4t}+\dfrac{1}{3}e^{-t} \\ \dfrac{4}{3}e^{-4t}-\dfrac{4}{3}e^{-t} & \dfrac{4}{3}e^{-4t}-\dfrac{1}{3}e^{-t} \end{bmatrix}\mathbf{x}(0)$$

Note that a partial fraction expansion can be performed easily through MATLAB®. Consider the following rational function with simple poles

$$H(s) = \frac{Q(s)}{P(s)} = \frac{s^m + q_{m-1}s^{m-1} + \cdots + q_1 s + q_0}{(s-p_1)(s-p_2)\cdots(s-p_n)}, \quad n \geq m$$

Then the "residue" command in MATLAB® will perform the partial fraction expansion of the form

$$H(s) = \frac{r_1}{(s-p_1)} + \frac{r_2}{(s-p_2)} + \cdots + \frac{r_n}{(s-p_n)}$$

where r_i, $i = 1, \ldots n$ are the residues of the partial fraction expansion. Consider, for example, the following rational function

$$\frac{s+5}{s^2+5s+4}$$

Then use the following MATLAB® commands:

```
num = [1 5];
den = [1 5 4];
[z,p,k] = residue(num,den)
```

This will produce z = [− 0.3333; 1.3333], p = [− 4; − 1] and k = []. In a similar way, the residues of all other three elements of the matrix can be computed. Note that residues for multiple poles can also be determined using the "residue" command. Please refer to MATLAB® help.

5.2.2 Properties of State Transition Matrix

There are certain properties that are used while determining the response of a system. Some of the properties are:

1) $\boldsymbol{\varphi}(0) = e^{\mathbf{A} \times 0} = \mathbf{I}$
2) $\boldsymbol{\varphi}^{-1}(t) = \left(e^{\mathbf{A}t}\right)^{-1} = e^{-\mathbf{A}t} = \boldsymbol{\varphi}(-t)$
3) $\boldsymbol{\varphi}(t_1 + t_2) = e^{\mathbf{A}(t_1 + t_2)} = e^{\mathbf{A}t_1} e^{\mathbf{A}t_2} = \boldsymbol{\varphi}(t_1)\boldsymbol{\varphi}(t_2)$
4) $\boldsymbol{\varphi}(t_2 - t_1)\, \boldsymbol{\varphi}(t_1 - t_0) = e^{\mathbf{A}(t_2 - t_1)} e^{\mathbf{A}(t_1 - t_0)} = e^{\mathbf{A}(t_2 - t_0)} = \boldsymbol{\varphi}(t_2 - t_0)$

Note that, for two square matrices **A** and **B**, $e^{\mathbf{A}+\mathbf{B}} = e^{\mathbf{A}} e^{\mathbf{B}}$ only when $\mathbf{AB} = \mathbf{BA}$. Therefore, properties 3 and 4 are valid. The fourth property is very useful as it implies that a state transition process can be divided into two sequential transitions. This property is used for some power electronic applications discussed in Section 5.11. In general, however, the transition process can be broken into any number of parts. Example 5.2 illustrates this.

Example 5.2 Consider the homogeneous state equation of Example 5.1. Let the initial condition at $t_0 = 0$ be $\mathbf{x}(0) = [3 \quad 5]^T$. It is assumed that $t_1 = 1$ seconds and $t_2 = 2$ seconds in property 4. Then we have

$$\mathbf{x}(t_1) = \boldsymbol{\varphi}(t_1)\mathbf{x}(0) = \left.\begin{bmatrix} -\frac{1}{3}e^{-4t} + \frac{4}{3}e^{-t} & -\frac{1}{3}e^{-4t} + \frac{1}{3}e^{-t} \\ \frac{4}{3}e^{-4t} - \frac{4}{3}e^{-t} & \frac{4}{3}e^{-4t} - \frac{1}{3}e^{-t} \end{bmatrix}\right|_{t=1} \begin{bmatrix} 3 \\ 5 \end{bmatrix} = \begin{bmatrix} 2.036 \\ -1.889 \end{bmatrix}$$

Also

$$\mathbf{x}(t_2) = \boldsymbol{\varphi}(t_2 - t_1)\mathbf{x}(t_1) = \left.\begin{bmatrix} -\frac{1}{3}e^{-4t} + \frac{4}{3}e^{-t} & -\frac{1}{3}e^{-4t} + \frac{1}{3}e^{-t} \\ \frac{4}{3}e^{-4t} - \frac{4}{3}e^{-t} & \frac{4}{3}e^{-4t} - \frac{1}{3}e^{-t} \end{bmatrix}\right|_{t=1} \begin{bmatrix} 2.036 \\ -1.889 \end{bmatrix} = \begin{bmatrix} 0.766 \\ -0.763 \end{bmatrix}$$

Or else $\mathbf{x}(t_2)$ can be directly computed from

$$\mathbf{x}(t_2) = \boldsymbol{\varphi}(t_2)\mathbf{x}(0) = \left.\begin{bmatrix} -\frac{1}{3}e^{-4t} + \frac{4}{3}e^{-t} & -\frac{1}{3}e^{-4t} + \frac{1}{3}e^{-t} \\ \frac{4}{3}e^{-4t} - \frac{4}{3}e^{-t} & \frac{4}{3}e^{-4t} - \frac{1}{3}e^{-t} \end{bmatrix}\right|_{t=2} \begin{bmatrix} 3 \\ 5 \end{bmatrix} = \begin{bmatrix} 0.766 \\ -0.763 \end{bmatrix}$$

In a similar way, a chain of solutions of the state equation can be used to arrive at the final solution for any given time instant.

Furthermore, from property 2 above, we can traverse back in time to find the values of the state variables, such as

$$\mathbf{x}(t_1) = \boldsymbol{\varphi}(t_2 - t_1)^{-1}\mathbf{x}(t_2) = \begin{bmatrix} 2.036 \\ -1.889 \end{bmatrix}$$

and

$$\mathbf{x}(0) = \boldsymbol{\varphi}(t_1)^{-1}\mathbf{x}(t_1) = \begin{bmatrix} 3 \\ 5 \end{bmatrix}$$

Since the state transition matrix satisfies the homogeneous state equation, it represents the free response of the system. It describes the response of the system that is excited by the initial condition only. As can be seen from (5.15), this matrix defines the transition from the initial time $t = 0$ to any finite time when the input is zero.

5.2.3 State Transition Equation

In this section, the solution of nonhomogeneous LTI systems is discussed, where the forcing function is not zero. Consider a state equation of the form (5.5), which need not be in the state variable canonical form. Taking Laplace transform on both sides of (5.5), we get

$$\begin{aligned} s\mathbf{X}(s) - \mathbf{x}(0) &= \mathbf{A}\mathbf{X}(s) + \mathbf{B}U(s) \\ \Rightarrow \mathbf{X}(s) &= (s\mathbf{I} - \mathbf{A})^{-1}[\mathbf{x}(0) + \mathbf{B}U(s)] \end{aligned} \tag{5.19}$$

The inverse Laplace transform of (5.19) is

$$\begin{aligned} \mathbf{x}(t) &= \mathcal{L}^{-1}\{(s\mathbf{I} - \mathbf{A})^{-1}[\mathbf{x}(0) + \mathbf{B}U(s)]\} \\ &= \boldsymbol{\varphi}(t)\mathbf{x}(0) + \mathcal{L}^{-1}\{(s\mathbf{I} - \mathbf{A})^{-1}\mathbf{B}U(s)\} \end{aligned}$$

Using the convolution integral, the above equation is written in time domain as

$$\mathbf{x}(t) = \boldsymbol{\varphi}(t)\mathbf{x}(0) + \int_0^t e^{\mathbf{A}(t-\tau)}\mathbf{B}u(\tau)d\tau = \boldsymbol{\varphi}(t)\mathbf{x}(0) + \int_0^t e^{\mathbf{A}\tau}\mathbf{B}u(t-\tau)d\tau \tag{5.20}$$

Note that (5.20) is defined with respect to the initial values at time $t = 0$. However, as evident from Example 5.2, this equation can be written between any two arbitrary time intervals t_1 and t_2 as

$$\mathbf{x}(t_2) = \boldsymbol{\varphi}(t_2 - t_1)\mathbf{x}(t_1) + \int_{t_1}^{t_2} e^{\mathbf{A}(t_2-\tau)}\mathbf{B}u(\tau)d\tau \tag{5.21}$$

5.3 Solution of State Equation of a Discrete-time System

The solution of a state Eq. (5.11) is akin to solving an nth-order difference equation. Equation (5.11) is expanded for the different values of k as

$$\begin{aligned} \mathbf{x}(1) &= \mathbf{A}\mathbf{x}(0) + \mathbf{B}u(0) \\ \mathbf{x}(2) &= \mathbf{A}\mathbf{x}(1) + \mathbf{B}u(1) \\ &= \mathbf{A}^2\mathbf{x}(0) + \mathbf{A}\mathbf{B}u(0) + \mathbf{B}u(1) \\ \mathbf{x}(3) &= \mathbf{A}\mathbf{x}(2) + \mathbf{B}u(2) \\ &= \mathbf{A}^3\mathbf{x}(0) + \mathbf{A}^2\mathbf{B}u(0) + \mathbf{A}\mathbf{B}u(1) + \mathbf{B}u(2) \end{aligned}$$

Repeating the sequence, we get

$$\mathbf{x}(k) = \mathbf{A}^k\mathbf{x}(0) + \sum_{i=0}^{k-1}\mathbf{A}^{k-i-1}\mathbf{B}u(i) \tag{5.22}$$

5.3.1 State Transition Matrix

Comparing (5.22) when the input is zero with (5.15), the state transition matrix of a discrete-time system is written as

$$\boldsymbol{\varphi}(k) = \mathbf{A}^k \tag{5.23}$$

It can be readily seen that

$$\boldsymbol{\varphi}(k+1) = \mathbf{A}^{k+1} = \mathbf{A}\boldsymbol{\varphi}(k)$$

which is of the same form as that given in (5.14).

The properties of the state transition matrix are also like those of the continuous-time systems, i.e.

$$\boldsymbol{\varphi}(0) = \mathbf{A}^{-0} = \mathbf{I}$$

$$\boldsymbol{\varphi}^{-1}(k) = \left(\mathbf{A}^k\right)^{-1} = \mathbf{A}^{-k} = \boldsymbol{\varphi}(-k)$$

$$\boldsymbol{\varphi}(k_1 + k_2) = \mathbf{A}^{(k_1 + k_2)} = \mathbf{A}^{k_1}\mathbf{A}^{k_2} = \boldsymbol{\varphi}(k_1)\boldsymbol{\varphi}(k_2)$$

$$\boldsymbol{\varphi}(k_2 - k_1)\,\boldsymbol{\varphi}(k_1 - k_0) = \mathbf{A}^{(k_2 - k_1)}\mathbf{A}^{(k_1 - k_0)} = \mathbf{A}^{(t_2 - t_0)} = \boldsymbol{\varphi}(k_2 - k_0)$$

5.3.2 Computation of State Transition Matrix

The simplest way of computing a state transition matrix is through an inverse z-transform. Consider the discrete-time homogeneous state equation of the form

$$\mathbf{x}(k+1) = \mathbf{A}\mathbf{x}(k) \tag{5.24}$$

Taking the z-transform of both sides of (5.24), we get

$$z\mathbf{X}(z) - z\mathbf{x}(0) = \mathbf{A}\mathbf{X}(z)$$

The solution of the above equation is

$$\mathbf{X}(z) = (z\mathbf{I} - \mathbf{A})^{-1}z\mathbf{x}(0)$$

Therefore, the state transition matrix is given by

$$\boldsymbol{\varphi}(k) = \mathcal{Z}^{-1}\left[z(z\mathbf{I} - \mathbf{A})^{-1}\right] \tag{5.25}$$

Example 5.3 Consider a homogeneous discrete-time state equation of the form

$$\mathbf{x}(k+1) = \begin{bmatrix} 0 & 1 \\ -4 & -5 \end{bmatrix}\mathbf{x}(k)$$

Then

$$\mathbf{F}(z) = z(z\mathbf{I} - \mathbf{A})^{-1} = z\begin{bmatrix} z & -1 \\ 4 & z+5 \end{bmatrix}^{-1} = \frac{z}{(z+4)(z+1)}\begin{bmatrix} z+5 & 1 \\ -4 & z \end{bmatrix}$$

Using partial fraction expansion such as is given in Example 5.1, we can write

$$\frac{\mathbf{F}(z)}{z} = \frac{1}{z+4}\begin{bmatrix} -\frac{1}{3} & -\frac{1}{3} \\ \frac{4}{3} & \frac{4}{3} \end{bmatrix} + \frac{1}{z+1}\begin{bmatrix} \frac{4}{3} & \frac{1}{3} \\ \frac{4}{3} & -\frac{1}{3} \end{bmatrix}$$

Therefore, the state transition matrix of the discrete-time system is

$$\begin{aligned} \boldsymbol{\varphi}(k) = \mathcal{Z}^{-1}[\mathbf{F}(z)] &= \frac{1}{3}\frac{z}{z+4}\begin{bmatrix} -1 & -1 \\ 4 & 4 \end{bmatrix} + \frac{1}{3}\frac{z}{z+1}\begin{bmatrix} 4 & 1 \\ -4 & -1 \end{bmatrix} \\ &= \frac{1}{3}\begin{bmatrix} -1 & -1 \\ 4 & 4 \end{bmatrix}(-4)^k + \frac{1}{3}\begin{bmatrix} 4 & 1 \\ -4 & -1 \end{bmatrix}(-1)^k \end{aligned}$$

5.3.3 Discretization of a Continuous-time System

Consider the state space description of (5.5). Choosing a sampling time of T seconds, it has a discrete-time form of

$$\mathbf{x}(k+1) = \mathbf{F}\mathbf{x}(k) + \mathbf{G}u(k) \tag{5.26}$$

For this, we assume that $u(t)$ is constant over two successive sampling intervals, i.e. $u(t) = u(kT)$ for $kT \le t < k(T+1)$. This implies that $u(kT)$ is the output of a zero-order hold circuit. From (5.20), we get the solution of (5.5) for the instant $k(T+1)$ as

$$\begin{aligned} \mathbf{x}[(k+1)T] &= e^{\mathbf{A}(k+1)T}\mathbf{x}(0) + \int_0^{(k+1)T} e^{\mathbf{A}(kT+T-\tau)}\mathbf{B}u(\tau)d\tau \\ &= e^{A(k+1)T}\mathbf{x}(0) + e^{\mathbf{A}(k+1)T}\int_0^{(k+1)T} e^{-\mathbf{A}\tau}\mathbf{B}u(\tau)d\tau \end{aligned} \tag{5.27}$$

Furthermore, the solution of (5.5) for the instant kT is

$$\mathbf{x}(kT) = e^{\mathbf{A}kT}\mathbf{x}(0) + e^{\mathbf{A}kT}\int_0^{kT} e^{-\mathbf{A}\tau}\mathbf{B}u(\tau)d\tau \tag{5.28}$$

Multiplying (5.28) by $e^{\mathbf{A}t}$ and subtracting from (5.27), the following equation is obtained

$$\mathbf{x}[(k+1)T]-e^{\mathbf{A}T}\mathbf{x}(kT) = e^{\mathbf{A}(k+1)T}\int_{kT}^{(k+1)T} e^{-\mathbf{A}\tau}\mathbf{B}u(\tau)d\tau = e^{\mathbf{A}T}\int_{kT}^{(k+1)T} e^{\mathbf{A}(kT-\tau)}\mathbf{B}u(\tau)d\tau \tag{5.29}$$

Define $\tau = kT + \lambda$ such that $d\tau = \lambda\ d\tau$. Also, this implies that when $\tau = kT$, $\lambda = 0$ and when $\tau = (k+1)T$, $\lambda = T$. Since the input remains constant at $u(kT)$ between the instants kT and $k(T+1)$, we can rewrite (5.29) as

$$\mathbf{x}[(k+1)T] - e^{\mathbf{A}T}\mathbf{x}(kT) = e^{\mathbf{A}T}\int_0^T e^{-\mathbf{A}\lambda}\mathbf{B}u(kT)d\lambda = \int_0^T e^{\mathbf{A}(T-\lambda)}\mathbf{B}u(kT)d\lambda$$

Again define $\lambda = T - t$ such that $d\lambda = -dt$. Also, as t changes from 0 to T, λ changes from T to 0. Thus, the above equation can be written as

$$\mathbf{x}[(k+1)T] = e^{\mathbf{A}T}\mathbf{x}(kT) + \left(\int_0^T e^{\mathbf{A}t}\mathbf{B}dt\right)u(kT) \tag{5.30}$$

Dropping the notation T and comparing (5.26) with (5.30), we have

$$\begin{aligned}\mathbf{F} &= e^{\mathbf{A}T}\\ \mathbf{G} &= \left(\int_0^T e^{\mathbf{A}t}dt\right)\mathbf{B}\end{aligned} \tag{5.31}$$

Example 5.4 Consider the following continuous-time state equation

$$\dot{\mathbf{x}} = \mathbf{A}\mathbf{x} + \mathbf{B}u = \begin{bmatrix} 0 & 1 \\ -4 & -5 \end{bmatrix}\mathbf{x} + \begin{bmatrix} 2 \\ 1 \end{bmatrix}u$$

Its discrete-time equivalent for a sampling time of 1 second is determined using MATLAB®. For this, we shall use the "c2d" command:

```
% Continuous time

A = [0 1;-4 -5];
B = [2;1];

% Discrete-time equivalent

T = 1;
[F,G] = c2d(A,B,T);
```

This will result in the following matrices

$$\mathbf{F} = \begin{bmatrix} 0.4844 & 0.1165 \\ -0.4661 & -0.0982 \end{bmatrix}, \quad \mathbf{G} = \begin{bmatrix} 1.6509 \\ -0.9147 \end{bmatrix}$$

5.4 Relation Between State Space Form and Transfer Function

In this section, the relation between state space form and transfer function representation is discussed, for both continuous-time and discrete-time systems.

5.4.1 Continuous-time System

The transfer function of a SISO is defined as the Laplace transform of the output divided by the Laplace transform of the input of an LTI system, given that the initial conditions are zero. Therefore, the transfer function for a SISO system can be computed by assuming that the initial condition $\mathbf{x}(0)$ in (5.19) is zero. This gives

$$\mathbf{X}(s) = (s\mathbf{I} - \mathbf{A})^{-1}\mathbf{B}U(s) \tag{5.32}$$

Consider a general output equation in which the output depends both on states as well as on the input, given by

$$y = \mathbf{C}\mathbf{x} + Du \tag{5.33}$$

The Laplace transform of the output in (5.33) is

$$Y(s) = \mathbf{C}\mathbf{X}(s) + DU(s)$$

Combining the last equation with (5.32), we get

$$\begin{aligned} Y(s) &= \mathbf{C}\mathbf{X}(s) = \mathbf{C}\left[(s\mathbf{I} - \mathbf{A})^{-1}\mathbf{B}U(s)\right] + DU(s) \\ \Rightarrow \frac{Y(s)}{U(s)} &= G(s) = \mathbf{C}(s\mathbf{I} - \mathbf{A})^{-1}\mathbf{B} + D \end{aligned} \tag{5.34}$$

Example 5.5 Consider a state space description of the form

$$\begin{aligned} \dot{\mathbf{x}} &= \begin{bmatrix} 0 & 1 \\ -4 & -5 \end{bmatrix}\mathbf{x} + \begin{bmatrix} 0 \\ 1 \end{bmatrix}u \\ y &= \begin{bmatrix} 1 & 0 \end{bmatrix}\mathbf{x} \end{aligned}$$

From Example 5.1, we know

$$(s\mathbf{I}-\mathbf{A})^{-1}=\frac{1}{(s+4)(s+1)}\begin{bmatrix} s+5 & 1 \\ -4 & s \end{bmatrix}$$

Therefore, from (5.34), the transfer function is

$$G(s)=\mathbf{C}(s\mathbf{I}-\mathbf{A})^{-1}\mathbf{B}=\frac{1}{(s+4)(s+1)}[1 \quad 0]\begin{bmatrix} s+5 & 1 \\ -4 & s \end{bmatrix}\begin{bmatrix} 0 \\ 1 \end{bmatrix}=\frac{1}{(s+4)(s+1)}$$

Note that it is easy to calculate a transfer function using MATLAB®. This is shown here.

```
% State space system
A = [0 1;-4 -5];
B = [0;1];
C = [1 0];
D = 0;

% Transfer function
[num,den] = ss2tf(A,B,C,D);
```

This will produces num = [0 0 1] and den = [1 5 4], which are essentially the polynomials in the decreasing order of s. On the other hand, to obtain a state space form from the transfer function, the command “tf2ss” can be used, which is of the form [A,B,C,D] = tf2ss(num,den).

The state space representation of a transfer function is not unique. In fact, there are three different ways of decomposing the transfer function from the state and output equations, if the system does not have any repeated roots. The decomposition techniques are not discussed here (for details see [1]). However, Example 5.6 illustrates this.

Example 5.6 Consider the transfer function

$$\frac{Y(s)}{U(s)}=\frac{s^2+8s+12}{s^2+5s+4}=\frac{(s+6)(s+2)}{(s+4)(s+1)}$$

- Direct decomposition: In this form, the transfer function is expanded in terms of s^{-1}. The ordinary differential equations (ODEs) are then obtained while keeping in mind that $s^{-1}={}^{d}\!/_{dt}$. The following state space form is then obtained

$$\dot{\mathbf{x}}=\begin{bmatrix} -5 & -4 \\ 1 & 0 \end{bmatrix}\mathbf{x}+\begin{bmatrix} 1 \\ 0 \end{bmatrix}u$$
$$y=[3 \quad 8]\mathbf{x}+u$$

- Cascade decomposition: In this form, the transfer function is expanded in cascade form as

$$\frac{Y(s)}{U(s)} = \frac{(s+6)}{(s+4)} \times \frac{(s+2)}{(s+1)}$$

Following this, two separate sets of ordinary differential equations (ODEs) are formed by expanding in terms of s^{-1}. This gives the state space equation

$$\dot{\mathbf{x}} = \begin{bmatrix} -4 & 0 \\ 2 & -1 \end{bmatrix} \mathbf{x} + \begin{bmatrix} 1 \\ 1 \end{bmatrix} u$$
$$y = [2 \quad 1]\mathbf{x} + u$$

- Parallel decomposition: In this form, the transfer function is expanded in parallel form as

$$\frac{Y(s)}{U(s)} = 1 + \frac{3s+8}{(s+4)(s+1)} = 1 + \frac{4/3}{(s+4)} + \frac{5/3}{(s+1)}$$

From this we get the following state space form

$$\dot{\mathbf{x}} = \begin{bmatrix} -4 & 0 \\ 0 & -1 \end{bmatrix} \mathbf{x} + \begin{bmatrix} 1 \\ 1 \end{bmatrix} u$$
$$y = \begin{bmatrix} \frac{4}{3} & \frac{5}{3} \end{bmatrix} \mathbf{x} + u$$

Using (5.34), it can be easily verified that all three forms produce the desired transfer function.

5.4.2 Discrete-time System

A discrete-time system is defined by

$$\begin{aligned} \mathbf{x}(k+1) &= \mathbf{F}\mathbf{x}(k) + \mathbf{G}u(k) \\ y(k) &= \mathbf{C}\mathbf{x}(k) + Du(k) \end{aligned} \tag{5.35}$$

The transfer function will then be

$$\frac{Y(z)}{U(z)} = \mathbf{C}(z\mathbf{I} - \mathbf{F})^{-1}\mathbf{G} + D \tag{5.36}$$

The state space analysis and design essentially depend on the matrix properties. Since these properties do not change whether the system is continuous or discrete, only continuous-time systems are considered in Sections 5.5–5.8. Feedback control design, however, is treated separately.

5.5 Eigenvalues and Eigenvectors

Equation (5.34) can be expanded as

$$\frac{Y(s)}{U(s)} = \mathbf{C}(s\mathbf{I}-\mathbf{A})^{-1}\mathbf{B} + D = \frac{\mathbf{C} \times adj(s\mathbf{I}-\mathbf{A}) \times \mathbf{B} + |sI-\mathbf{A}| \times D}{|s\mathbf{I}-\mathbf{A}|} \tag{5.37}$$

Equating the denominator to zero, the characteristic equation of the system is obtained as

$$|s\mathbf{I}-\mathbf{A}| = 0 \tag{5.38}$$

The solution of the characteristic equation results in the system poles. These are also called the eigenvalues of the matrix **A**.

5.5.1 Eigenvalues

The eigenvalues of the matrix $\mathbf{A} \in \Re^{n\times n}$ are given by the solution of the characteristic equation $|s\mathbf{I} - \mathbf{A}| = 0$. Let these eigenvalues be $\lambda_1, \lambda_2, ..., \lambda_n$. Then there are certain properties of eigenvalues that are of importance for control system design. These are:

1) Eigenvalues of the matrix **A** remain unchanged under any linear transformation. For a system of the form given by (5.5), let $\mathbf{x} = \mathbf{Pz}$, where $\mathbf{P} \in \Re^{n\times n}$ and is nonsingular. Then

$$\dot{\mathbf{x}} = \mathbf{P}\dot{\mathbf{z}} = \mathbf{Ax} + \mathbf{B}u = \mathbf{AP}^{-1}\mathbf{z} + \mathbf{B}u$$

Hence

$$\dot{\mathbf{z}} = \mathbf{P}^{-1}\mathbf{APz} + \mathbf{P}^{-1}\mathbf{B}u$$

The eigenvalues of the transformed system are given by the solution of the characteristic equation

$$|s\mathbf{I} - \mathbf{P}^{-1}\mathbf{AP}| = 0$$
$$\Rightarrow |s\mathbf{P}^{-1}\mathbf{P} - \mathbf{P}^{-1}\mathbf{AP}| = |\mathbf{P}^{-1}(s\mathbf{I}-\mathbf{A})\mathbf{P}| = |\mathbf{P}^{-1}\mathbf{P}||s\mathbf{I}-\mathbf{A}| = |s\mathbf{I}-\mathbf{A}| = 0$$

In other words, the eigenvalues of the matrix $\mathbf{P}^{-1}\mathbf{AP}$ are the same as the eigenvalues of the matrix **A**.

2) The trace of the matrix **A** is summation of the eigenvalues, i.e.

$$Tr(\mathbf{A}) = \lambda_1 + \lambda_2 + \cdots + \lambda_n$$

3) The determinant of the matrix **A** is the product of the eigenvalues, i.e.

$$|\mathbf{A}| = \lambda_1 \times \lambda_2 \times \cdots \times \lambda_n$$

4) If **A** is nonsingular, then $1/\lambda_1, 1/\lambda_2, \ldots, 1/\lambda_n$ are the eigenvalues of the matrix $\mathbf{A}^{-1}$.
5) If the matrix **A** is real and symmetric, then its eigenvalues are all real.
6) If a matrix $\mathbf{B} \in \Re^{n \times n}$ is such that

$$|s\mathbf{I} - \mathbf{AB}| = |s\mathbf{I} - \mathbf{BA}|$$

then the eigenvalues of **AB** are the same as that of **BA**.

5.5.2 Eigenvectors

Assume that a square matrix has simple (nonrepeated) eigenvalues. It can then be diagonalized in a form in which the diagonal elements are the eigenvalues using a matrix formed using the eigenvectors that are associated with these eigenvalues. In case the matrix has repeated eigenvalues, it can also nearly be diagonalized in a form which is called the Jordan form using the eigenvectors. Let $\boldsymbol{\gamma}_i$ be an n dimensional column vector that satisfies the equation

$$\lambda_j \boldsymbol{\gamma}_j = \mathbf{A}\boldsymbol{\gamma}_j \tag{5.39}$$

where λ_j is the jth eigenvalue of the matrix **A**. Then $\boldsymbol{\gamma}_i$ is called the right eigenvector of the matrix **A**, associated with the eigenvalue λ_j. This is called the right eigenvector since it is placed on the right of the matrix **A**.

Alternatively, the n dimensional row vector $\boldsymbol{\rho}_j$ that satisfies the equation

$$\lambda_j \boldsymbol{\rho}_j = \boldsymbol{\rho}_j \mathbf{A} \tag{5.40}$$

is called the left eigenvector of the matrix **A**, and is associated with the eigenvalue λ_j. This is called the left eigenvector since it is placed on the left of the matrix **A**. Transposing both sides of (5.40), we get

$$\lambda_j \boldsymbol{\rho}_j^T = \left(\boldsymbol{\rho}_j \mathbf{A}\right)^T = \mathbf{A}^T \boldsymbol{\rho}_j^T \tag{5.41}$$

This implies that the transpose of the left eigenvector is the right eigenvector of the transpose of the matrix **A**. In general, the right eigenvector is simply called the eigenvector and is used for the matrix properties discussed here.

Some of the important properties of eigenvectors are:

1) The rank of the matrix $(\lambda_j \mathbf{I} - \mathbf{A})$, where $\lambda_1, \lambda_2, \ldots, \lambda_n$ are the distinct eigenvalues of the matrix **A**, is $n - 1$.

2) Post-multiplying the matrix $(\lambda_j\mathbf{I} - \mathbf{A})$ by $\boldsymbol{\gamma}_j$ it can be observed that the matrix $(\lambda_j\boldsymbol{\gamma}_j - \mathbf{A}\boldsymbol{\gamma}_j)$ is not full rank and hence it is not invertible.
3) If the matrix **A** has a set of n distinct eigenvalues $\lambda_1, \lambda_2, ..., \lambda_n$, then the set of eigenvectors $\boldsymbol{\gamma}_j$, $j = 1, 2, ..., n\gamma_j$ are linearly independent, i.e.

$$\alpha_1\boldsymbol{\gamma}_1 + \alpha_2\boldsymbol{\gamma}_2 + \cdots + \alpha_n\boldsymbol{\gamma}_n \neq 0, \text{for nonzero } \alpha_j, \ \ j = 1, 2, \cdots, n$$

4) If $\boldsymbol{\gamma}_j$ is an eigenvector of the matrix **A**, then $\beta\boldsymbol{\gamma}_j$ is also an eigenvector of **A**, where β is an arbitrary scalar.

Example 5.7 Consider the matrix

$$\mathbf{A} = \begin{bmatrix} 0 & 1 & 0 \\ 0 & 0 & 1 \\ -40 & -38 & -11 \end{bmatrix}$$

It can be easily seen that

$$|s\mathbf{I} - \mathbf{A}| = s^3 + 11s^2 + 38s + 40 = (s + 2)(s + 4)(s + 5)$$

This means that the eigenvalues of the matrix **A** are $-2, -4$ and -5.

Define an eigenvector $\boldsymbol{\gamma}_1$ associated with the eigenvalue -2 as

$$\boldsymbol{\gamma}_1 = \begin{bmatrix} \gamma_{11} \\ \gamma_{12} \\ \gamma_{13} \end{bmatrix}$$

Then, from (5.39) we have

$$-2\begin{bmatrix} \gamma_{11} \\ \gamma_{12} \\ \gamma_{13} \end{bmatrix} = \begin{bmatrix} 0 & 1 & 0 \\ 0 & 0 & 1 \\ -40 & -38 & -11 \end{bmatrix} \times \begin{bmatrix} \gamma_{11} \\ \gamma_{12} \\ \gamma_{13} \end{bmatrix} = \begin{bmatrix} \gamma_{12} \\ \gamma_{13} \\ -40\gamma_{11} - 38\gamma_{12} - 11\gamma_{13} \end{bmatrix}$$

From the second property of the eigenvector given above, it is obvious that there cannot be any unique solution of the above set of equations. Let us define $\gamma_{11} = 1$. Then

$$\gamma_{12} = -2\gamma_{11} = -2$$
$$\gamma_{13} = -2\gamma_{12} = 4$$

It can be easily verified that this choice satisfies the equation form from the third row. Hence, the eigenvector $\boldsymbol{\gamma}_1$ is

$$\boldsymbol{\gamma}_1 = \begin{bmatrix} 1 \\ -2 \\ 4 \end{bmatrix}$$

In a similar way, we can determine

$$\boldsymbol{\gamma}_2 = \begin{bmatrix} 1 \\ -4 \\ 16 \end{bmatrix} \text{ associated with } -4 \text{ and } \boldsymbol{\gamma}_3 = \begin{bmatrix} 1 \\ -5 \\ 25 \end{bmatrix} \text{ associated with } -5.$$

5.6 Diagonalization of a Matrix Using Similarity Transform

A procedure through which the matrix **A** given in (5.5) can be diagonalized is discussed in this section. Diagonal matrices have a significant influence on the control design as can be seen in Sections 5.7 and 5.8. The diagonalization method is simple when the matrix has distinct (nonrepeated) eigenvalues. However, in the case of repeated eigenvalues, an almost diagonal form, called the Jordan canonical form is derived.

5.6.1 Matrix with Distinct Eigenvalues

Assume that the matrix $\mathbf{A} \in \Re^{n\times n}$ has distinct eigenvalues $\lambda_1, \lambda_2, ..., \lambda_n$. Let the respective eigenvectors be $\boldsymbol{\gamma}_1, \boldsymbol{\gamma}_2, ..., \boldsymbol{\gamma}_n$. Let us now form a matrix **P** as

$$\mathbf{P} = [\boldsymbol{\gamma}_1 \quad \boldsymbol{\gamma}_2 \quad \cdots \quad \boldsymbol{\gamma}_n] \tag{5.42}$$

Now we shall use the similarity transform of the form

$$\boldsymbol{\Lambda} = \mathbf{P}^{-1}\mathbf{A}\mathbf{P} \tag{5.43}$$

From (5.42), we get

$$\mathbf{AP} = [\mathbf{A}\boldsymbol{\gamma}_1 \quad \mathbf{A}\boldsymbol{\gamma}_2 \quad \cdots \quad \mathbf{A}\boldsymbol{\gamma}_n]$$

Using (5.39), the above equation is rewritten as

$$\mathbf{AP} = [\boldsymbol{\gamma}_1 \quad \boldsymbol{\gamma}_2 \quad \cdots \quad \boldsymbol{\gamma}_n] \begin{bmatrix} \lambda_1 & & & \\ & \lambda_2 & & \\ & & \ddots & \\ & & & \lambda_n \end{bmatrix} = \mathbf{P} \begin{bmatrix} \lambda_1 & & & \\ & \lambda_2 & & \\ & & \ddots & \\ & & & \lambda_n \end{bmatrix}$$

Therefore, substituting the above equation in (5.43), we have

$$\Lambda = \mathbf{P}^{-1}\mathbf{A}\mathbf{P} = \begin{bmatrix} \lambda_1 & & & \\ & \lambda_2 & & \\ & & \ddots & \\ & & & \lambda_n \end{bmatrix} \tag{5.44}$$

Example 5.8 Consider the matrix of Example 5.7, where the eigenvalues and the eigenvectors are also calculated. Then the transformation matrix of (5.2) is

$$\mathbf{P} = \begin{bmatrix} 1 & 1 & 1 \\ -2 & -4 & -5 \\ 4 & 16 & 25 \end{bmatrix}$$

It can be easily verified that

$$\Lambda = \mathbf{P}^{-1}\mathbf{A}\mathbf{P} = \begin{bmatrix} -2 & & \\ & -4 & \\ & & -5 \end{bmatrix}$$

Now, defining $\lambda_1 = -2, \lambda_2 = -4$, and $\lambda_3 = -5$, the transformation matrix $\mathbf{P}$ is in the form

$$\mathbf{P} = \begin{bmatrix} 1 & 1 & 1 \\ -2 & -4 & -5 \\ (-2)^2 & (-4)^2 & (-5)^2 \end{bmatrix} = \begin{bmatrix} 1 & 1 & 1 \\ \lambda_1 & \lambda_2 & \lambda_3 \\ \lambda_1^2 & \lambda_2^2 & \lambda_3^2 \end{bmatrix}$$

In general, if the matrix $\mathbf{A}$ with n distinct eigenvalues is in the controllable canonical form, then it can be diagonalized using the so-called Vandermonde matrix, given by

$$\mathbf{P} = \begin{bmatrix} 1 & 1 & \cdots & 1 \\ \lambda_1 & \lambda_2 & \cdots & \lambda_n \\ \lambda_1^2 & \lambda_2^2 & \cdots & \lambda_n^2 \\ \vdots & \vdots & \ddots & \vdots \\ \lambda_1^{n-1} & \lambda_2^{n-1} & \cdots & \lambda_n^{n-1} \end{bmatrix} \tag{5.45}$$

It is, however, not necessary for the matrix $\mathbf{A}$ to be in the controllable canonical form for it to be diagonalized. Example 5.9 illustrates this concept.

Example 5.9 Consider the matrix

$$\mathbf{A} = \begin{bmatrix} -89 & -77 & -49 \\ 90 & 78 & 50 \\ 0 & -1 & 0 \end{bmatrix}$$

The eigenvalues of the matrix are given by $\lambda_1 = -5$, $\lambda_2 = -4$, and $\lambda_3 = -2$. Using the same procedure as in Example 5.7, the following eigenvectors are obtained

$$\boldsymbol{\gamma}_1 = \begin{bmatrix} -5.1667 \\ 5 \\ 1 \end{bmatrix}, \boldsymbol{\gamma}_2 = \begin{bmatrix} -4.2 \\ 4 \\ 1 \end{bmatrix}, \text{and } \boldsymbol{\gamma}_3 = \begin{bmatrix} -2.3333 \\ 2 \\ 1 \end{bmatrix}$$

It can be easily verified that the matrix **P** formed by the eigenvectors will diagonalize the matrix **A**, given by $\mathbf{A} = diag([-5 \quad -4 \quad -2])$.

The eigenvalues and eigenvectors can be obtained by using the command "eig" in MATLAB®. For example, the command "eig(A)" will produce the three eigenvalues. On the other hand, the command "[V, U] = eig(A)" will generate two 3×3 matrices, where V is transformation matrix **P**, given in (5.42), and U is the diagonal matrix $\boldsymbol{\Lambda}$, given in (5.44). Note that MATLAB® produces normalized eigenvectors, where the sum of the square of each element of the vector is 1. This, however, is not a problem since any eigenvector, when multiplied by a scalar, still remains an eigenvector, as per property 4 of the eigenvectors. Using the MATLAB® command, the following eigenvectors are obtained (subscript M is used to indicate that these are produced by MATLAB®).

$$\boldsymbol{\gamma}_{1M} = \begin{bmatrix} 0.7218 \\ -0.6888 \\ -0.1378 \end{bmatrix}, \boldsymbol{\gamma}_{2M} = \begin{bmatrix} 0.7136 \\ -0.6796 \\ -0.1699 \end{bmatrix}, \text{and } \boldsymbol{\gamma}_{3M} = \begin{bmatrix} 0.7220 \\ -0.6189 \\ -0.3094 \end{bmatrix}$$

Note that by diving $\boldsymbol{\gamma}_{1M}$ by -0.1378, $\boldsymbol{\gamma}_{2M}$ by -0.1699, and $\boldsymbol{\gamma}_{3M}$ by -0.3094 we get the eigenvectors $\boldsymbol{\gamma}_1$, $\boldsymbol{\gamma}_2$, and $\boldsymbol{\gamma}_3$ respectively calculated above, which is in accordance with property 4 of eigenvectors.

Alternatively, the command "[V, U, W] = eig(A)" will produce a matrix **V** containing the right eigenvectors, the diagonal matrix $\boldsymbol{\Lambda}$ (U), and a matrix **W** (W) containing the left eigenvector, which is given by

$$\mathbf{W} = \begin{bmatrix} 0.6919 & 0.6918 & 0.6952 \\ 0.6458 & 0.6534 & 0.6759 \\ 0.3229 & 0.3075 & 0.2331 \end{bmatrix}$$

5.6.2 Matrix with Repeated Eigenvalues

To explain the principle, let us assume that the matrix $\mathbf{A} \in \Re^{5\times5}$ has the eigenvalues of $\lambda_1, \lambda_1, \lambda_1, \lambda_4,$ and λ_5. This means that it has three distinct eigenvalues $\lambda_1, \lambda_4, \lambda_5$ and two repeated eigenvalues. The eigenvectors for the distinct eigenvalues then are

$$\begin{aligned} \lambda_1\boldsymbol{\gamma}_1 &= \mathbf{A}\boldsymbol{\gamma}_1 \\ \lambda_4\boldsymbol{\gamma}_4 &= \mathbf{A}\boldsymbol{\gamma}_4 \\ \lambda_5\boldsymbol{\gamma}_5 &= \mathbf{A}\boldsymbol{\gamma}_5 \end{aligned} \tag{5.46}$$

The other two eigenvectors are given by the solutions of

$$\begin{aligned} \mathbf{A}\boldsymbol{\gamma}_2 &= \lambda_1\boldsymbol{\gamma}_2 + \boldsymbol{\gamma}_1 \\ \mathbf{A}\boldsymbol{\gamma}_3 &= \lambda_1\boldsymbol{\gamma}_3 + \boldsymbol{\gamma}_2 \end{aligned} \tag{5.47}$$

The transformation matrix $\mathbf{P}$ of (5.42) is now formed such that

$$\boldsymbol{\Lambda} = \mathbf{P}^{-1}\mathbf{A}\mathbf{P} = \mathbf{P}^{-1}\mathbf{A}[\boldsymbol{\gamma}_1 \quad \boldsymbol{\gamma}_2 \quad \boldsymbol{\gamma}_3 \quad \boldsymbol{\gamma}_4 \quad \boldsymbol{\gamma}_5]$$

From (5.46) and (5.47), the above equation can be written as

$$\begin{aligned} \boldsymbol{\Lambda} = \mathbf{P}^{-1}\mathbf{A}\mathbf{P} &= \mathbf{P}^{-1}[\lambda_1\boldsymbol{\gamma}_1 \quad \lambda_1\boldsymbol{\gamma}_2 + \boldsymbol{\gamma}_1 \quad \lambda_1\boldsymbol{\gamma}_3 + \boldsymbol{\gamma}_2 \quad \lambda_3\boldsymbol{\gamma}_4 \quad \lambda_5\boldsymbol{\gamma}_5] \\ &= \mathbf{P}^{-1}\mathbf{P}\begin{bmatrix} \lambda_1 & 1 & & & \\ & \lambda_1 & 1 & & \\ & & \lambda_1 & & \\ & & & \lambda_4 & \\ & & & & \lambda_5 \end{bmatrix} \end{aligned}$$

The matrix $\boldsymbol{\Lambda}$ is written in the Jordan form as

$$\boldsymbol{\Lambda} = \mathbf{P}^{-1}\mathbf{A}\mathbf{P} = \begin{bmatrix} \mathbf{J}_1 & & \\ & \mathbf{J}_2 & \\ & & \mathbf{J}_3 \end{bmatrix} \tag{5.48}$$

where

$$\mathbf{J}_1 = \begin{bmatrix} \lambda_1 & 1 & \\ & \lambda_1 & 1 \\ & & \lambda_1 \end{bmatrix}, \quad \mathbf{J}_2 = \lambda_4, \quad \mathbf{J}_3 = \lambda_5$$

The matrices $\mathbf{J}_1$, $\mathbf{J}_2$, and $\mathbf{J}_3$ are called the Jordan blocks. The Jordan canonical form has the following properties:

1) The number of Jordan blocks is equal to the number of distinct eigenvalues, i.e. each distinct eigenvalue is associated with one Jordan block.

2) All the elements on the main diagonal are the eigenvalues of **A**.
3) All the elements below the main diagonal are zero.
4) The element immediately above all but one of a group of repeated eigenvalues is 1.
5) The total number of 1s in the Jordan form is the difference between the total number of eigenvalues and the total number of Jordan blocks.

Example 5.10 Consider the matrix

$$A = \begin{bmatrix} 0 & 6 & -5 \\ 1 & 0 & 2 \\ 3 & 2 & 4 \end{bmatrix}$$

The eigenvalues of the matrix are 2, 1, and 1. Let us define $\lambda_1 = 2,\ \lambda_2 = \lambda_3 = 1$. Then

$$\boldsymbol{\gamma}_1 = \begin{bmatrix} 2 \\ -1 \\ 2 \end{bmatrix} \text{ and } \boldsymbol{\gamma}_2 = \begin{bmatrix} 1 \\ -3/7 \\ -5/7 \end{bmatrix}$$ From (5.47), we get

$$\mathbf{A}\boldsymbol{\gamma}_3 = \lambda_1\boldsymbol{\gamma}_3 + \boldsymbol{\gamma}_2 \Rightarrow \boldsymbol{\gamma}_3 = \begin{bmatrix} 1 \\ -22/49 \\ -46/49 \end{bmatrix}$$

Verify that the transformation matrix $\mathbf{P} = [\boldsymbol{\gamma}_1 \quad \boldsymbol{\gamma}_2 \quad \boldsymbol{\gamma}_3]$ will lead to the Jordan form

$$\boldsymbol{\Lambda} = \begin{bmatrix} 2 & & \\ & 1 & 1 \\ & & 1 \end{bmatrix}$$

5.7 Controllability of LTI Systems

The controllability of a system is defined by the ability of one or more of the inputs to influence all the state variables such that a certain objective can be achieved in a finite time. Intuitively, one can say that, if one of the state variables is independent of the control inputs, then this particular state variable cannot be driven. This is discussed with the help of the diagonalization method explained in Section 5.6. Consider the state equation of the form (5.5), on which the linear transformation $\mathbf{x} = \mathbf{P}\mathbf{z}$ has been used to obtain

$$\dot{\mathbf{z}} = \mathbf{P}^{-1}\mathbf{A}\mathbf{P}\mathbf{z} + \mathbf{P}^{-1}\mathbf{B}\mathbf{u} = \boldsymbol{\Lambda}\mathbf{z} + \boldsymbol{\Gamma}\mathbf{u} \tag{5.49}$$

In (5.49), $\mathbf{\Lambda}$ is a diagonal matrix that contains the n distinct eigenvalues. Assume that the system has two control inputs. Then, (5.49) can be written as

$$\begin{bmatrix} \dot{z}_1 \\ \dot{z}_2 \\ \dot{z}_3 \\ \vdots \\ \dot{z}_n \end{bmatrix} = \begin{bmatrix} \lambda_1 & & & & \\ & \lambda_3 & & & \\ & & \lambda_3 & & \\ & & & \ddots & \\ & & & & \lambda_n \end{bmatrix} \begin{bmatrix} z_1 \\ z_2 \\ z_3 \\ \vdots \\ z_n \end{bmatrix} + \begin{bmatrix} \tau_{11} & \tau_{12} \\ \tau_{21} & \tau_{22} \\ \tau_{31} & \tau_{32} \\ \vdots & \vdots \\ \tau_{n1} & \tau_{n2} \end{bmatrix} \begin{bmatrix} u_1 \\ u_2 \end{bmatrix}$$

Let us now assume that the third row of the matrix $\mathbf{\Gamma}$ has only zero elements, i.e. $\tau_{31} = \tau_{32} = 0$. Then, none of the control inputs will have any influence on the third mode (z_3) that is associated with the eigenvalue λ_3. The system in this case will be called uncontrollable. Additionally, if λ_3 has positive real parts, then the response of the third mode will grow asymptotically for any nonzero initial condition. The system in this case will not be stabilizable.

Consider the case of a Jordan form in which the matrix $\mathbf{\Lambda}$ is given by (5.48). Again, assuming that there are two inputs, the following transformed state equation can be written

$$\begin{bmatrix} \dot{z}_1 \\ \dot{z}_2 \\ \dot{z}_3 \\ \dot{z}_4 \\ \dot{z}_5 \end{bmatrix} = \begin{bmatrix} \lambda_1 & 1 & & & \\ & \lambda_1 & 1 & & \\ & & \lambda_1 & & \\ & & & \lambda_4 & \\ & & & & \lambda_5 \end{bmatrix} \begin{bmatrix} z_1 \\ z_2 \\ z_3 \\ z_4 \\ z_5 \end{bmatrix} + \begin{bmatrix} \tau_{11} & \tau_{12} \\ \tau_{21} & \tau_{22} \\ \tau_{31} & \tau_{32} \\ \tau_{41} & \tau_{41} \\ \tau_{51} & \tau_{52} \end{bmatrix} \begin{bmatrix} u_1 \\ u_2 \end{bmatrix} \tag{5.50}$$

Obviously, the system will not be controllable when $\tau_{31} = \tau_{32} = 0$ or $\tau_{41} = \tau_{42} = 0$ or $\tau_{51} = \tau_{52} = 0$. Now consider the case where $\tau_{31} \neq 0$ and $\tau_{32} \neq 0$. What happens when $\tau_{11} = \tau_{12} = 0$ or $\tau_{21} = \tau_{22} = 0$? To explain, we expand the first two rows of (5.50) to get

$$\dot{z}_1 = \lambda_1 z_1 + z_2 + \tau_{11} u_1 + \tau_{12} u_2$$
$$\dot{z}_2 = \lambda_1 z_2 + z_3 + \tau_{21} u_1 + \tau_{22} u_2$$

When $\tau_{21} = \tau_{22} = 0$, even if the control inputs do not directly reach the state z_2, it can still be indirectly influenced by the control inputs through z_3. Similarly, when $\tau_{11} = \tau_{12} = 0$, z_1 can be influenced by z_2. From this discussion, we can conclude that, for a system in Jordan canonical form to be controllable, the rows of the matrix $\mathbf{\Gamma}$ associated with the last element in each Jordan block cannot be zero, i.e. $\tau_{31} \neq 0$ and $\tau_{32} \neq 0$ or $\tau_{41} \neq 0$ and $\tau_{42} \neq 0$, or $\tau_{51} \neq 0$ and $\tau_{52} \neq 0$.

The question then is, "Should a system be converted into its diagonal form before checking for the controllability?" The answer is negative. However, before the controllability test condition is discussed, we shall briefly discuss the implication of the Cayley–Hamilton theorem.

5.7.1 Implication of Cayley–Hamilton Theorem

The Cayley–Hamilton theorem states that a matrix must satisfy its own characteristic equation. Therefore, if the matrix $\mathbf{A} \in \Re^{n \times n}$ has the characteristic equation of

$$|s\mathbf{I} - \mathbf{A}| = s^n + a_{n-1}s^{n-1} + \cdots + a_1 s + a_0 = 0$$

then

$$\mathbf{A}^n + a_{n-1}\mathbf{A}^{n-1} + \cdots + a_1\mathbf{A} + a_0\mathbf{I} = 0 \quad (5.51)$$

Rearranging (5.51), we write

$$\mathbf{A}^n = -a_{n-1}\mathbf{A}^{n-1} - a_{n-2}\mathbf{A}^{n-2} - \cdots - a_1\mathbf{A} - a_0\mathbf{I}$$

Post-multiplying the above equation by $\mathbf{A}$, the following equation is obtained

$$\begin{aligned}\mathbf{A}^{n+1} &= \mathbf{A}^n \times \mathbf{A} = -a_{n-1}\mathbf{A}^n - a_{n-2}\mathbf{A}^{n-1} - \cdots - a_1\mathbf{A}^2 - a_0\mathbf{A} \\ &= -a_{n-1}\left(-a_{n-1}\mathbf{A}^{n-1} - a_{n-2}\mathbf{A}^{n-2} - \cdots - a_1\mathbf{A} - a_0\mathbf{I}\right) - a_{n-2}\mathbf{A}^{n-1} - \cdots - a_1\mathbf{A}^2 - a_0\mathbf{A} \\ &= \left(a_{n-1}^2 - a_{n-2}\right)\mathbf{A}^{n-1} + \left(a_{n-1}a_{n-2} - a_{n-3}\right)\mathbf{A}^{n-2} + \cdots + \left(a_{n-1}a_1 - a_0\right)\mathbf{A} + a_{n-1}a_0\mathbf{I}\end{aligned}$$

In a similar way, all the higher powers of the matrix which are greater than or equal to n can be expressed by a matrix polynomial of the order of $n - 1$.

5.7.2 Controllability Test Condition

Consider a state equation of the form given in (5.5). The solution of the equation, assuming that the initial condition is zero, is given from (5.20) as

$$\mathbf{x}(t) = \int_0^t e^{\mathbf{A}(t-\tau)}\mathbf{B}u(\tau)dt$$

Expanding the matrix exponential, we get

$$\mathbf{x}(t) = \mathbf{B}\int_0^t u(\tau)dt + \mathbf{AB}\int_0^t (t-\tau)u(\tau)dt + \mathbf{A}^2\mathbf{B}\int_0^t \frac{(t-\tau)^2}{2!}u(\tau)dt + \cdots \quad (5.52)$$

Therefore, the terminal state $\mathbf{x}(t)$ is in the linear subspace spanned by the column vectors of infinite sequence matrices $\mathbf{B}$, $\mathbf{AB}$, $\mathbf{A}^2\mathbf{B}$, ... Again, from the Cayley–Hamilton theorem, we know that $\mathbf{A}^k$, $k \geq 0$, can always be expressed in terms of $\mathbf{A}$, $\mathbf{A}^2$, ..., $\mathbf{A}^{n-1}$, thus (5.52) is rewritten as

$$\mathbf{x}(t) = \begin{bmatrix} \mathbf{B} & \mathbf{AB} & \mathbf{A}^2\mathbf{B} & \cdots & \mathbf{A}^{n-1}\mathbf{B} \end{bmatrix} \boldsymbol{\Omega} = \mathbf{S}\boldsymbol{\Omega} \tag{5.53}$$

where $\boldsymbol{\Omega}$ is a vector that contains the combination of the integrals on the right-hand side of (5.52). Note that the dimension of the matrix $\mathbf{S}$ is $n \times (n \times m)$, m being the number of inputs. The aim is to check if all the elements of the state vector $\mathbf{x}(t)$ can be determined uniquely from (5.53). This is only possible if the matrix $\mathbf{S}$ has n linearly independent columns. In other words, the rank of the matrix must be equal to n.

Example 5.11 Consider a system in phase variable canonical form as given by

$$\mathbf{A} = \begin{bmatrix} 0 & 1 & 0 & 0 \\ 0 & 0 & 1 & 0 \\ 0 & 0 & 0 & 1 \\ -1 & -1 & -3 & -4 \end{bmatrix}, \quad \mathbf{B} = \begin{bmatrix} 0 \\ 0 \\ 0 \\ 1 \end{bmatrix}$$

Then we have

$$S = \begin{bmatrix} B & AB & A^2B & A^3B \end{bmatrix} = \begin{bmatrix} 0 & 0 & 0 & 1 \\ 0 & 0 & 1 & -4 \\ 0 & 1 & -4 & 13 \\ 1 & -4 & 13 & -42 \end{bmatrix}$$

Note that $|\mathbf{S}| = 1$, i.e. the system is full rank and hence controllable. Also, it can be observed that all the elements in the secondary diagonal of the matrix $\mathbf{S}$ are 1. Hence a system in phase variable canonical form is always controllable. This is the reason such systems are also said to be in controllable canonical form.

Example 5.12 Let the state and input matrices of a system be given by

$$\mathbf{A} = \begin{bmatrix} 0 & 1 & 0 \\ 0 & 0 & \alpha \\ 0 & 0 & -1 \end{bmatrix}, \quad \mathbf{B} = \begin{bmatrix} 0 \\ \beta \\ 1 \end{bmatrix}$$

A relation between α and β must be determined so that the system is controllable. For this, the controllability matrix is

$$\mathbf{S} = \begin{bmatrix} \mathbf{B} & \mathbf{AB} & \mathbf{A}^2\mathbf{B} \end{bmatrix} = \begin{bmatrix} 0 & \beta & \alpha \\ \beta & \alpha & -\alpha \\ 1 & -1 & 1 \end{bmatrix}$$

Then

$$\begin{aligned}|\mathbf{S}| &= -\beta(\beta+\alpha)+\alpha(-\beta-\alpha)\\ &= -\beta^2-\alpha\beta-\alpha\beta-\alpha^2 = -(\alpha+\beta)^2\end{aligned}$$

Therefore, for nonsingular **S**, we must have $\alpha \neq -\beta$.

5.8 Observability of LTI Systems

An LTI system state is said to be observable if it can be uniquely determined from the input and output $u(t), y(t), t_0 \leq t_f$ and the matrices **A**, **B**, **C**, and **D** for any finite time t_f. In other words, an LTI system is said to be observable if every state of the system can be estimated from the knowledge of the system matrices, output measurements, and input data. To test for the observability, consider an LTI system that is given by (5.5) and (5.7). Then it is said to be observable if the matrix

$$\mathbf{O} = \begin{bmatrix}\mathbf{C}\\ \mathbf{CA}\\ \mathbf{CA}^2\\ \vdots\\ \mathbf{CA}^{n-1}\end{bmatrix} \tag{5.54}$$

has a full rank. The proof of this follows the same pattern as the proof for the controllability condition. Consider now the transformation given by (5.49), which diagonalizes the system. The output equation is of the form

$$y = \mathbf{C}x = \mathbf{CPz} = \mathbf{\Theta z} \tag{5.55}$$

Then the observability matrix will be given by

$$\mathbf{L} = \begin{bmatrix}\mathbf{\Theta}\\ \mathbf{\Theta\Lambda}\\ \mathbf{\Theta\Lambda}^2\\ \vdots\\ \mathbf{\Theta\Lambda}^{n-1}\end{bmatrix}$$

It can be seen from the above equation that matrix **L** will have full rank if all the elements of the matrix $\mathbf{\Theta}$ are not zero. This means none of the columns should be zero. In a similar way, it can be shown that, for the system to be observable, the first column of each Jordan block must be nonzero.

Example 5.13 Let a state space system be characterized by the matrices

$$\mathbf{A} = \begin{bmatrix} 0 & 1 & 0 \\ 0 & 0 & 1 \\ -2 & -5 & -4 \end{bmatrix}, \quad \mathbf{B} = \begin{bmatrix} 0 \\ 0 \\ 1 \end{bmatrix}, \quad \mathbf{C} = [2 \quad 3 \quad 1]$$

Since the system is in the phase variable canonical form, it is obviously controllable. To check for the observability, we use (5.54) to get

$$\mathbf{O} = \begin{bmatrix} \mathbf{C} \\ \mathbf{CA} \\ \mathbf{CA}^2 \end{bmatrix} = \begin{bmatrix} 2 & 3 & 1 \\ -2 & -3 & -1 \\ 2 & 3 & 1 \end{bmatrix}$$

The matrix **O** has only one independent row, and therefore has a rank of 1. Thus, the system is not observable. Note that the transfer function of the system is given by

$$\frac{Y(s)}{U(s)} = \mathbf{C}(s\mathbf{I} - \mathbf{A})^{-1}\mathbf{B} = \frac{s^2 + 3s + 2}{s^3 + 4s^2 + 5s + 2} = \frac{(s+1)(s+2)}{(s+1)^2(s+2)}$$

This implies that two poles get canceled by two zeros. Does this mean that if there are pole-zero cancelations the system will not be observable? Let us consider another example.

Example 5.14 Let a state space system be characterized by the matrices

$$\mathbf{A} = \begin{bmatrix} -1 & 1 & -1 \\ 0 & -1 & -1 \\ 0 & 0 & -2 \end{bmatrix}, \quad \mathbf{B} = \begin{bmatrix} 1 \\ 1 \\ 1 \end{bmatrix}, \quad \mathbf{C} = [1 \quad 0 \quad 0]$$

To check for controllability, we use (5.53) to get

$$\mathbf{S} = \begin{bmatrix} \mathbf{B} & \mathbf{AB} & \mathbf{A}^2\mathbf{B} \end{bmatrix} = \begin{bmatrix} 1 & -1 & 1 \\ 1 & -2 & 4 \\ 1 & -2 & 4 \end{bmatrix}$$

Since the matrix **S** has two independent rows, it has a rank of 2. Now to check for observability, we use (5.54) to get

$$\mathbf{O} = \begin{bmatrix} \mathbf{C} \\ \mathbf{CA} \\ \mathbf{CA}^2 \end{bmatrix} = \begin{bmatrix} 1 & 0 & 0 \\ -1 & 1 & -1 \\ 1 & -2 & 2 \end{bmatrix}$$

This matrix has two independent columns, and therefore also has a rank of 2. Since both the matrices **S** and **O** do not have full rank, the system is neither controllable nor observable. The transfer function of the system is given by

$$\frac{Y(s)}{U(s)} = \mathbf{C}(s\mathbf{I}-\mathbf{A})^{-1}\mathbf{B} = \frac{s^2+3s+2}{s^3+4s^2+5s+2} = \frac{(s+1)(s+2)}{(s+1)^2(s+2)}$$

This implies that one of the poles at the location of −1 gets canceled by the zero at the same location.

Examples 5.13 and 5.14 illustrate that a system is either not controllable, not observable, or both uncontrollable and unobservable if there are pole-zero cancelations in the transfer function. This can be a critical problem in a power system or in a power electronic circuit where several blocks are connected together to form a composite system. In such a case, it is quite likely that a pole-zero cancelation between blocks may occur. In such an event, careful consideration must be given to controller or observer design.

5.9 Pole Placement Through State Feedback

The pole technique pertains to the placement of closed-loop poles through feedback. In the transfer function domain, the poles are placed through a negative feedback of the output. In a similar fashion, the system states can be fed back to place the closed-loop poles in desired locations. The feedback control law for a SISO system is given by

$$u = -\mathbf{K}\mathbf{x} + K_P y_{ref} \tag{5.56}$$

where $\mathbf{K}$ is the feedback gain matrix strictly containing real numbers, y_{ref} is the reference, and K_P is a proportional gain. Substituting (5.56) in (5.5), we have

$$\dot{\mathbf{x}} = (\mathbf{A}-\mathbf{B}\mathbf{K})\mathbf{x} + \mathbf{B}K_P y_{ref} \tag{5.57}$$

From (5.57) and (5.7), the closed-loop transfer function of the system is obtained as

$$\frac{Y(s)}{Y_{ref}(s)} = \mathbf{C}(s\mathbf{I}-\mathbf{A}+\mathbf{B}\mathbf{K})^{-1}\mathbf{B}K_P \tag{5.58}$$

From (5.58), we find that the closed-loop poles are given by the solution of the equation

$$|s\mathbf{I}-(\mathbf{A}-\mathbf{B}\mathbf{K})| = 0 \tag{5.59}$$

This implies that the eigenvalues of the matrix $(\mathbf{A}-\mathbf{B}\mathbf{K})$ are the closed-loop poles.

Now consider a system that is in phase variable canonical form with

$$\mathbf{A} = \begin{bmatrix} 0 & 1 & 0 & \cdots & 0 & 0 \\ 0 & 0 & 1 & \cdots & 0 & 0 \\ \vdots & \vdots & \vdots & \ddots & \vdots & \vdots \\ 0 & 0 & 0 & \cdots & 0 & 1 \\ -a_0 & -a_1 & -a_2 & \cdots & -a_{n-2} & -a_{n-1} \end{bmatrix}, \quad \mathbf{B} = \begin{bmatrix} 0 \\ 0 \\ \vdots \\ 0 \\ 1 \end{bmatrix}$$

Let the feedback gain matrix of a SISO system be given by

$$\mathbf{K} = [k_1 \;\; k_2 \;\; \cdots \;\; k_{n-1} \;\; k_n]$$

Note that

$$\mathbf{BK} = \begin{bmatrix} 0 & 0 & \cdots & 0 & 0 \\ 0 & 0 & \cdots & 0 & 0 \\ 0 & 0 & \cdots & 0 & 0 \\ \vdots & \vdots & \ddots & \vdots & \vdots \\ -k_1 & k_2 & \cdots & -k_{n-1} & -k_n \end{bmatrix}$$

Therefore

$$\mathbf{A}-\mathbf{BK} = \begin{bmatrix} 0 & 1 & 0 & \cdots & 0 & 0 \\ 0 & 0 & 1 & \cdots & 0 & 0 \\ \vdots & \vdots & \vdots & \ddots & \vdots & \vdots \\ 0 & 0 & 0 & \cdots & 0 & 1 \\ -a_0-k_1 & -a_1-k_2 & -a_2-k_3 & \cdots & -a_{n-2}-k_{n-1} & -a_{n-1}-k_n \end{bmatrix}$$

Hence, we get the following characteristic equation of the closed-loop system

$$\Delta_{CL}(s) = |s\mathbf{I}-(\mathbf{A}-\mathbf{BK})| = s^n + (a_{n-1}+k_n)s^{n-1} + \cdots + (a_1+k_2)s + (a_0+k_1) = 0 \tag{5.60}$$

The closed-loop poles can therefore be arbitrarily placed by the choice of $k_1, k_2, ..., k_n$.

Example 5.15 Let a state space system be characterized by the matrices

$$\mathbf{A} = \begin{bmatrix} 0 & 1 & 0 \\ 0 & 0 & 1 \\ -2 & -5 & -4 \end{bmatrix}, \quad \mathbf{B} = \begin{bmatrix} 0 \\ 0 \\ 1 \end{bmatrix}$$

Since this is in the phase variable canonical form, it is controllable. Define $K = [k_1 \;\; k_2 \;\; k_3]$. The characteristic equation of the closed-loop system is obtained from (5.60) as

$$|s\mathbf{I} - (\mathbf{A} - \mathbf{BK})| = s^3 + (4 + k_3)s^2 + (5 + k_2)s + (2 + k_1) = 0$$

It is desired that the dominant poles of the closed-loop system have a damping ratio (ξ) of 0.8 and an undamped natural frequency (ω_n) of 4 rad/s. The third pole is placed at -10, which is far away from the imaginary axis. The closed-loop characteristic equation is then

$$\Delta_{CL}(s) = (s^2 + 2\xi\omega_n s + \omega_n^2)(s + 10) = s^3 + 16.4s^2 + 80s + 160 = 0$$

Then, comparing the last two equations, we get $k_1 = 158$, $k_2 = 75$, and $k_3 = 12.4$. The closed-loop transfer function (5.58) is then given by

$$\frac{Y(s)}{Y_{ref}(s)} = \frac{K_P}{s^3 + 16.4s^2 + 80s + 160}$$

The DC gain of the system is $K_P/160$. Figure 5.1 shows the step response of the system for two values of the gain K_P. It can be seen that, as this gain increases, the steady state error reduces. However, this is not a particularly attractive solution. A more suitable approach is to increase the system type using an integral controller, which is presented in Section 5.9.1.

Example 5.15 shows that the closed-loop eigenvalues can be arbitrarily placed provided that the system is in phase variable canonical form. There is a process by which any controllable system can be converted into a phase variable form. The pole placement method using Ackerman's formula is based on this phase variable conversion technique [1]. These days, however, it is easy to accomplish pole placement in MATLAB® using the command "K = place(A,B,P)," where A and B are the system matrices and P a vector containing the desired eigenvalue locations. This command is valid when there are no repeated eigenvalues. For example, if the desired roots of $\Delta_{CL}(s)$ are -10, $-3.2 \pm j2.4$, then the vector P is written as

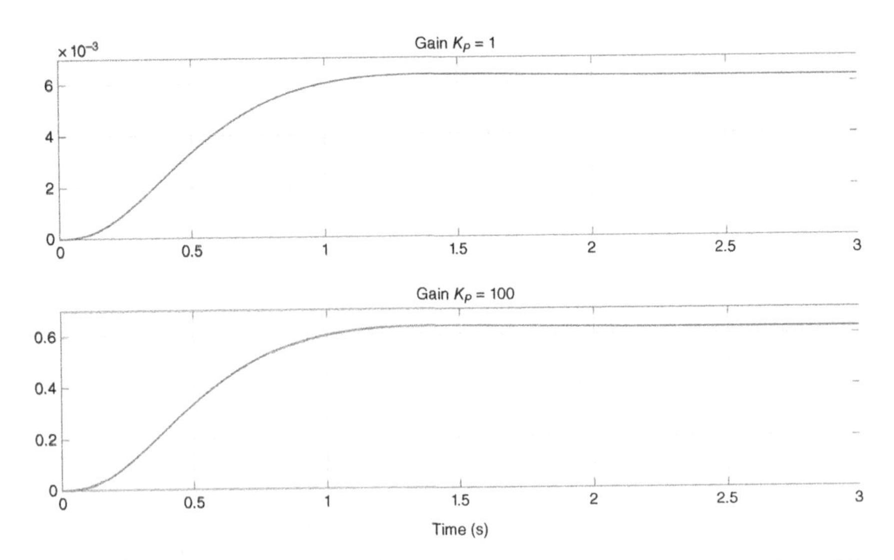

Figure 5.1 Step response of the system of Example 5.15 with two different values of the proportional gain.

$$P = \begin{bmatrix} -10 \\ -3.2 + 2.4i \\ -3.2 - 2.4i \end{bmatrix}$$

5.9.1 Pole Placement with Integral Action

As Example 5.15 shows, a state feedback controller can guarantee stable operation and can place the poles in desirable locations. It cannot, however, ensure proper reference tracking. To eliminate the steady state error, an integral controller is introduced, which has the following form

$$w = K_I \int \left(y_{ref} - y\right) dt \Rightarrow \dot{w} = K_I \left(y_{ref} - y\right) \tag{5.61}$$

An extended state vector is then defined as $\mathbf{x}_e = [\mathbf{x} \quad w]^T$. Then combining (5.5) and (5.7) with (5.61), we have

$$\dot{\mathbf{x}}_e = \begin{bmatrix} \mathbf{A} & 0 \\ -K_I\mathbf{C} & 0 \end{bmatrix} \mathbf{x}_e + \begin{bmatrix} \mathbf{B} \\ 0 \end{bmatrix} u + \begin{bmatrix} 0 \\ K_I \end{bmatrix} y_{ref} = \mathbf{A}_e \mathbf{x}_e + \mathbf{B}_e u + \begin{bmatrix} 0 \\ K_I \end{bmatrix} y_{ref} \tag{5.62}$$

Then the pole placement with integral control is given by

$$u = -\mathbf{K}\mathbf{x}_\mathbf{e} = -[\mathbf{K}_1 \quad K_2]\mathbf{x}_e = -[\mathbf{K}_1 \quad K_2]\begin{bmatrix} \mathbf{x} \\ z \end{bmatrix} = -\mathbf{K}_1\mathbf{x} - K_2 w \tag{5.63}$$

where $\mathbf{K}_1$ is a vector that is the gain associated with $\mathbf{x}$ and K_2 the scalar gain that multiplies w. Substituting u from (5.63) in (5.62), we get

$$\dot{\mathbf{x}}_e = \begin{bmatrix} \mathbf{A} - \mathbf{B}\mathbf{K}_1 & -\mathbf{B}K_2 \\ -K_I\mathbf{C} & 0 \end{bmatrix} \mathbf{x}_e + \begin{bmatrix} 0 \\ K_I \end{bmatrix} y_{ref} \tag{5.64}$$

The closed-loop transfer function is then given by

$$\frac{Y(s)}{Y_{ref}(s)} = [\mathbf{C} \quad 0]\left(s\mathbf{I} - \begin{bmatrix} \mathbf{A} - \mathbf{B}\mathbf{K}_1 & -\mathbf{B}K_2 \\ -K_I\mathbf{C} & 0 \end{bmatrix}\right)\begin{bmatrix} 0 \\ K_I \end{bmatrix} \tag{5.65}$$

The block diagram of the system is shown in Figure 5.2.

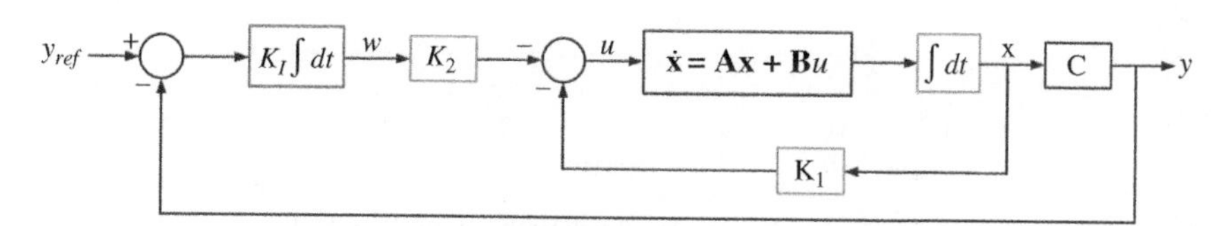

Figure 5.2 Continuous-time state feedback with integral control.

Example 5.16 Let us continue with Example 5.15. For the stated feedback control with integral action, the following parameters are chosen

$$K_I = 150, \quad P(s) = (s + 10)(s + 3.2 - j2.4)(s + 3.2 + j2.4)(s + 20)$$

where $P(s)$ is a polynomial defining the closed-loop poles. The first three of the desired closed-loop poles are placed at the same locations as given in Example 5.15. The matrices for the extended state space system of (5.62) are given by

$$\mathbf{A}_e = \begin{bmatrix} 0 & 1 & 0 & 0 \\ 0 & 0 & 1 & 0 \\ -2 & -5 & -4 & 0 \\ -150 & 0 & 0 & 0 \end{bmatrix}, \quad \mathbf{B}_e = \begin{bmatrix} 0 \\ 0 \\ 1 \\ 0 \end{bmatrix}$$

The gain matrix and the closed-loop transfer function are given by

$$\mathbf{K} = \begin{bmatrix} 1.758 \times 10^3 & 403 & 32.4 & -21.33 \end{bmatrix}$$

$$\frac{Y(s)}{Y_{ref}(s)} = \frac{3200}{s^4 + 36.4s^3 + 408s^2 + 1760s + 3200}$$

It can be seen that the closed-loop system has a DC gain of 1. The step response of the system is shown in Figure 5.3, where the system settles within 1.5 seconds and the output tracks the input accurately.

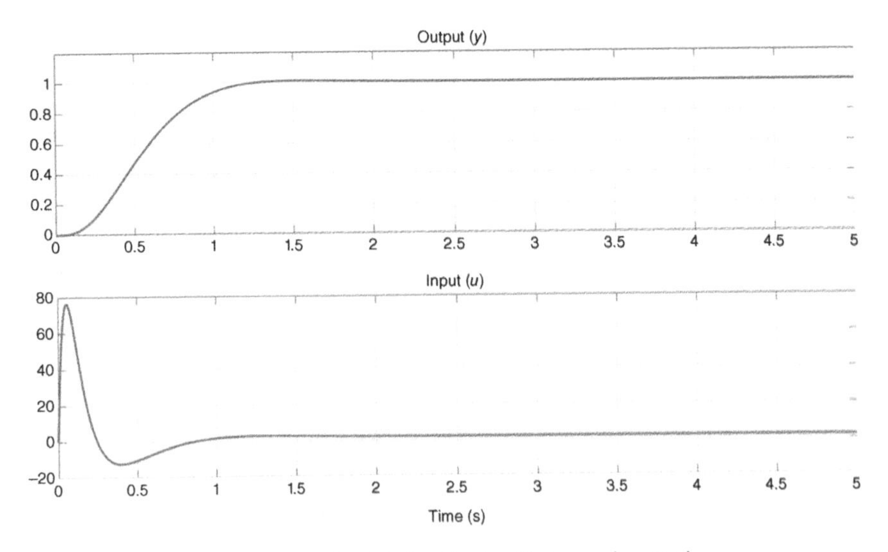

Figure 5.3 System response with state feedback with integral control.

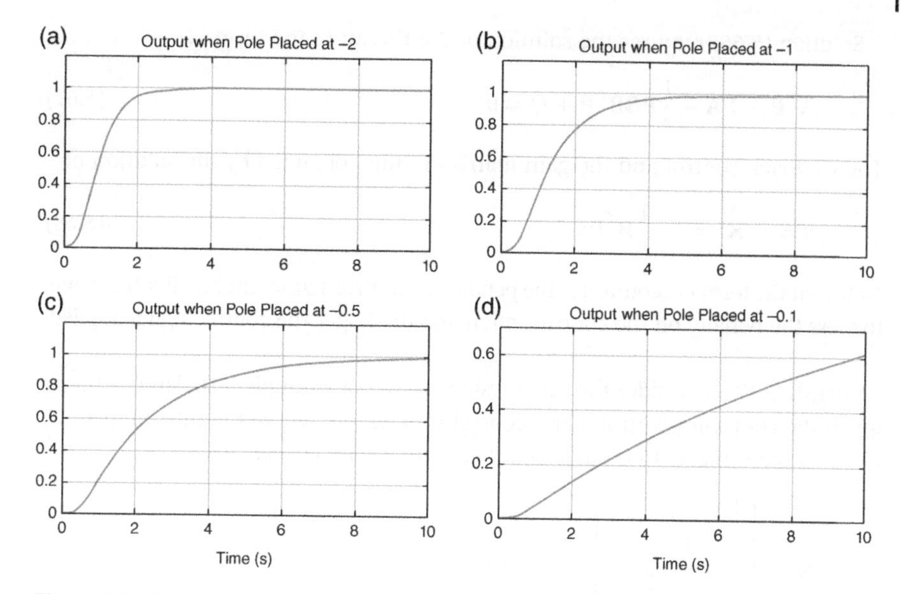

Figure 5.4 System response with different locations of the fourth pole. (a) Output when Pole Placed at −2, (b) Output when Pole Placed at −1, (c) Output when Pole Placed at −0.5, (d) Output when Pole Placed at −0.1.

With the first three poles remaining in the same locations and the integral gain remaining the same, the location of the fourth pole is varied. Figure 5.4 shows the output response with four different pole locations. As the fourth pole moves closer to the origin, the gain K_2 becomes smaller and the impact of the integral controller reduces. Therefore, the system response starts to become sluggish. It is thus important to choose the pole location carefully.

5.9.2 Linear Quadratic Regulator (LQR)

In a linear quadratic regulator (LQR) for the system of (5.5), a performance index of the form given below is chosen

$$J = \int_0^\infty \left(\mathbf{x}^T \mathbf{Q} \mathbf{x} + \mathbf{u}^T \mathbf{R} \mathbf{u}\right) dt$$

where both **Q** and **R** are positive-definite weighting matrices. Note that for a single-input system the performance index is

$$J = \int_0^\infty \left(\mathbf{x}^T \mathbf{Q} \mathbf{x} + r u^2\right) dt \tag{5.66}$$

where r is a positive scalar.

Solution (5.66) requires the solution of the Riccati of the form [2]

$$\mathbf{A}^T\mathbf{P} + \mathbf{P}\mathbf{A} - \frac{1}{r}\mathbf{P}\mathbf{B}\mathbf{B}^T\mathbf{P} + \mathbf{Q} = 0 \tag{5.67}$$

The feedback control and the gain matrix are then obtained by the solution of

$$u = -\mathbf{K}\mathbf{x} = -\frac{1}{r}\mathbf{B}^T\mathbf{P}\mathbf{x} \tag{5.68}$$

Note that the term r accounts for the penalty on control action: the smaller the value, the less the penalty on the control action and the larger the control signal applied.

Example 5.17 Consider the same system as that of Example 5.16. Here, a linear quadratic controller with integral control is designed, where the integral gain K_I remains the same at 150. The weighting matrices are chosen as

$$\mathbf{Q} = \begin{bmatrix} 10 & & & \\ & 1 & & \\ & & 1 & \\ & & & 1 \end{bmatrix}, \quad r = 1$$

The gain is obtained using the "K = lqr(Ae,Be,Q,r)" command in MATLAB® with these matrices.

The step response of the system is shown in Figure 5.5a, while the control input is shown in Figure 5.5b. It can be seen that the steady state error is zero due to the

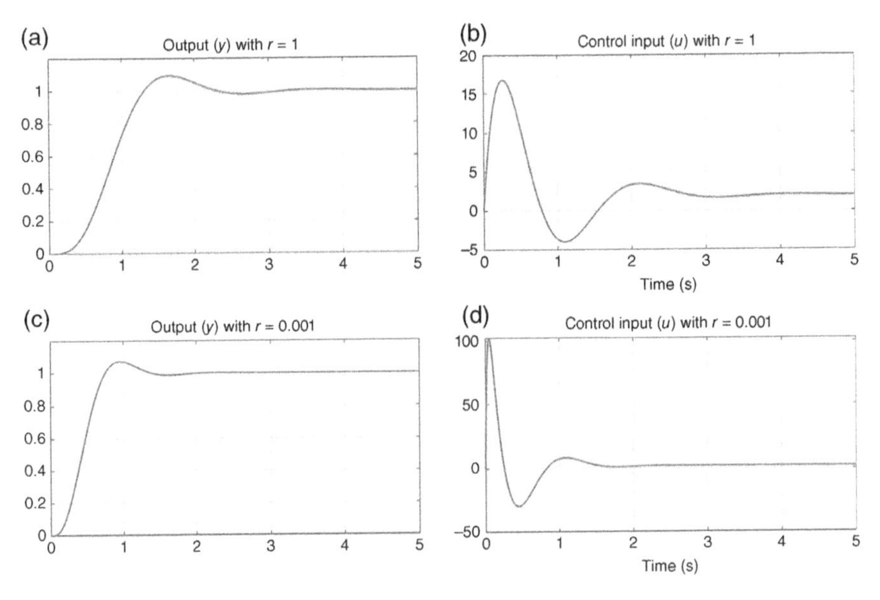

Figure 5.5 System response with LQR for two values of control weighting. (a) Output (y) with $r = 1$, (b) Control input (u) with $r = 1$, (c) Output (y) with $r = 0.001$, (d) Control input (u) with $r = 0.001$.

integral action, while the peak of the control input is around 17. The control weighting r is now changed to 0.001. The results are shown in Figure 5.5c,d. In this case, the output requires less time to settle, albeit at the expense of a much larger control action at the beginning.

5.9.3 Discrete-time State Feedback with Integral Control

The discrete-time design is also similar to that of the continuous-time design. A discrete-time integral control can be written as

$$\begin{aligned} e(k) &= y_r(k) - y(k) \\ w(k) &= w(k-1) + K_I e(k) = w(k-1) + K_I y_r(k) - K_I y(k) \end{aligned} \tag{5.69}$$

Then defining the extended state vector as $\mathbf{x}_e(k) = [\mathbf{x}(k) \quad w(k-1)]^T$, the extended state space matrix can be written using (5.26) and (5.7) as

$$\mathbf{x}_e(k+1) = \begin{bmatrix} \mathbf{F} & 0 \\ -K_I\mathbf{C} & 1 \end{bmatrix} \mathbf{x}_e(k) + \begin{bmatrix} \mathbf{G} \\ 0 \end{bmatrix} u(k) + \begin{bmatrix} 0 \\ K_I \end{bmatrix} y_{ref}(k) \tag{5.70}$$

We can then obtain the feedback gain matrix either through pole placement using "place" or through discrete-time LQR using the "dlqr" command in MATLAB®.

5.10 Observer Design (Full Order)

We have seen in the state feedback control design that the measurements of all the states are required. If, however, all the states are not accessible, it is necessary to estimate (or observe) them from the information contained in the output and input variables. The structure of feedback control with the observer is shown in Figure 5.6.

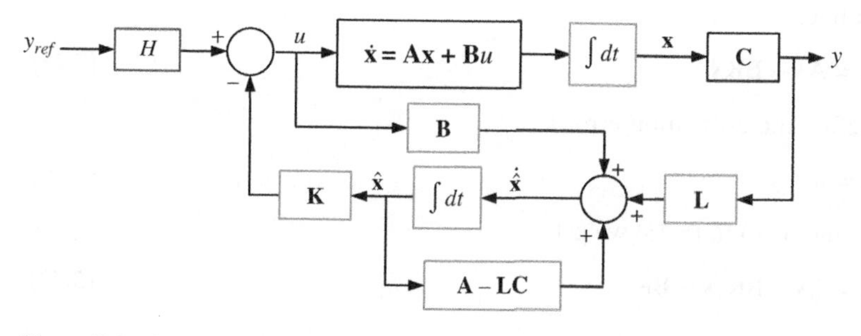

Figure 5.6 Structure of feedback controller with observer.

The state observer must be such that the observed state $\hat{\mathbf{x}}$ is as close as possible to the actual state $\mathbf{x}$. The observer should be such that the error between the observed state and the actual state should exponentially become zero, i.e.

$$\dot{\hat{\mathbf{x}}} - \dot{\mathbf{x}} = \mathbf{W}(\hat{\mathbf{x}} - \mathbf{x}) \tag{5.71}$$

The matrix $\mathbf{W}$ should be chosen such that its eigenvalues have large negative real parts. The closed-loop observer equation is given by

$$\dot{\hat{\mathbf{x}}} = (\mathbf{A} - \mathbf{LC})\hat{\mathbf{x}} + \mathbf{B}u + \mathbf{L}y \tag{5.72}$$

where $\mathbf{L}$ is the observer gain matrix. Eq. (5.72) can be rewritten as

$$\dot{\hat{\mathbf{x}}} = (\mathbf{A} - \mathbf{LC})\hat{\mathbf{x}} + \mathbf{B}u + \mathbf{LC}\mathbf{x} = \mathbf{A}\hat{\mathbf{x}} + \mathbf{B}u + \mathbf{LC}(\mathbf{x} - \hat{\mathbf{x}})$$

Subtracting the state Eq. (5.5) from the above equation, we get

$$\dot{\hat{\mathbf{x}}} - \dot{\mathbf{x}} = (\mathbf{A} - \mathbf{LC})\,(\hat{\mathbf{x}} - \mathbf{x}) = \mathbf{W}(\hat{\mathbf{x}} - \mathbf{x}) \tag{5.73}$$

It can be proved that, if the system is observable, the eigenvalues of the matrix $\mathbf{W}$ can be arbitrarily placed.

5.10.1 Separation Principle

So, if all the states of the system are not measurable, the observed states in the state feedback control law need to be used. This will require two sets of pole placement algorithms. The question is then how should the poles of the observer be placed such that they do not interfere with the control action? To answer this question, let us consider the state and output equations of the forms given by (5.5) and (5.7) respectively. The observer is given by (5.72). First, we assume that the reference input y_{ref} in Figure 5.6 is zero. Then the control law uses the estimated values of the states, i.e.

$$u = -\mathbf{K}\hat{\mathbf{x}} \tag{5.74}$$

We then have

$$\dot{\mathbf{x}} = \mathbf{A}\mathbf{x} - \mathbf{BK}\hat{\mathbf{x}} \tag{5.75}$$

Let us define the estimation error as

$$\mathbf{e} = \mathbf{x} - \hat{\mathbf{x}} \tag{5.76}$$

Substituting (5.76) in (5.75) we get

$$\dot{\mathbf{x}} = (\mathbf{A} - \mathbf{BK})\mathbf{x} - \mathbf{BK}e \tag{5.77}$$

Now (5.76) is combined with (5.5), (5.7), and (5.72) to give

$$\dot{\mathbf{e}} = \dot{\mathbf{x}} - \dot{\hat{\mathbf{x}}} = \mathbf{A}\mathbf{x} + \mathbf{B}u - (\mathbf{A} - \mathbf{L}\mathbf{C})\,\hat{\mathbf{x}} - \mathbf{B}u - \mathbf{L}\mathbf{C}\mathbf{x} = (\mathbf{A} - \mathbf{L}\mathbf{C})\,\mathbf{e} \tag{5.78}$$

Let us define an extended state vector of the form $\mathbf{x}_e = [\mathbf{x} \quad \mathbf{e}]^T$. Then, combining (5.77) and (5.78), we have

$$\dot{\mathbf{x}}_e = \begin{bmatrix} \mathbf{A} - \mathbf{B}\mathbf{K} & -\mathbf{B}\mathbf{K} \\ \mathbf{0} & \mathbf{A} - \mathbf{L}\mathbf{C} \end{bmatrix} \mathbf{x}_e \tag{5.79}$$

It is useful to note that if

$$\mathbf{H} = \begin{bmatrix} \mathbf{A} & \mathbf{B} \\ \mathbf{C} & \mathbf{D} \end{bmatrix}$$

then

$$|\mathbf{H}| = |\mathbf{D}|\left|\mathbf{A} - \mathbf{B}\mathbf{D}^{-1}\mathbf{C}\right| = |\mathbf{A}|\left|\mathbf{D} - \mathbf{C}\mathbf{A}^{-1}\mathbf{B}\right| \tag{5.80}$$

Using the identity (5.80), we get the characteristic equation of (5.79) as

$$\left| s\mathbf{I} - \begin{bmatrix} \mathbf{A} - \mathbf{B}\mathbf{K} & -\mathbf{B}\mathbf{K} \\ \mathbf{0} & \mathbf{A} - \mathbf{L}\mathbf{C} \end{bmatrix} \right| = \begin{vmatrix} s\mathbf{I} - \mathbf{A} + \mathbf{B}\mathbf{K} & \mathbf{B}\mathbf{K} \\ \mathbf{0} & s\mathbf{I} - \mathbf{A} + \mathbf{L}\mathbf{C} \end{vmatrix} = 0 \tag{5.81}$$

$$\Rightarrow |s\mathbf{I} - \mathbf{A} + \mathbf{B}\mathbf{K}| \times |s\mathbf{I} - \mathbf{A} + \mathbf{L}\mathbf{C}| = 0$$

This implies that the controller and the observer can be designed independently. They do not necessarily interfere with the performance of each other provided that the eigenvalues are well separated.

Note that, for the derivation above, we have assumed that the reference input y_{ref} in Figure 5.6 is zero. This assumption, however, can be relaxed and the control law including the reference input given by

$$u = -\mathbf{K}\hat{\mathbf{x}} + Hy_{ref} \tag{5.82}$$

where H is a feedforward gain. This gain must be chosen such that the DC gain of the system is 1. Including the reference input in (5.82), the system state space Eq. (5.79) is then rewritten as

$$\dot{\mathbf{x}}_e = \begin{bmatrix} \mathbf{A} - \mathbf{B}\mathbf{K} & -\mathbf{B}\mathbf{K} \\ \mathbf{0} & \mathbf{A} - \mathbf{L}\mathbf{C} \end{bmatrix} \mathbf{x}_e + \begin{bmatrix} \mathbf{B} \\ 0 \end{bmatrix} Hy_{ref} = \mathbf{A}'\mathbf{x}_e + \mathbf{B}'Hy_{ref} \tag{5.83}$$

Therefore, the closed-loop transfer function can then be written as

$$\frac{Y(s)}{Y_{rer}(s)} = H[\mathbf{C} \quad [0 \quad 0 \quad 0]](s\mathbf{I} - \mathbf{A}')^{-1}\mathbf{B}' \tag{5.84}$$

Example 5.18 Consider the following system

$$\dot{\mathbf{x}} = \begin{bmatrix} 0 & 1 & 0 \\ 0 & 0 & 1 \\ -0.1 & -0.3 & -0.5 \end{bmatrix}\mathbf{x} + \begin{bmatrix} 0 \\ 0 \\ 1 \end{bmatrix} u = \mathbf{A}\mathbf{x} + \mathbf{B}u$$

$$y = \begin{bmatrix} 1 & 0 & 0 \end{bmatrix}\mathbf{x} = \mathbf{C}\mathbf{x}$$

It is desired that the closed-loop poles be placed at $-1, \ -0.5 \pm j0.4$. The observer poles should be placed at locations which are far away from these closed-loop poles and the locations are chosen as $-25, \ -30$, and -35. The system is both controllable and observable.

The "place" command in MATLAB® will be used to compute that controller and observer gain matrices **K** and **L**. This command is of the form "K = place(A,B, P1)," where

$$\mathbf{P}_1^T = \begin{bmatrix} -1 & -0.5 + j0.4 & -0.5 - j0.4 \end{bmatrix}$$

which produces the gain matrix of

$$\mathbf{K} = \begin{bmatrix} 0.31 & 1.11 & 1.5 \end{bmatrix}$$

The observer gain matrix **L** will place the eigenvalues of $\mathbf{A} - \mathbf{LC}$ at desired locations. To use the place command, we note that $(\mathbf{A} - \mathbf{LC})^T = \mathbf{A}^T - \mathbf{C}^T\mathbf{L}^T$. Therefore, the command "L = place(A′,C′,P2)" is used, where

$$\mathbf{P}_2^T = \begin{bmatrix} -25 & -30 & -35 \end{bmatrix}$$

which produces the gain matrix of

$$\mathbf{L} = \begin{bmatrix} 89.5 & 2.63 \times 10^3 & 2.49 \times 10^4 \end{bmatrix}^T$$

Using these gain matrices in (5.84), to have a DC gain of 1, the parameter H must be chosen as 0.41. The system response is shown in Figure 5.7. It can be seen that the output settles to 1 and the estimation errors become 0.

5.11 Control of DC-DC Converter

In this section, the state space modeling and control of DC-DC converters is discussed. Only the boost converters are considered – the models of buck or buck-boost converters and their associated control can also be derived following the procedures described in this section. The schematic diagram of a boost converter is shown in Figure 5.8a. It is assumed that the converter operates in continuous conduction mode (CCM). Then the equivalent circuits when the switch is closed and when the switch is open are shown in Figure 5.8b and Figure 5.8c respectively.

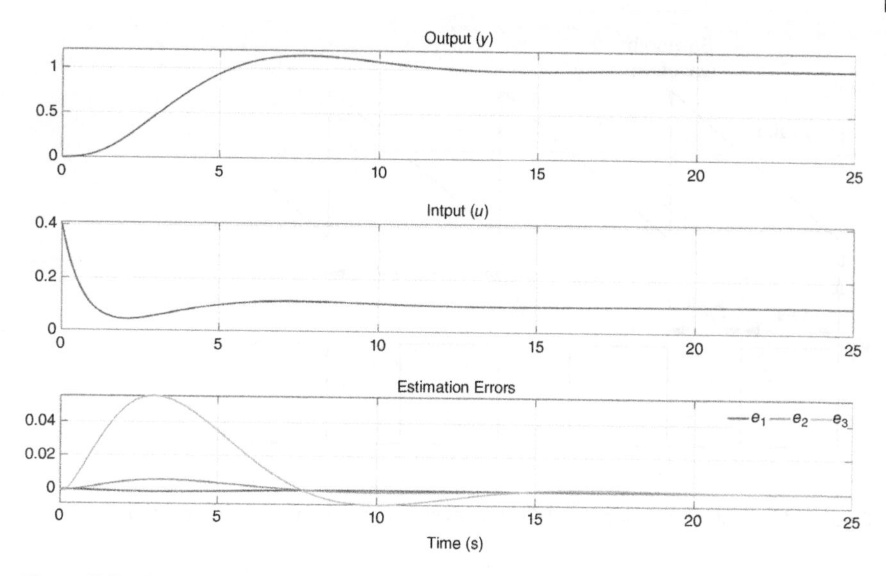

Figure 5.7 System response with state feedback control and observer.

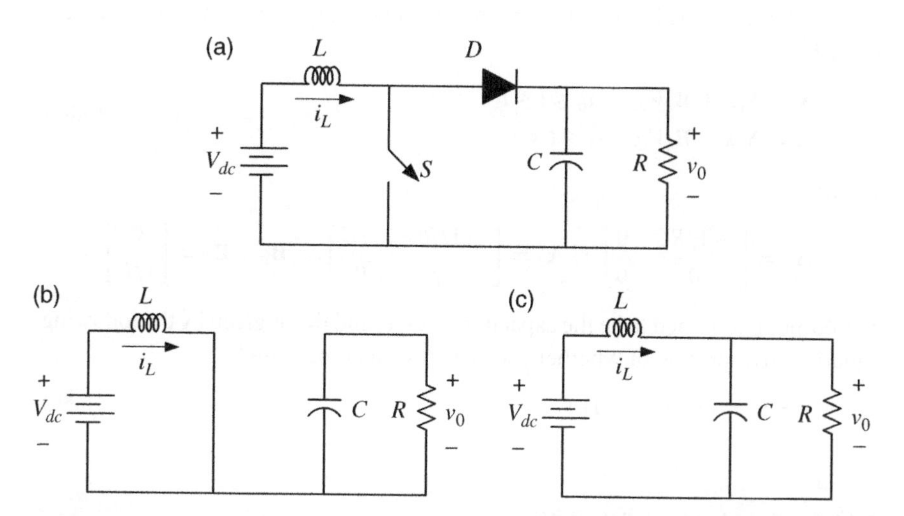

Figure 5.8 (a) The schematic diagram of a boost converter and its equivalent circuit when (b) the switch is closed and (c) the switch is open.

As discussed in Chapter 4, these converters are controlled by their duty ratio (d, $0 < d < 1$). The switching pulses for these converters are generated by comparing a sawtooth waveform of frequency $f(=1/T)$ with the duty ratio, as shown in Figure 5.9, where the switch on and off periods are also indicated.

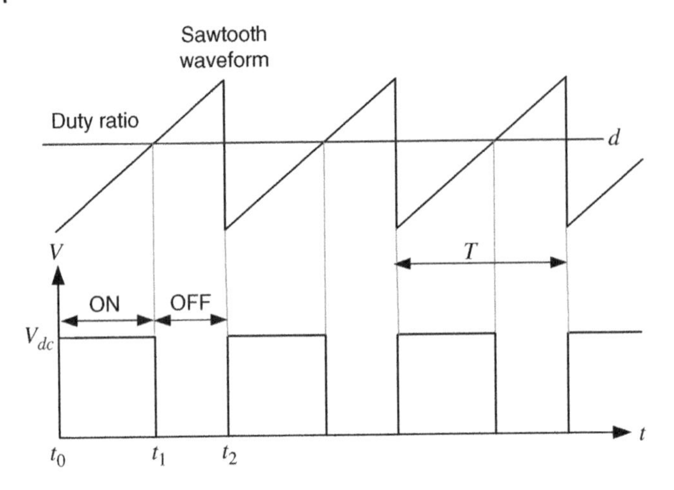

Figure 5.9 Switching scheme of DC-DC converters based on duty ratio control.

Let us define a state vector as $\mathbf{x} = \begin{bmatrix} v_0 & i_L \end{bmatrix}^T$. Then the state space equation when the switch is closed (Figure 5.8b) and when it is open (Figure 5.8c) are given respectively by

$$\begin{aligned} \dot{\mathbf{x}} &= \mathbf{A}_1\mathbf{x} + \mathbf{B}_1 V_{dc}, \quad t_0 \le t \le t_1 \\ \dot{\mathbf{x}} &= \mathbf{A}_2\mathbf{x} + \mathbf{B}_2 V_{dc} \quad t_1 \le t < t_2 \end{aligned} \tag{5.85}$$

where

$$\mathbf{A}_1 = \begin{bmatrix} -1/RC & 0 \\ 0 & 0 \end{bmatrix}, \quad \mathbf{A}_2 = \begin{bmatrix} -1/RC & 1/C \\ -1/L & 0 \end{bmatrix}, \quad \mathbf{B}_1 = \mathbf{B}_2 = \begin{bmatrix} 0 \\ 1/L \end{bmatrix}$$

The output is assumed to be the capacitor voltage, and this is given by the following equation, irrespective of whether the switch is open or closed.

$$y = v_0 = \begin{bmatrix} 1 & 0 \end{bmatrix}\mathbf{x} = \mathbf{H}\mathbf{x} \tag{5.86}$$

5.11.1 Steady State Calculation

From Figure 5.9, the following equations are obtained with respect to the duty ratio d and the switching time $T(=1/f)$

$$\begin{aligned} t_2 - t_0 &= T \\ t_1 - t_0 &= dT \\ t_2 - t_1 &= (1-d)T \end{aligned} \tag{5.87}$$

For the boost converter equations of (5.85), the solutions of the state equations are given by

$$
\begin{aligned}
\mathbf{x}(t_1) &= e^{\mathbf{A}_1 dT}\mathbf{x}(t_0) + \left\{\int_0^{dT} e^{\mathbf{A}_1(dT-\tau)}d\tau\right\}\mathbf{B}_1 V_{dc} \\
\mathbf{x}(t_2) &= e^{\mathbf{A}_2(1-d)T}\mathbf{x}(t_1) + \left\{\int_0^{(1-d)T} e^{\mathbf{A}_2(T-dT-\tau)}d\tau\right\}\mathbf{B}_2 V_{dc}
\end{aligned} \tag{5.88}
$$

In the above equations, the matrix $\mathbf{A}_1$ is singular and therefore noninvertible. Fortunately, this matrix is a diagonal, whose exponential is a diagonal matrix containing the exponentials of each diagonal element, i.e.

$$
e^{\mathbf{A}_1(dT-\tau)} = \begin{bmatrix} e^{-(dT-\tau)/RC} & 0 \\ 0 & 1 \end{bmatrix}
$$

The integral terms in (5.88) are then given by

$$
\mathbf{G}_1 = \int_0^{dT} e^{\mathbf{A}_1(dT-\tau)}d\tau = \begin{bmatrix} -RC\, e^{-(dT-\tau)/RC} & 0 \\ 0 & \tau \end{bmatrix}\Bigg|_0^{dT} = \begin{bmatrix} RC\, e^{-dT/RC} - RC & 0 \\ 0 & dT \end{bmatrix}
$$

$$
\mathbf{G}_2 = \int_0^{(1-d)T} e^{\mathbf{A}_2(T-dT-\tau)}d\tau = -\mathbf{A}_2^{-1}\left(e^{\mathbf{A}_2(T-dT)}\Big|_0^{(1-d)T}\right) = -\mathbf{A}_2^{-1}\left(\mathbf{I} - e^{\mathbf{A}_2(T-dT)}\right)
$$

Substituting the above two equations in (5.88), we have

$$
\begin{aligned}
\mathbf{x}(t_1) &= e^{\mathbf{A}_1 dT}\mathbf{x}(t_0) + \mathbf{G}_1\mathbf{B}_1 V_{dc} \\
\mathbf{x}(t_2) &= e^{\mathbf{A}_2(1-d)T}\mathbf{x}(t_1) + \mathbf{G}_2\mathbf{B}_2 V_{dc}
\end{aligned} \tag{5.89}
$$

Note that $\mathbf{B}_1 = \mathbf{B}_2$. Therefore, by combining equations given by (5.89), the following composite equation is obtained

$$
\begin{aligned}
\mathbf{x}(t_2) &= e^{\mathbf{A}_2(1-d)T}\left[e^{\mathbf{A}_1 dT}\mathbf{x}(t_0) + \mathbf{G}_1\mathbf{B}_1 V_{dc}\right] + \mathbf{G}_2\mathbf{B}_2 V_{dc} \\
&= e^{\mathbf{A}_2(1-d)T}e^{\mathbf{A}_1 dT}\mathbf{x}(t_0) + \left(e^{\mathbf{A}_2(1-d)T}\mathbf{G}_1 + \mathbf{G}_2\right)\mathbf{B}_1 V_{dc}
\end{aligned} \tag{5.90}
$$

When the system is in the steady state at a time T_n, we expect $\mathbf{x}(T_n + t_2) = \mathbf{x}(T_n + t_0)$. Using this identity, the steady state equation of a boost converter is obtained from (5.90) as

$$
\mathbf{x}_{ss} = \left(\mathbf{I} - e^{\mathbf{A}_2(1-d)T}e^{\mathbf{A}_1 dT}\right)^{-1}\left(e^{\mathbf{A}_2(1-d)T}\mathbf{G}_1 + \mathbf{G}_2\right)\mathbf{B}_1 V_{dc} \tag{5.91}
$$

There is, however, a simpler numerical method of obtaining the steady state equation using "c2d" command in MATLAB®, as given here:

```
% Define the duty ratio d and the cycle time T.

[F1,G1] = c2d(A1,B1,d*T);
[F2,G2] = c2d(A2,B2,(1-d)*T);

xss = inv(eye(2)-F2*F1)*(F2*G1+G2)*Vdc
```

Example 5.19 Consider a boost converter with the following system parameters

$$L = 1\,\text{mH}, C = 100\,\mu\text{F}, R = 10\,\Omega, V_{dc} = 10\,\text{V},\ \ f = 20\,\text{kHz}$$

We then have

$$\mathbf{A}_1 = \begin{bmatrix} -1000 & 0 \\ 0 & 0 \end{bmatrix}, \quad \mathbf{A}_2 = \begin{bmatrix} -1000 & 10000 \\ -1000 & 0 \end{bmatrix}, \quad \mathbf{B}_1 = \mathbf{B}_2 = \begin{bmatrix} 0 \\ 1000 \end{bmatrix}$$

Let us choose the duty ratio as 0.5. Then the steady state condition obtained using the MATLAB® code given above is

$$\mathbf{x}_{ss}(t_0) = \mathbf{x}_{ss}(t_2) = \begin{bmatrix} 20.24\,\text{V} \\ 3.87\,\text{A} \end{bmatrix}$$

The steady state waveforms for two cycles are shown in Figure 5.10. The steady state waveforms are calculated at the beginning of a switching circle (i.e. at t_0 or t_2).

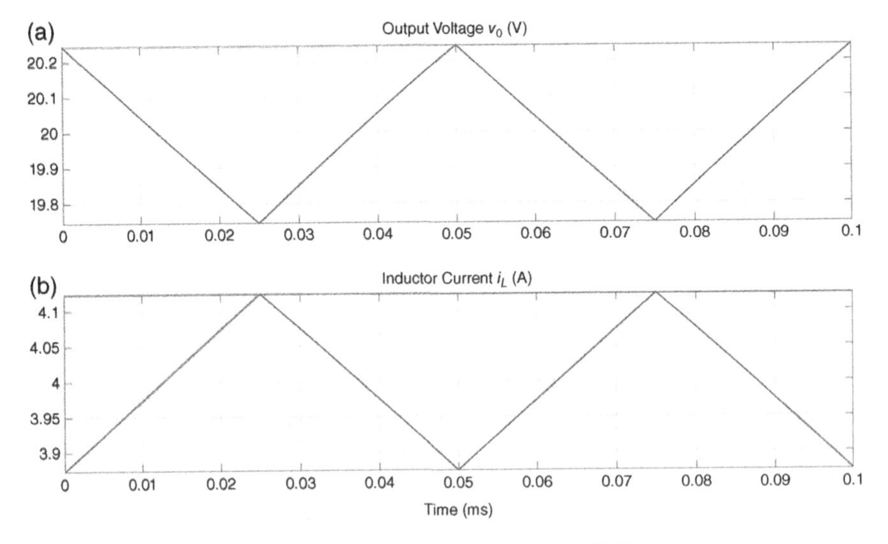

Figure 5.10 Steady state voltage and current waveforms of a boost converter.

Therefore, the peak of the capacitor voltage and the trough of the inductor current occur at this instant, as is evident from this figure. The trough of the capacitor voltage and the peak of the inductor current can also be calculated from (5.89) as

$$\mathbf{x}_{ss}(t_1) = e^{\mathbf{A}_1 dT}\mathbf{x}_{ss}(t_0) + \mathbf{G}_1\mathbf{B}_1 V_{dc} = \begin{bmatrix} 19.74\text{ V} \\ 4.12\text{ A} \end{bmatrix}$$

It can be seen from Figure 5.10 that the trough of the capacitor voltage and the peak of the inductor current match the values calculated above.

Example 5.19 demonstrates the open-loop control of the boost converter, where the switch is turned on and off based on a fixed duty ratio. However, as is mentioned in Chapter 4, for feedback control design an averaging method and a linearized model are required. These are explained in Section 5.11.2.

5.11.2 Linearized Model of a Boost Converter

The averaged model will be determined using the state space averaging method proposed by Middlebrook and Cúk in 1976 [3]. Consider the boost converter model given by (5.85). Then the averaged model is given by

$$\dot{\mathbf{x}} = [\mathbf{A}_1 d + \mathbf{A}_2(1-d)]\mathbf{x} + [\mathbf{B}_1 d + \mathbf{B}_2(1-d)]V_{dc} \tag{5.92}$$

It is assumed that the DC voltage V_{dc} is constant and has no variation. Since $\mathbf{B}_1 = \mathbf{B}_2$, the averaged model can be simplified as

$$\dot{\mathbf{x}} = [\mathbf{A}_1 d + \mathbf{A}_2(1-d)]\mathbf{x} + \mathbf{B}_1 V_{dc} \tag{5.93}$$

Note that (5.93) is nonlinear due to the product of the duty ratio and the state vector. This will thus have to be linearized.

Equation (5.93) is linearized around the operating point $\mathbf{x}_0$ and d_0. For this, the perturbation equations are written as

$$\mathbf{x} = \mathbf{x}_0 + \Delta\mathbf{x}, \quad d = d_o + \Delta d \tag{5.94}$$

where $\Delta\mathbf{x}$ and Δd are the perturbed variables. Noting that V_{dc} is constant, (5.93) can be linearized as

$$\begin{aligned}\Delta\dot{\mathbf{x}} &= [\mathbf{A}_1 d_0 + \mathbf{A}_2(1-d_0)]\Delta\mathbf{x} + (\mathbf{A}_1 - \mathbf{A}_2)\mathbf{x}_0\Delta d \\ &= \mathbf{A}\Delta\mathbf{x} + \mathbf{B}\Delta d\end{aligned} \tag{5.95}$$

where

$$\mathbf{A} = \mathbf{A}_1 d_0 + \mathbf{A}_2(1-d_0), \quad \mathbf{B} = (\mathbf{A}_1 - \mathbf{A}_2)\mathbf{x}_0$$

The output equation is then given from (5.86) by

$$\Delta y = \mathbf{H}\Delta\mathbf{x} \tag{5.96}$$

5.11.3 State Feedback Control of a Boost Converter

The schematic diagram of a state feedback with integral control of a boost converter is shown in Figure 5.11. The control law, including an integral action, is given by

$$\begin{aligned} d &= d_0 + \Delta d \\ \Delta d &= k_1 \Delta v_0 + k_2 \Delta i_L + k_3 K_I \int \left(\Delta V_{ref} - \Delta v_0\right) dt \end{aligned} \tag{5.97}$$

Since the controller is designed based on the linear model given in (5.95) and (5.96), the first step in this process is to calculate the steady state values using (5.91). Since the controller is designed based on the averaged variable, the measured values of the capacitor voltage and the inductor current are averaged over one cycle. These averaged values are then subtracted from the steady state values as per (5.94), as shown in Figure 5.11. These are then multiplied by their respective gains and added to the output of the integral controller.

When a change in the capacitor voltage is required, V_{ref} is changed. The steady state value of the capacitor voltage is subtracted from V_{ref} to obtain ΔV_{ref}. The error $\Delta V_{ref} - \Delta v_0$ is then passed through the integral controller and multiplied by the gains and added with the other two components to form Δd. This is then added to d_0. Before, the duty ratio is given as an input to the pulse width modulator, a limiter is used to restrict d between, say, $0.05 \le d \le 0.95$.

Example 5.20 Consider the boost converter of Example 5.19, where the duty ratio is chosen as 0.5. From Example 5.19, we find the average steady state quantities as

$$\mathbf{x}_{ss} = \begin{bmatrix} v_{0ss} \\ i_{Lss} \end{bmatrix} = d \times \mathbf{x}_{ss}(t_0) + (1-d) \times \mathbf{x}_{ss}(t_2) = \begin{bmatrix} 19.994\ \text{V} \\ 3.998\ \text{A} \end{bmatrix}$$

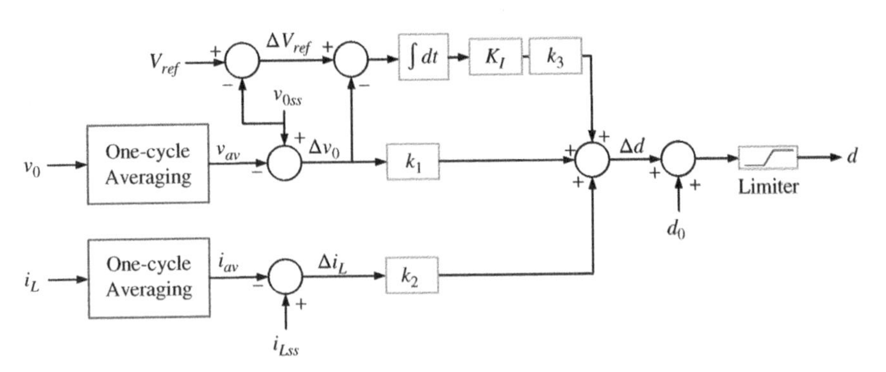

Figure 5.11 Feedback control structure of DC-DC converter.

The averaged matrices are computed from (5.95) as

$$\mathbf{A} = \begin{bmatrix} -1000 & 5000 \\ -500 & 0 \end{bmatrix}, \quad \mathbf{B} = \begin{bmatrix} -3.87 \\ 2.02 \end{bmatrix} \times 10^4$$

A linear quadratic regular with an integral controller is designed with the following parameters

$$K_I = 1000, \quad \mathbf{Q} = \begin{bmatrix} 10 & & \\ & 1 & \\ & & 1 \end{bmatrix}, \quad r = 0.1$$

The resultant gain matrix is

$$\mathbf{K} = [k_1 \quad k_2 \quad k_3] = [2.78 \quad 24.8 \quad -3.16]$$

The converter is started from rest at $t = 0$ seconds, with a duty ratio of 0.5 where the desired output voltage is V_{0ss} and the output resistance is $10\,\Omega$. Then at $t = 0.075$ seconds, the desired output voltage is changed to 30 V. Subsequently, at $t = 0.125$ seconds, the output impedance is changed to $6.67\,\Omega$. The results are shown in Figure 5.12. The duty ratio increases to 0.667 as the output voltage

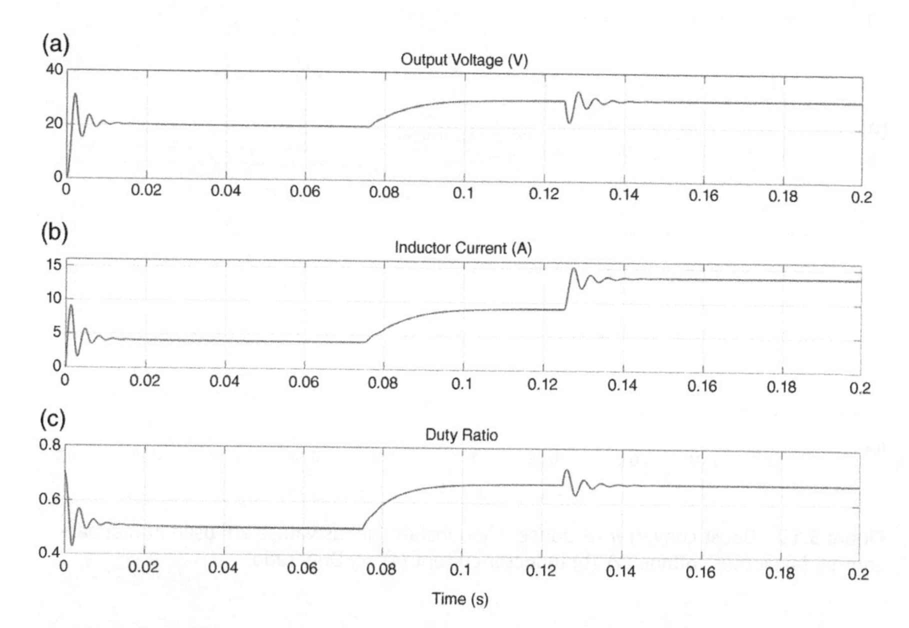

Figure 5.12 Boost converter response when averaged values are used in feedback control. (a) Output voltage (V), (b) Inductor current (A), (c) Duty ratio.

increases to 30 V. However, the duty ratio remains unchanged when the load changes, albeit for an initial transient. It might be argued that the averaging process increases the computational requirements. To alleviate this, the instantaneous values of output voltage and the inductor current are instead used for the control law computation. This is shown in Figure 5.13. It can be seen that the duty ratio has a significant ripple in this case: it chatters between 0.66 and 0.67 after 0.125 seconds.

Only a state feedback controller of the form $\Delta d = -\mathbf{Kx}$ is employed without the integral controller. The LQR gain matrix is computed using $\mathbf{Q} = diag([100 \quad 1])$ and $r = 0.1$. This produces the gain matrix of $\mathbf{K} = [4.5 \quad 69.29]$. The system response for the same voltage reference of $V_{Oss} = 19.994$ V and load change at 0.125 seconds is shown in Figure 5.14. Since the simple state feedback is essentially a proportional controller, the change in the reference voltage is not tracked by the controller. The controller is not even able to maintain the desired output voltage following the load change. This proves the superiority of the state feedback controller with integral control.

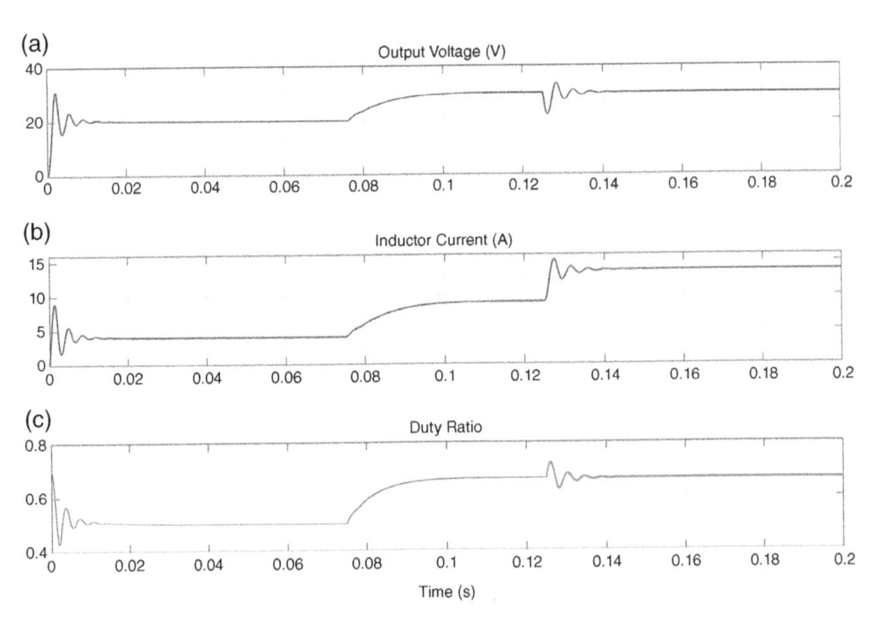

Figure 5.13 Boost converter response when instantaneous values are used in feedback control. (a) Output voltage (V), (b) Inductor current (A), (c) Duty ratio.

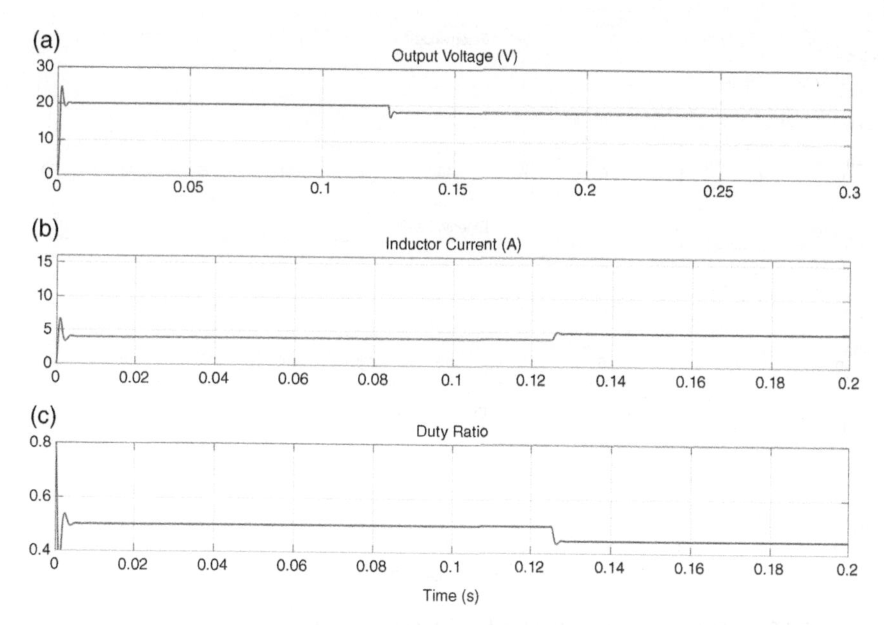

Figure 5.14 Boost converter response with state feedback control without the integral controller. (a) Output voltage (V), (b) Inductor current (A), (c) Duty ratio.

Example 5.21 In this example, the limits of stable operation for the boost controller designed in Example 5.20 are investigated. With the controller gains obtained from Example 5.20, the eigenvalues of the system are computed as the resistance R changes from 20 to 0.1 Ω. The eigenvalues are all real, as shown in Figure 5.15. However, all of them have negative real parts, indicating a stable operation. Eigenvalues-1 and 2 are very stable, while Eigenvalue-3 moves closer to the imaginary axis as the load resistance decreases. This eigenvalue is located at −6.29 for $R = 0.1\ \Omega$.

With the system operating at the steady state with a duty ratio of 0.5, the load resistance is changed to 0.91 Ω at 0.25 seconds. The results are shown in Figure 5.16. It can be seen that output voltage, following an initial transient, returns back to 20 V. The duty ratio, however, sees a marginal increase from 0.5 to 0.525. Consequently, the inductor current becomes 46 A. Nevertheless, the system remains stable, as is expected from the eigenvalue analysis. However, whether such large inductor current is acceptable from the practical design point of view is debatable.

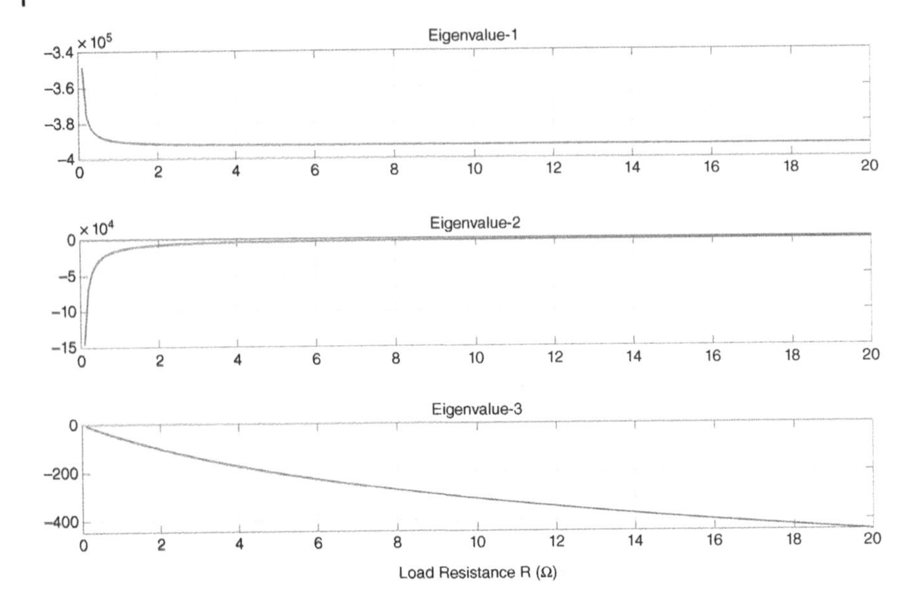

Figure 5.15 Eigenvalue plot as the load resistance changes.

5.12 Concluding Remarks

In this chapter, we discuss state space analysis and the design of linear systems. These are very important in the analysis and control design of both power systems and power electronic circuits. These are extensively used in the following chapters. Even though only LTI systems are discussed here, these techniques are also extensively used in linear controller design for nonlinear systems and for the eigenvalue analysis of interconnected power electronic circuits, such as microgrids, as well as for power system stability analysis. An example of the use of a linearization method for a single-machine, infinite-bus (SMIB) system, compensated by a thyristor-controlled series capacitor, is pointed out in the Notes and References. There are several such examples that can be found in the literature.

Also, the state feedback control design for a DC-DC boost converter is presented in Section 5.11. The state feedback controller is designed in the linearized domain, which is obtained using the state space averaging method. Despite using the linear approximation, it can been shown that the controller can stabilize the system for a wide variation in both output voltage and resistance.

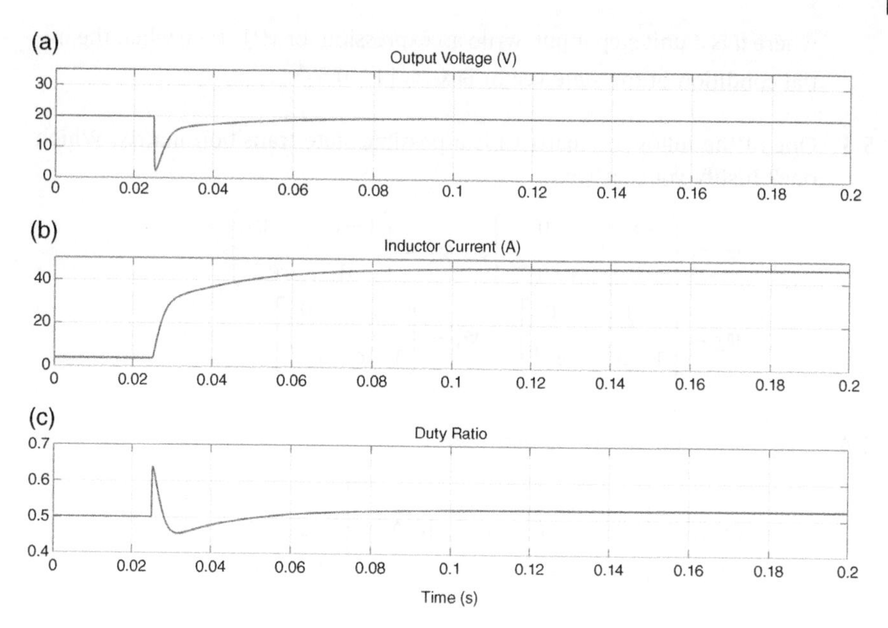

Figure 5.16 Boost converter response for a large change in the output resistance. (a) Output voltage (V), (b) Inductor current (A), (c) Duty ratio.

Problems

5.1 Defining a state vector as $\mathbf{x} = [i_1 \quad i_2 \quad v_c]^T$, find the state space model of the circuit shown in Figure P5.1.

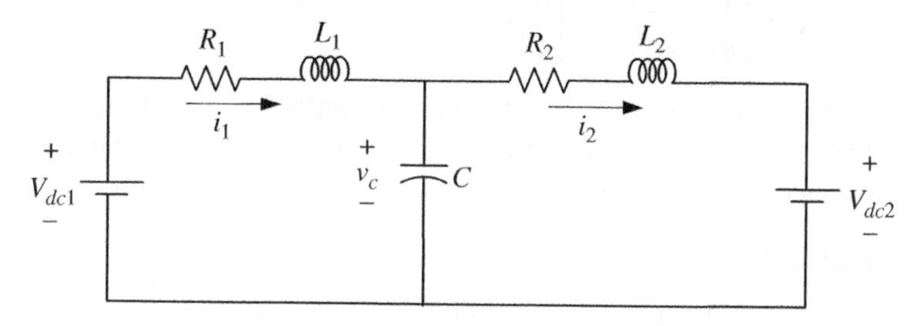

Figure P5.1 The circuit diagram of Problem 5.1.

5.2 Given the following LTI system

$$\dot{\mathbf{x}} = \begin{bmatrix} -2 & 1 \\ 0 & -1 \end{bmatrix}\mathbf{x} + \begin{bmatrix} 1 \\ 1 \end{bmatrix}u$$

where u is a unit step input, write an expression for $y(t)$, $t > 0$ when the initial condition of the state vector is $\mathbf{x}_0 = \begin{bmatrix} 1 & 0.5 \end{bmatrix}^T$.

5.3 One of the following matrices is a possible state transition matrix. Which one? Justify your answer.

$$\boldsymbol{\varphi}_1 = \begin{bmatrix} -e^{-t} & 0 \\ 0 & 1-e^{-t} \end{bmatrix}, \quad \boldsymbol{\varphi}_2 = \begin{bmatrix} 1-e^{-t} & 0 \\ 1 & e^{-t} \end{bmatrix},$$

$$\boldsymbol{\varphi}_3 = \begin{bmatrix} 1 & 0 \\ 1-e^{-t} & e^{-t} \end{bmatrix}, \quad \boldsymbol{\varphi}_4 = \begin{bmatrix} 1 & 0 \\ 1-e^{t} & e^{-t} \end{bmatrix}$$

5.4 Diagonalize the following matrices

$$\text{(a) } \mathbf{A} = \begin{bmatrix} -1 & -4 & 0 \\ 1 & -1 & 0 \\ 4 & 2 & -3 \end{bmatrix}, \text{(b) } \mathbf{A} = \begin{bmatrix} 3 & -2 & 0 \\ -2 & 3 & 0 \\ 0 & 0 & 5 \end{bmatrix}$$

5.5 Convert the following matrix in Jordan form

$$\mathbf{A} = \begin{bmatrix} 0 & 1 & 0 \\ 0 & 0 & 1 \\ -25 & -35 & -11 \end{bmatrix}$$

5.6 Consider the LTI system

$$\dot{\mathbf{x}} = \begin{bmatrix} 1 & -3 \\ 8 & 0 \end{bmatrix} \mathbf{x} + \begin{bmatrix} 0 \\ 1 \end{bmatrix} u$$

It is to be controlled by a state feedback controller of the form $u = -\mathbf{K}\mathbf{x}$, where the feedback gain matrix is given by $\mathbf{K} = \begin{bmatrix} k_1 & k_2 \end{bmatrix}$. Find the constraints on k_1 and k_2 for stable operation.

5.7 Consider the LTI system

$$\dot{\mathbf{x}} = \begin{bmatrix} 0 & 1 & 0 \\ 0 & 0 & 1 \\ 0 & -3 & -2 \end{bmatrix} \mathbf{x} + \begin{bmatrix} 0 \\ 0 \\ 1 \end{bmatrix} u$$

It is a state feedback controller of the form $u = -\mathbf{K}\mathbf{x}$, where the feedback gain matrix is given by $\mathbf{K} = \begin{bmatrix} k & 7 & 5 \end{bmatrix}$. Determine the value of k for stability.

5.8 For the LTI system given below, determine the conditions on $\beta_1, \beta_2, \gamma_1$, and γ_2 such that the system is both controllable and observable.

$$\dot{\mathbf{x}} = \begin{bmatrix} 1 & 1 \\ 0 & 1 \end{bmatrix}\mathbf{x} + \begin{bmatrix} \beta_1 \\ \beta_2 \end{bmatrix}u, \quad y = [\gamma_1 \quad \gamma_2]\mathbf{x}$$

5.9 Consider the LTI of Problem 5.8, in which $\beta_1 = \beta_2 = \gamma_1 = 1$ and $\gamma_2 = 0$. Design a state feedback controller such that the closed-loop system has a damping ratio of 0.707 and an undamped natural frequency of 5 rad/s.

5.10 An LTI system is given by

$$\dot{\mathbf{x}} = \begin{bmatrix} 0 & 1 \\ 0 & 0 \end{bmatrix}\mathbf{x} + \begin{bmatrix} 0 \\ 1 \end{bmatrix}u, \quad y = [1 \quad 0]\mathbf{x}$$

(a) Design a state feedback controller such that closed-loop characteristic equation with the controller is given by $s^2 + 8s + 32 = 0$.
(b) Design an observer such that the characteristic equation of the observer is $(s + 10)^2 = 0$.
(c) Compute the characteristic equation of the overall system.

5.11 For the boost converter of Examples 5.19 and 5.20, discretize the system with a sampling time of $T(=1/f)$ and duty ratio of 0.5. Determine the gain matrix of a state feedback with integral controller such that the closed-loop poles are placed at $-0.5 \pm j0.5$, 0.1. Choose the integral gain given of 0.1.

5.12 Determine the linearized state space average model of the buck converter given in Example 4.2, where the duty ratio is chosen as 0.5.

5.13 The schematic diagram of a buck-boost converter is shown in Figure P5.13. Draw the equivalent circuits and write the state space equation when (a) the switch is closed and (b) when the switch is open.

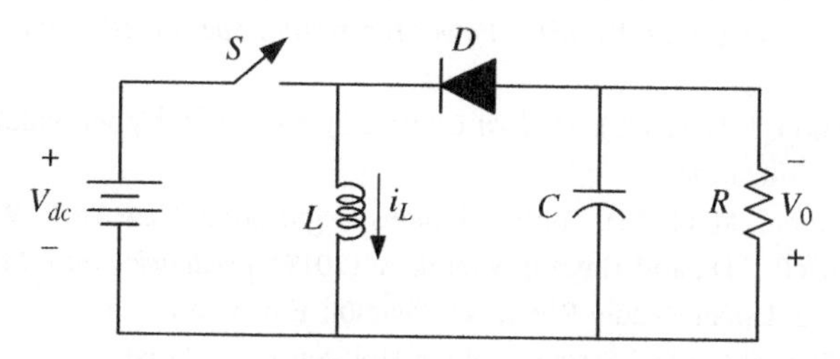

Figure P5.13 Schematic diagram of a buck-boost converter.

5.14 For the buck-boost converter of Problem 5.13, determine the linearized state space average model for a duty ratio of 0.5 and the following converter parameters.

$$L = 4\,\text{mH}, C = 250\,\mu\text{F}, R = 10\,\Omega, V_{dc} = 100\,\text{V}, \text{and}\, f = 1/T = 10\,\text{kHz}$$

Notes and References

This chapter assumes that the reader has prior knowledge of classical and digital control systems. However, there are some wonderful texts that can be used for brushing up the background materials. My personal favorites are books by Kuo [1] (I prefer the fifth or sixth editions, which are not readily available anymore) and Ogata [2]. These texts probably really stand out from the other texts. There are other very worthy books that can also be considered [4–6]. Digital control is covered in considerable depth in [7, 8]. There are, however, many other books that cover these subjects in detail. The linearization of a thyristor-controlled series compensator (TCSC) SMIB system and the TCSC controller design are discussed in [9, 10]. Apart from the linearization of the circuit, the sensitivity of the shift in the zero-crossing of the line current is included in the model, which makes the model very accurate. It is shown that the predicted output and the actual output matches very closely. The book by Mohan, Undeland, and Robbins [11] is an excellent text that presents the detailed analysis of DC-DC converters. The state space averaging is first presented in [3] and has been widely used in a variety of problems since then. The derivation of a discrete-time model for DC-DC converters, based on which the steady state model is derived here, is presented in [12].

1 Kuo, B.C. (1987). *Automatic Control System*, 5e. Englewood Cliffs, NJ: Prentice Hall.
2 Ogata, K. (2010). *Modern Control Engineering*, 5e. Englewood Cliffs, NJ: Prentice Hall.
3 Middlebrook, R.D. and Cúk, S. (1976). A general unified approach to modelling switching inverter power stages. In: *IEEE Power Electronics Specialists' Conference*, 18–34. IEEE.
4 Dorf, R.C. and Bishop, R.H. (2005). *Modern Control Systems*, 10e. Upper Saddle River, NJ: Pearson Education.
5 Nagrath, I.J. and Gopal, M. (1975). *Control Systems Engineering*. New York: Wiley.
6 Franklin, G.F., Powell, J.D., and Emami-Naeini, A. (2019). *Feedback Control of Dynamic Systems*, 8e. Upper Saddle River, NJ: Pearson Education.
7 Kuo, B.C. (1980). *Digital Control System*. Tokyo: Holt-Saunders Japan.
8 Franklin, G.F. and Powell, J.D. (1998). *Digital Control of Dynamic Systems*, 3e. Englewood Cliffs, NJ: Prentice Hall.

9 Ghosh, A. and Ledwich, G. (1995). A discrete-time model of thyristor controlled series compensators. *Electric Power Syst. Res.* 33: 211–218.

10 Ghosh, A. and Ledwich, G. (1995). Modelling and control of thyristor-controlled series compensators. *Proc. IEE Gener. Trans. & Dist.* 142 (3): 297–304.

11 Mohan, N., Undeland, T.E., and Robbins, W.R. (2003). *Power Electronics*, 3e. Hoboken, NJ: Wiley.

12 Ghosh, A. and Ledwich, G. (1995). Modelling and control of switch-mode DC-DC converters using state transition matrices. *Int. J. Electron.* 79 (1): 113–127.

6

Discrete-time Control

In this chapter, discrete-time control using difference equation is discussed. In Chapter 3, we discuss z-transform and how a transfer function can be converted into a difference equation. These will now be used for designing a controller for single-input, single-output (SISO) systems. Our discussion focuses on prediction-based controllers, under the assumption that dynamic systems have disturbances that are stochastic in nature. However, these controllers can be used for deterministic systems as well.

The basic premise of predictive control is simple. It involves calculating a sequence of future control signals that will minimize a set of cost functions over a finite prediction horizon. However, the theory behind this is not simple, neither is its application. Even though the concept has been used by industry for a long time, it has been restricted to processes that have much lower time constants, such as chemical processes. The basic drawback of implementing such types of control is that they are computation intensive. However, with the modern-day fast processors, such controllers are now under the realm of the possibility of being used in fast dynamic systems like power converters.

In this chapter, both prediction base controllers and their application in adaptive control are discussed. Our discussion is limited only to SISO systems in the discrete-time domain. However, predictive controllers can be developed for multivariable systems or for continuous-time systems. Moreover, prediction-based controllers are often developed for systems that contain random noise. It will therefore be assumed that the systems are stochastic in nature.

Control of Power Electronic Converters with Microgrid Applications, First Edition.
Arindam Ghosh and Firuz Zare.

Published 2023 by John Wiley & Sons, Inc.

6.1 Minimum Variance (MV) Prediction and Control

The first step in developing a predictive control is to construct a model. Several models based on difference equations are available in the literature. Their use depends on the behavior of the plant under discussion. We begin our discussion with a summary of the models.

6.1.1 Discrete-time Models for SISO Systems

A generalized SISO model for a linear system with stochastic input is given by

$$\begin{aligned} A(z^{-1})y(k) &= z^{-d}B(z^{-1})u(k) + C(z^{-1})e(k) \\ &= B(z^{-1})u(k-d) + C(z^{-1})e(k) \end{aligned} \tag{6.1}$$

where $y(k)$ is the output; $u(k)$ is the input; $e(k)$ is a noise signal, which is considered white noise; and d is an integer input delay or dead time. Note that the second line of (6.1) is written using the shifting theorem of the z-transform. The polynomials are given in the backward shift operator z^{-1} by

$$\begin{aligned} A(z^{-1}) &= 1 + a_1z^{-1} + a_2z^{-2} + \cdots + a_{n_a}z^{-n_a} \\ B(z^{-1}) &= b_0 + b_1z^{-1} + b_2z^{-2} + \cdots + b_{n_b}z^{-n_b} \\ C(z^{-1}) &= 1 + c_1z^{-1} + c_2z^{-2} + \cdots + c_{n_c}z^{-n_c} \end{aligned} \tag{6.2}$$

From a causality condition, the polynomial orders will have to obey the following conditions

$$n_a \geq n_b + d \text{ and } n_a \geq n_c \tag{6.3}$$

The system of (6.1) is usually called an autoregressive moving average process with an exogenous input, or ARMAX, process. This is the most general form that is used in stochastic control literature. Using this general model, different models can be derived. These are listed in Table 6.1. Note that, even if the input $e(k)$ of

Table 6.1 List of different stochastic processes.

Process name	Characteristics	Equation
Autoregressive (AR)	$B(z^{-1}) = 0$ and $C(z^{-1}) = 1$	$A(z^{-1})y(k) = e(k)$
Moving average (MA)	$A(z^{-1}) = 1$ and $B(z^{-1}) = 0$	$y(k) = C(z^{-1})e(k)$
Autoregressive moving average (ARMA)	$B(z^{-1}) = 0$	$A(z^{-1})y(k) = C(z^{-1})e(k)$
Autoregressive with exogenous input (ARX)	$C(z^{-1}) = 1$	$A(z^{-1})y(k) = z^{-d}B(z^{-1})u(k) + e(k)$

an MA process is an uncorrelated white noise, the output $y(k)$ is not uncorrelated and it is called a colored noise [1].

6.1.2 MV Prediction

Consider the ARMAX process given by (6.1) and (6.2). In the MV control, we choose the performance index

$$J = E[y(k+d) - y_r(k+d)]^2 \tag{6.4}$$

and minimize it to obtain $u(k)$, where $y_r(k)$ is the reference input. The performance index of (6.4), however, cannot be minimized in a straightforward manner. To minimize the square of the error between the reference input and system output at time $k+d$, the control action must be taken at time k. The delay term adds d samples between the control signals and their effect on the output. At time instant k only the values of the system output up until that instant are known. The future values of $y(k+i)$, $i = 1, \ldots d$ are not yet measured. Therefore, a process is required to predict the value of $y(k+d)$, given the measurements up to time k.

Replacing k by $k+d$, we rewrite (6.1) as

$$y(k+d) = \frac{B(z^{-1})}{A(z^{-1})}u(k) + \frac{C(z^{-1})}{A(z^{-1})}e(k+d) \tag{6.5}$$

Expanding the second rational term on the right-hand side of (6.5) we get

$$\frac{C(z^{-1})}{A(z^{-1})} = F(z^{-1}) + z^{-d}\frac{G(z^{-1})}{A(z^{-1})} \tag{6.6}$$

where $F(z^{-1})$ and $G(z^{-1})$ respectively are the quotient and remainder polynomials, given by

$$\begin{aligned} F(z^{-1}) &= 1 + f_1 z^{-1} + \cdots + f_{d-1} z^{-d+1} \\ G(z^{-1}) &= g_0 + g_1 z^{-1} + \cdots + g_{n_g} z^{-n_g} \end{aligned} \tag{6.7}$$

Equation (6.6) is rewritten in a form that is called the Diophantine equation (or Aryabhata identity) as

$$C(z^{-1}) = A(z^{-1})F(z^{-1}) + z^{-d}G(z^{-1}) \tag{6.8}$$

To have nonredundant coefficients, the order of the polynomial $C(z^{-1})$ in (6.8) is bounded by

$$n_c \leq (n_a + d - 1) \tag{6.9}$$

Therefore, to have any nonredundant coefficients on the right-hand side of (6.9), the order of the polynomial $G(z^{-1})$ should be calculated from

$$\begin{aligned} n_a + d - 1 &= d + n_g \\ \Rightarrow n_g &= n_a - 1 \end{aligned} \tag{6.10}$$

Let us now define

$$\begin{aligned} x(k+d) &= \frac{C(z^{-1})}{A(z^{-1})} e(k+d) \\ &= F(z^{-1})\ e(k+d) + \frac{G(z^{-1})}{A(z^{-1})} e(k) \end{aligned} \tag{6.11}$$

Note that $e(k)$ in (6.11) is a random noise and hence its future values cannot be predicted. Therefore, the best d-step-ahead MV prediction can then be obtained from the observation of the values until time k [2], i.e.

$$\hat{x}(k+d|k) = \frac{G(z^{-1})}{A(z^{-1})} e(k) \tag{6.12}$$

The term $\hat{x}(k+d|k)$ indicates that it is the prediction of x at instant $k+d$ given the measurements up to time k. Equation (6.11) can also be written as

$$e(k) = \frac{A(z^{-1})}{C(z^{-1})} x(k)$$

Substituting the above equation in (6.12), we have

$$\hat{x}(k+d|k) = \frac{G(z^{-1})}{C(z^{-1})} x(k) \tag{6.13}$$

Therefore, the poles of the predictor are roots of the polynomial $C(z^{-1})$. Subtracting (6.12) from (6.11), the following error in prediction is obtained

$$\tilde{x}(k+d|k) = x(k+d) - \hat{x}(k+d|k) = F(z^{-1})\ e(k+d) \tag{6.14}$$

Assume that $e(k)$ is an uncorrelated sequence with zero mean. Then the prediction error $E[F(z^{-1})e(k+d)]$ is also zero mean. Therefore, the variance of the prediction error is given by

$$\begin{aligned} E[\tilde{x}(k+d|k)]^2 &= E[x(k+d) - \hat{x}(k+d|k)]^2 \\ &= E\left[F(z^{-1})e(k+d) + \frac{G(z^{-1})}{A(z^{-1})} e(k) - \hat{x}(k+d|k)\right]^2 \end{aligned}$$

The above equation can be expanded as

$$E[\tilde{x}(k+d|k)]^2 = E\left[F(z^{-1})\,e(k+d)\right]^2 + E\left[\frac{G(z^{-1})}{A(z^{-1})}e(k) - \hat{x}(k+d|k)\right]^2 + E\left\{2F(z^{-1})\left[\frac{G(z^{-1})}{A(z^{-1})}e(k+d)e(k) - e(k+d)\hat{x}(k+d|k)\right]\right\}$$

The second term on the right-hand side can be eliminated using (6.12). Therefore

$$E[\tilde{x}(k+d|k)]^2 = E[F(z^{-1})\,e(k+d)]^2 + E\left\{2F(z^{-1})\left[\frac{G(z^{-1})}{A(z^{-1})}e(k+d)e(k) - e(k+d)\hat{x}(k+d|k)\right]\right\} \tag{6.15}$$

In (6.15), since $e(k)$ is uncorrelated, $E[e(k+d)e(k)] = 0$. Also, using the same argument, we get the following by substituting (6.12)

$$E[e(k+d)\hat{x}(k+d|k)] = \frac{G(z^{-1})}{A(z^{-1})}E[e(k+d)e(k)] = 0$$

Thus, the last term on the right-hand side of (6.15) is equal to zero. Then the variance can be written as

$$E[\tilde{x}(k+d|k)]^2 = E\left[F(z^{-1})\,e(k+d)\right]^2 \tag{6.16}$$

Since $e(k)$ is uncorrelated and has a mean of zero, (6.16) can be directly obtained from (6.14). If the uncorrelated noise $e(k)$ has a variance of σ_e^2, (6.16) can be expanded as

$$E[\tilde{x}(k+d|k)]^2 = \left[1 + f_1^2 + \cdots + f_d^2\right]\sigma_e^2 \tag{6.17}$$

Example 6.1 We shall design a three-step-ahead MV predictor for the process

$$x(k) = \frac{C(z^{-1})}{A(z^{-1})} = \frac{1 - 1.4z^{-1} + 0.5z^{-2}}{1 - 1.2z^{-1} + 0.4z^{-2}}e(k)$$

where $e(k) \sim N(0,\ \sigma_e^2)$, i.e. $e(k)$ is a normal (Gaussian) random process with zero mean and a variance of σ_e^2. The polynomials $F(z^{-1})$ and $G(z^{-1})$ are then given from (6.7) as

$$F(z^{-1}) = 1 + f_1 z^{-1} + f_2 z^{-2}$$
$$G(z^{-1}) = g_0 + g_1 z^{-1}$$

Therefore, the Diophantine equation of (6.8) is written for this case as

$$
\begin{aligned}
1-1.4z^{-1}+0.5z^{-2} &= \left(1-1.2z^{-1}+0.4z^{-2}\right)\left(1+f_1z^{-1}+f_2z^{-2}\right) \\
&\quad + z^{-3}\left(g_0+g_1z^{-1}\right) \\
&= 1-1.2z^{-1}+0.4z^{-2}+f_1z^{-1}-1.2f_1z^{-2}+0.4f_1z^{-3} \\
&\quad + f_2z^{-2}-1.2f_2z^{-3} \\
&\quad 0.4f_2z^{-4}+g_0z^{-3}+g_1z^{-4}
\end{aligned}
$$

Equating the coefficients of the negative powers of z, we have

$$
\begin{aligned}
&z^{-1} \rightarrow -0.2 = f_1 \\
&z^{-2} \rightarrow 0.1 = -1.2f_1+f_2 \Rightarrow f_2 = 0.1+1.2f_1 = -0.14 \\
&z^{-3} \rightarrow 0 = 0.4f_1-1.2f_2+g_0 \Rightarrow g_0 = -0.4f_1+1.2f_2 = -0.088 \\
&z^{-4} \rightarrow 0 = 0.4f_2+g_1 \Rightarrow g_1 = -0.4f_2 = 0.056
\end{aligned}
$$

i.e. $f_1 = -0.2, \quad f_2 = -0.14, \quad g_0 = -0.088, \quad g_1 = 0.056$

The variance of the prediction error is then

$$
E[\tilde{x}(k+3|k)]^2 = \left[1+f_1^2+f_2^2\right]\sigma_e^2 = 1.0596\sigma_e^2
$$

6.1.3 MV Control Law

By replacing the last term on the right-hand side of (6.5) by (6.11), a d-step-ahead prediction of the output can be written as

$$
\hat{y}(k+d|k) = \frac{B(z^{-1})}{A(z^{-1})}u(k) + \hat{x}(k+d|k)
$$

Substituting (6.12) in the above equation, we have

$$
\hat{y}(k+d|k) = \frac{B(z^{-1})}{A(z^{-1})}u(k) + \frac{G(z^{-1})}{A(z^{-1})}e(k) \tag{6.18}
$$

The noise term $e(k)$ can be eliminated from (6.18) by substituting (6.1) into (6.18), i.e.

$$
\begin{aligned}
\hat{y}(k+d|k) &= \frac{B(z^{-1})}{A(z^{-1})}u(k) + \frac{G(z^{-1})}{A(z^{-1})}\left[\frac{A(z^{-1})}{C(z^{-1})}y(k) - z^{-d}\frac{B(z^{-1})}{C(z^{-1})}u(k)\right] \\
&= \frac{G(z^{-1})}{C(z^{-1})}y(k) + \frac{B(z^{-1})}{A(z^{-1})C(z^{-1})}\left[C(z^{-1}) - z^{-d}G(z^{-1})\right]u(k)
\end{aligned}
$$

Substitution of (6.8) in the expression of $\hat{y}(k+d|k)$ yields the following prediction equation

$$\hat{y}(k+d|k) = \frac{G(z^{-1})}{C(z^{-1})}y(k) + \frac{B(z^{-1})F(z^{-1})}{C(z^{-1})}u(k) \tag{6.19}$$

Replacing the actual value of $y(k+d)$ in (6.4) by its predicted value from (6.19), the performance index of (6.4) is given as

$$\begin{aligned} J &= E[\hat{y}(k+d|k) - y_r(k+d)]^2 \\ &= \left[\frac{G(z^{-1})}{C(z^{-1})}y(k) + \frac{B(z^{-1})F(z^{-1})}{C(z^{-1})}u(k) - y_r(k+d)\right]^2 \end{aligned} \tag{6.20}$$

Note that (6.20) contains the terms which are known or measured. Since all the values are deterministic, the expectation operator is no longer needed. Expanding the right-hand side of (6.20), we have

$$\begin{aligned} J = &\left[\frac{G(z^{-1})}{C(z^{-1})}y(k) - y_r(k+d)\right]^2 + \left[\frac{B(z^{-1})F(z^{-1})}{C(z^{-1})}u(k)\right]^2 \\ &+ 2\left[\frac{G(z^{-1})}{C(z^{-1})}y(k) - y_r(k+d)\right] \times \left[\frac{B(z^{-1})F(z^{-1})}{C(z^{-1})}u(k)\right] \end{aligned}$$

The performance index can then be minimized by taking its derivative with respect to $u(k)$ and equating it to zero. It is to be noted that both $y(k)$ and $y_r(k+d)$ are independent of $u(k)$. Then the minimization of the above equation with respect to $u(k)$ results in

$$\begin{aligned} \frac{\partial J}{\partial u(k)} &= 2\left[\frac{B(z^{-1})F(z^{-1})}{C(z^{-1})}\right]^2 u(k) + 2\left[\frac{G(z^{-1})}{C(z^{-1})}y(k) - y_r(k+d)\right]\frac{B(z^{-1})F(z^{-1})}{C(z^{-1})} \\ &= \frac{B(z^{-1})F(z^{-1})}{C(z^{-1})}u(k) + \frac{G(z^{-1})}{C(z^{-1})}y(k) - y_r(k+d) = 0 \end{aligned}$$

Rearranging the above equation, we get the MV control law as

$$u(k) = \frac{1}{B(z^{-1})F(z^{-1})}\left[C(z^{-1})y_r(k+d) - G(z^{-1})y(k)\right] \tag{6.21}$$

Substitution of (6.21) into (6.1) yields

$$A(z^{-1})y(k) = \frac{z^{-d}}{F(z^{-1})}\left[C(z^{-1})y_r(k+d) - G(z^{-1})y(k)\right] + C(z^{-1})e(k)$$

This is rearranged as

$$\begin{aligned} \left[A(z^{-1})F(z^{-1}) + z^{-d}G(z^{-1})\right]y(k) = &\, z^{-d}C(z^{-1})y_r(k+d) \\ &+ F(z^{-1})C(z^{-1})e(k) \end{aligned}$$

Again, substituting (6.8) in the above equation, the closed-loop input–output relation is obtained as

$$y(k) = y_r(k) + F(z^{-1})\, e(k) \tag{6.22}$$

It is interesting to note from (6.22) that all the closed-loop poles are placed at the origin. The tracking error is then given by

$$\tilde{y}(k) = y(k) - y_r(k) = F(z^{-1})\, e(k) \tag{6.23}$$

For a zero-mean noise with a variance of σ_e^2, the variance of the tracking error is the same as that given in (6.17).

Example 6.2 Consider an ARMAX process of (6.1) with $d = 3$ and

$$A(z^{-1}) = 1 - 1.2z^{-1} + 0.4z^{-2}, \quad B(z^{-1}) = 0.5 + 0.2z^{-1},$$
$$C(z^{-1}) = 1 - 1.4z^{-1} + 0.5z^{-2}$$

The polynomials $F(z^{-1})$ and $G(z^{-1})$ are calculated in Example 6.1 as

$$F(z^{-1}) = 1 - 0.2z^{-1} - 0.14z^{-2}$$
$$G(z^{-1}) = -0.088 + 0.056z^{-1}$$

Then

$$B(z^{-1})F(z^{-1}) = 0.5 + 0.1z^{-1} - 0.11z^{-2} - 0.028z^{-3}$$

Therefore, the MV control law is given by

$$u(k) = \frac{1}{0.5}\{y_r(k+3) - 1.4y_r(k+2) + 0.5y_r(k+1) + 0.088y(k)$$
$$-0.056y(k-1) - 0.1u(k-1) + 0.11u(k-2) + 0.028u(k-3)\}$$

The system noise input is assumed to be $e(k) \sim N(0,\ 10^{-3})$. The system output and input are shown in Figure 6.1. At the beginning y_r is chosen as a unit step and at the 50th sample. This is changed to twice the unit step. Note that a three-step-ahead prediction of the reference output is required from the control equation. Since the reference is held constant at the previous value for three samples, there is no change in the output for the first three samples after the reference change. However, the system settles within 10 samples thereafter.

6.1.4 One-step-ahead Control

Consider an ARX model with a unit delay, given by

$$y(k) = z^{-1}\frac{B(z^{-1})}{A(z^{-1})}u(k) + \frac{1}{A(z^{-1})}e(k) \tag{6.24}$$

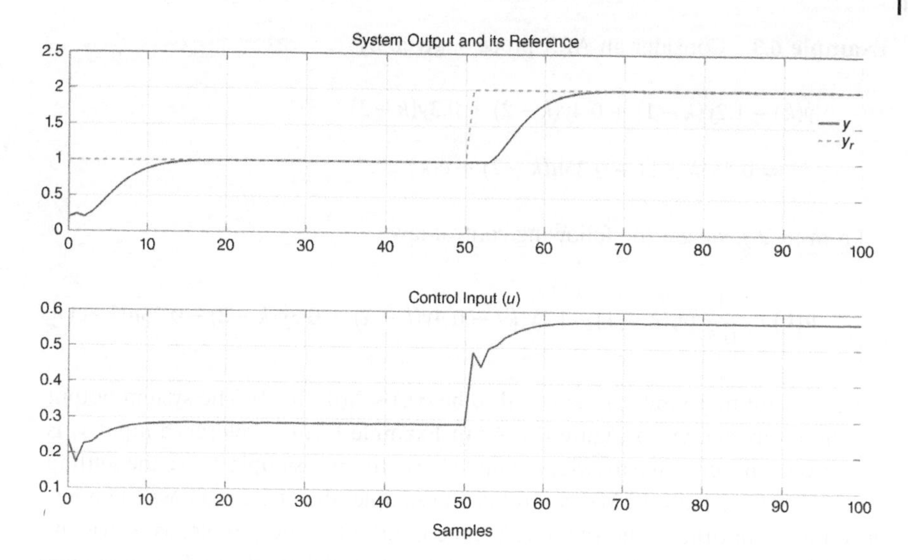

Figure 6.1 System output and input for the three-step-ahead MV control of Example 6.1.

Since $F(z^{-1}) = 1$, the Diophantine equation of (6.8) then becomes

$$1 = A(z^{-1}) + z^{-1}G(z^{-1}) \tag{6.25}$$

Therefore, the one-step-ahead control law is derived from (6.21) as

$$\begin{aligned} u(k) &= \frac{1}{B(z^{-1})}\left[y_r(k+1) - G(z^{-1})y(k)\right] \\ &= \frac{1}{B(z^{-1})}\left[y_r(k+1) + \left(1 - A(z^{-1})\right)y(k)\right] \end{aligned} \tag{6.26}$$

Substituting (6.26) in (6.24), the closed-loop control system is given by

$$A(z^{-1})y(k+1) = y_r(k+1) + \left[1 - A(z^{-1})\right]y(k) + e(k+1)$$

Note that

$$A(z^{-1})y(k+1) - \left[1 - A(z^{-1})\right]y(k) = y(k+1)$$

Therefore, the closed-loop system can be expressed as

$$y(k+1) = y_r(k+1) + e(k+1) \tag{6.27}$$

This implies that the one-step-ahead MV control predicts the output one step ahead of the current sample and then applies the control signal such that the output follows the reference input.

Example 6.3 Consider an ARX process, given by

$$y(k) - 1.2y(k-1) + 0.4y(k-2) + 0.3y(k-3)$$
$$= 0.5u(k-1) + 0.25u(k-2) + e(k)$$

From (6.26), we get the following control law

$$u(k) = \frac{1}{0.5}\,[y_r(k+1) - 1.2y(k) + 0.4y(k-1) + 0.3y(k-2) - 0.25u(k-1)]$$

The system noise input is assumed to be $e(k) \sim N(0,\ 10^{-3})$. The system output and input are shown in Figure 6.2. As in Example 6.1, the reference input y_r is changed from a unit step to twice the unit step at the 50th sample. Since the settling time of a one-step-ahead MV control is one sample, the output follows the reference the output then follows the reference input after one sample, as is evident from Figure 6.2a. This, however, comes at a cost of high control effort, as shown in Figure 6.2.

The MV control has a very desirable control action. However, it has one major disadvantage, as is illustrated by Example 6.4.

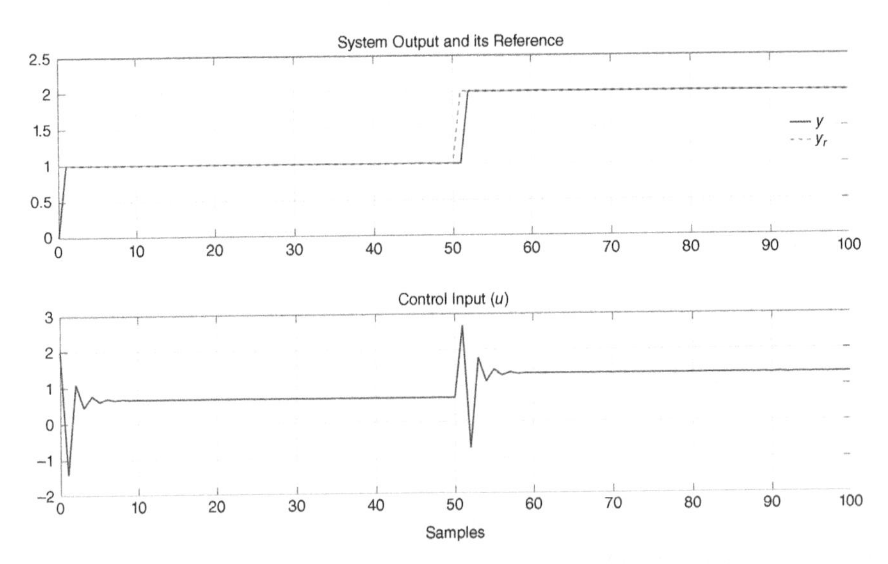

Figure 6.2 System output and input for the one-step-ahead MV control of Example 6.2.

Example 6.4 Consider an ARX process given by

$$y(k) - 1.2y(k-1) + 0.4y(k-2) + 0.3y(k-3)$$
$$= 0.5u(k-1) + 1.75u(k-2) + e(k)$$

This implies that the polynomial $A(z^{-1})$ is the same as that of Example 6.3 and

$$B(z^{-1}) = 0.5 + 1.75z^{-1}$$

The system step response is shown in Figure 6.3. It can be seen that the output follows the reference for about 30 samples despite having a very large control action. But eventually the control action overflows, thus resulting in an unbounded output. In a physical system there will be limiters in the actuator that will prevent the input signal from being excessively large and this might result in a bang-bang response, as shown in Figure 6.4, where $u(k)$ is restricted to ± 10.

The zero of the system of this example is at −6.5, i.e. outside the unit circle. This is called a nonminimum phase condition. From (6.26) it is obvious that the poles of the controller are the roots of the polynomial $B(z^{-1})$, i.e. the open loop system zeros. Therefore, for a nonminimum phase system, control action is unstable and produces an excessive control action. This obviously is not at all desirable. There are methods in which a suboptimal control is obtained by factoring the polynomial $B(z^{-1})$ into two polynomials: one containing system zeros that are inside the unit

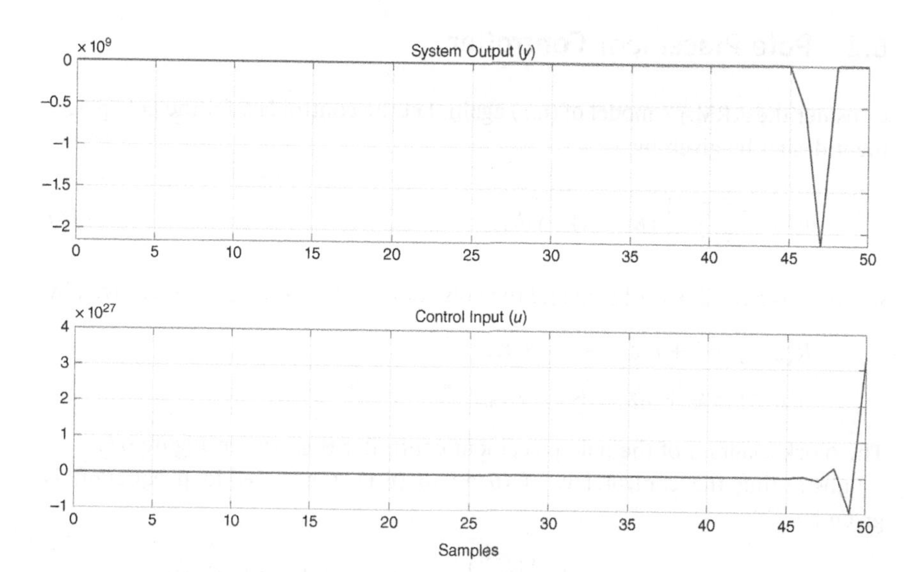

Figure 6.3 System output and input for unrestricted control of Example 6.6.

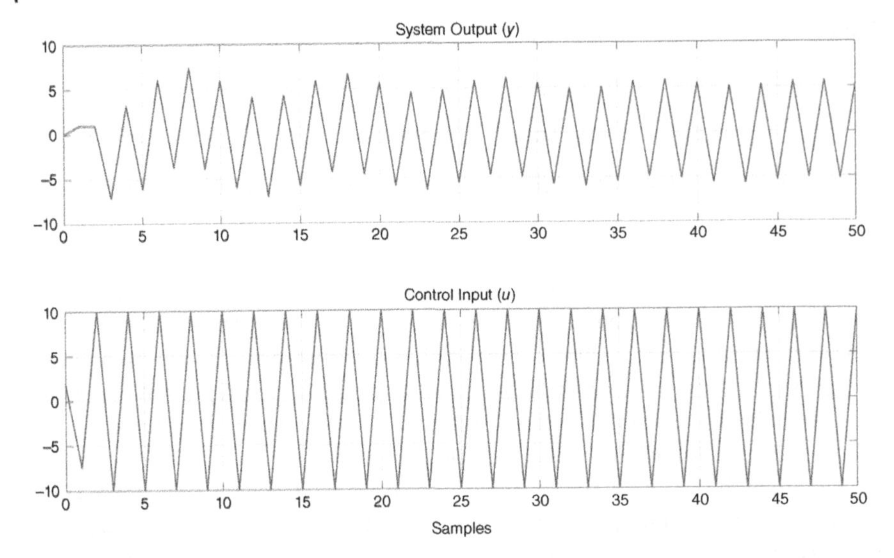

Figure 6.4 System output and input for restricted bang-bang control of Example 6.6.

circle and the other containing zeros that are outside the unit circle. The controller is then synthesized using the polynomial with stable zeros [1, 3].

6.2 Pole Placement Controller

Consider the ARMAX model of (6.1) again. Let the control law for the pole placement design be given by

$$u(k) = \frac{S(z^{-1})}{R(z^{-1})}\{K_c y_r(k) - y(k)\} \tag{6.28}$$

where K_c is a feedforward gain and the polynomials $R(z^{-1})$ and $S(z^{-1})$ are given by

$$R(z^{-1}) = 1 + r_1 z^{-1} + \cdots + r_{n_r} z^{-n_r}$$
$$S(z^{-1}) = s_0 + s_1 z^{-1} + \cdots + s_{n_s} z^{-n_s}$$

The block diagram of the pole placement controller is shown in Figure 6.5.

Substituting the control law of (6.28) in (6.1), the closed-loop equation is given by

$$A(z^{-1})y(k) = z^{-d}B(z^{-1})\frac{S(z^{-1})}{R(z^{-1})}[K_c y_r(k) - y(k)] + C(z^{-1})e(k)$$

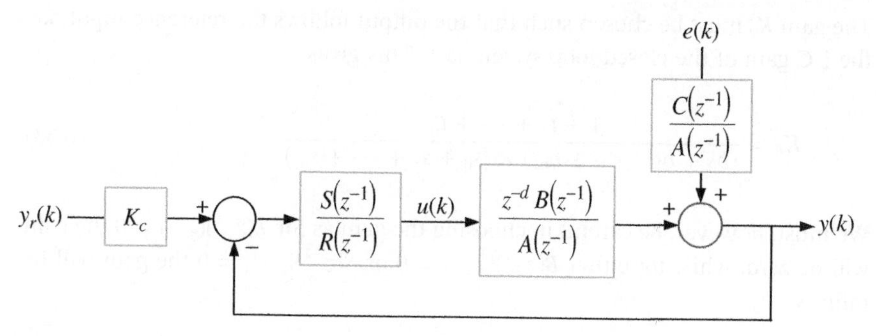

Figure 6.5 Block diagram of the pole placement controller.

Rearranging the above equation, we get

$$y(k) = \frac{z^{-d}K_cB(z^{-1})S(z^{-1})y_r(k) + C(z^{-1})R(z^{-1})e(k)}{A(z^{-1})R(z^{-1}) + z^{-d}B(z^{-1})S(z^{-1})} \tag{6.29}$$

The poles of the closed-loop system are given by the solution of the characteristic equation

$$A(z^{-1})R(z^{-1}) + z^{-d}B(z^{-1})S(z^{-1}) = 0$$

In the pole placement design, the controller polynomials are obtained through the solution of the Diophantine equation

$$A(z^{-1})R(z^{-1}) + z^{-d}B(z^{-1})S(z^{-1}) = T(z^{-1}) \tag{6.30}$$

where $T(z^{-1}) = 0$ defines the desired characteristics equation of the closed-loop system. This can be denoted as

$$T(z^{-1}) = 1 + t_1z^{-1} + \cdots + t_{n_t}z^{-n_t}$$

From (6.30), we can stipulate the following for the polynomial orders

$$n_a + n_r = d + n_b + n_s \geq n_t \tag{6.31}$$

Neglecting the noise term in (6.29), the steady state value of the output to a step input is given by

$$\begin{aligned} y_{ss}(k) &= \left.\frac{K_cB(z^{-1})S(z^{-1})}{T(z^{-1})}\right|_{z=1} \\ &= \frac{K_c \times (b_0 + b_1 + \cdots + b_{n_b}) \times (s_0 + s_1 + \cdots + s_{n_s})}{1 + t_1 + \cdots + t_{n_t}} y_r(k) \end{aligned} \tag{6.32}$$

The gain K_c must be chosen such that the output follows the reference input, i.e. the DC gain of the closed-loop system is 1. This gives

$$K_c = \frac{1 + t_1 + \cdots + t_{n_t}}{(b_0 + b_1 + \cdots + b_{n_b}) \times (s_0 + s_1 + \cdots + s_{n_s})} \tag{6.33}$$

We must, however, be careful in choosing the gain as for $T(z^{-1})|_{z=1} = 0$ the gain will be zero, while for either $B(z^{-1})|_{z=1} = 0$ or $S(z^{-1})|_{z=1} = 0$ the gain will be infinity.

Example 6.5 Consider an ARMAX model of Example 6.2, where the polynomials are

$$A(z^{-1}) = 1 + a_1 z^{-1} + a_2 z^{-2}, \quad B(z^{-1}) = b_0 + b_1 z^{-1},$$
$$C(z^{-1}) = 1 - 1.4z^{-1} + 0.5z^{-2}$$

where $a_1 = -1.2$, $a_2 = 0.4$, $b_0 = 0.5$, $b_1 = 0.2$. Also, the delay d is again chosen as 3. Note that the open-loop poles are located at $-0.6 \pm j1.908$. It is desired that the closed-loop poles will be placed at $-0.1 \pm j0.2$ such that

$$T(z^{-1}) = 1 + t_1 z^{-1} + t_2 z^{-2} = 1 + 0.2z^{-1} + 0.05z^{-2}$$

Choosing

$$S(z^{-1}) = s_0 + s_1 z^{-1}$$

the following polynomial orders are obtained from (6.31)

$$n_r = d + n_b + n_s - n_a = 3$$

Thus

$$R(z^{-1}) = 1 + r_1 z^{-1} + r_2 z^{-2} + r_3 z^{-3}$$

Therefore, the Diophantine equation of (6.30) is given as

$$(1 + a_1 z^{-1} + a_2 z^{-2})(1 + r_1 z^{-1} + r_2 z^{-2} + r_3 z^{-3})$$
$$+ z^{-3}(b_0 + b_1 z^{-1})(s_0 + s_1 z^{-1}) = (1 + t_1 z^{-1} + t_2 z^{-2})$$

Expanding and equating the negative powers of z, the above equation is written in matrix vector form as

$$\begin{array}{l} z^{-1} \Rightarrow \\ z^{-2} \Rightarrow \\ z^{-3} \Rightarrow \\ z^{-4} \Rightarrow \\ z^{-5} \Rightarrow \end{array} \begin{bmatrix} 1 & 0 & 0 & 0 & 0 \\ a_1 & 1 & 0 & 0 & 0 \\ a_2 & a_1 & 1 & b_0 & 0 \\ 0 & a_2 & a_1 & b_1 & b_0 \\ 0 & 0 & a_2 & 0 & b_1 \end{bmatrix} \begin{bmatrix} r_1 \\ r_2 \\ r_3 \\ s_0 \\ s_1 \end{bmatrix} = \begin{bmatrix} t_1 - a_1 \\ t_2 - a_2 \\ 0 \\ 0 \\ 0 \end{bmatrix}$$

Substituting the coefficient of the polynomials $A(z^{-1})$ and $B(z^{-1})$ and solving, we get

$$S(z^{-1}) = 0.7877 - 0.4523z^{-1}$$
$$R(z^{-1}) = 1 + z^{-1} + 0.85z^{-2} + 0.2262z^{-3}$$

The gain is calculated from (6.33) as $K_c = 6.62$. The control law is therefore

$$\varepsilon(k) = K_c y_{ref}(k) - y(k)$$
$$u(k) = s_0\varepsilon(k) + s_1\varepsilon(k-1) - r_1 u(k-1) - r_2 u(k-2) - r_3 u(k-3)$$

The system output and input are shown in Figure 6.6, where the reference input is changed from 1 to 2 at 0.5 seconds. It can be seen that the response is closer to

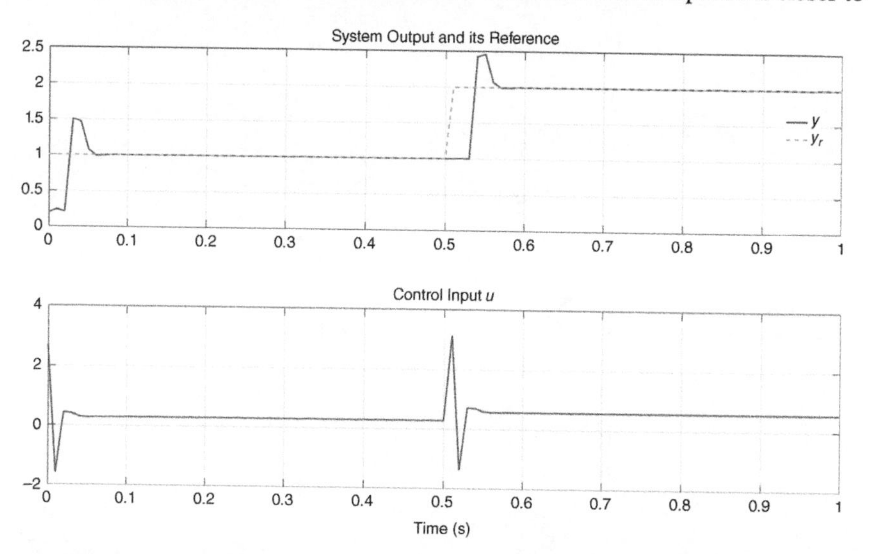

Figure 6.6 System output and input for the three-step-ahead pole placement control of Example 6.5.

that of the MV controller since the poles are placed near the origin. However, an arbitrary pole placement can lead to excessive control action. In order to avoid this, the open-loop poles can be shifted to more stable locations using the pole shift control technique.

6.2.1 Pole Shift Control

Consider a system that has a pair of complex conjugate poles and one real pole, located respectively at $\alpha \pm j\beta$ and $-\sigma$. The position of the poles in the complex plane is shown in Figure 6.7, where it has been assumed that all the poles lie within the unit circle. Note that, in this figure, only one of the complex conjugate pair has been shown. The polynomial $A(z^{-1})$ is then given by

$$\begin{aligned} A(z^{-1}) &= (1+\sigma z^{-1}) \times (1-\alpha z^{-1} + j\beta z^{-1}) \times (1-\alpha z^{-1} - j\beta z^{-1}) \\ &= 1 + (\sigma - 2\alpha) z^{-1} + (\alpha^2 + \beta^2 - 2\alpha\sigma) z^{-2} + (\alpha^2 + \beta^2)\, \sigma z^{-3} \end{aligned} \tag{6.34}$$

Defining $A(z^{-1}) = 1 + a_1 z^{-1} + a_2 z^{-2} + a_3 z^{-3}$, and comparing with (6.34), we get

$$a_1 = (\sigma - 2\alpha), \quad a_2 = (\alpha^2 + \beta^2 - 2\alpha\sigma), \quad a_3 = (\alpha^2 + \beta^2)\, \sigma$$

Also note from Figure 6.7 that $\varphi = \tan^{-1}(\beta/\alpha)$.

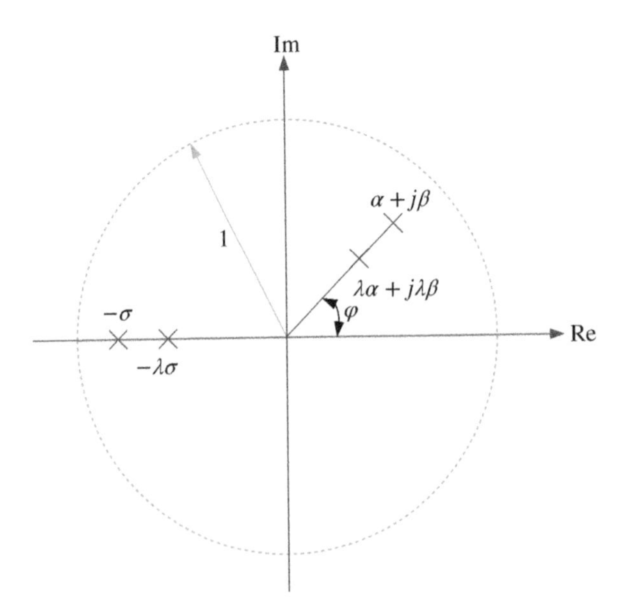

Figure 6.7 Radial shifting of poles in the z-plane.

In the pole shift control, the closed-loop poles are determined by multiplying open-loop poles by a factor λ, $0 < \lambda < 1$. Then the following polynomial defines the closed-loop poles.

$$\begin{aligned} T(z^{-1}) &= A(\lambda z^{-1}) = (1 + \lambda\sigma z^{-1}) \times (1 - \lambda\alpha z^{-1} + j\lambda\beta z^{-1}) \\ &\quad \times (1 - \lambda\alpha z^{-1} - j\lambda\beta z^{-1}) \\ &= 1 + (\sigma - 2\alpha)\lambda z^{-1} + (\alpha^2 + \beta^2 - 2\alpha\sigma)\lambda^2 z^{-2} + (\alpha^2 + \beta^2)\sigma\lambda^3 z^{-3} \end{aligned} \tag{6.35}$$

Comparing (6.35) with (6.34), we find that

$$T(z^{-1}) = 1 + \lambda a_1 z^{-1} + \lambda^2 a_2 z^{-2} + \lambda^3 a_3 z^{-3}$$

The closed-loop pole locations are also shown in Figure 6.7. Note that, as the value of λ is reduced, the closed-loop poles radially move toward the origin, with the complex conjugate pair staying on the line joining the origin with the pole at location $\alpha \pm j\beta$. This implies that angle φ will remain unchanged. Also, the real pole will move toward the origin along the real axis. Therefore, in the pole shift control, the closed-loop poles are determined by radially shifting the open-loop poles by a factor λ $(0 < \lambda < 1)$ to more stable locations [4]. In general, the closed loop is then defined by the polynomial

$$T(z^{-1}) = A(\lambda z^{-1}) = 1 + \lambda a_1 z^{-1} + \cdots + \lambda^{n_a} t_{n_a} z^{-n_a} \tag{6.36}$$

Since $n_a = n_t$, condition (6.31) is satisfied.

Example 6.6 Consider the ARMAX model of Example 6.5. Here the procedure for the determination polynomials $R(z^{-1})$ and $S(z^{-1})$ remains the same, except that the polynomial $T(z^{-1})$ now is

$$T(z^{-1}) = 1 - 1.2\lambda z^{-1} + 0.4\lambda^2 z^{-2}$$

The system output and input for $\lambda = 0.9$ and 0.5 are shown in Figure 6.8, where the reference input is changed from 1 to 2 at 0.5 seconds. It can be seen that, as the closed-loop poles move closer to the origin, the system response becomes faster (Figure 6.8c,d).

Example 6.7 Let us consider the nonminimum phase system of Example 6.4, given by

$$\begin{aligned} y(k) &= 1.2y(k-1) - 0.4y(k-2) - 0.3y(k-3) + 0.5u(k-1) \\ &\quad + 1.75u(k-2) + e(k) \end{aligned}$$

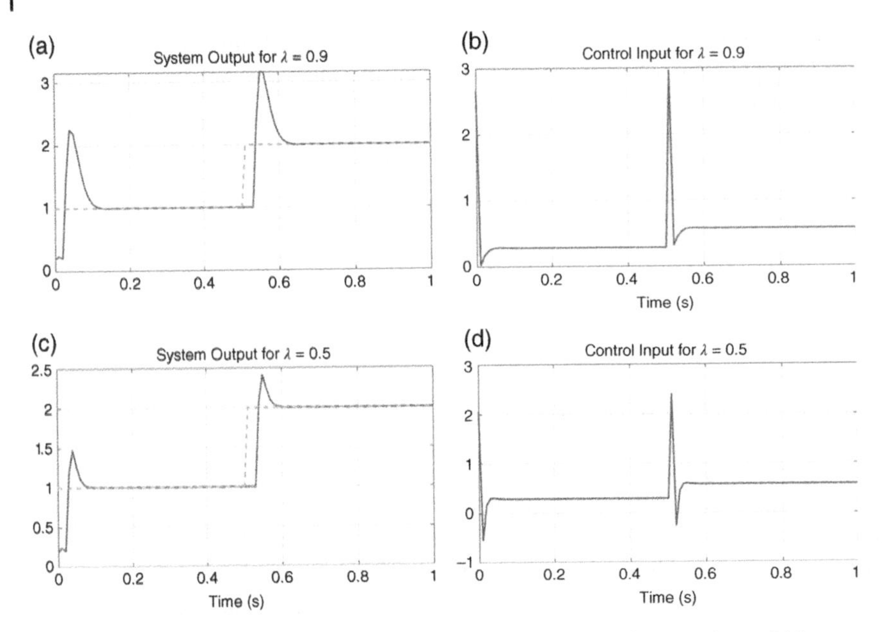

Figure 6.8 System response with pole shift control for two different values of λ in Example 6.6. (a) System Output for $\lambda = 0.9$, (b) Control Input for $\lambda = 0.9$, (c) System Output for $\lambda = 0.5$, (d) Control Input for $\lambda = 0.5$.

Let the polynomial $R(z^{-1})$ be given by

$$R(z^{-1}) = 1 + r_1 z^{-1}$$

Then, from (6.31), the following order of the polynomial $S(z^{-1})$ is obtained

$$n_s = n_a + n_r - d - n_b = 2$$

Therefore, the polynomial $S(z^{-1})$ is defined as

$$S(z^{-1}) = s_0 + s_1 z^{-1} + s_2 z^{-2}$$

The Diophantine equation is written as

$$(1 + a_1 z^{-1} + a_2 z^{-2} + a_3 z^{-3})(1 + r_1 z^{-1}) + z^{-1}(b_0 + b_1 z^{-1})(s_0 + s_1 z^{-1} + s_2 z^{-2})$$
$$= (1 + \lambda a_1 z^{-1} + \lambda^2 a_2 z^{-2} + \lambda^3 a_2 z^{-3})$$

Let us choose a pole shift factor of 0.7 such that

$$T(z^{-1}) = 1 - 0.84 z^{-1} + 0.196 z^{-2} + 0.103 z^{-3}$$

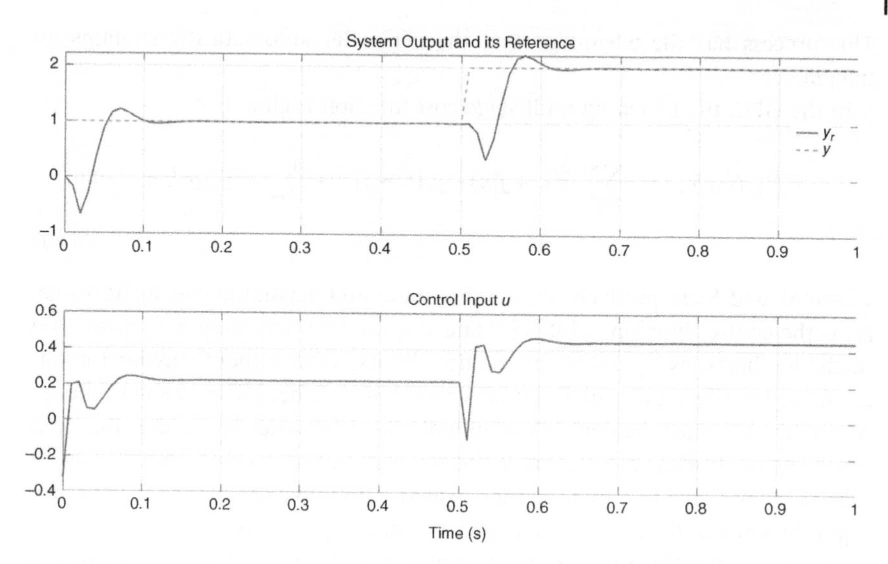

Figure 6.9 System response with pole shift control for the nonminimum phase system of Example 6.7.

Then the polynomials $R(z^{-1})$ and $S(z^{-1})$ the gain K_c are calculated as

$$R(z^{-1}) = 1 + 0.294z^{-1}$$

$$S(z^{-1}) = 0.132 - 0.165z^{-1} - 0.05z^{-2}$$

$$K_c = -2.44$$

Note that, since $T(z^{-1})|_{z=1} > 0$, $B(z^{-1})|_{z=1} > 0$, and $S(z^{-1})|_{z=1} < 0$, the gain K_c must be negative. The system response is shown in Figure 6.9. The closed-loop system is stable, unlike that of the case of MV.

6.3 Generalized Predictive Control (GPC)

GPC was first introduced by D. W. Clarke [5–7]. Since then, it has become very popular and is discussed in several books [3, 8–10]. Generally, a model that is known as the controlled autoregressive integrated moving average, or CARIMA, process is used for this controller design. This is given by

$$A(z^{-1})y(k) = z^{-d}B(z^{-1})u(k-1) + \frac{e(k)}{\Delta} \tag{6.37}$$

where

$$\Delta = 1 - z^{-1} \tag{6.38}$$

This process has the advantage that the controller automatically contains an integrator.

In the GPC, the following multistage cost function is chosen

$$J(N_1, N_2, N_u) = \sum_{j=N_1}^{N_2} [\hat{y}(k+j|k) - y_r(k+j)]^2 + \sum_{j=1}^{N_u} \lambda(j)[\Delta u(k+j-1)]^2 \tag{6.39}$$

where N_1 and N_2 respectively are the minimum and maximum costing horizons, N_u is the control horizon, while $\delta(j)$ and $\lambda(j)$ are two weighting sequences. The prediction horizons N_1 and N_2 are the time limits of the output following the reference input. When the system contains a dead time d, there is no reason to choose N_1 less than d since the control action will affect the system only after the dead time. The main idea is to minimize the cost function to compute future control sequences in such a way that the future plant output follows the future reference input. In some of the applications, the reference trajectory is known a priori, for instance in a repetitive process such as a chemical process. However, in power electronic circuits, this is often not true. A smooth first-order approximation of the output toward a known reference input can be formed, as discussed in [8].

Comparing (6.37) with (6.1), we find that

$$C(z^{-1}) = \frac{1}{\Delta} = \frac{1}{1 - z^{-1}}$$

The Diophantine equation of (6.8) is then modified as

$$1 = \Delta A(z^{-1})F(z^{-1}) + z^{-j}G(z^{-1}) \tag{6.40}$$

where

$$\begin{aligned} F(z^{-1}) &= f_0 + f_1 z^{-1} + \cdots + f_{j-1} z^{-j+1} \\ G(z^{-1}) &= g_0 + g_1 z^{-1} + \cdots + g_{n_g} z^{-n_g} \end{aligned} \tag{6.41}$$

Note that, in this case, the Diophantine equation does not depend on d but on the prediction horizon value j. The order of $\Delta A(z^{-1})$ is $n_a + 1$. Therefore, to have non-redundant coefficients, the order of the polynomial G is obtained by modifying (6.10) as

$$\begin{aligned} &(n_a + 1) + (j - 1) = j + n_g \\ &\Rightarrow n_g = n_a \end{aligned} \tag{6.42}$$

Multiplication of (6.37) by $\Delta F(z^{-1})z^j$ results in

$$\Delta A(z^{-1})F(z^{-1})y(k+j) = \Delta F(z^{-1})B(z^{-1})u(k+j-d-1) + F(z^{-1})e(k+j) \tag{6.43}$$

Substituting (6.40) in (6.43), we get

$$[1-z^{-j}G(z^{-1})]y(k+j) = F(z^{-1})B(z^{-1})\Delta u(k+j-d-1) + F(z^{-1})e(k+j)$$

The above equation can be rewritten as

$$y(k+j) = G(z^{-1})y(k) + F(z^{-1})B(z^{-1})\Delta u(k+j-d-1) + F(z^{-1})e(k+j)$$

Since the future values of $e(k)$ cannot be predicted a priori, the output prediction equation is then

$$\hat{y}(k+j|k) = G(z^{-1})y(k) + H(z^{-1})\Delta u(k+j-d-1) \tag{6.44}$$

where $H(z^{-1}) = F(z^{-1})B(z^{-1})$.

From (6.41) notice that $F(z^{-1})$ depends on the prediction horizon and changes with the value of j. Consequently, the solution of the Diophantine equation of (6.40) changes with j. Therefore, we need to evaluate $F(z^{-1})$ and $G(z^{-1})$ as j changes. Let us denote these polynomials as $F_j(z^{-1})$ and $G_j(z^{-1})$ for $j = 1, 2, \ldots$, such that

$$\left.\begin{array}{l} F_j(z^{-1}) = 1 + f_{j,1}z^{-1} + \cdots + f_{j,j-1}z^{-j+1} \\ G_j(z^{-1}) = g_{j,0} + g_{j,1}z^{-1} + \cdots + g_{j,n_a}z^{-n_a} \end{array}\right\} j = 1, 2, \ldots \tag{6.45}$$

Now consider the following example.

Example 6.8 Consider a CARIMA process of (6.37) with

$$A(z^{-1}) = 1 - 0.9z^{-1} \text{ and } B(z^{-1}) = 0.5 + 0.2z^{-1}$$

It can be seen that $d = 0$. Choosing $N_1 = 1$ and $N_2 = N_u = 3$, we define

$$A_\delta(z^{-1}) = \Delta A(z^{-1}) = 1 + a_{\delta 1}z^{-1} + a_{\delta 2}z^{-2}$$
$$G_j(z^{-1}) = g_{j,0} + g_{j,1}z^{-1}$$

where $a_{\delta 1} = -1.9$ and $a_{\delta 2} = 0.9$.

For $j = 1$, we have

$$F_1(z^{-1}) = 1$$

Therefore, the Diophantine equation of (6.40) is

$$1 = 1 + a_{\delta 1}z^{-1} + a_{\delta 2}z^{-2} + g_{1,0}z^{-1} + g_{1,1}z^{-2}$$

The solution of the above equation produces

$$G_1(z^{-1}) = -a_{\delta 1} - a_{\delta 2}z^{-1}$$

For $j = 2$, we have

$$F_2(z^{-1}) = 1 + f_{2,1}z^{-1}$$

and the Diophantine equation is

$$1 = (1 + f_{2,1}z^{-1})(1 + a_{\delta 1}z^{-1} + a_{\delta 1}z^{-2}) + g_{2,0}z^{-2} + g_{2,1}z^{-3}$$

The above equation can be rewritten as

$$1 = 1 + (f_{2,1} + a_{\delta 1})z^{-1} + (a_{\delta 2} + a_{\delta 1}f_{2,1} + g_{2,0})z^{-2} + (a_{\delta 2}f_{2,1} + g_{2,1})z^{-3}$$

Solving this, we get $f_{2,1} = -a_{\delta 1}$ and therefore

$$F_2(z^{-1}) = 1 - a_{\delta 1}z^{-1} \text{ and } G_2(z^{-1}) = -(a_{\delta 2} - a_{\delta 1}^2) + a_{\delta 1}a_{\delta 2}z^{-1}$$

Similarly, for $j = 3$, we have

$$F_3(z^{-1}) = 1 + f_{3,1}z^{-1} + f_{3,2}z^{-2}$$

Therefore, the Diophantine equation of (6.40) results in

$$1 = (1 + f_{3,1}z^{-1} + f_{3,2}z^{-2})(1 + a_{\delta 1}z^{-1} + a_{\delta 2}z^{-2}) + g_{3,0}z^{-3} + g_{3,1}z^{-4}$$

The above equation is expanded as

$$\begin{aligned} 1 = 1 &+ (f_{3,1} + a_{\delta 1})z^{-1} + (a_{\delta 2} + a_{\delta 1}f_{3,1} + f_{3,2})z^{-2} \\ &+ (a_{\delta 2}f_{3,1} + a_{\delta 1}f_{3,2} + g_{3,1})z^{-3} + (a_{\delta 2}f_{3,2} + g_{3,2})z^{-4} \end{aligned}$$

Solving this we get $f_{3,1} = -a_{\delta 1}$, $f_{3,2} = -a_{\delta 2} + a_{\delta 1}^2$ and therefore

$$F_3(z^{-1}) = 1 - a_{\delta 1}z^{-1} + (a_{\delta 1}^2 - a_{\delta 2})z^{-2} \text{ and } G_3(z^{-1}) = -(a_{\delta 2}f_{3,1} + a_{\delta 1}f_{3,2}) - a_{\delta 2}f_{3,2}z^{-1}$$

The polynomials F and G can then be summarized as

$$\begin{aligned} F_1(z^{-1}) &= 1 & G_1(z^{-1}) &= 1.9 - 0.9z^{-1} \\ F_2(z^{-1}) &= 1 + 1.9z^{-1} & G_2(z^{-1}) &= 2.71 - 1.71z^{-1} \\ F_3(z^{-1}) &= 1 + 1.9z^{-1} + 2.71z^{-3} & G_3(z^{-1}) &= 3.439 - 2.439z^{-1} \end{aligned}$$

Once polynomials F_j are obtained, the polynomials H_j are obtained as

$$\begin{aligned} H_1(z^{-1}) &= B(z^{-1})F_1(z^{-1}) = 0.5 + 0.2z^{-1} \\ H_2(z^{-1}) &= B(z^{-1})F_2(z^{-1}) = 0.5 + 1.15z^{-1} + 0.38z^{-2} \\ H_3(z^{-1}) &= B(z^{-1})F_3(z^{-1}) = 0.5 + 1.15z^{-1} + 1.735z^{-2} + 0.542z^{-3} \end{aligned}$$

Note that the polynomials in the above example are derived assuming a relatively smaller prediction horizon. However, there are processes (e.g. chemical processes) that require a much larger prediction horizon. In that case, the polynomial can also be derived recursively, which is discussed in [9].

As in the case of d-step-ahead MV control, a set of control input signals over the control horizon N must be obtained for the minimization of (6.39). Noting that the system contains a dead time of d, we define the horizons as

$$\begin{aligned} N_1 &= d + 1 \\ N_2 &= d + N \\ N_u &= N \end{aligned} \tag{6.46}$$

The following optimal predictions for the control horizon are then obtained from (6.44)

$$\begin{aligned} \hat{y}(k+d+1|k) &= G_{d+1}(z^{-1})y(k) + H_{d+1}(z^{-1})\Delta u(k) \\ \hat{y}(k+d+2|k) &= G_{d+2}(z^{-1})y(k) + H_{d+2}(z^{-1})\Delta u(k+1) \\ &\vdots \\ \hat{y}(k+d+N|k) &= G_{d+N}(z^{-1})y(k) + H_{d+N}(z^{-1})\Delta u(k+N-1) \end{aligned} \tag{6.47}$$

Now consider the polynomials H obtained in Example 6.7. From these, we can write

$$\begin{aligned} H_1(z^{-1})\Delta u(k) &= 0.5\Delta u(k) + 0.2\Delta u(k-1) \\ H_2(z^{-1})\Delta u(k+1) &= 0.5\Delta u(k+1) + 1.15\Delta u(k) + 0.38\Delta u(k-1) \\ H_3(z^{-1})\Delta u(k+2) &= 0.5\Delta u(k+2) + 1.15\Delta u(k+1) \\ &\quad + 1.73\Delta u(k+1) + 0.542\Delta u(k-1) \end{aligned}$$

Let us define

$$H_3(z^{-1}) = h_0 + h_1 z^{-1} + h_2 z^{-2} + h_3 z^{-3}$$

where

$$h_0 = 0.5,\ h_1 = 1.15,\ h_2 = 1.73, \text{ and } h_3 = 0.542.$$

Then we can write

$$\begin{aligned} H_1(z^{-1})\Delta u(k) &= h_0\Delta u(k) + 0.2\Delta u(k-1) \\ H_2(z^{-1})\Delta u(k+1) &= h_0\Delta u(k+1) + h_1\Delta u(k) + 0.38\Delta u(k-1) \\ H_3(z^{-1})\Delta u(k+2) &= h_0\Delta u(k+2) + h_1\Delta u(k+1) + h_2\Delta u(k+1) \\ &\quad + h_3\Delta u(k-1) \end{aligned}$$

such that

$$\begin{aligned} H_1(z^{-1}) &= h_0 + (H_1(z^{-1}) - h_0)z \\ H_2(z^{-1}) &= B(z^{-1})F_2(z^{-1}) = 0.5 + 1.15z^{-1} + 0.38z^{-2} \\ H_3(z^{-1}) &= B(z^{-1})F_3(z^{-1}) = 0.5 + 1.15z^{-1} + 1.735z^{-2} + 0.542z^{-3} \end{aligned}$$

The equations above can be written in the following matrix–vector form

$$\begin{bmatrix} H_1(z^{-1})\Delta u(k) \\ H_2(z^{-1})\Delta u(k+1) \\ H_3(z^{-1})\Delta u(k+2) \end{bmatrix} = \begin{bmatrix} h_0 & 0 & 0 \\ h_1 & h_0 & 0 \\ h_2 & h_1 & h_0 \end{bmatrix} \begin{bmatrix} \Delta u(k) \\ \Delta u(k+1) \\ \Delta u(k+2) \end{bmatrix} + \mathbf{\Gamma}\Delta u(k-1)$$

where

$$\mathbf{\Gamma} = \begin{bmatrix} 0.2 \\ 0.36 \\ 0.488 \end{bmatrix} = \begin{bmatrix} (H_1(z^{-1}) - h_0)z \\ (H_2(z^{-1}) - h_0 - h_1 z^{-1})z^2 \\ (H_2(z^{-1}) - h_0 - h_1 z^{-1} - h_2 z^{-2})z^3 \end{bmatrix}$$

Therefore, the last term on the right-hand side of (6.47) can be written as

$$\begin{bmatrix} H_{d+1}(z^{-1})\Delta u(k) \\ H_{d+2}(z^{-1})\Delta u(k+1) \\ \vdots \\ H_{d+N}(z^{-1})\Delta u(k+N-1) \end{bmatrix} = \begin{bmatrix} h_0 & 0 & \cdots & 0 \\ h_1 & h_0 & \cdots & 0 \\ \vdots & \vdots & \vdots & \vdots \\ h_{N-1} & h_{N-2} & \cdots & h_0 \end{bmatrix} \begin{bmatrix} \Delta u(k) \\ \Delta u(k+1) \\ \vdots \\ \Delta u(k+N-1) \end{bmatrix} + \mathbf{\Gamma}\Delta u(k-1) \tag{6.48}$$

where

$$\mathbf{\Gamma} = \begin{bmatrix} (H_{d+1}(z^{-1}) - h_0)z \\ (H_{d+2}(z^{-1}) - h_0 - h_1 z^{-1})z^2 \\ \vdots \\ (H_{d+N}(z^{-1}) - h_0 - h_1 z^{-1} - \cdots - h_{N-1} z^{-N+1})z^N \end{bmatrix}$$

The following vectors and matrix can then be defined as

$$\hat{\mathbf{Y}} = \begin{bmatrix} \hat{y}(k+d+1|k) \\ \hat{y}(k+d+2|k) \\ \vdots \\ \hat{y}(k+d+N|k) \end{bmatrix}, \quad \mathbf{\Delta u} = \begin{bmatrix} \Delta u(k) \\ \Delta u(k+1) \\ \vdots \\ \Delta u(k+N-1) \end{bmatrix},$$

$$\mathbf{H} = \begin{bmatrix} h_0 & 0 & \cdots & 0 \\ h_1 & h_0 & \cdots & 0 \\ \vdots & \vdots & \vdots & \vdots \\ h_{N-1} & h_{N-2} & \cdots & h_0 \end{bmatrix}, \quad \mathbf{f} = \begin{bmatrix} G_{d+1}(z^{-1}) \\ G_{d+2}(z^{-1}) \\ \vdots \\ G_{d+N}(z^{-1}) \end{bmatrix} y(k) + \mathbf{\Gamma}\Delta u(k-1)$$

Then, using (6.47), (6.48) can be rewritten in a compact form as

$$\widehat{\mathbf{Y}} = \mathbf{H}\Delta\mathbf{u} + \mathbf{f} \tag{6.49}$$

Defining

$$\mathbf{Y_r} = [y_r(k+d+1)\ y_r(k+d+2)\ \cdots\ y_r(k+d+N)]^T$$

the objective function (6.39) is given as

$$J = (\mathbf{H}\Delta\mathbf{u} + \mathbf{f} - \mathbf{Y_r})^T(\mathbf{H}\Delta\mathbf{u} + \mathbf{f} - \mathbf{Y_r}) + \lambda\Delta\mathbf{u}^T\Delta\mathbf{u} \tag{6.50}$$

By taking the derivative of J with respect to $\mathbf{U}$ and equating it to zero, the control law is obtained as

$$\Delta\mathbf{u} = -\left(\mathbf{H}^T\mathbf{H} + \lambda\mathbf{I}\right)^{-1}\mathbf{H}^T(\mathbf{f} - \mathbf{Y_r}) \tag{6.51}$$

Example 6.9 We shall continue with Example 6.8 for which we have already derived polynomials F, G, and H. Therefore, we have

$$\mathbf{H} = \begin{bmatrix} 0.5 & 0 & 0 \\ 1.15 & 0.5 & 0 \\ 1.735 & 1.15 & 0.5 \end{bmatrix}, \quad \mathbf{f} = \begin{bmatrix} 1.9y(k) - 0.9y(k-1) + 0.2\Delta u(k-1) \\ 2.71y(k) - 1.71y(k-1) + 0.38\Delta u(k-1) \\ 3.439y(k) - 2.439y(k-1) + 0.542\Delta u(k-1) \end{bmatrix}$$

Let us choose $\lambda = 0.75$. Then

$$\left(\mathbf{H}^T\mathbf{H} + \lambda\mathbf{I}\right)^{-1}\mathbf{H}^T = \begin{bmatrix} 0.2061 & 0.2597 & 0.1667 \\ -0.2143 & -0.0191 & 0.2597 \\ -0.0556 & -0.2143 & 0.2061 \end{bmatrix}$$

Note that only the first row of (6.51) is required to calculate $\Delta u(k)$. Therefore, the control law is given by

$$\begin{aligned} \Delta u(k) &= -0.2061 \times [\mathbf{f}(1) - y_r(k+1)] - 0.2597[\mathbf{f}(2) - y_r(k+2)] \\ &\quad -0.1667[\mathbf{f}(3) - y_r(k+3)] \\ &= -1.6685y(k) + 1.036y(k-1) - 0.2302\Delta u(k-1) \\ &\quad +0.2061y_r(k+1) + 0.2597y_r(k+2) + 0.1667y_r(k+3) \end{aligned}$$

Since $\Delta u(k) = u(k) - u(k-1)$, the above equation can be rewritten as

$$\begin{aligned} u(k) = &-1.6685y(k) + 1.036y(k-1) + 0.7698u(k-1) + 0.2302u(k-2) \\ &+ 0.2061y_r(k+1) + 0.2597y_r(k+2) + 0.1667y_r(k+3) \end{aligned}$$

The system response to a unit step is shown in Figure 6.10, assuming that the system starts from rest.

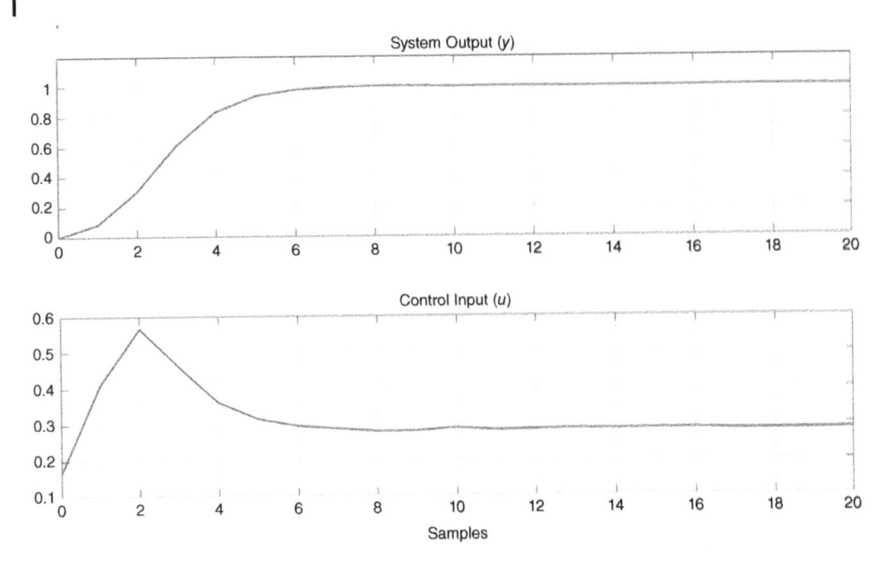

Figure 6.10 System response with GPC of Example 6.7.

The control law above is now defined as

$$B'(z^{-1})u(k) = A'(z^{-1})y(k) + C'(z^{-1})y_r(k+3)$$

where

$$\begin{aligned} A'(z^{-1}) &= -1.6685 + 1.036z^{-1} \\ B'(z^{-1}) &= 1 - 0.7698z^{-1} - 0.2302z^{-2} \\ C'(z^{-1}) &= 0.2061 + 0.2597z^{-1} + 0.1667z^{-2} \end{aligned}$$

Substituting this in the plant model, we get

$$A(z^{-1})y(k+1) = B(z^{-1})u(k) = \frac{B(z^{-1})}{B'(z^{-1})}\left[A'(z^{-1})y(k) + C'(z^{-1})y_r(k+3)\right]$$

$$\left[A(z^{-1})B'(z^{-1}) - B(z^{-1})A'(z^{-1})z^{-1}\right]y(k+1) = B(z^{-1})C'(z^{-1})y_r(k+3)$$

The solution of the above equation yields

$$\begin{aligned} &(1 - 0.8355z^{-1} + 0.2782z^{-2})y(k+1) \\ &\quad = (0.1031 + 0.1711z^{-1} + 0.1353z^{-2} + 0.0333z^{-3})y_r(k+3) \end{aligned}$$

The DC gain of the above equation is 1, as is with all GPC control designs due to the presence of Δ term with the controller. However, the closed-loop system behavior depends on the choice of parameters such as N_1, N_2, N_u, and λ. This is hard to predict [8].

6.3.1 Simplified GPC Computation

To compute GPC control law, the Diophantine equation must be solved to obtain the matrix **H** and the vector **f**, which is not a trivial task for longer control horizons. Note that from Example 6.9 we only needed to compute the first column of the matrix **H**, as this is a lower triangular matrix, where the elements of the first column are propagated in the triangular elements. The matrix **H** is composed of the plant step response elements and can be computed from [9]

$$h_j = -\sum_{i=1}^{j} a_i h_{j-1} + \sum_{i=0}^{j-1} b_i, \quad h_k = 0, \quad \forall k < 0 \tag{6.52}$$

From Examples 6.8 and 6.9, we know that $A(z^{-1}) = 1 - 0.9z^{-1}$ and $B(z^{-1}) = 0.5 + 0.2z^{-1}$, such that $a_1 = -0.9$, $b_0 = 0.5$, and $b_1 = 0.2$. Then

$$h_0 = b_0 = 0.5$$
$$h_1 = -a_1 h_0 + b_0 + b_1 = 0.9 \times 0.5 + 0.5 + 0.2 = 1.15$$
$$h_2 = -a_1 h_1 + b_0 + b_1 = 0.9 \times 1.15 + 0.5 + 0.2 = 1.735$$

The system output equation is

$$y(k) = 0.9y(k-1) + 0.5u(k-1) + 0.2u(k-2)$$

Shifting the above equation by one sampling instant, we get

$$y(k+1) = 0.9y(k) + 0.5u(k) + 0.2u(k-1)$$

Subtraction of $y(k)$ from $y(k+1)$ produces

$$y(k+1) - y(k) = 0.9y(k) - 0.9y(k-1) + 0.5[u(k) - u(k-1)] + 0.2[u(k-1) - u(k-2)]$$

Rearranging the above equation and noting that $\Delta u(k) = u(k) - u(k-1)$, we have

$$y(k+1) = 1.9y(k) - 0.9y(k-1) + 0.5\Delta u(k) + 0.2\Delta u(k-1)$$

The control increments will now be considered before the instant k occurs to compute the elements of **f**. Therefore, we discard $\Delta u(k)$ in the above equation to define

$$f(k+1) = 1.9y(k) - 0.9y(k-1) + 0.2\Delta u(k-1)$$

In a similar way, the following equations are obtained

$$\begin{aligned} f(k+2) &= 1.9f(k+1) - 0.9y(k) = 1.9 \times [1.9y(k) - 0.9y(k-1) + 0.2\Delta u(k-1)] \\ &\quad - 0.9y(k) \\ &= 2.71y(k) - 1.71y(k-1) + 0.38\Delta u(k-1) \end{aligned}$$

$$\begin{aligned} f(k+3) &= 1.9f(k+2) - 0.9f(k+1) \\ &= 3.439y(k) - 2.439y(k-1) + 0.542\Delta u(k-1) \end{aligned}$$

Defining $\mathbf{f} = [f(k+1) \quad f(k+2) \quad f(k+3)]^T$, we get the same results as those given in Example 6.9.

6.4 Adaptive Control

An adaptive quantity is able to adapt to the changes in its environment. In control literature, an adaptive controller is defined as "a controller with adjustable parameters and a mechanism for adjusting the parameters" [3]. A SISO system of the form shown in Figure 6.11 is considered in this section. The input–output relationship is then given by

$$Y(z) = \frac{1}{1 + G_1(z)G_C(z)}[G_1(z)G_C(z)Y_r + G_2(z)E(z)] \tag{6.53}$$

where y_r is the reference input and e is a disturbance input. The stability of the closed-loop system will be governed by the zeros of the polynomial $1 + G_1(z)G_C(z) = 0$.

The purpose of the control is to minimize the error $e_y(k) = y_r(k) - y(k)$, as $k \to \infty$, as well as to provide a well-damped transient behavior. For a deterministic $e(k)$, the feedback control can be designed by minimizing a performance index, such as the sum of the square of $e_y(k)$ as k varies from 0 to ∞. However, in adaptive control literature, the term $e(k)$ is often assumed to be stochastic and therefore a stochastic optimal control problem must be considered.

In general, two types of adaptive controls are reported in the literature: model reference adaptive control (MRAC) and self-tuning control or regulator. These are

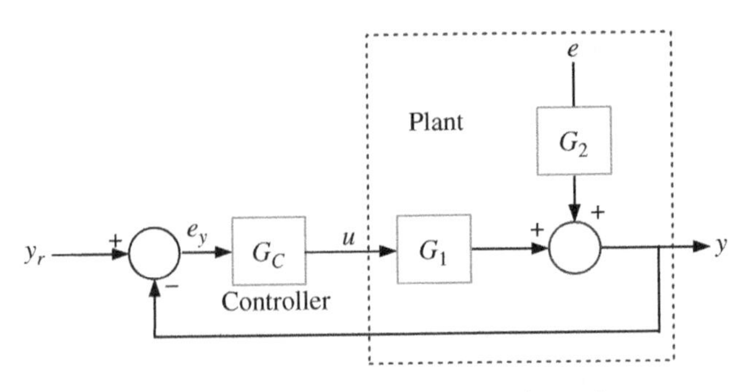

Figure 6.11 Typical SISO system with a disturbance input.

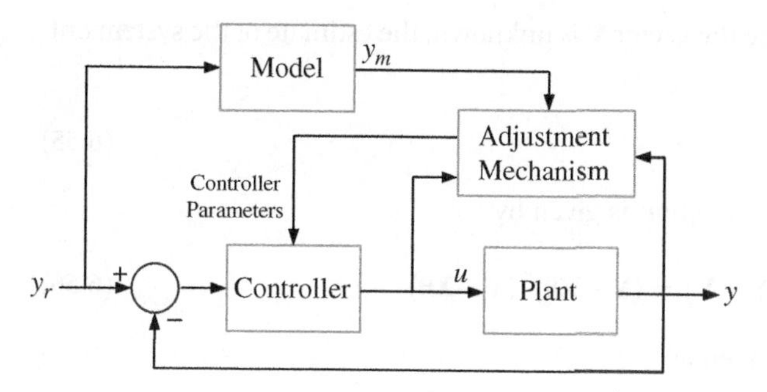

Figure 6.12 Model reference adaptive control.

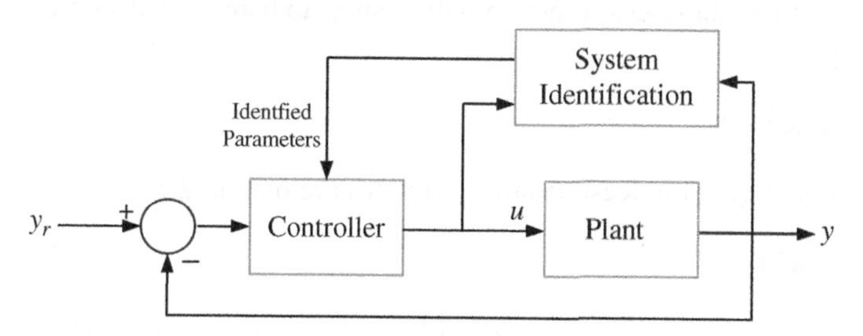

Figure 6.13 Self-tuning adaptive control.

respectively shown in Figures 6.12 and 6.13. In MRAC, the performance specification is given in terms of a reference model with an output y_m and input y_r [10]. The adjustment mechanism then determines the controller parameters such that the error between y_m and y asymptotically decreases. A self-tuning control, however, is a direct method which includes a block called system identification. This block estimates the transfer function parameters of the plant. The controller is then designed based on the identified parameters. While the identification process is almost the same for different types of control design, the control strategies vary. In this chapter, we concentrate on different types of self-tuning controllers.

6.5 Least-squares Estimation

Consider a batch model, given by

$$\mathbf{Y} = \mathbf{X}\boldsymbol{\theta} + \mathbf{V} \tag{6.54}$$

where $\mathbf{Y} \in \Re^n$ is the observation vector, $\boldsymbol{\theta} \in \Re^m$ is the parameter vector, $\mathbf{V} \in \Re^n$ is a vector random observation errors, and the matrix $\mathbf{X} \in \Re^{n \times m}$ is called the observation matrix. Since the vector $\mathbf{V}$ is unknown, the estimate of the system output is given by

$$\hat{\mathbf{Y}} = \mathbf{X}\boldsymbol{\theta} \tag{6.55}$$

The least-squares cost function is given by

$$J_{LS} = \left(\mathbf{Y} - \hat{\mathbf{Y}}\right)^T \left(\mathbf{Y} - \hat{\mathbf{Y}}\right) = (\mathbf{Y} - \mathbf{X}\boldsymbol{\theta})^T (\mathbf{Y} - \mathbf{X}\boldsymbol{\theta}) \tag{6.56}$$

Equation (6.56) is expanded as

$$J_{LS} = \mathbf{Y}^T\mathbf{Y} - \boldsymbol{\theta}^T\mathbf{X}^T\mathbf{Y} - \mathbf{Y}^T\mathbf{X}\boldsymbol{\theta} + \boldsymbol{\theta}^T\mathbf{X}^T\mathbf{X}\boldsymbol{\theta}$$

Taking a derivative of the above cost function with respect to $\boldsymbol{\theta}$ and equating the result to zero, we get

$$-2\mathbf{X}^T\mathbf{Y} + \mathbf{X}^T\mathbf{X}\hat{\boldsymbol{\theta}} = 0$$

Rearranging the above equation, least-squares estimates are obtained as

$$\hat{\boldsymbol{\theta}} = \left(\mathbf{X}^T\mathbf{X}\right)^{-1}\mathbf{X}^T\mathbf{Y} \tag{6.57}$$

Example 6.10 An experiment is performed to determine gravitational acceleration by dropping a steel ball from a tall building. Assuming that there is no wind drag, the length of the fall is measured at equally spaced 0.1-second intervals. However, these measurements are not accurate, and they are corrupted by sensor noise. The length of fall (l) is obtained from the following equation

$$l = \frac{1}{2}gt^2 + e$$

where g is the gravitational acceleration and e is a zero-mean sensor noise that has a variance of 1. The measured length obtained from the above equation and the actual length with the value of g being 9.8 m/s^2 are shown in Figure 6.14, where the measurements are taken every 100 ms.

Let us define a vector containing the length of the fall as $\mathbf{L}$ and another vector containing the terms $t^2/2$ as $\mathbf{T}$. Then from (6.57), the gravitational acceleration can be estimated as

$$\hat{g} = \frac{\mathbf{T}^T\mathbf{L}}{\mathbf{T}^T\mathbf{T}} \text{ m/s}^2$$

Table 6.2 lists the values of the estimates for different total number of samples, from which it is evident that the estimated value becomes more accurate as the number of samples increases.

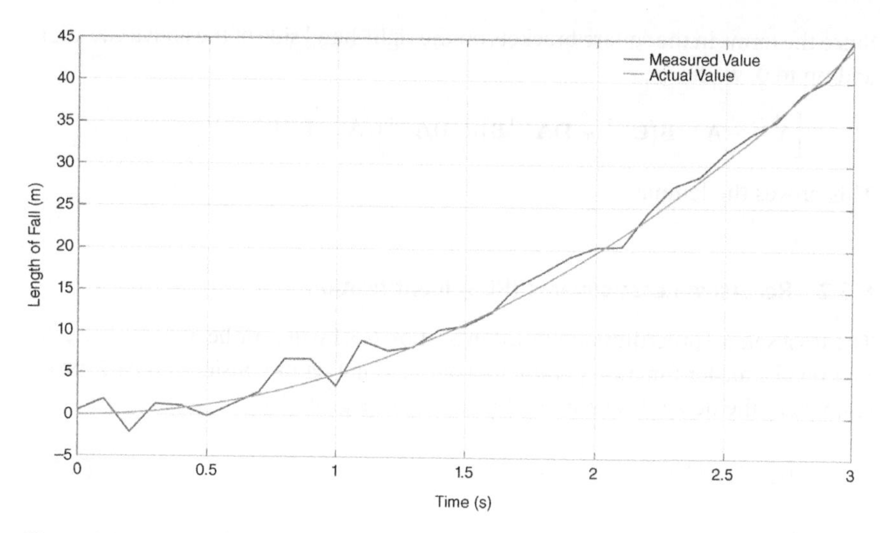

Figure 6.14 The actual length of fall and the measured length of fall.

Table 6.2 Estimated values for different total number of samples.

Time (s)	Total number of samples	Estimate $\hat{g}$ (m/s^2)
2	21	10.447
5	51	9.805
10	101	9.7996
50	501	9.8

6.5.1 Matrix Inversion Lemma

Given two nonsingular matrices, **A** and **C**, the inverse of the matrix **A** + **BCD** is given by

$$(\mathbf{A}+\mathbf{BCD})^{-1}=\mathbf{A}^{-1}-\mathbf{A}^{-1}\mathbf{B}\left(\mathbf{C}^{-1}+\mathbf{DA}^{-1}\mathbf{B}\right)^{-1}\mathbf{DA}^{-1} \tag{6.58}$$

The proof of the lemma is rather simple. We post-multiply the right-hand side of (6.58) by the matrix **A** + **BCD** to get

$$\begin{aligned}
&\left[\mathbf{A}^{-1}-\mathbf{A}^{-1}\mathbf{B}\left(\mathbf{C}^{-1}+\mathbf{DA}^{-1}\mathbf{B}\right)^{-1}\mathbf{DA}^{-1}\right](\mathbf{A}+\mathbf{BCD})\\
&=\mathbf{I}+\mathbf{A}^{-1}\mathbf{BCD}-\mathbf{A}^{-1}\mathbf{B}\left(\mathbf{C}^{-1}+\mathbf{DA}^{-1}\mathbf{B}\right)^{-1}\mathbf{D}\\
&\quad-\mathbf{A}^{-1}\mathbf{B}\left(\mathbf{C}^{-1}+\mathbf{DA}^{-1}\mathbf{B}\right)^{-1}\mathbf{DA}^{-1}\mathbf{BCD}\\
&=\mathbf{I}+\mathbf{A}^{-1}\mathbf{B}\left(\mathbf{C}^{-1}+\mathbf{DA}^{-1}\mathbf{B}\right)^{-1}\left[\left(\mathbf{C}^{-1}+\mathbf{DA}^{-1}\mathbf{B}\right)\mathbf{CD}-\mathbf{D}-\mathbf{DA}^{-1}\mathbf{BCD}\right]\\
&=\mathbf{I}+\mathbf{A}^{-1}\mathbf{B}\left(\mathbf{C}^{-1}+\mathbf{DA}^{-1}\mathbf{B}\right)^{-1}\left[\mathbf{D}+\mathbf{DA}^{-1}\mathbf{BCD}-\mathbf{D}-\mathbf{DA}^{-1}\mathbf{BCD}\right]
\end{aligned}$$

Since the terms in the square brackets on the right-hand side of the above equation add up to 0, we have

$$\left[\mathbf{A}^{-1} - \mathbf{A}^{-1}\mathbf{B}\left(\mathbf{C}^{-1} + \mathbf{D}\mathbf{A}^{-1}\mathbf{B}\right)^{-1}\mathbf{D}\mathbf{A}^{-1}\right](\mathbf{A} + \mathbf{BCD}) = \mathbf{I}$$

This proves the lemma.

6.5.2 Recursive Least-squares (RLS) Identification

The least square procedure discussed in Section 6.5.1 will now be used for the estimation of transfer function parameters online. First let us consider an ARX process, where the delay d is assumed to be 1. This is given by

$$A\left(z^{-1}\right)y(k) = z^{-1}B\left(z^{-1}\right)u(k) + e(k) \tag{6.59}$$

Equation (6.59) can be written in a difference equation form as

$$\begin{aligned} y(k) = & -a_1 y(k-1) - a_2 y(k-2) + \cdots - a_{n_a} y(k-n_a) \\ & + b_0 u(k-1) + b_1 u(k-2) + \cdots + b_{n_b} u(k-n_b) + e(k) \end{aligned} \tag{6.60}$$

Defining a parameter vector $\boldsymbol{\theta}$ and a regression vector $\boldsymbol{\varphi}$ as

$$\begin{aligned} \boldsymbol{\theta}^T &= [a_1 \quad a_2 \quad \cdots \quad a_{n_a} \quad b_0 \quad b_1 \quad \cdots \quad b_{n_b}] \\ \boldsymbol{\varphi}^T(k) &= [-y(k-1) \quad -y(k-2) \quad \cdots \quad -y(k-n_a) \\ & \qquad u(k-1) \quad u(k-1) \quad \cdots \quad u(k-n_b)] \end{aligned} \tag{6.61}$$

Equation (6.60) is written in a compact form as

$$y(k) = \boldsymbol{\theta}^T\boldsymbol{\varphi}(k) + e(k) = \boldsymbol{\varphi}^T(k)\boldsymbol{\theta} + e(k) \tag{6.62}$$

In (6.60), the measured and control input values on the right-hand side are known at time instant $k-1$. However, since $e(k)$ is uncorrelated, its value at instant k is unknown. Therefore, $y(k)$ in (6.62) will be replaced by its estimate, which is given by

$$\hat{y}(k|k-1) = \boldsymbol{\theta}^T\boldsymbol{\varphi}(k) = \boldsymbol{\varphi}^T(k)\boldsymbol{\theta} \tag{6.63}$$

In the least squares, the following cost function is chosen

$$J_{LS} = \frac{1}{k}\sum_{i=1}^{k}\{y(k) - \hat{y}(k|k-1)\}^2 = \frac{1}{k}\sum_{i=1}^{k}\{y(k) - \boldsymbol{\varphi}^T(k)\boldsymbol{\theta}\}^2 \tag{6.64}$$

The cost function is then minimized with respect to the parameter vector to obtain its estimate. Let us define the following two vectors

$$\mathbf{Y}(k) = \begin{bmatrix} y(1) \\ y(2) \\ \vdots \\ y(k) \end{bmatrix}, \quad \mathbf{X}(k) = \begin{bmatrix} \boldsymbol{\varphi}^T(1) \\ \boldsymbol{\varphi}^T(2) \\ \vdots \\ \boldsymbol{\varphi}^T(k) \end{bmatrix} \tag{6.65}$$

Then the least-squares cost function is written as

$$J_{LS} = \frac{1}{k}[Y(k) - \mathbf{X}(k)\boldsymbol{\theta}]^T[\mathbf{Y}(k) - \mathbf{X}(k)\boldsymbol{\theta}] \tag{6.66}$$

Note that (6.66) has a similar form to (6.56). Therefore, the cost function of (6.66) is minimized by taking its derivative with respect to $\boldsymbol{\theta}$ and equating it to zero to obtain the estimates in the same fashion as (6.57). The estimates at time k are then given by

$$\hat{\boldsymbol{\theta}}(k) = \left[\mathbf{X}^T(k)\mathbf{X}(k)\right]^{-1}\mathbf{X}^T(k)\mathbf{Y}(k) \tag{6.67}$$

Note that the inverse of the matrix product given inside the square brackets can be obtained by using the matrix inversion lemma of (6.58).

Let us define a matrix $\mathbf{P} \in \Re^{(n+m)\times(n+m)}$ as

$$\mathbf{P}(k) = \left[\mathbf{X}^T(k)\mathbf{X}(k)\right]^{-1} \tag{6.68}$$

Noting from (6.65) that the matrix $\mathbf{X}(k)$ can be written as

$$\mathbf{X}(k) = \begin{bmatrix} \mathbf{X}(k-1) \\ \boldsymbol{\varphi}^T(k) \end{bmatrix}$$

$\mathbf{P}(k)$ is given by

$$\mathbf{P}(k) = \left[\mathbf{X}^T(k-1)\mathbf{X}(k-1) + \boldsymbol{\varphi}(k)\boldsymbol{\varphi}^T(k)\right]^{-1} = \left[\mathbf{P}(k-1) + \boldsymbol{\varphi}(k)\boldsymbol{\varphi}^T(k)\right]^{-1} \tag{6.69}$$

We shall now apply the matrix inversion lemma of (6.58) to (6.69) by denoting

$\mathbf{A} = \mathbf{P}^{-1}(k-1)$, $\mathbf{B} = \boldsymbol{\varphi}(k)$, $\mathbf{C} = 1$, and $\mathbf{D} = \varphi^T(k)$

The resultant $\mathbf{P}$ matrix at time k is given by the following recursive relationship

$$\mathbf{P}(k) = \mathbf{P}(k-1) - \frac{\mathbf{P}(k-1)\boldsymbol{\varphi}(k)\boldsymbol{\varphi}^T(k)\mathbf{P}(k-1)}{1 + \boldsymbol{\varphi}^T(k)\mathbf{P}(k-1)\boldsymbol{\varphi}(k)} \tag{6.70}$$

Let us now define a gain matrix as

$$\mathbf{K}(k) = \frac{\mathbf{P}(k-1)\boldsymbol{\varphi}(k)}{1 + \boldsymbol{\varphi}^T(k)\mathbf{P}(k-1)\boldsymbol{\varphi}(k)} \tag{6.71}$$

Then (6.70) can be rewritten as

$$\mathbf{P}(k) = \left[\mathbf{I} - \mathbf{K}(k)\boldsymbol{\varphi}^T(k)\right]\mathbf{P}(k-1) \tag{6.72}$$

Substituting (6.68) into (6.67), the estimates at time k are given by

$$\hat{\boldsymbol{\theta}}(k) = \mathbf{P}(k)\mathbf{X}^T(k)\mathbf{Y}(k) \tag{6.73}$$

Again $\mathbf{Y}(k)$ can be written as

$$\mathbf{Y}(k) = \begin{bmatrix} \mathbf{Y}(k-1) \\ y(k) \end{bmatrix}$$

Therefore

$$\mathbf{X}^T(k)\mathbf{Y}(k) = \begin{bmatrix} \mathbf{X}(k-1) \\ \boldsymbol{\varphi}^T(k) \end{bmatrix}^T \begin{bmatrix} \mathbf{Y}(k-1) \\ y(k) \end{bmatrix} = \mathbf{X}^T(k-1)\mathbf{Y}(k-1) + \boldsymbol{\varphi}(k)y(k)$$

Substitution of the above equation and (6.72) in (6.73) results in

$$\begin{aligned} \hat{\boldsymbol{\theta}}(k) &= \left[\mathbf{I} - \mathbf{K}(k)\boldsymbol{\varphi}^T(k)\right]\mathbf{P}(k-1)\left[\mathbf{X}^T(k-1)\mathbf{Y}(k-1) + \boldsymbol{\varphi}(k)y(k)\right] \\ &= \mathbf{P}(k-1)\mathbf{X}^T(k-1)\mathbf{Y}(k-1) + \mathbf{P}(k-1)\boldsymbol{\varphi}(k)y(k) \\ &\quad - \mathbf{K}(k)\boldsymbol{\varphi}^T(k)\mathbf{P}(k-1)\left[\mathbf{X}^T(k-1)\mathbf{Y}(k-1) + \boldsymbol{\varphi}(k)y(k)\right] \end{aligned} \tag{6.74}$$

Now, the following two equations are obtained from (6.73) and (6.71)

$$\begin{aligned} \hat{\boldsymbol{\theta}}(k-1) &= \mathbf{P}(k-1)\mathbf{X}^T(k-1)\mathbf{Y}(k-1) \\ \mathbf{P}(k-1)\boldsymbol{\varphi}(k) &= \mathbf{K}(k) + \mathbf{K}(k)\boldsymbol{\varphi}^T(k)\mathbf{P}(k-1)\boldsymbol{\varphi}(k) \end{aligned}$$

Substituting the above two equations in (6.74), we get

$$\begin{aligned} \hat{\boldsymbol{\theta}}(k) = \hat{\boldsymbol{\theta}}(k-1) &+ \mathbf{K}(k)y(k) + \mathbf{K}(k)\boldsymbol{\varphi}^T(k)\mathbf{P}(k-1)\boldsymbol{\varphi}(k)y(k) \\ &- \mathbf{K}(k)\boldsymbol{\varphi}^T(k)\hat{\boldsymbol{\theta}}(k-1) - \mathbf{K}(k)\boldsymbol{\varphi}^T(k)\mathbf{P}(k-1)\boldsymbol{\varphi}(k)y(k) \end{aligned}$$

Solving and rearranging the above equation, the recursive least-squares (RLS) estimates are obtained as

$$\hat{\boldsymbol{\theta}}(k) = \hat{\boldsymbol{\theta}}(k-1) + \mathbf{K}(k)\left[y(k) - \hat{\boldsymbol{\theta}}^T(k-1)\boldsymbol{\varphi}(k)\right] \tag{6.75}$$

The step-by-step recursive relationships for RLS are given in Table 6.3. In the equation of $\mathbf{K}(k)$, the term $0 < \gamma < 1$ is taken as a forgetting factor that is used to discount the old measurements. The implication of this is discussed in [11].

Table 6.3 RLS algorithm.

Initial conditions:

$\hat{\boldsymbol{\theta}}(0) = 0, \quad \mathbf{P}(0) = \alpha\mathbf{I}, \textit{for } \alpha >> 0$

Then, at every k, execute the following equations sequentially

$$\varepsilon(k) = y(k) - \hat{\boldsymbol{\theta}}^T(k-1)\boldsymbol{\varphi}(k)$$

$$\mathbf{K}(k) = \frac{\mathbf{P}(k-1)\boldsymbol{\varphi}(k)}{\gamma + \boldsymbol{\varphi}^T(k)\mathbf{P}(k-1)\boldsymbol{\varphi}(k)}$$

$$\hat{\boldsymbol{\theta}}(k) = \hat{\boldsymbol{\theta}}(k-1) + \mathbf{K}(k)\varepsilon(k)$$

$$\mathbf{P}(k) = [\mathbf{I} - \mathbf{K}(k)\boldsymbol{\varphi}^T(k)]\mathbf{P}(k-1)$$

Example 6.11 Consider an ARX system given by

$$y(k) = 1.2y(k-1) - 0.4y(k-2) - 0.3y(k-3) + 0.5u(k-1) + 0.25u(k-2) + e(k)$$

where $e(k)$ is a normal, zero-mean Gaussian random process with a variance of 0.001, i.e. $e(k) \sim N(0, 0.001)$. For a unit step input, the system output is shown in Figure 6.15. The identified and actual parameters are shown in Figure 6.16a–e. The estimates are fairly accurate. The error $\varepsilon(k) = y(k) - \hat{\boldsymbol{\theta}}^T(k-1)\boldsymbol{\varphi}(k)$ is called

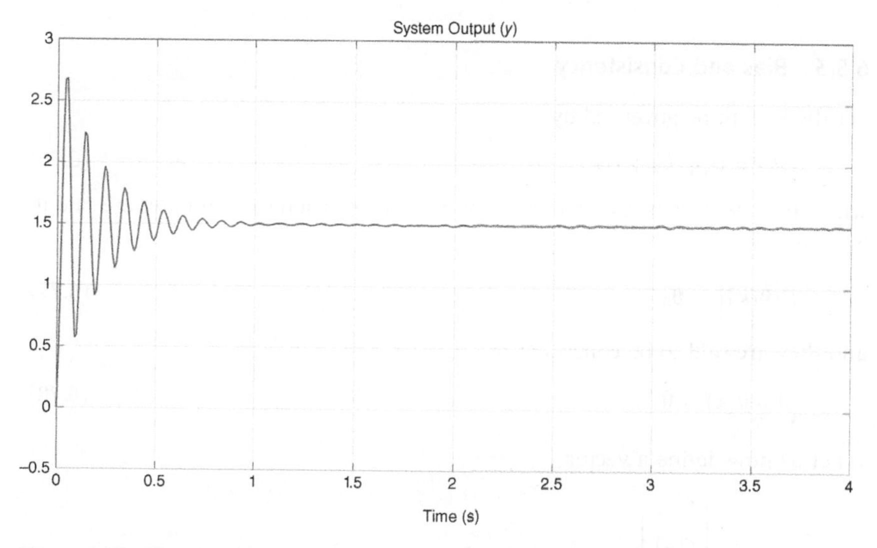

Figure 6.15 The system response to a unit step input.

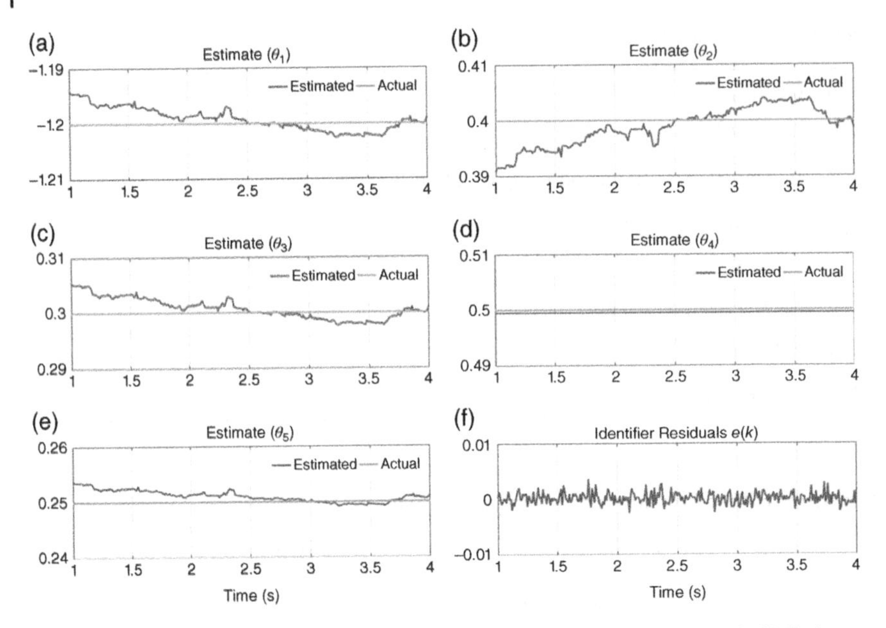

Figure 6.16 Identified parameters and identifier residuals. (a) Estimate (θ_1), (b) Estimate (θ_2), (c) Estimate (θ_3), (d) Estimate (θ_4), (e) Estimate (θ_5), (f) Identifier Residuals $e(k)$.

the identifier residuals. This is shown in Figure 6.16f. Note that, as the identifier converges, the residuals become equal to $e(k)$.

6.5.3 Bias and Consistency

Let the system be governed by

$$y(k) = \boldsymbol{\theta}_0^T \boldsymbol{\varphi}(k) + e(k) \tag{6.76}$$

where $\boldsymbol{\theta}_0$ is the true parameter vector of the system. Then the estimates are said to be unbiased if

$$E\left[\hat{\boldsymbol{\theta}}(k)\right] = \boldsymbol{\theta}_0 \tag{6.77}$$

and they are said to be consistent if

$$\lim_{k\to\infty} \hat{\boldsymbol{\theta}}(k) = \boldsymbol{\theta}_0 \tag{6.78}$$

Let us now define a vector

$$\boldsymbol{\Sigma}(k) = \begin{bmatrix} e(1) \\ e(2) \\ \vdots \\ e(k) \end{bmatrix}$$

Then the vector $\mathbf{Y}(k)$ is given as

$$\mathbf{Y}(k) = \mathbf{X}(k)\boldsymbol{\theta}_0 + \boldsymbol{\Sigma}(k) \tag{6.79}$$

Using (6.79), (6.67) is rewritten as

$$\begin{aligned}\hat{\boldsymbol{\theta}}(k) &= \left[\mathbf{X}^T(k)\mathbf{X}(k)\right]^{-1}\mathbf{X}^T(k)[\mathbf{X}(k)\boldsymbol{\theta}_0 + \boldsymbol{\Sigma}(k)] \\ &= \boldsymbol{\theta}_0 + \left[\mathbf{X}^T(k)\mathbf{X}(k)\right]^{-1}\mathbf{X}^T(k)\boldsymbol{\Sigma}(k)\end{aligned} \tag{6.80}$$

As $E[\mathbf{X}^T(k)\mathbf{X}(k)]^{-1}$ cannot be equal to zero, clearly the estimates are unbiased only when

$$E\left[\mathbf{X}^T(k)\mathbf{E}(k)\right] = 0$$

This implies that

$$E[\boldsymbol{\varphi}(1)e(1) + \boldsymbol{\varphi}(2)e(2) + \cdots + \boldsymbol{\varphi}(k)e(k)] = 0 \tag{6.81}$$

Equation (6.81) is satisfied when the expected value of each individual product term is zero, i.e.

$$E[\boldsymbol{\varphi}(k)e(k)] = 0, \quad \forall k \tag{6.82}$$

Note that the vector $\boldsymbol{\varphi}(k)$ contains the outputs $y(k-1)$, $y(k-2)$, ..., $y(k-n_a)$, which will, in turn, contain the noise terms $e(k-1)$, $e(k-2)$, ..., $e(k-n_a)$. Additionally, this vector also contains the inputs $u(k-1)$, $u(k-2)$, ..., $u(k-n_b)$. Therefore, to have unbiased estimates, the condition of (6.82) must be satisfied and this can only be guaranteed when the following two statements are true:

- If the input $u(k)$ is independent of $e(k)$.
- If the noise $e(k)$ has a mean of zero and is uncorrelated.

Example 6.12 Now consider the same system as in Example 6.11, except that we have chosen an ARMAX model, given by

$$\begin{aligned}y(k) = 1.2y(k-1) - 0.4y(k-2) - 0.3y(k-3) + 0.5u(k-1) \\ + 0.25u(k-2) + C(z^{-1})e(k)\end{aligned}$$

where $e(k) \sim N(0,\ 0.001)$ and

$$C\left(z^{-1}\right) = 1 + 0.2z^{-1} - 0.3z^{-2}$$

Note that the noise term here can be expressed as

$$\chi(k) = C\left(z^{-1}\right)e(k)$$

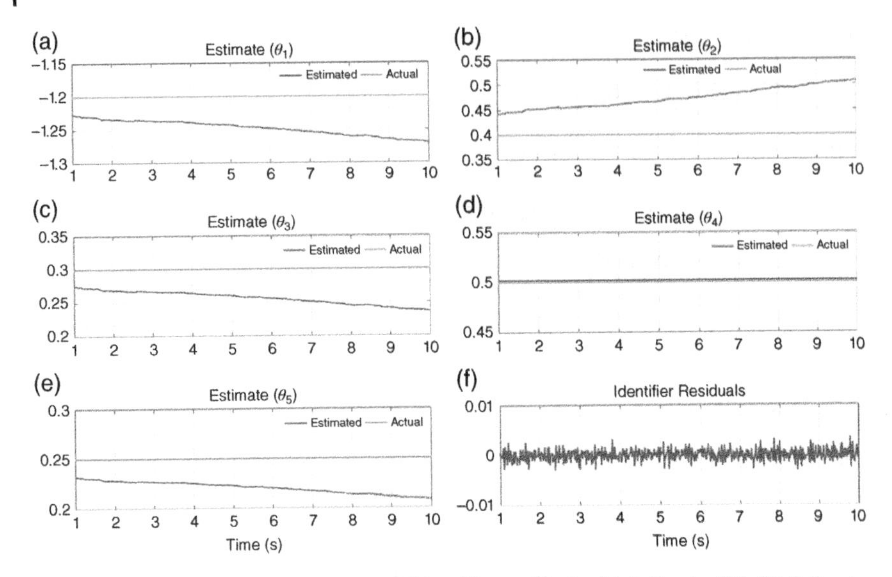

Figure 6.17 Identified parameters and identifier residuals. (a) Estimate (θ_1), (b) Estimate (θ_2), (c) Estimate (θ_3), (d) Estimate (θ_4), (e) Estimate (θ_5), (f) Identifier Residuals.

This is a moving average process and therefore $\chi(k)$ is correlated, as is mentioned in Section 6.1.1.

The identified parameters are shown in Figure 6.17a–e, while the identifier residuals are shown in Figure 6.17f. It can be seen that the estimates, except for θ_4, drift from their nominal values. This drift becomes more pronounced as the time progresses. This means that the estimates are biased due to the presence of correlated noise.

6.6 Self-tuning Controller

A self-tuning controller, the schematic diagram of which is shown in Figure 6.13, is based on the certainty equivalence principle of stochastic control theory. As we have seen in Example 6.12, the identified parameters do not converge to the true parameter values using RLS identification due to the presence of the correlated noise. Under the certainty equivalence principle, the control law is designed assuming that the identified parameters are the actual system parameters [12].

6.6.1 MV Self-tuning Control

The self-tuning controller that was first proposed in [12] shows a remarkable result: if the identified parameters converge, then the controller converges to a MV control law. The main results can be summarized into two theorems.

- The first theorem states that the autocovariance of the output and the cross-covariance between the input and the output are zero using ergodic assumption.
- The results of the first theorem are then used in the second theorem, which states that if the parameter estimation converges and if the polynomials A and B have no common factor then the control law will converge to a MV control.

To illustrate this, let us consider Examples 6.13 and 6.14.

Example 6.13 Consider the ARMAX system of Example 6.12, where $e(k) \sim N(0, 0.001)$. We now assume that the system parameters are unknown and are estimated by an RLS identifier. An MV controller is designed based on the estimated parameters. The results are shown in Figure 6.18, in which the reference input is changed from 1 to 2 at 0.5 seconds. It can be seen that the output response is exactly like that of a known parameter MV controller. Also, the system parameters have converged to those given in polynomials A and B as if polynomial C is nonexistent. Therefore, the controller is not influenced by the colored noise and forces the identifier to converge to the correct parameters.

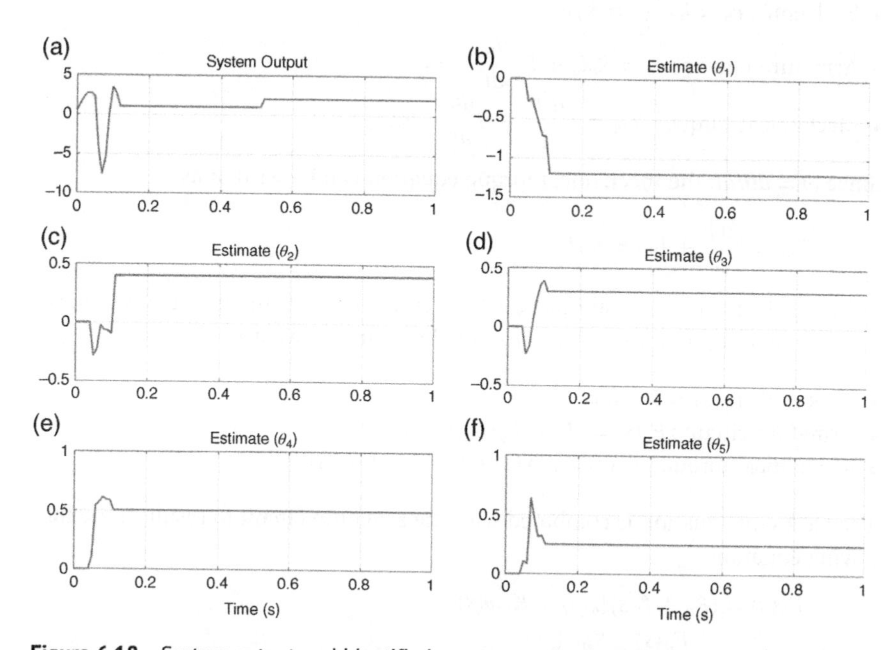

Figure 6.18 System output and identified parameters for the MV self-tuning controller of Example 6.13. (a) System Output, (b) Estimate (θ_1), (c) Estimate (θ_2), (d) Estimate (θ_3), (e) Estimate (θ_4), (f) Estimate (θ_5).

Example 6.14 This example considers the speed control of a DC motor, the parameters of which are defined as

R_a	Armature resistance (Ω)
L_a	Armature inductance (H)
i_a	Armature current (A)
i_f	Field current (A)
e_a	Input voltage (V)
e_b	Back emf (V)
T_m	Motor torque (N-m)
ω	Angular velocity of the rotor (rad/s)
J	Equivalent inertia of the rotor and load referred to the motor shaft (kg-m^2)
B	Viscous damping coefficient of the rotor and load referred to the motor shaft (N-s/rad)
K_b	Electromotive force (EMF) constant (V-s/rad)
K_T	Motor torque constant (N/A)

The governing equations of the DC motor are written as [13]

- Back emf : $e_b = K_b \dfrac{d\theta}{dt} = K_b \omega$
- Armature circuit : $e_a = R_a i_a + L_a \dfrac{di_a}{dt} + e_b$
- Mechanical torque : $T_m = J\dfrac{d^2\theta}{dt^2} + B\dfrac{d\theta}{dt} = K_T i_a$

Since $\omega = d\theta/dt$, the mechanical torque equation can be written as

$$T_m = J\frac{d\omega}{dt} + B\omega = K_T i_a$$

Taking the Laplace transform of the back emf, armature circuit, and mechanical torque equations and assuming zero initial conditions, we get

- Back emf: $E_b(s) = K_b \omega(s)$
- Armature circuit: $E_a(s) = (R_a + L_a s) I_a(s) + E_b(s)$
- Mechanical torque: $T_m(s) = sJ\omega(s) + B\omega(s) = K_T I_a(s)$

The back emf equation is combined with the armature circuit to result in the following equation

$$E_a(s) = (R_a + L_a s) I_a(s) + K_b \omega(s)$$
$$\Rightarrow I_a(s) = \frac{E_a(s) - K_b \omega(s)}{(R_a + L_a s)}$$

Again, the following expression of the armature current can be written from the torque equation

$$T_m(s) = sJ\omega(s) + B\omega(s) = K_T I_a(s)$$
$$\Rightarrow I_a(s) = \frac{sJ\omega(s) + B\omega(s)}{K_T}$$

Equating the last two equations, we have

$$I_a(s) = \frac{E_a(s) - K_b\omega(s)}{(R_a + L_a s)} = \frac{sJ\omega(s) + B\omega(s)}{K_T}$$
$$\Rightarrow K_T E_a(s) = [sJ\omega(s) + B\omega(s)] \times (R_a + L_a s) + K_b K_T \omega(s)$$

The transfer function of the motor is then

$$\frac{\omega(s)}{E_a(s)} = \frac{K_T}{(sJ + B)(R_a + L_a s) + K_b K_T}$$

$$= \frac{K_T}{L_a J s^2 + (JR_a + BL_a)s + BR_a + K_b K_T}$$

$$= \frac{K}{s^2 + \alpha_1 s + \alpha_2}$$

where

$$K = \frac{K_T}{L_a J}, \quad \alpha_1 = \frac{JR_a + BL_a}{L_a J}, \quad \alpha_2 = \frac{BR_a + K_b K_T}{L_a J}$$

The parameters chosen for this study are [14]:

$$R_a = 1.9, \quad L_a = 0.05, \quad J = 0.1, \quad B = 0.0045, \quad K_T = 0.5, \quad K_b = 0.5$$

A sampling time of 1 ms is chosen. This gives an ARX model of

$$(1 - 1.9626z^{-1} + 0.9627z^{-2})y(k) = (0.4937 + 0.4875z^{-1}) \times 10^{-4}u(k) + e(k)$$

Here $e(k) \sim N(0,\ 0.001)$ is assumed to be the measurement noise. At the beginning, it has been assumed that only rough values of the motor parameters are known. Based on these values the motor model is assumed to be

$$(1 - 1.72z^{-1} + 0.62z^{-2})y(k) = (0.5 + 0.4z^{-1}) \times 10^{-3}u(k) + e(k)$$

A MV controller is designed based on these parameter values where a reference speed is set as 157 rad/s. The identification process has been started at the beginning with a constant input voltage of 100 V. At 0.5 seconds, the self-tuning regulator is switched on. The results are shown in Figure 6.19. It is obvious that, as soon as the self-tuning controller is switched on, the speed converges to the reference

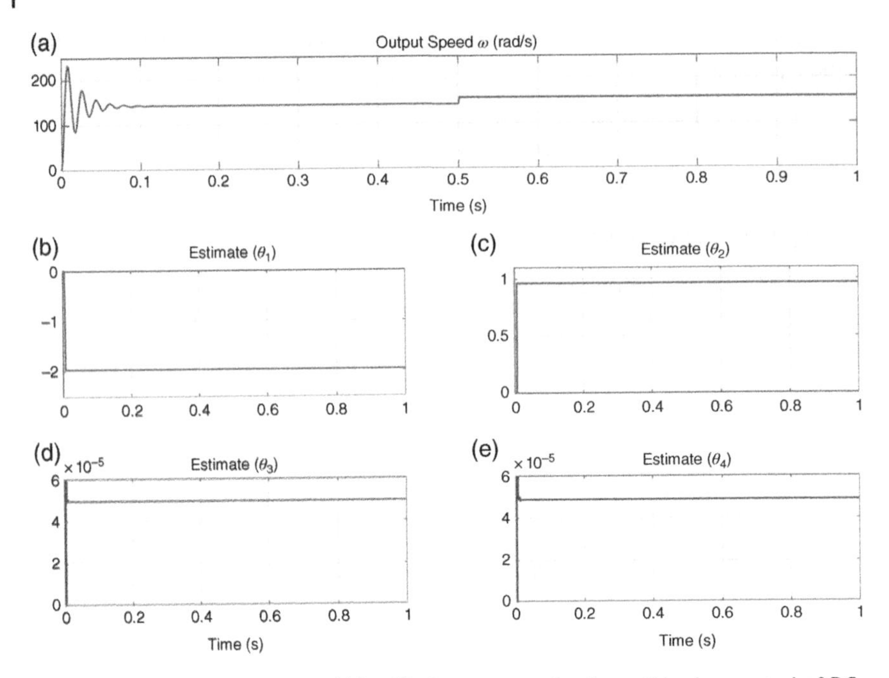

Figure 6.19 System output and identified parameters for the self-tuning control of DC motor of Example 6.14. (a) Output speed ω (rad/s), (b) Estimate (θ_1), (c) Estimate (θ_2), (d) Estimate (θ_3), (e) Estimate (θ_4).

input. Since the identified parameters have already converged, the change in speed is smooth.

6.6.2 Pole Shift Self-tuning Control

The original self-tuning regulator was proposed for MV-based controllers. Since then, other controllers which synthesized the control law based on identified parameters are generically called self-tuning controllers. The pole placement self-tuning controller was proposed by Wellstead et al. [17], who show that this type of controller can accommodate variable time delays. Obviously, the pole placement design is nonoptimal as compared to optimal MV controllers. However, it is more robust in the sense that it can be used for nonminimum phase plants easily. A self-tuning pole shift control for a nonminimum phase system is discussed in Example 6.15

Example 6.15 In this example, we consider a nonminimum phase system that is governed by the ARMAX process of

$$\begin{aligned} y(k) = 1.2y(k-1) - 0.4y(k-2) - 0.3y(k-3) + 0.5u(k-1) + 1.75u(k-2) \\ + e(k) + 0.2e(k-1) - 0.3e(k-2) \end{aligned}$$

The same control design as in Example 6.7 is now followed, except that the estimated parameters using RLS are used instead of the actual parameters to compute the controller polynomials of

$$R(z^{-1}) = 1 + r_1 z^{-1}$$
$$S(z^{-1}) = s_0 + s_1 z^{-1} + s_2 z^{-2}$$

The Diophantine equation can be written in matrix vector form by replacing the actual parameters with the identified parameters as

$$\begin{bmatrix} 1 & \hat{b}_0 & 0 & 0 \\ \hat{a}_1 & \hat{b}_1 & \hat{b}_0 & 0 \\ \hat{a}_2 & 0 & \hat{b}_1 & \hat{b}_0 \\ \hat{a}_3 & 0 & 0 & \hat{b}_1 \end{bmatrix} \begin{bmatrix} r_1 \\ s_0 \\ s_1 \\ s_2 \end{bmatrix} = \begin{bmatrix} \hat{a}_1(\lambda - 1) \\ \hat{a}_2(\lambda^2 - 1) \\ \hat{a}_3(\lambda^3 - 1) \\ 0 \end{bmatrix}$$

The feedforward gain is computed as

$$K_C = \frac{1 + \hat{a}_1(\lambda - 1) + \hat{a}_2(\lambda^2 - 1) + \hat{a}_3(\lambda^3 - 1)}{\left(\hat{b}_0 + \hat{b}_1\right) \times (s_0 + s_1 + s_2)}$$

A pole shift factor (λ) of 0.9 is chosen to form the closed-loop pole locations. The system output response and the identified parameters are shown in Figure 6.20. The reference input is assumed to be 1 at the beginning and has been changed to 2 at 0.5 seconds. At the beginning, there are large transients. However, as the estimated parameters converge, the tracking performance improves. In fact, the steady state tracking is perfect after the change in the reference since the estimated parameters have converged to their actual values by this time.

6.6.3 Self-tuning Control of Boost Converter

Consider the boost converter of Example 5.19. With the state space linearized model of (5.95), and the switching frequency of 20 kHz (i.e. sampling time of 0.05 ms), the discrete-time input–output relationship is written in difference equation form for the parameters given in Example 5.20. The state space averaged model is first converted into its equivalent discrete-time equivalent for a sampling time of 0.05 ms using the "c2d" command in MATLAB®. The following difference equation is then obtained using the "ss2tf" command.

$$(1 - 1.945z^{-1} + 0.95z^{-2})y(k) = (-1.76z^{-1} + 2.01z^{-2})u(k) + e(k)$$

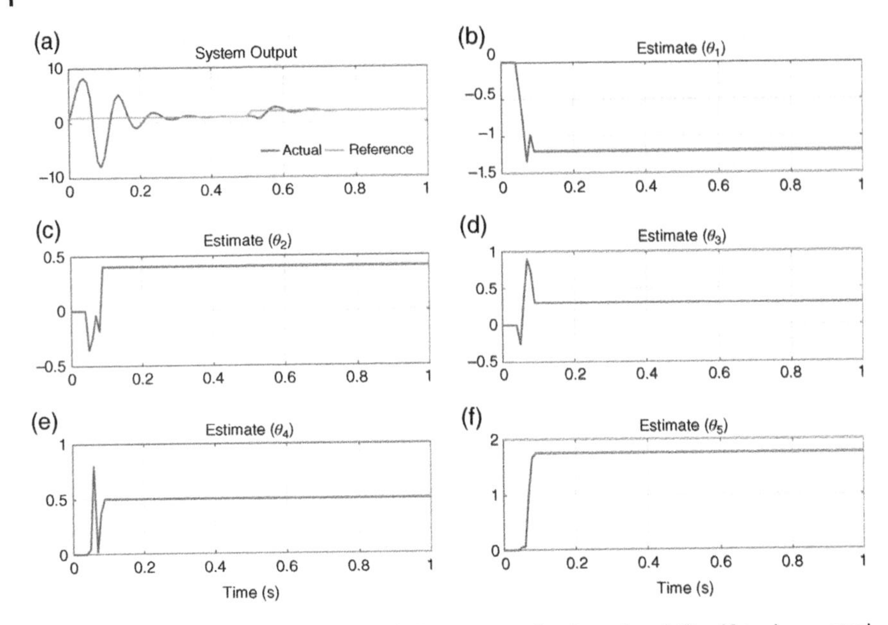

Figure 6.20 System output and identified parameters for the poles shift self-tuning control of Example 6.15. (a) System Output, (b) Estimate (θ_1), (c) Estimate (θ_2), (d) Estimate (θ_3), (e) Estimate (θ_4), (f) Estimate (θ_5).

The system zero is placed at 1.14, which indicates that the system is a nonminimum phase. An MV controller cannot be used; therefore, a pole shift controller is designed instead.

Let us write the difference equation in terms of the identified parameters as

$$\left(1 + \hat{a}_1 z^{-1} + \hat{a}_2 z^{-2}\right) y(k) = \left(\hat{b}_0 z^{-1} + \hat{b}_1 z^{-2}\right) u(k) + e(k)$$

Also let us define the controller polynomials as

$$R\left(z^{-1}\right) = 1 + r_1 z^{-1} \text{ and } S\left(z^{-1}\right) = s_0 + s_1 z^{-1}$$

Then the coefficients of the polynomials can be obtained from

$$\begin{bmatrix} 1 & \hat{b}_0 & 0 \\ \hat{a}_1 & \hat{b}_1 & \hat{b}_0 \\ \hat{a}_2 & 0 & \hat{b}_1 \end{bmatrix} \begin{bmatrix} r_1 \\ s_0 \\ s_1 \end{bmatrix} = \begin{bmatrix} \hat{a}_1(\lambda - 1) \\ \hat{a}_2\left(\lambda^2 - 1\right) \\ 0 \end{bmatrix}$$

The controller is then designed using (6.28) for pole shift factor (λ) of 0.0995.

At the beginning, it is assumed that the load of $R = 10\ \Omega$ is known and therefore the initial parameter vector and the covariance matrix are chosen as

$$\hat{\boldsymbol{\theta}}(0) = [\,-1.945 \quad 0.95 \quad -1.76 \quad 2.01\,]^T, \quad \mathbf{P}(0) = \mathbf{I}_4 \times 10^{10}$$

The load is changed to $R = 5\ \Omega$ at 0.01 seconds, to $R = 12.5\ \Omega$ at 0.03 seconds, and to $R = 1\ \Omega$ at 0.06 seconds. The desired output voltage is kept at 20 V. The output voltage, inductor current, and duty ratio are shown in Figure 6.21. Despite large changes in the load, the controller is able to regulate to the desired output voltage, even though the ripples increase as the load resistance decreases, as can be seen from Figure 6.21a. The inductor current, shown in Figure 6.21b, obviously increases/decreases with the decrease/increase in the load resistance. The duty ratio remains in the vicinity of 0.5 barring transients during load changes, as shown in Figure 6.21c.

The estimated parameters are shown in Figure 6.22. We must keep in mind that the controller is designed based on a linear model of the converter. Therefore, the estimated parameters must try to adapt during the load changes such that the control law is computed based on these parameters can force the output to follow the reference voltage. It is obvious that this has been achieved.

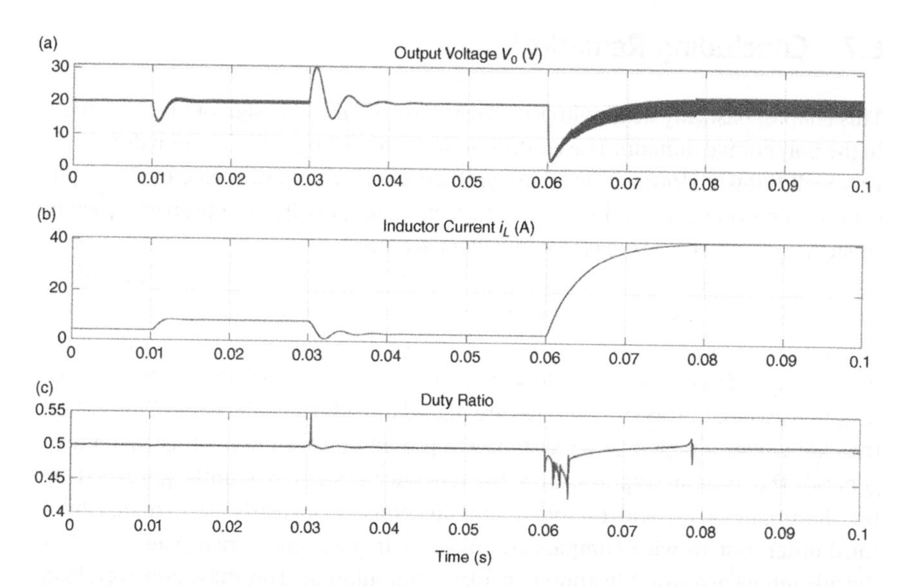

Figure 6.21 Boost converter (a) output voltage, (b) inductor current, and (c) duty ratio during load changes.

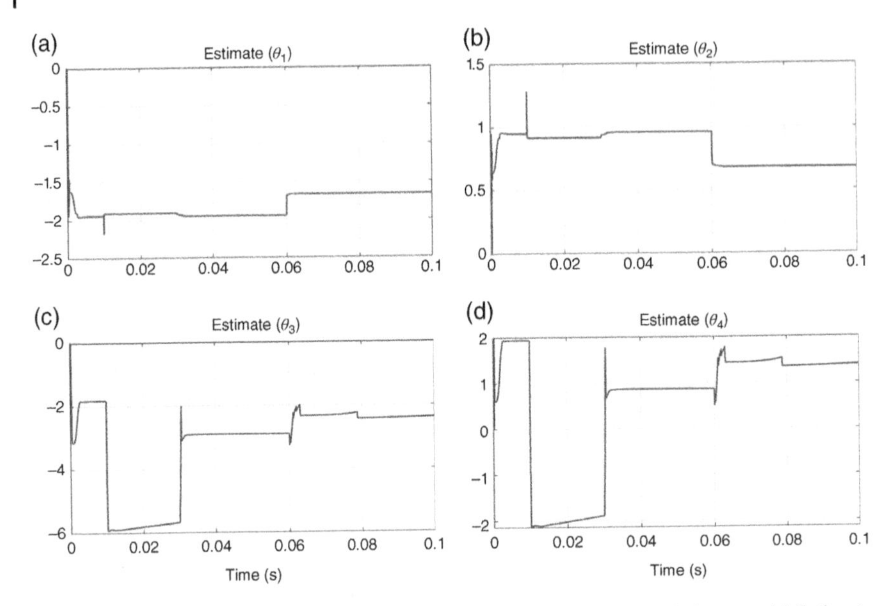

Figure 6.22 Estimated parameters of the boost converter during load changes. (a) Estimate (θ_1), (b) Estimate (θ_2), (c) Estimate (θ_3), (d) Estimate (θ_4).

6.7 Concluding Remarks

This chapter basically deals with the discrete-time control design of linear systems in the polynomial domain. The controller design in the transfer function domain is very well known. However, in many problems involving power electronics applications in power systems, the system characteristics are often not properly known. In such cases, prediction-based controllers are needed.

There are, however, several approaches to design prediction based adaptive controllers. We have restricted ourselves to a certain class, which is based on the certainty equivalence principle. Of these, MV controllers have excellent performance in tracking reference signals. However, they have the problem associated with instability when the open-loop zeros are placed outside the unit circle. This is often the case where a higher-order system is represented by a lower-order system. In general, the system response of a higher-order system is mainly governed by the dominant poles, and therefore their response can nearly be matched by a third-order system with complex conjugate pair poles and a real pole. This has the advantage of a simplification of control computation. This may, however, lead to a nonminimum phase behavior, as discussed in [4]. A pole placement controller can, however, remain stable under such conditions.

Finally, predictive controllers are another important class that has been used extensively in power converter applications. They can be applied to a variety of systems that may include constraints and nonlinearities. Even though the design is somewhat complicated, the resultant controller is easy to implement on a microprocessor. However, this will require a high amount of computation, which, given the availability of high-speed microcontrollers or a digital signal processor (DSP) system, will not be a problem. These controllers are fairly robust and very suitable for interconnected systems.

Problems

6.1 Consider the batch model given by (6.54). Find the weighted least-squares estimates of $\boldsymbol{\theta}$ that minimize the cost function

$$J_{WLS} = \left(\mathbf{Y} - \hat{\mathbf{Y}}\right)^T \mathbf{W} \left(\mathbf{Y} - \hat{\mathbf{Y}}\right)$$

where $\mathbf{W}$ is a positive definite symmetric matrix.

6.2 Design an MV controller for the ARMAX process, given by

$$y(k) - y(k-1) + 0.5y(k-2) = u(k-2) + 0.5u(k-2) + e(k) + 0.8e(k-1) + 0.25(k-2)$$

6.3 Consider the ARMAX process, given by

$$y(k) - 0.25y(k-1) + 0.5y(k-2) = \beta u(k-2) + e(k) + 0.5e(k-1)$$

(a) Design an MV controller assuming $\beta = 1$.
(b) Assuming $y_r = 0$, the controller is computed from

$$u(k) = s_1 y(k) + s_2 y(k-1) + r_1 u(k-1)$$

Then find a set of possible convergence points of s_1, s_2, and r_1 when $\beta = 1.5$.

6.4 For the ARMAX process of Problem 6.2, design a pole placement controller that will place the closed-loop poles at $T(z^{-1}) = 1 - 0.5z^{-1} + 0.5z^{-2} = 0$.

6.5 For the ARMAX process of Problem 6.2, design a pole shift controller with a pole shift factor of $\lambda = 0.7$.

6.6 Consider the ARMAX process

$$y(k) + ay(k-1) = bu(k-1) + e(k) + ce(k-1)$$

We use an RLS method to estimate the parameter α in the model

$$y(k) = -\alpha y(k-1) + \beta u(k-1) + \xi(k)$$

where β is known. Assuming $y_r = 0$, the control signal is generated from

$$u(k) = \frac{\alpha}{\beta} y(k)$$

Then show that an MV controller with

$$\alpha = \frac{a-c}{b}\beta$$

is a possible value of the parameter convergence.

6.7 Consider the buck-boost converter of Problem 5.14. Discretize the state space average model with a sampling time of 0.1 ms. Then design a pole shift controller with a pole shift factor of $\lambda = 0.75$.

Notes and References

Karl Johan Åström was the Chair of the Department of Automatic Control at Lund University in Sweden. Due to his pioneering contributions to the theory and application of adaptive control, he was awarded the IEEE Medal of Honor, which is the highest recognition by IEEE, in 1996. Most of what is discussed in this chapter is primarily due to his contribution and that of Rudolf E. Kalman (also an IEEE Medal of Honor winner in 1974) before that. Even though we did not discuss the Kalman filter in this chapter, the fundamental concepts of estimation theory owe a great deal to him.

The books written by Åström and his colleague Wittenmark are very thorough and immensely enjoyable to read [1–3]. Also, the concept of self-tuning control with the remarkable property based on the certainty equivalence principle was proposed by Åström and Wittenmark in [12]. Since then, other forms of self-tuning control such as one based on optimal control [15, 16] or based on pole placement [17] have been proposed.

There are several classes of predictive control that are generally categorized under the common name model predictive control [9]. General discussions on GPC can also be found in [3] and [10]. GPCs and model predictive controllers (MPCs) have gained much attention in power electronic applications. DC-DC converter control using GPC is reported in [18], where the converter discontinuous mode of operation is also considered.

1 Åström, K.J. and Wittenmark, B. (1990). *Computer-Controlled Systems: Theory and Design*, 2e. Englewood Cliffs, NJ: Prentice Hall.

2 Åström, K.J. (1970). *Introduction to Stochastic Control Theory*. New York: Academic Press.

3 Åström, K.J. and Wittenmark, B. (1994). *Adaptive Control*, 2e. Englewood Cliffs, NJ: Prentice Hall.

4 Ghosh, A., Ledwich, G., Malik, O.P., and Hope, G.S. (1984). Power system stabilizers based on adaptive control techniques. *IEEE Trans. Power Appl. & Syst.* PAS-103: 1983–1989.

5 Clarke, D.W., Mohtadi, C., and Tuffs, P.S. (1987). Generalized predictive control: part I: the basic algorithm. *Automatica* 23 (2): 137–148.

6 Clarke, D.W., Mohtadi, C., and Tuffs, P.S. (1987). Generalized predictive control: part II: extensions and interpretations. *Automatica* 23 (2): 149–160.

7 Clarke, D.W. and Mohtadi, C. (1989). Properties of generalized predictive control. *Automatica* 25 (6): 859–875.

8 Bitmead, R.R., Gevers, M., and Wertz, V. (1990). *Adaptive Optimal Control: The Thinking Man's GPC*. Englewood Cliffs, NJ: Prentice Hall.

9 Camacho, E.F. and Bordons, C. (2007). *Model Predictive Control*. London: Springer.

10 Landau, I.D., Lozano, R., Saad, M.M., and Karimi, A. (2011). *Adaptive Control: Algorithms, Analysis and Applications*, 2e. London: Springer.

11 Ljung, L. and Söderström, T. (1986). *Theory and Practice of Recursive Identification*. Cambridge, MA: MIT Press.

12 Åström, K.J. and Wittenmark, B. (1976). On self tuning regulators. *Automatica* 9: 185–199.

13 Kuo, B.C. (1987). *Automatic Control System*, 5e. Englewood Cliffs, NJ: Prentice Hall.

14 Dupuis, A., Ghribi, M. and Kaddouri, A. (2004). Multiobjective genetic estimation of DC motor parameters and load torque. IEEE International Conference on Industrial Technology (ICIT), Hammamet, Tunisia.

15 Clarke, D.W. and Gawthrop, P.J. (1975). Self-tuning controller. *Proc. IEE* 122 (9): 929–934.

16 Clarke, D.W. and Gawthrop, P.J. (1979). Self-tuning control. *Proc. IEE* 126 (6): 633–640.

17 Wellstead, P.E., Prager, D., and Zanker, P. (1979). Pole placement self-tuning regulator. *Proc. IEE* 126 (8): 781–787.

18 Karamanakos, P., Geyer, T., and Manias, S. (2014). Direct voltage control of DC–DC boost converters using enumeration-based model predictive control. *IEEE Trans. Power Electron.* 29 (2): 968–978.

7

DC-AC Converter Modulation Techniques

In this chapter, AC-DC converter structures and their modulation (switching) techniques are discussed. In general, DC-AC converters are of two types: current source converter (CSC) and voltage source converter (VSC). The power circuits are slightly different (type and configuration of the semiconductor switches), but the energy conversion technique is the same. However, the main difference is how they are supplied on their DC sides. A CSC is supplied by a DC current source, while a DC input voltage supplies a VSC. The DC side of the CSC is realized by a controlled DC source that is connected to a large inductor in series. The presence of the inductor in a CSC makes it reliable and fault tolerant as the inductor can limit the rate of change in current during a fault. However, CSCs have higher losses and size due to the presence of the inductors. The VSCs, on the other hand, are supplied by either DC voltage sources (e.g. battery storage, photovoltaic, fuel cell, etc.) or by DC storage capacitors. These types of converters are more popular than CSCs and are more important for renewable energy integration. Henceforth, our discussion will be restricted solely to VSCs only.

There are several converter switching techniques that are available in the literature and in practice. Only three of them are discussed here: hysteresis control, sinusoidal pulse width modulation (SPWM), and space vector pulse width modulation (SVPWM). How VSCs and the modulation techniques produce an output voltage or current in the open loop is also discussed. In practice, however, the converters have output filters, and their feedback control systems are designed by taking into consideration these output filters. Different types of controllers for both voltage and current control are discussed in Chapters 8–10. Excellent texts (for example [1–2]) covering different aspects of power electronics can be referred to complement the discussions presented in this chapter.

High-voltage power converters are utilized in many high- and medium-power applications such as motor drives, grid connected inverters, and power quality

Control of Power Electronic Converters with Microgrid Applications, First Edition.
Arindam Ghosh and Firuz Zare.

Published 2023 by John Wiley & Sons, Inc.

management units in industrial and commercial systems. As an example, the demand for high-power renewable energy systems has been increasing and solar and wind farms have been utilized in low- and medium-distribution networks based on power electronics technology. A direct connection of conventional power converters (two-level H-bridge inverters) to medium-voltage distribution networks is not possible, due to the maximum blocking voltage rating of semiconductor switches. As a result, a multilevel power converter structure or a transformer-based topology is required for these applications. In Section 7.5, three main multi-level converter topologies, as well as modular multilevel converter, are discussed as alternatives for high- and medium-voltage applications. Pulse width modulation (PWM) of multilevel converters is also discussed.

7.1 Single-phase Bridge Converter

The schematic diagram of a single-phase converter is shown in Figure 7.1a. It is called an H-bridge converter as it looks like the eighth letter of the English alphabet (H). The converter contains four switches $S_1 - S_4$, each comprising a power semiconductor device and an antiparallel diode, as shown in Figure 7.1b. Modern VSCs employ insulated gate bipolar transistors (IGBTs) or metal oxide semiconductor field effect transistors (MOSFET) as they can carry fairly large current and have fast switching characteristics and low losses. The wide bandgap devices like silicon carbide (SiC) are still in their infancy. The converter switching frequency will drastically change when such devices are readily available for high-power applications.

The converter of Figure 7.1a supplies a passive RL load. It is connected between the two legs of the converter, which is supplied by a DC source with a voltage of V_{dc}. The switches of each leg usually have complementary values, e.g. when S_1 is on, S_4 is off, and vice versa. Furthermore, when the switches S_1 and S_2 are on, S_3 and S_4 are off. Similarly, when S_3 and S_4 are on, S_1 and S_2 are off. However, a small time delay is provided between the turning off of one pair of switches and the turning on of the other pair. This period, called the blanking period, is provided to prevent the DC source from being short circuited through the legs. Consider, for example, the transition when S_3 and S_4 are turned off and S_1 and S_2 are turned on. During this period, if switch S_1 gets turned on before switch S_4 turns off completely, these two switches will connect the two leads of the DC source directly. It is therefore mandatory that switch S_4 turns off completely before switch S_1 is turned on. To ensure this, the blanking period (deadtime) is used. The continuity of the current during the blanking period is maintained by the antiparallel diodes.

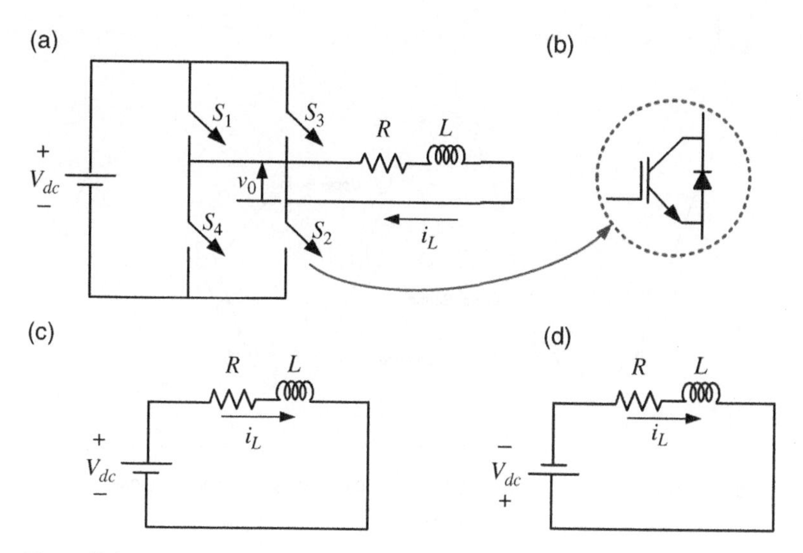

Figure 7.1 Single-phase VSC circuit: (a) equivalent representation, (b) switch representation, (c) equivalent circuit when S_1 and S_2 are on, and (d) when S_3 and S_4 are on.

The equivalent circuit of the converter when the switches S_1 and S_2 are on is shown in Figure 7.1c, while when S_3 and S_4 being on is shown in Figure 7.1d. Neglecting the blanking period, the current flowing through the circuit is given by

$$\frac{d}{dt}i_L = -\frac{R}{L}i_L + \frac{1}{L}V_{dc}\,u \tag{7.1}$$

where u is the variable that takes on the values of ± 1 as per

$$u = \begin{cases} +1 & \text{when } S_1 \text{ and } S_2 \text{ are on} \\ -1 & \text{when } S_3 \text{ and } S_4 \text{ are on} \end{cases} \tag{7.2}$$

7.1.1 Hysteresis Current Control

Hysteresis control is the simplest form of current control. In this, the H-bridge converter is required to track a reference current (i_{Lref}). An upper and lower band are chosen around this reference current, as shown in Figure 7.2. When the current touches the lower band, a positive voltage is applied across the load by switching S_1 and S_2 on. Conversely, when the current reaches the upper band, a negative voltage is applied across the load by switching S_3 and S_4 on. The switching logic is given for a small positive scalar h by

$$\begin{aligned} &\text{If } i_L \geq i_{Lref} + h \text{ then } u = -1 \\ &\quad \text{elseif } i_L \leq i_{Lref} - h \text{ then } u = +1 \end{aligned} \tag{7.3}$$

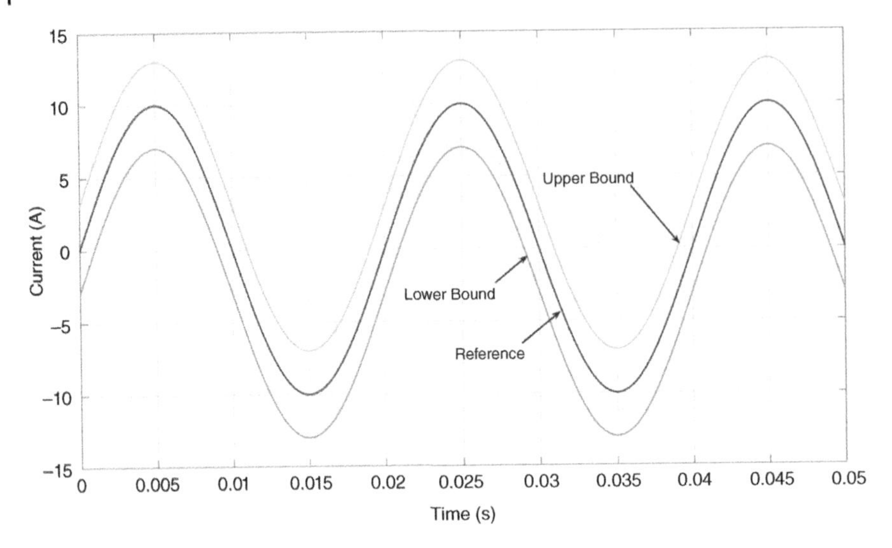

Figure 7.2 Reference current and upper and lower hysteresis bands.

Example 7.1 Consider the circuit shown in Figure 7.1a with the following parameters

- DC input voltage: $V_{dc} = 550$ V
- Sinusoidal reference current: Peak value = 400 A, frequency = 10 Hz, i.e. $i_{Lref} = 400\sin(20\pi t)$ A
- RL load: $0.42 + j0.121\ \Omega$
- Hysteresis band: $h = 10$ A.

The steady state waveforms of the output current, its reference, and upper and lower hysteresis bands are shown in Figure 7.3a. The corresponding converter output voltage is shown in Figure 7.3b.

The output current for two cycles is shown in Figure 7.4a. It can be seen that the current follows the reference faithfully, in spite of the ripples due to the hysteresis band. Now, the load impedance is increased five times such that the load is $2.1 + j0.605\ \Omega$. The result of the current tracking is shown in Figure 7.4b, from which it is obvious that the current fails to track the reference due to the presence of the larger inductor. With the same load, the DC voltage is now increased to $V_{dc} = 1500$ V. The reference current tracking becomes accurate for this increased voltage, as can be seen from Figure 7.4c. However, as the supply voltage increases, the rate of change di_L/dt increases as well.

The behavior of the current tracking vis-à-vis the load impedance is shown in Figure 7.5. As the value of the load inductor increases, the rate of change of the

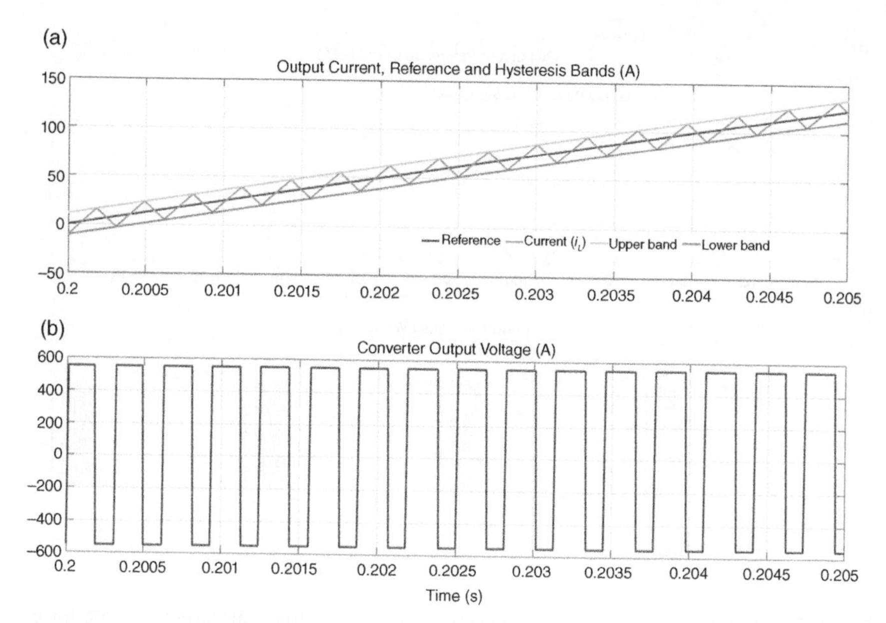

Figure 7.3 Steady state voltage and current waveforms of an H-bridge converter. (a) Output current, reference and hysteresis bands (A), (b) Converter output voltage (A).

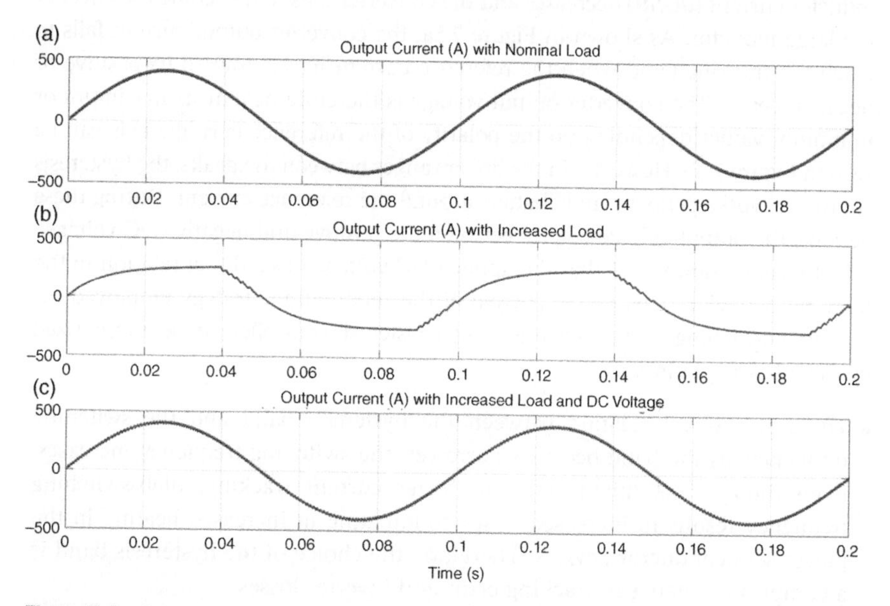

Figure 7.4 Current tracking performance of H-bridge converter with different load and DC voltage. (a) Output current (A) with nominal load, (b) Output current (A) with increased load, (c) Output current (A) with increased load and DC voltage.

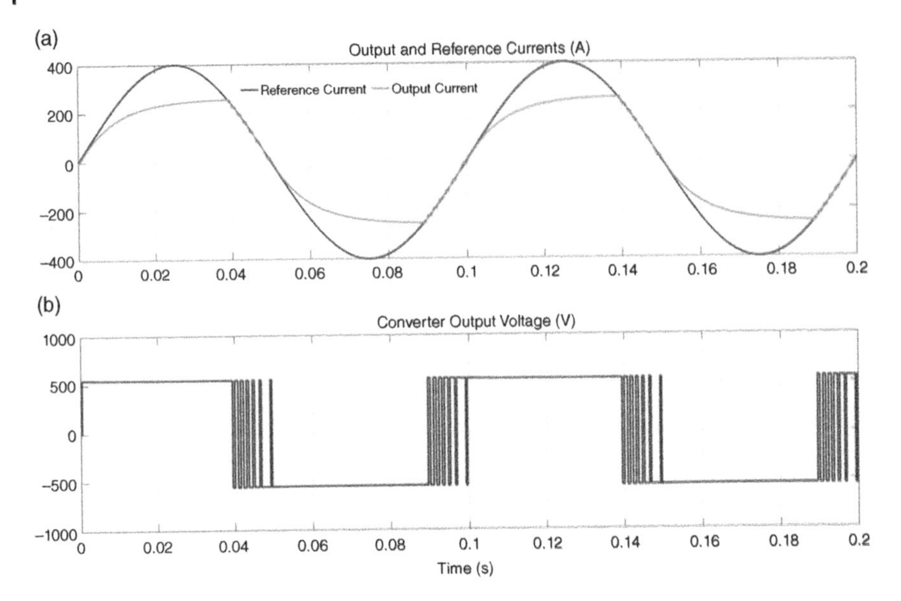

Figure 7.5 Converter output current and voltage with a high load. (a) Output and reference current (A), (b) Converter output voltage (V).

inductor current (di_L/dt) decreases and the converter fails to overcome the effect of the large inductor. As shown in Figure 7.5a, the converter output current fails to reach the tracking band when the reference current moves toward its positive or negative peaks. The converter output voltage is therefore held at its maximum or minimum values depending on the polarity of the reference current. This can be seen in Figure 7.5b. However, in the linear ranges between the peaks, the hysteresis controller works perfectly and chatters around the reference current. During these periods, the output voltage toggles between the positive and negative DC voltage.

Note from Example 7.1 that the applied DC voltage has a direct relation to the tracking behavior. This is irrespective of the modulation strategy employed for converter switching. The characteristics of a hysteresis controller can be summarized by the following points:

- There is a direct relation between the hysteresis band and the switching frequency: as the band becomes narrower, the switching frequency increases. Even though this might result in better current tracking, high-switching frequency results in increased losses culminating in increased heating in the power semiconductor devices. Therefore, the choice of the hysteresis band is a compromise between tracking error and inverter losses.
- For the same value of the hysteresis band, the switching frequency decreases as the value of the inductor increases, since the rate of change in the inductor current (di_L/dt) decreases with the increase in the value of the inductor. As we can see

in Example 7.1, the tracking with a larger inductor becomes inferior compared to a smaller inductor for the same value of the hysteresis band and input DC voltage.

- The tracking performance can be improved by increasing di_L/dt. For a large inductor, this can be achieved by increasing the DC voltage.
- The switching frequency of this controller is not fixed, as it varies based on the reference signal. Thus, in some applications, a filter designed for harmonic mitigation and system stability needs to be considered for the overall system.

Hysteresis current control techniques with adaptive hysteresis band or different topologies such as a single-phase, a three-phase, or a multilevel converter are discussed in [3–5].

7.1.2 Bipolar Sinusoidal Pulse Width Modulation (SPWM)

In this section, the SPWM of the H-bridge converter is discussed. For SPWM, the switching signals are generated by the intersections of a high-frequency triangular carrier wave and a sinusoidal modulating signal, the frequency of which is the desired frequency of the output voltage. Consider the H-bridge inverter of Figure 7.1a. In a bipolar SPWM, the switching signals are generated as per

$$\begin{aligned}&\text{If } V_{MS} \geq V_{TC} \text{ then } u = +1\\ &\quad \text{elseif } V_{MS} < V_{TC} \text{ then } u = -1\end{aligned} \tag{7.4}$$

where V_{MS} is the modulating signal and V_{TC} is the triangular carrier waveform. We define the modulation ratio or index (m_a), which is the ratio of the peaks of these two waveforms, as

$$m_a = \frac{V_{MS}^{peak}}{V_{TC}^{peak}} \tag{7.5}$$

The modulation index determines if the signal is under- or overmodulated. If $m_a < 1$, the signal is undermodulated, while it is overmodulated when $m_a > 1$. One example of an undermodulated operation is shown in Figure 7.6, where the modulation index is chosen as $m_a = 0.75$. The overmodulation case is shown in Figure 7.7 for $m_a = 1.5$.

It is also obvious from Figures 7.6 and 7.7 that the output voltage (v_0) (see Figure 7.6b) will contain harmonics. The fundamental component (V_{01}) of this voltage depends on the modulation index. The root mean square (rms) value of the fundamental component of the output voltage is given, for the undermodulated case, by [1]

$$V_{01} = m_a \times \frac{V_{dc}}{\sqrt{2}}, \quad m_a \leq 1 \tag{7.6}$$

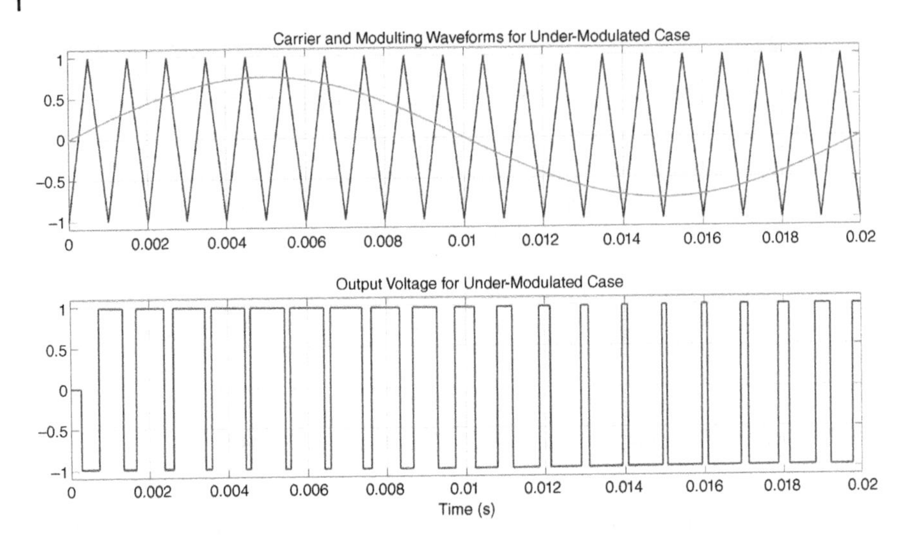

Figure 7.6 Undermodulated bipolar SPWM and converter output voltage.

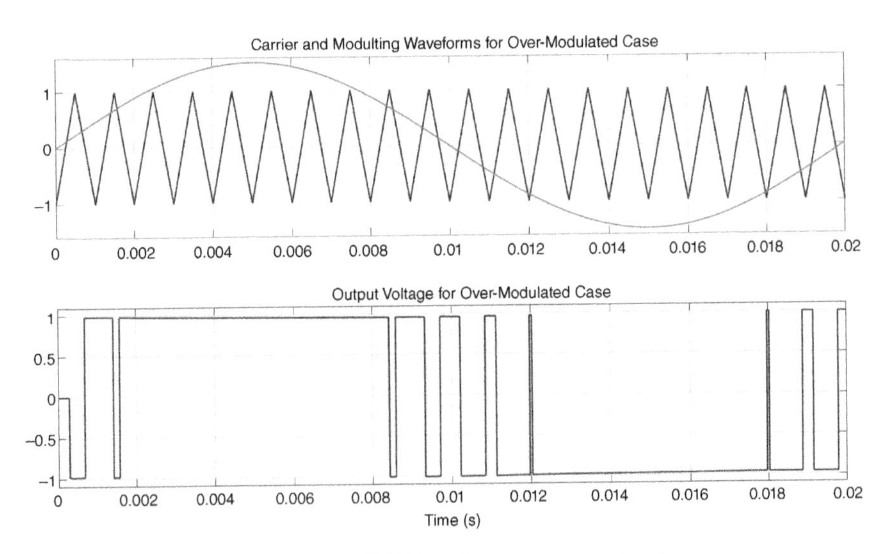

Figure 7.7 Overmodulated bipolar SPWM and converter output voltage.

For the overmodulated case, the rms value of the fundamental voltage is [1]

$$\frac{V_{dc}}{\sqrt{2}} < V_{01} < \frac{4}{\pi}\frac{V_{dc}}{\sqrt{2}}, \quad m_a > 1 \tag{7.7}$$

Now consider the case where the carrier waveform has a frequency of 15 kHz and a normalized peak of 1 per unit (pu) for a base voltage of 400 V. The modulating signal has a frequency of 50 Hz and a peak of 0.75 pu, i.e. the desired output of 300 V. The DC bus voltage is chosen as $V_{dc} = 550$ V, i.e. 1.375 pu. From (7.6), we have

$$m_a = 0.75$$

$$V_{01} = m_a \times \frac{V_{dc}}{\sqrt{2}} = 0.75 \times \frac{1.375}{\sqrt{2}} = 0.7292 \text{ pu}$$

Therefore, the fundamental rms output voltage is $0.7292 \times 400 = 291.68$ V, as shown in Figure 7.8a. For the same carrier waveform and DC voltage, a 50 Hz modulating waveform is chosen that has a peak of 1.5 pu (i.e. 600 V). Obviously, the modulating index in this case is 1.5. Then from (7.7), we have

$$\frac{1.375}{\sqrt{2}} < V_{01} < \frac{4}{\pi}\frac{1.375}{\sqrt{2}} \Rightarrow 0.9723 < V_{01} < 1.370 \text{ pu}$$

This implies that the fundamental rms of the output voltage will be between $0.9723 \times 400 = 388.91$ V and $1.2379 \times 400 = 495.17$ V. In Figure 7.8b, this voltage is plotted, where its value is found to be 453.1 V.

7.1.3 Unipolar Sinusoidal Pulse Width Modulation

In this method, the switches of the two legs of an H-bridge converter of Figure 7.1a are not switched simultaneously. The first step in this process is to generate two modulating signals, where one is the negative of the other. This implies that these two signals are phase shifted by 180°. Let us denote them as V_{MS} and $-V_{MS}$. These are then compared with the triangular carrier signal V_{TC}. The switching law is given by the following relations with respect to Figure 7.1a.

$$\begin{aligned}
&\text{If } V_{MS} \geq V_{TC} \text{ then } S_1 \text{ is on; otherwise off}\\
&\text{If } V_{MS} < V_{TC} \text{ then } S_4 \text{ is on; otherwise off}\\
&\text{If } -V_{MS} \geq V_{TC} \text{ then } S_3 \text{ is on; otherwise off}\\
&\text{If } -V_{MS} < V_{TC} \text{ then } S_2 \text{ is on; otherwise off}
\end{aligned} \tag{7.8}$$

Figure 7.9 (a) shows the carrier and the modulating waveforms, while Figure 7.9b shows the output voltage waveform. The rms fundamental voltage remains the same as those given in (7.6) and (7.7) respectively for the under- and overmodulating signals.

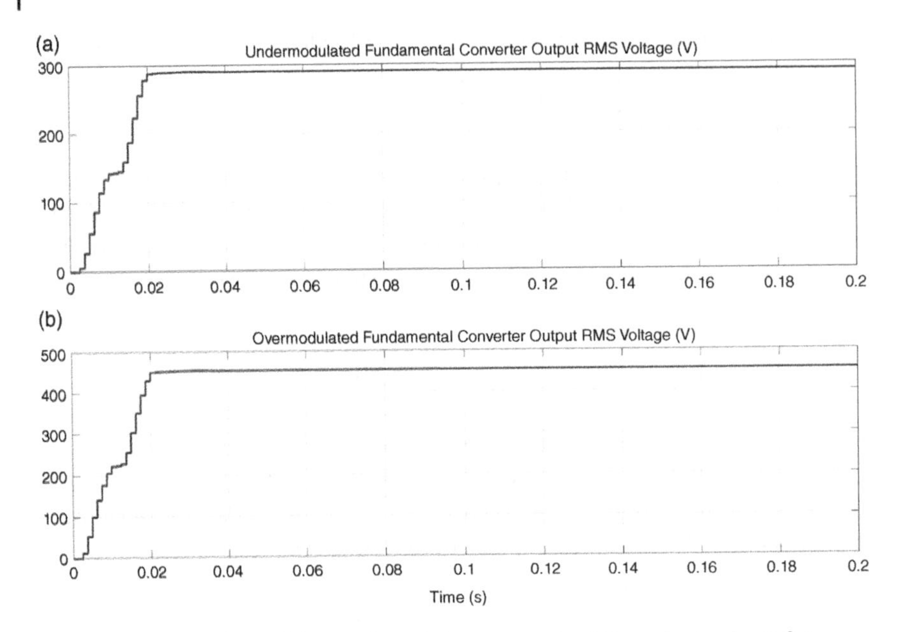

Figure 7.8 Rms values of the fundamental components of the output voltage for (a) $m_a = 0.75$ and (b) $m_a = 1.5$.

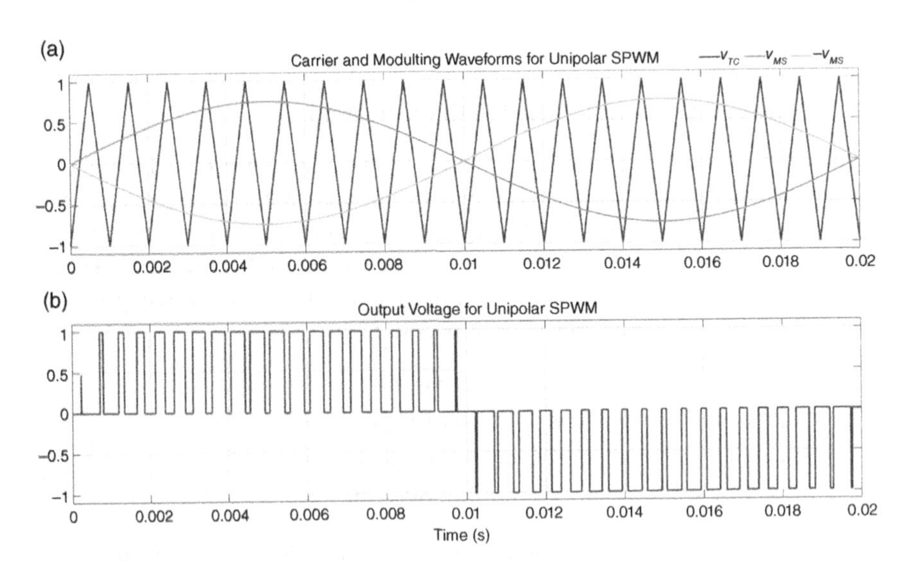

Figure 7.9 Undermodulated unipolar SPWM and converter output voltage. (a) Carrier and Modulting Waveforms for Unipolar SPWM, (b), Output Voltage for Unipolar SPWM.

7.2 SPWM of Three-phase Bridge Converter

The schematic diagram of a three-phase full-bridge converter is shown in Figure 7.10a. This contains six switches S_1 to S_6, each consisting of a power semiconductor device and an antiparallel diode. As in the case of the H-bridge converter, the switches of each leg are complementary (e.g. when S_1 is on, S_4 is off, and vice versa). This is the most common form of three-phase converter available, where the DC bus voltage is equal to V_{dc}, as shown in Figure 7.10a. However, this has the disadvantage that the algebraic sum of the three output currents must be zero, i.e. $i_a + i_b + i_c = 0$, i.e. the zero-sequence current must be zero. There is an alternative way to connect the DC link, as shown in Figure 7.10b. In this, the DC bus is split into two, with a center point N. The DC bus voltage is still equal to V_{dc}. But it now contains two DC sources, each equal to $V_{dc}/2$. The center point N can be connected to protective earth to circulate common mode and electromagnetic interference (EMI) currents. In a special case, when N is connected to the load neutral, it provides a path for the unbalanced (zero-sequence) current to flow. Hence, the three legs of the converter can be treated separately, and each leg can be controlled independent of the other two legs.

Two other possible three-phase converter structures are shown in Figure 7.11, each of which provides a path for the zero-sequence load current to circulate. The VSC of Figure 7.11a consists of three H-bridge converters. The output of each H-bridge contains a single-phase transformer that provides a galvanic isolation and prevents the DC bus from being short circuited [6]. The secondary side of the three transformers are star-connected, and this point can be connected to the load neutral (N). An alternate converter structure containing four legs is shown in Figure 7.10b. This converter was first proposed in [7–8] and was subsequently used for compensating unbalanced loads in [9–10]. It is evident that the center point fourth leg is connected to the load neutral (N). The current through this path

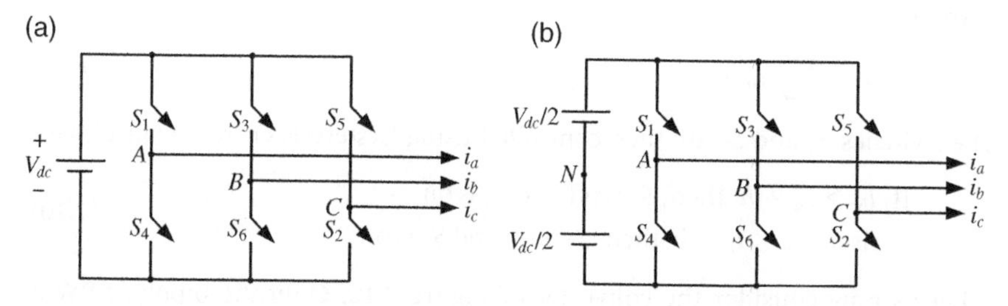

Figure 7.10 Three-phase converter circuits: (a) bridge converter and (b) neutral-clamped converter.

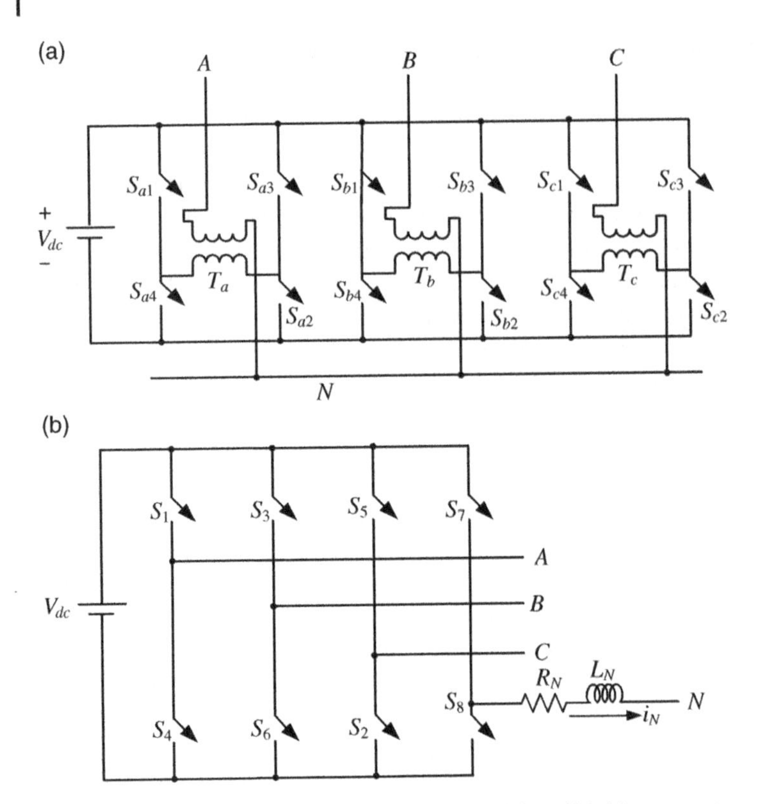

Figure 7.11 Structure of (a) transformer couple of three H-bridge converters and (b) a four-leg VSC.

is i_N and this current is used to cancel the zero-sequence component of the load current. The switches S_7 and S_8 are controlled to cancel the zero-sequence current. Therefore, the reference current for the fourth leg is the negative sum of the load currents, i.e.

$$i_N^* = -\sum i_{Load} \tag{7.9}$$

The switches S_7 and S_8 are then controlled using hysteresis current control, i.e.

$$\begin{aligned} &\text{If } i_N \geq i_N^* + h \text{ then } S_7 \text{ is off and } S_8 \text{ is on} \\ &\text{elseif } i_N \leq i_N^* - h \text{ then } S_7 \text{ is on and } S_8 \text{ is off} \end{aligned} \tag{7.10}$$

Let us now consider the converter of Figure 7.10. Only the bipolar SPWM of this converter is considered here, where a triangular waveform is compared with three reference sinusoidal waveforms that have equal magnitude and are phase displaced by 120°. The switching signals for each leg of the converter are then generated following the same logic as given in (7.4). For the undermodulated case, the rms value of the fundamental line-to-neutral voltage is given by [1]

$$V_{aN} = m_a \frac{V_{dc}}{2\sqrt{2}}, \quad m_a < 1 \tag{7.11}$$

Example 7.2 Consider the circuit of Figure 7.12, in which the converter is connected to a balanced back emf through an RL circuit. The system parameters are:

- DC voltage (V_{dc}) = 600 V
- back emf = 415 V (L-L rms)
- frequency = 50 Hz
- $R = 5\ \Omega$
- $L = 11.6$ mH.

A base voltage of 400 V is chosen, with the assumption that the peak of the triangular carrier waveform is 1.0 pu. The peak of the modulating signal is 315 V (i.e. 0.7875 pu). Therefore, the modulation index is also 0.7875. Now the DC voltage is 1.5 pu. Therefore, from (7.11), the rms value of the line-to-neutral voltage is given by $V_{an} = 0.7875 \times 1.5/2\sqrt{2} = 0.4176$ pu. This translates into $0.4176 \times 400 = 167$, as shown in Figure 7.13b, while their phase angles are shown in Figure 7.13c. The output voltages are balanced, i.e. they have the same magnitude and are phase displaced by 120°.

The output voltage waveforms will contain harmonics that are dependent on the frequency of the carrier waveforms, and these voltages will produce current harmonics. Phase a of the output current is shown in Figure 7.14 for three different frequencies of the triangular waveform (f_{TC}). It can be seen that the ripples in the current decrease as the frequency increases.

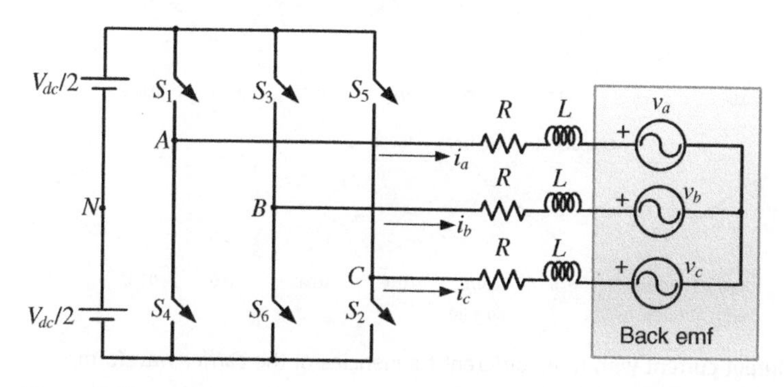

Figure 7.12 Converter connected to a back emf through an RL impedance.

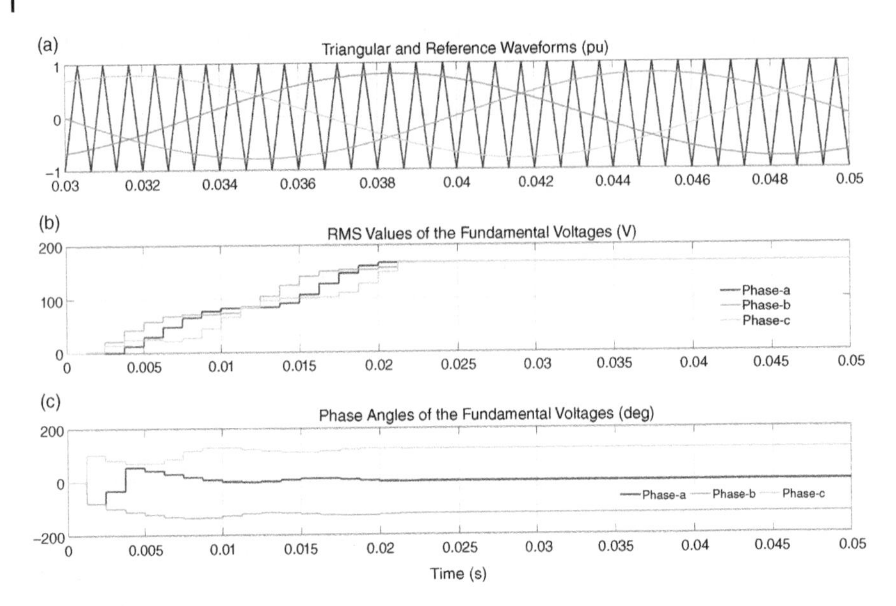

Figure 7.13 SPWM control of a three-phase converter: (a) carrier and reference waveforms, (b) rms fundamental voltage, and (c) their phase angles.

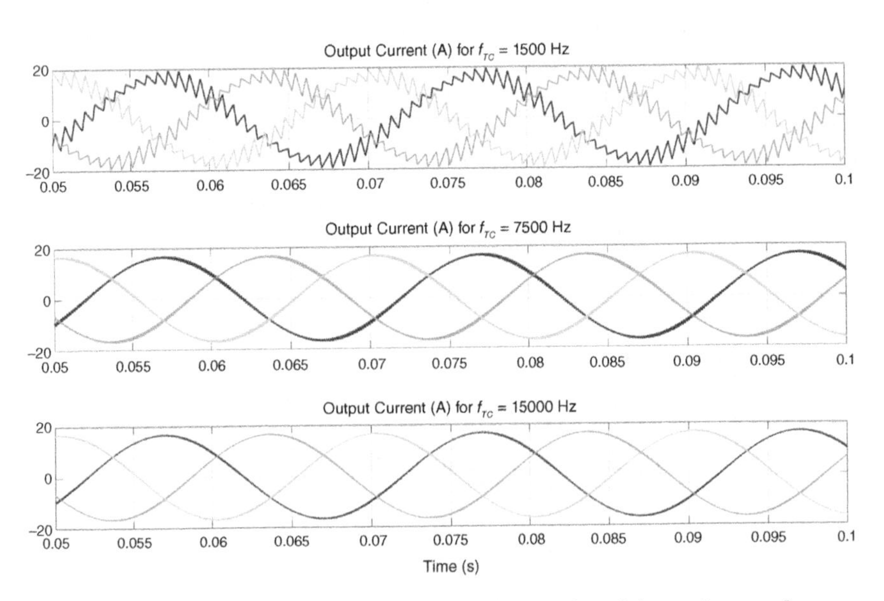

Figure 7.14 Output current with three different frequencies of the carrier waveform.

7.3 Space Vector Modulation (SVM)

SVM is essentially an averaging technique which takes into consideration that a three-phase inverter has only eight switch states. Figure 7.15 shows a three-phase converter and its eight switching states. Noting that the top and bottom switches of each leg are complementary, there are two states for each leg. Therefore, for the three independent legs of the inverter, a total combination of $2^3 = 8$ states can be obtained. These are also shown in Figure 7.15.

The voltage vectors for the eight switch states are depicted in Figure 7.16. Each vector is displaced from its adjacent vectors by 60°. The entire voltage vector space is divided into six regions. Each region is the triangular area between two contiguous vectors, e.g. Region-1 is the space between vectors V_1 and V_2, and so on. Note that

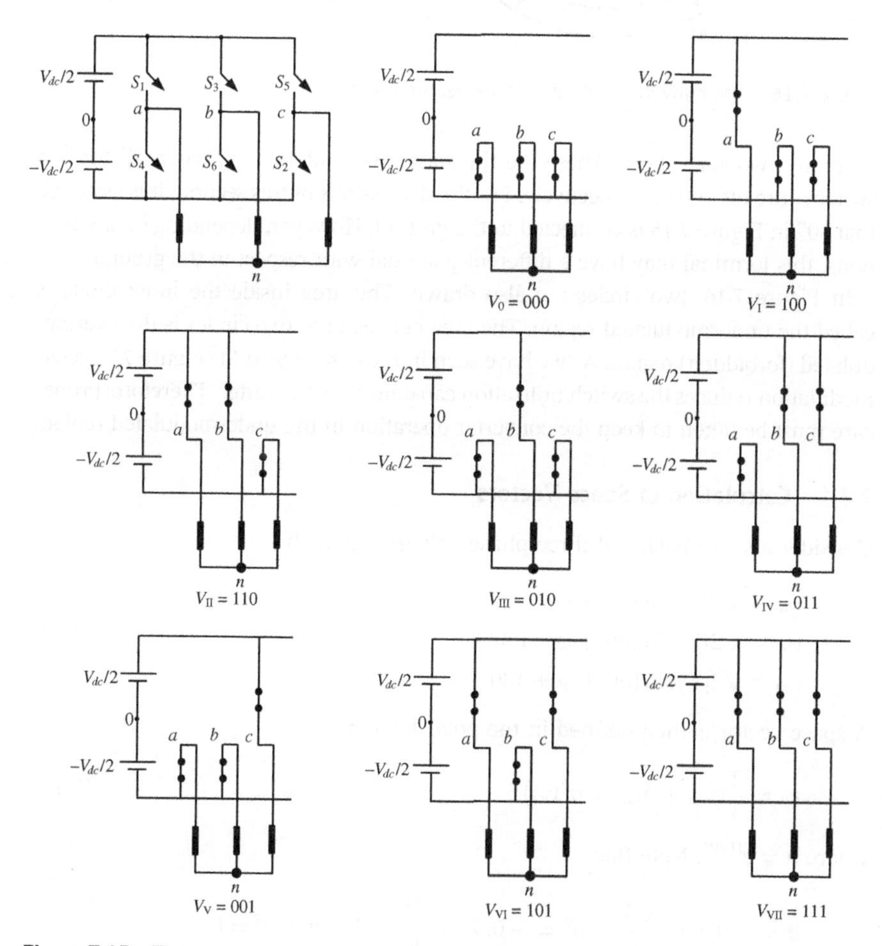

Figure 7.15 The converter structure and the switching states.

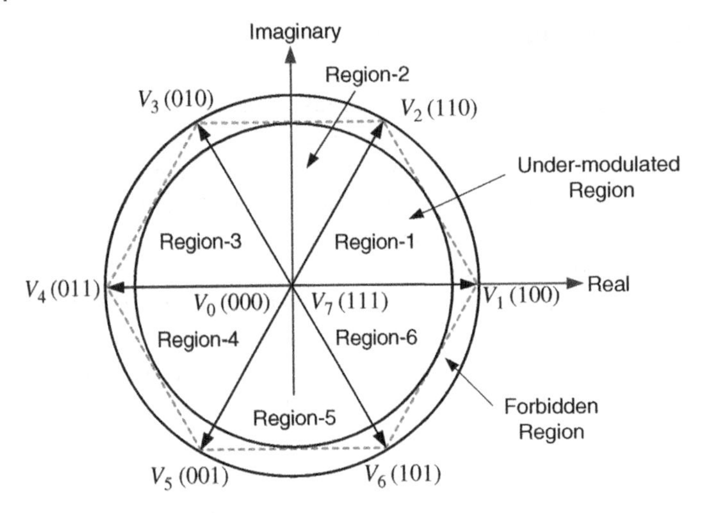

Figure 7.16 The converter output voltage vector space.

there are two zero states. These are the states 000 and 111, i.e. when all the top switches are off and on respectively. For the discussions of this section, it is assumed that "0" in Figure 7.15 is connected to the ground. However, depending on applications, this terminal may have a different potential with respect to the ground.

In Figure 7.16, two circles are also drawn. The area inside the inner circle is called the undermodulated region. The area between the two circles is the overmodulated (forbidden) region. As we have seen in the case of SPWM (Figure 7.7), overmodulation reduces the switch utilization capability of an inverter. Therefore, proper care must be taken to keep the converter operation in the undermodulated region.

7.3.1 Calculation of Space Vectors

Consider a set of balanced three-phase voltages, given by

$$\begin{aligned} v_{a0} &= \sqrt{2}|V|\sin(\omega t + \varphi) \\ v_{b0} &= \sqrt{2}|V|\sin(\omega t + \varphi - 120^\circ) \\ v_{c0} &= \sqrt{2}|V|\sin(\omega t + \varphi + 120^\circ) \end{aligned} \tag{7.12}$$

A space vector is then defined in the polar form as

$$v_P = \frac{2}{3}\left(v_{a0} + a v_{b0} + a^2 v_{c0}\right) \tag{7.13}$$

where $a = e^{j120^\circ}$. Note that

$$a = -0.5 + j\frac{\sqrt{3}}{2}, \quad a^2 = -0.5 - j\frac{\sqrt{3}}{2}, \quad 1 + a + a^2 = 0$$

Using (7.12), the last two terms on the right-hand side of (7.13) can be expanded as

$$
\begin{aligned}
av_{b0} &= \frac{2\sqrt{2}}{3}|V|a\left[-0.5\sin(\omega t+\varphi)-\frac{\sqrt{3}}{2}\cos(\omega t+\varphi)\right]\\
a^2v_{c0} &= \frac{2\sqrt{2}}{3}|V|a^2\left[-0.5\sin(\omega t+\varphi)+\frac{\sqrt{3}}{2}\cos(\omega t+\varphi)\right]
\end{aligned}
$$

Adding these two terms, we have

$$
\begin{aligned}
av_{b0}+a^2v_{c0} &= \frac{2\sqrt{2}}{3}|V|\left[-(a+a^2)0.5\sin(\omega t+\varphi)-(a-a^2)\frac{\sqrt{3}}{2}\cos(\omega t+\varphi)\right]\\
&= \frac{2\sqrt{2}}{3}|V|\left[0.5\sin(\omega t+\varphi)-j\sqrt{3}\times\frac{\sqrt{3}}{2}\cos(\omega t+\varphi)\right]\\
&= \frac{2\sqrt{2}}{3}|V|\left[0.5\sin(\omega t+\varphi)-j\frac{3}{2}\cos(\omega t+\varphi)\right]
\end{aligned}
$$

Substituting these in (7.13), the voltage vector v_P is expressed as

$$
\begin{aligned}
v_P &= \frac{2\sqrt{2}}{3}|V|\left[\frac{3}{2}\sin(\omega t+\varphi)-j\frac{3}{2}\cos(\omega t+\varphi)\right]\\
&= -j\sqrt{2}|V|[\cos(\omega t+\varphi)+j\sin(\omega t+\varphi)] = \sqrt{2}|V|e^{j(\omega t+\varphi-90^\circ)}
\end{aligned}
\tag{7.14}
$$

Therefore, the vector has a constant magnitude and rotates counterclockwise with a speed of ω.

7.3.2 Common Mode Voltage

The voltage between the star point of the load and the midpoint of the DC link is called the common mode voltage, v_{n0}. From Figure 7.15, we can write

$$
\begin{aligned}
v_{a0} &= v_{an}-v_{n0}\\
v_{b0} &= v_{bn}-v_{n0}\\
v_{c0} &= v_{cn}-v_{n0}
\end{aligned}
\tag{7.15}
$$

Adding the two sides of (7.15), we get

$$v_{a0}+v_{b0}+v_{c0} = v_{an}+v_{bn}+v_{cn}-3v_{n0}$$

Note that, for a balanced system, the sum $v_{an}+v_{bn}+v_{cn}$ will be equal to zero. Therefore, the common mode voltage is given as

$$v_{n0} = \frac{v_{a0}+v_{b0}+v_{c0}}{3} \tag{7.16}$$

The common mode voltage for the eight switching states is listed in Table 7.1.

Table 7.1 Eight switching states of a three-phase converter.

State Number	S_1	S_3	S_5	v_{a0}	v_{b0}	v_{c0}	v_{n0}
V_0	0	0	0	$-\frac{V_{dc}}{2}$	$-\frac{V_{dc}}{2}$	$-\frac{V_{dc}}{2}$	$-\frac{V_{dc}}{2}$
V_1	1	0	0	$\frac{V_{dc}}{2}$	$-\frac{V_{dc}}{2}$	$-\frac{V_{dc}}{2}$	$-\frac{V_{dc}}{6}$
V_2	1	1	0	$\frac{V_{dc}}{2}$	$\frac{V_{dc}}{2}$	$-\frac{V_{dc}}{2}$	$\frac{V_{dc}}{6}$
V_3	0	1	0	$-\frac{V_{dc}}{2}$	$\frac{V_{dc}}{2}$	$-\frac{V_{dc}}{2}$	$-\frac{V_{dc}}{6}$
V_4	0	1	1	$-\frac{V_{dc}}{2}$	$\frac{V_{dc}}{2}$	$\frac{V_{dc}}{2}$	$\frac{V_{dc}}{6}$
V_5	0	0	1	$-\frac{V_{dc}}{2}$	$-\frac{V_{dc}}{2}$	$\frac{V_{dc}}{2}$	$-\frac{V_{dc}}{6}$
V_6	1	0	1	$\frac{V_{dc}}{2}$	$-\frac{V_{dc}}{2}$	$\frac{V_{dc}}{2}$	$\frac{V_{dc}}{6}$
V_7	1	1	1	$\frac{V_{dc}}{2}$	$\frac{V_{dc}}{2}$	$\frac{V_{dc}}{2}$	$\frac{V_{dc}}{2}$

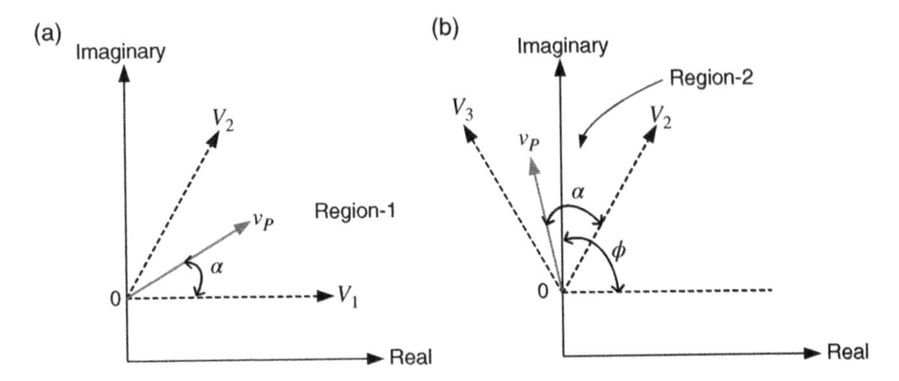

Figure 7.17 Placement of vector v_P in (a) Region-1 between vectors V_1 and V_2 and (b) Region-2 between vectors V_2 and V_3.

7.3.3 Timing Calculations

Now suppose at any given instant of time we want to recreate the vector v_P in Region-1, as shown in Figure 7.17a. This can be done by time averaging of the nearest inverter state vectors V_1 and V_2 and a zero vector V_0 or V_7. The averaging is done over a suitably chosen time interval T_s. This technique is known as space vector modulation [11]. Let us denote the converter voltages as V_i for states $i = 0$, 1, ..., 7. Then, if the converter spends time T_a in state V_1 and time T_b in state V_2, we get the following equation

$$v_P T_s = V_1 T_a + V_2 T_b + V_0 (T_s - T_a - T_b) \tag{7.17}$$

Let us assume that v_P has a magnitude of $|V_P|$ and an angle α with vector V_1 so that it is written as

$$v_P = |V_P| \angle \alpha = |V_P| (\cos\alpha + j \sin\alpha) \tag{7.18}$$

From (7.13) and Table 7.1, v_P in state V_1 is written as

$$v_P = \frac{2}{3}\left(\frac{V_{dc}}{2} - a\frac{V_{dc}}{2} - a^2\frac{V_{dc}}{2}\right) = \frac{V_{dc}}{3}(1 - a - a^2) = \frac{2V_{dc}}{3}$$

Similarly, v_P in state V_2 is

$$v_P = \frac{2}{3}\left(\frac{V_{dc}}{2} + a\frac{V_{dc}}{2} - a^2\frac{V_{dc}}{2}\right) = \frac{V_{dc}}{3}(1 + a - a^2) = \frac{2V_{dc}}{3}\left(0.5 + j\frac{\sqrt{3}}{2}\right)$$

Furthermore, v_P in state V_0 is equal to zero. Then (7.18) can be written in the form of (7.17) as

$$|V_P|(\cos\alpha + j\sin\alpha) T_s = \frac{2V_{dc}}{3}\left[T_a + \left(\frac{1}{2} + j\frac{\sqrt{3}}{2}\right) T_b + 0 \times (T_s - T_a - T_b)\right] \tag{7.19}$$

Equation (7.19) is resolved into real and imaginary parts as

$$\frac{3|V_P|}{2V_{dc}} T_s \cos\alpha = T_a + \frac{1}{2} T_b \tag{7.20}$$

$$\frac{3|V_P|}{2V_{dc}} T_s \sin\alpha = \frac{\sqrt{3}}{2} T_b \tag{7.21}$$

From (7.21), we have

$$\frac{T_b}{T_s} = \sqrt{3}\frac{|V_P|}{V_{dc}} \sin\alpha \tag{7.22}$$

Substitution of (7.22) in (7.20) results in

$$\frac{3|V_P|}{2V_{dc}} T_s \cos\alpha = T_a + \frac{\sqrt{3}}{2}\frac{|V_P|}{V_{dc}} T_s \sin\alpha$$

Rearranging (7.22), we get

$$T_a = \sqrt{3}\frac{|V_P|}{V_{dc}} T_s \left(\frac{\sqrt{3}}{2}\cos\alpha - \frac{1}{2}\sin\alpha\right) = \sqrt{3}\frac{|V_P|}{V_{dc}} T_s \sin(60^\circ - \alpha) \tag{7.23}$$

Equations (7.22) and (7.23) give the time in which the converter will be state vectors V_1 and V_2.

Let us now consider that the vector is in Region-2 between state vectors V_2 and V_3, as shown in Figure 7.16b, where the vector is defined as

$$v_P = |V_P|\angle\phi \tag{7.24}$$

If the converter spends time T_a in state V_2 and time T_b in state V_3 then, following the same process as before, we have

$$\begin{aligned}(|v_P|\angle\phi)T_s &= |v_P|\angle(\alpha + 60^\circ)T_s \\ &= \frac{2V_{dc}}{3}\angle 60^\circ\left[T_a + \left(\frac{1}{2} + j\frac{\sqrt{3}}{2}\right)T_b + 0\times(T_s - T_a - T_b)\right]\end{aligned} \tag{7.25}$$

This is similar to the expression given in (7.19) and hence the timing calculations of (7.22) and (7.23) will remain valid. However, comparing (7.19) with (7.25), we find that the positioning of the vector changes with the region in which it is placed. For example, in Region-3, (7.19) can be written as

$$|V_P|\angle(\alpha + 120^\circ)T_s = \frac{2V_{dc}}{3}\angle 120^\circ\left[T_a + \left(\frac{1}{2} + j\frac{\sqrt{3}}{2}\right)T_b + 0\times(T_s - T_a - T_b)\right] \tag{7.26}$$

We must therefore determine the angle α before the computation of the times T_a and T_b. The first step in the process is to determine the sector (region) in which the vector is placed by observing ϕ. An integer value is then assigned to the regions (or sectors) as

$$\text{Region-}i = i, \quad i = 1, \ldots, 6 \tag{7.27}$$

Once the regions are identified, the angle α is calculated based on the following formula

$$\alpha = \phi - (\text{Region-}i - 1)\times 60^\circ, \quad i = 1, \ldots, 6 \tag{7.28}$$

Substituting (7.28) in (7.22) and (7.23), the following timings are obtained

$$\frac{T_b}{T_s} = \sqrt{3}\frac{|V_P|}{V_{dc}}\sin\{\phi - (\text{Region-}i - 1)\times 60^\circ\}, \quad i = 1, \ldots, 6 \tag{7.29}$$

$$\frac{T_a}{T_s} = \sqrt{3}\frac{|V_P|}{V_{dc}} \sin\{60^\circ - \phi + (\text{Region-}i - 1) \times 60^\circ\}$$
$$= \sqrt{3}\frac{|V_P|}{V_{dc}} \sin(\text{Region-}i \times 60^\circ - \phi), \quad i = 1, ..., 6 \tag{7.30}$$

7.3.4 An Alternate Method for Timing Calculations

This section discusses an alternate method for timing calculation. Let us define the vector v_P as

$$v_P = v_d + jv_q \tag{7.31}$$

Assume that the vector is in Region-1, where the converter is required to spend time T_a in state V_1 and T_b in state V_2. Then, from (7.17), the following equation is obtained

$$\left(v_d + jv_q\right) T_s = \frac{2V_{dc}}{3} T_a + \frac{2V_{dc}}{3}\left(\frac{1}{2} + j\frac{\sqrt{3}}{2}\right) T_b \tag{7.32}$$

Equating the real and imaginary parts, the following equations are obtained

$$v_d \frac{3T_s}{2V_{dc}} = T_a + \frac{t_2}{2}$$
$$v_q \frac{3T_s}{2V_{dc}} = \frac{\sqrt{3}}{2} T_b$$

Solving the above two equations, we have

$$T_a = \frac{3T_s}{2V_{dc}}\left(v_d - \frac{1}{\sqrt{3}} v_q\right) \tag{7.33}$$

$$T_b = \sqrt{3}\frac{T_s}{V_{dc}} v_q \tag{7.34}$$

In a similar way, the converter operating times for the other regions are computed, and they are listed in Table 7.2. It is assumed that the converter spends time T_a in the first state of each sector and time T_b in the second state of the sector while moving in a counterclockwise direction.

Note that the time T_0 is determined from the following equation

$$T_0 = T_s - T_a - T_b \tag{7.35}$$

Table 7.2 Timing information for the six regions.

Region	Vector Placement	T_a	T_b
1	Between 0° and 60°	$\frac{3T_s}{2V_{dc}}\left(v_d - \frac{1}{\sqrt{3}}v_q\right)$	$\sqrt{3}\frac{T_s}{V_{dc}}v_q$
2	Between 60° and 120°	$\frac{3T_s}{2V_{dc}}\left(v_d + \frac{1}{\sqrt{3}}v_q\right)$	$\frac{3T_s}{2V_{dc}}\left(-v_d + \frac{1}{\sqrt{3}}v_q\right)$
3	Between 120° and 180°	$\sqrt{3}\frac{T_s}{V_{dc}}v_q$	$\frac{3T_s}{2V_{dc}}\left(-v_d - \frac{1}{\sqrt{3}}v_q\right)$
4	Between 180° and 240°	$\frac{3T_s}{2V_{dc}}\left(-v_d + \frac{1}{\sqrt{3}}v_q\right)$	$-\sqrt{3}\frac{T_s}{V_{dc}}v_q$
5	Between 240° and 300°	$\frac{3T_s}{2V_{dc}}\left(-v_d - \frac{1}{\sqrt{3}}v_q\right)$	$\frac{3T_s}{2V_{dc}}\left(v_d - \frac{1}{\sqrt{3}}v_q\right)$
6	Between 300° and 360°	$-\sqrt{3}\frac{T_s}{V_{dc}}v_q$	$\frac{3T_s}{2V_{dc}}\left(v_d + \frac{1}{\sqrt{3}}v_q\right)$

Example 7.3 Consider a converter with the following parameters

$$V_{dc} = 600\text{ V},\quad \sqrt{2}|V| = 315\text{ V},\quad \omega = 100\pi\text{ rad/s},\quad \varphi = -30^\circ$$

Using (7.12) and (7.13), the state vector v_P, evaluated at $t = 0.923$ seconds, is

$$v_P = 128.12 - j287.77 = 315\angle -66^\circ = 315\angle 294^\circ\text{ V}$$

This is obviously in Region-5, between 240° and 300°. Then from (7.30) and (7.29), we get

$$\frac{T_a}{T_s} = \sqrt{3}\frac{315}{600}\sin(294^\circ - 240^\circ) = 0.0951$$
$$\frac{T_b}{T_s} = \sqrt{3}\frac{315}{600}\sin(300^\circ - 294^\circ) = 0.7357$$

Now, for Region-5, $v_d = 128.12$ V and $v_q = -287.77$ V. Therefore, the following values are obtained from Table 7.2

$$\frac{T_a}{T_s} = \frac{1}{600}\left(-128.12 + \frac{1}{\sqrt{3}}287.77\right) = 0.0951$$
$$\frac{T_b}{T_s} = \frac{1}{600}\left(128.12 + \frac{1}{\sqrt{3}}287.77\right) = 0.7357$$

Thus, it can be concluded that both methods produce the same values. From (7.35), T_0/T_s is calculated as

$$\frac{T_0}{T_s} = 1 - \frac{T_a}{T_s} - \frac{T_b}{T_s} = 0.1693$$

7.3.5 Sequencing of Space Vectors

The SVPWM algorithm is carried out in the following steps:

1) Calculate v_P from (7.13) and determine $|V_P|$ and the angle ϕ.
2) From the angle ϕ, determine the sector in which the space vector lies at that instant.
3) Determine T_a and T_b respectively from (7.30) and (7.29) or from Table 7.3 and T_0 from (7.35).
4) Determine the switching sequence in which the upper and lower switches are turned on or off.

However, it is desired that the space vectors remain in the undermodulated region. For this, it must be ensured that the maximum absolute value of the space vector is bounded by [12]

$$|V_P|_{\max} = \frac{2V_{dc}}{3}\cos 30^{\circ} \tag{7.36}$$

The modulation index is then given by

$$m_a = \frac{\pi|V_P|}{2V_{dc}} \tag{7.37}$$

Then the maximum modulation index is calculated as 0.9069.

The advantage of SVPWM is that the number of switching (and hence switching losses) can be reduced using a special arrangement in which only one switching on each inverter leg occurs during the transition from one state to the next. This can only be achieved due to the use of the zero vectors V_0 and V_7. The switching patterns are shown in Figure 7.18. Consider the

Table 7.3 Switching information for the six regions.

Region	S_1	S_5	S_5
1	$T_a + T_b + T_0/2$	$T_b + T_0/2$	$T_0/2$
2	$T_a + T_0/2$	$T_a + T_b + T_0/2$	$T_0/2$
3	$T_0/2$	$T_a + T_b + T_0/2$	$T_b + T_0/2$
4	$T_0/2$	$T_a + T_0/2$	$T_a + T_b + T_0/2$
5	$T_b + T_0/2$	$T_0/2$	$T_a + T_b + T_0/2$
6	$T_a + T_b + T_0/2$	$T_0/2$	$T_a + T_0/2$

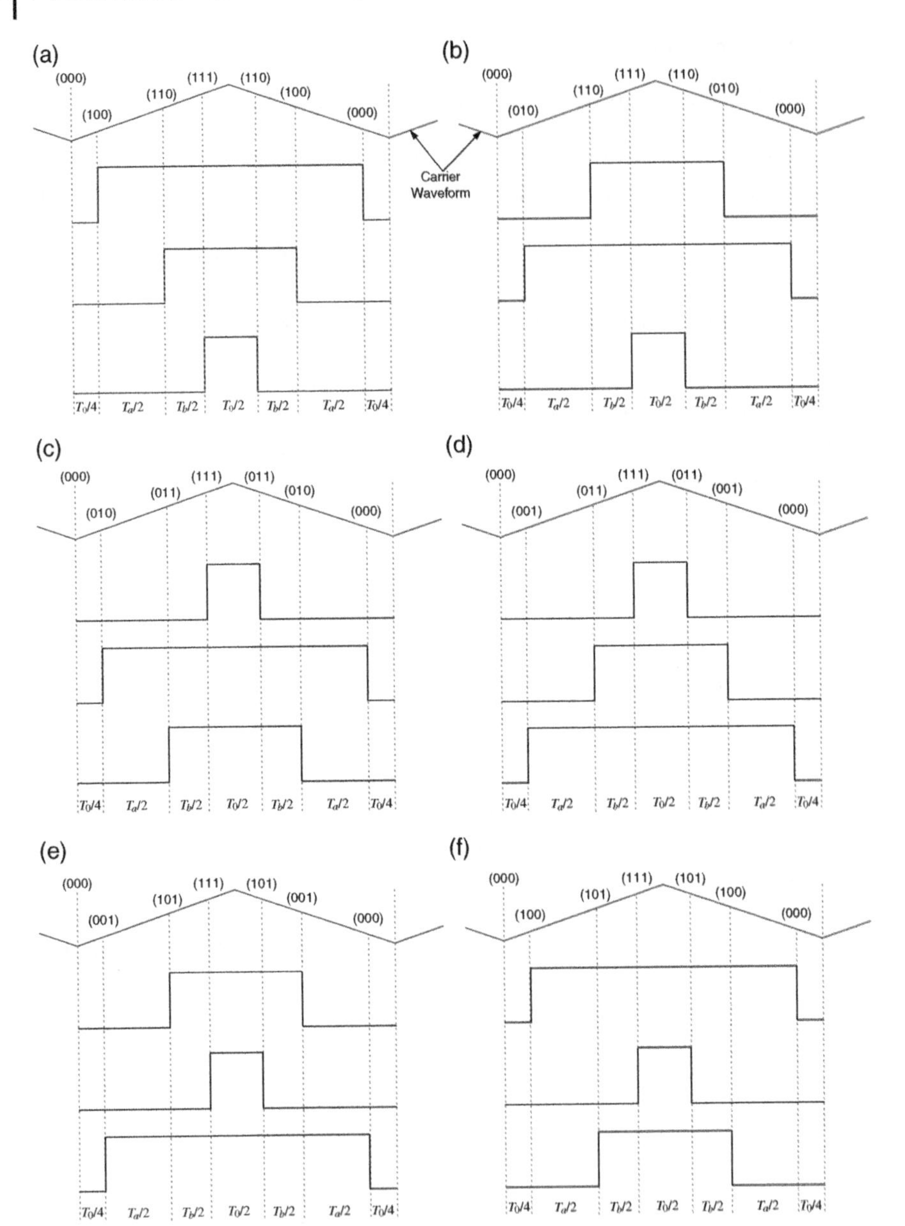

Figure 7.18 Switching patterns in six regions: (a) Region-1, (b) Region-2, (c) Region-3, (d) Region-4, (e) Region-5, and (f) Region-6.

switching pattern of Figure 7.18a. The converter transition pattern and the timing sequence used for this case are:

000	100	110	111	110	100	000
$T_0/4$	$T_a/2$	$T_b/2$	$T_0/2$	$T_b/2$	$T_a/2$	$T_0/4$

This is synchronized with a triangular carrier wave, as shown in Figure 7.18. The time duration of one full cycle of the carrier wave is T_s. The switching times of the upper switches for all the sectors are listed in Table 7.3. Note that the lower switches are complementary to the upper switches.

Example 7.4 Consider the same system as discussed in Example 7.2. Since the DC voltage is chosen as 600 V, $|V_P|_{max}$ is calculated from (7.36) as 346.41 V. The balanced sinusoidal modulating voltage has a peak of 315 V, and so $|V_P| = 315$ V. Therefore, the modulation index that is calculated from (7.37) is 0.8247, which is higher than the one obtained for SPWM. This proves the higher switch utilization for SVPWM. Next, a switching frequency of 5 kHz is chosen. Figure 7.19 shows the angle variation of the vector V_P and the corresponding regions (sectors). Figure 7.20 shows the switching signals and the output current. It can be seen that the output current has a higher magnitude than that in the case of SPWM due to better switch utilization. Also, the switching frequency is considerably lower in this case.

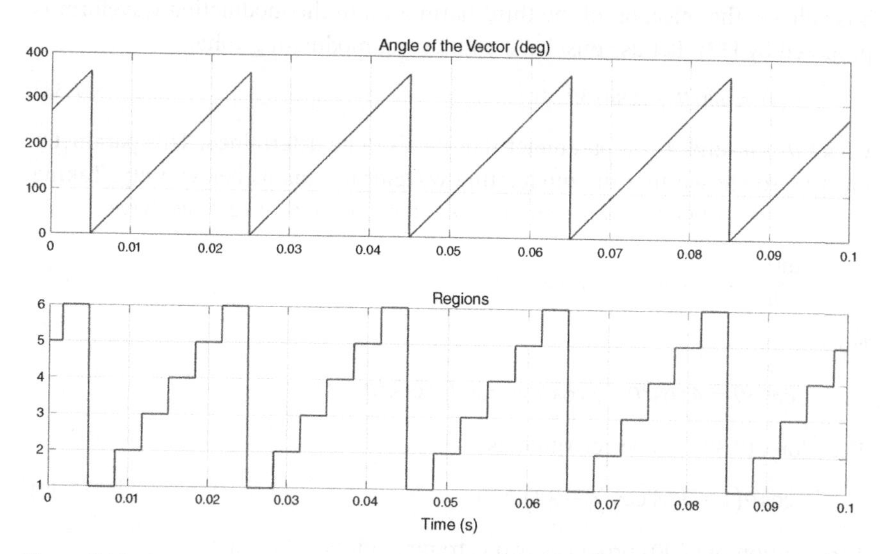

Figure 7.19 Variation in angle of the vector V_P and the computation of the regions in Example 7.4.

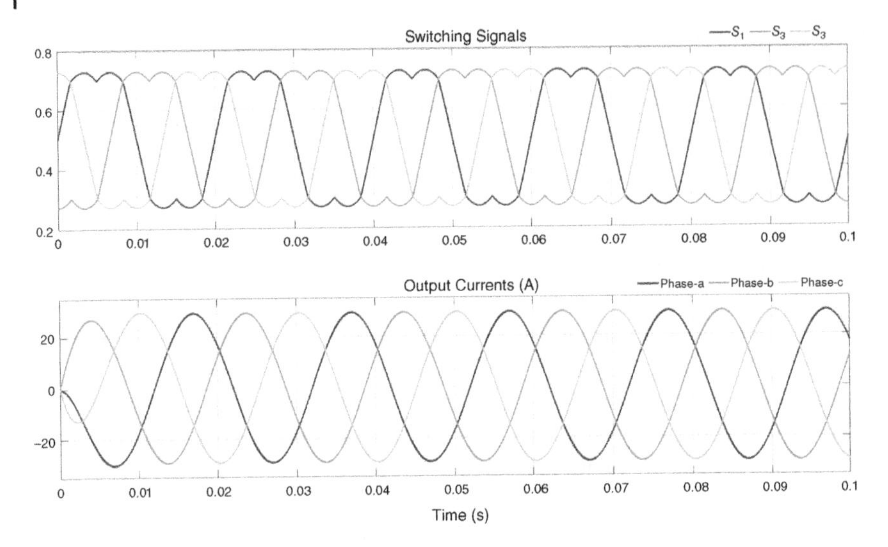

Figure 7.20 Switching signals and the output currents in Example 7.4.

7.4 SPWM with Third Harmonic Injection

To improve the performance of SPWM so that it can better utilize the available DC bus voltage, the injection of the third harmonics in the modulating waveform is proposed by [13]. Let us consider the following modulating voltage

$$v(t) = \sin\theta + A\sin 3\theta \tag{7.38}$$

where $\theta = \omega$ tand A is a parameter that needs to be determined. This parameter must be so chosen that the modulating waveform remains below unity. Taking the derivative of (7.36) with respect to θ and equating it to zero, we have

$$\frac{dv(t)}{d\theta} = \cos\theta + 3A\cos 3\theta = 0 \tag{7.39}$$

Now

$$\cos 3\theta = 4\cos^3\theta - 3\cos\theta = \cos\theta\left(4\cos^2\theta - 3\right)$$

Therefore, (7.39) can be rewritten as

$$\cos\theta\left(1 + 12A\cos^2\theta - 9A\right) = 0 \tag{7.40}$$

The solution of (7.40) produces two extremal points, which are

$$\cos\theta = 0 \tag{7.41}$$

$$\cos\theta = \sqrt{\frac{9A-1}{12A}} \tag{7.42}$$

Equation (7.41) is equivalent to

$$\sin\theta = 1 \tag{7.43}$$

Squaring both sides of (7.42), we get $\cos^2\theta = 9A - 1/12A$. Therefore, $\sin^2\theta$ can be written as

$$\sin^2\theta = 1 - \cos^2\theta = 1 - \frac{9A-1}{12A} = \frac{3A+1}{12A}$$

Therefore

$$\sin\theta = \sqrt{\frac{3A+1}{12A}} \tag{7.44}$$

We now substitute these values in (7.38) to determine the optimal value of A. Using the identity

$$\sin 3\theta = 3\sin\theta - 4\sin^3\theta = \sin\theta\left(3 - 4\sin^2\theta\right)$$

(7.38) is rewritten as

$$v(t) = \sin\theta\left(1 + 3A - 4A\sin^2\theta\right) \tag{7.45}$$

Substituting (7.43) in (7.45), we get

$$v(t) = 1 - A \tag{7.46}$$

Since (7.46) does not serve any useful purpose, we substitute (7.44) in (7.45) instead to get

$$\begin{aligned} v(t) &= \sqrt{\frac{3A+1}{12A}}\left(1 + 3A - 4A\frac{3A+1}{12A}\right) = \sqrt{\frac{3A+1}{12A}}\left[\frac{12A(3A+1) - 4A(3A+1)}{12A}\right] \\ &= 8A\left(\frac{3A+1}{12A}\right)^{\frac{3}{2}} \end{aligned} \tag{7.47}$$

The maximum value of $v(t)$ in (7.47) is obtained with respect to A as

$$\frac{dv(t)}{dA} = \sqrt{\frac{3A+1}{12A}}\left(2 - \frac{1}{3A}\right) = 0 \tag{7.48}$$

Equation (7.48) has two possible solutions, which are

$$A = \frac{1}{6} \text{ and } A = -\frac{1}{3} \tag{7.49}$$

Consider the waveform of phase-a, where the third harmonic component is in phase with the fundamental voltage. Then, during the negative peak, the modulating waveform will be more than the carrier waveform if we choose $A = -1/3$. As can be seen from Figure 7.21a, the modulating waveform is higher than the carrier waveform indicating an overmodulation. However, it remains in the undermodulated region if $A = 1/6$ is chosen, as shown in Figure 7.21b. Then (7.38) can be rewritten as

$$v(t) = \sin\theta + \frac{1}{6}\sin 3\theta \tag{7.50}$$

Substituting $A = 1/6$, θ can be evaluated from (7.44) as

$$\theta = \sin^{-1}\left(\sqrt{\frac{3/6 + 1}{12/6}}\right) = \sin^{-1}\left(\sqrt{\frac{1.5}{2}}\right) = 0.866 = \frac{\sqrt{3}}{2} \tag{7.51}$$

Thus, the positive and negative peaks will occur at 60°, 240°,.... The maximum output voltage at these instants will then be

$$|v(t)|_{\max} = \sin 60^\circ + \frac{1}{6}\sin 180^\circ = 0.866$$

To increase the output voltage even further, the peak of the modulating voltage can be made equal to 1 by using

$$v(t) = K\left(\sin\theta + \frac{1}{6}\sin 3\theta\right) \tag{7.52}$$

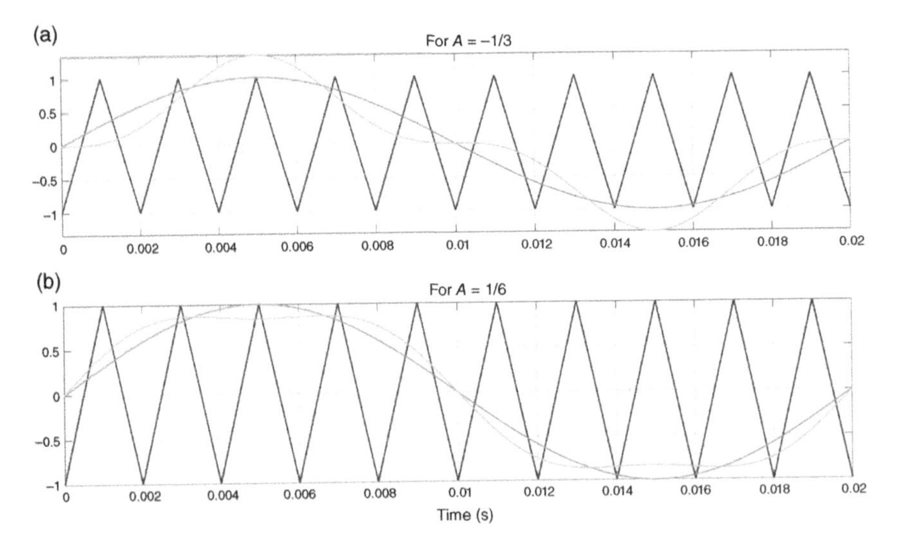

Figure 7.21 Modulating wave for two different values of A. (a) For $A = -1/3$ and (b) For $A = 1/6$.

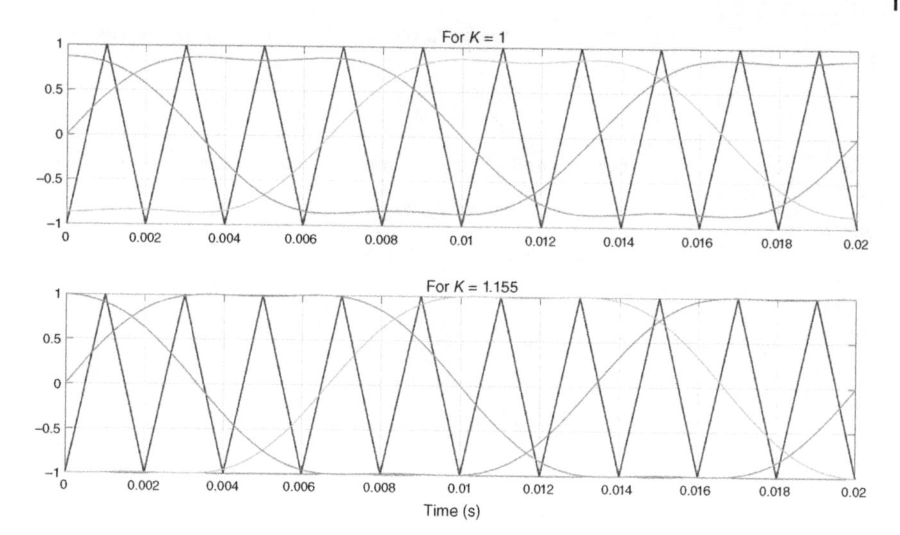

Figure 7.22 Modulating wave for two different values of K.

where the gain K is chosen as

$$K = \frac{1}{|v(t)|_{\max}} = 1.155$$

Figure 7.22 clearly shows that the switch utilization for $K = 1.155$ is better than that with $K = 1$.

Note that the triplen harmonics of the three phases have the same phase angles and, therefore, they are eliminated from the line to neutral voltages. The lowpass filtered version of these three-phase voltages are shown in Figure 7.23 for two different values of K ($K = 1$ and $K = 1.155$). The reference voltage is chosen as 400 V. The peak of the line to neutral voltage with $K = 1.155$ is higher than that with $K = 1$. Note that the converter bandwidth must be at least three times the desired output frequency in order to accommodate to the triplen harmonics [13].

7.5 Multilevel Converters

The first application of power electronic converters in power systems was in high-voltage direct current (HVDC) bulk power transmission, in which the converters were basically line frequency commutated. Most popular among these converters were 12-step converters, which were constructed using two 6-step converters that are phase shifted through magnetic connections. These were mainly thyristor-based

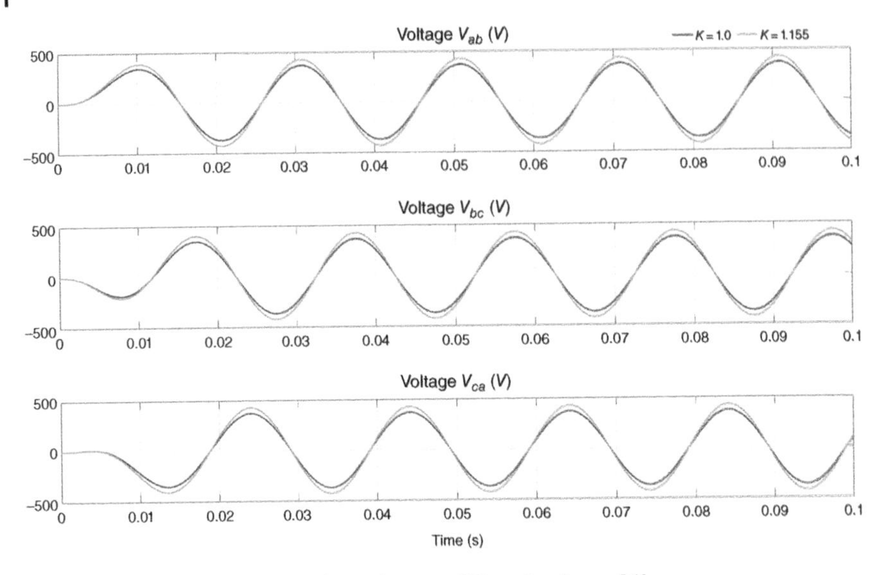

Figure 7.23 Line to neutral voltage for two different values of K.

devices. Later, gate turn-off thyristor (GTO) based devices were proposed for multi-step converters. However, with the improvement in the power ratings of IGBT switches, different types of multilevel converters have been proposed. These converters are capable of high-power operation. The structures of some multilevel converters are discussed in this section. It is, however, to be noted that the control of these converters is more involved than two-level converters.

Multilevel converters operate using a combination of a series connection of semiconductor switches such as IGBTs or MOSFETs with different DC voltage sources or capacitors to synthesize and generate different voltage levels which are either a low-frequency staircase or a high-frequency modulated voltage waveform. In real life applications, different energy sources such as batteries or photovoltaic panels can be considered as the DC voltage sources in various multilevel converter structures.

Figure 7.24a shows one leg of a general schematic of a multilevel converter with n voltage level based on several capacitors as voltage sources. In this example, the high-voltage DC source (V_{dc}) can be generated by a high-voltage rectifier and the capacitor voltages can be charged equally or unequally. In other cases, different low-voltage DC sources (batteries or solar panels) can be utilized to generate a high-voltage across the DC link and the load. Assuming that the voltage ripple across each capacitor is negligible and the capacitors are charged at defined voltage levels (depending on the multilevel topology and configuration), the output

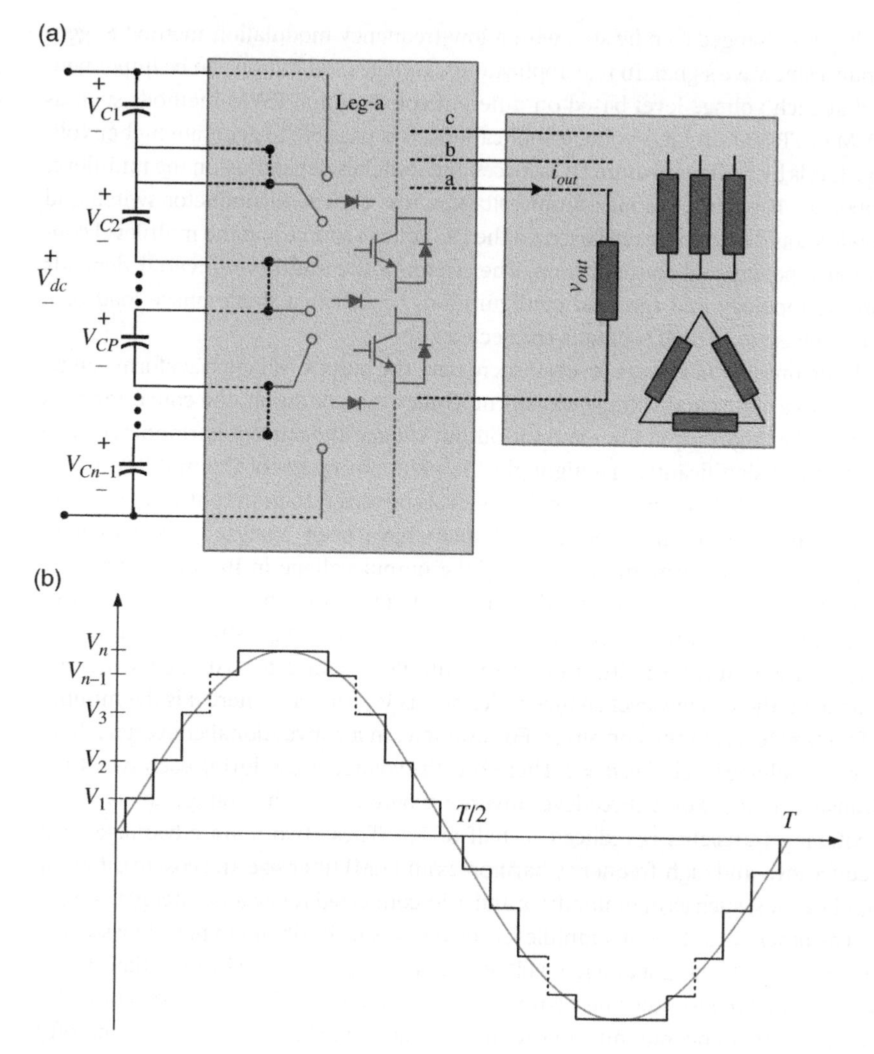

Figure 7.24 (a) Schematic diagram of multilevel converter and (b) output voltage waveform of one phase of the converter.

voltage can be generated based on the position of the switch in the leg and the configuration of the capacitors, in series with the load.

Figure 7.24b shows the output voltage of the multilevel converter for a single-phase system with different voltage levels utilizing n number of DC capacitors which can be charged equally or unequally [14–15]. In this case, the output voltages are generated without using a high-frequency PWM strategy, and the output voltage is changed step by step with a low-frequency modulation method to generate a sinewave signal. In real applications, voltage modulation can be implemented at each voltage level based on different conventional PWM methods such as SVM or SPWM. In high-voltage applications, it is possible to generate higher voltage levels by adding up more DC sources and switches depending on the multilevel topology. However, the maximum voltage across each semiconductor switch and diode should be considered based on the DC voltage source and the multilevel converter topology and configuration. The circuit of the multilevel inverter depends on the topology and the load configuration, i.e. either a single-phase load or a three-phase load (delta- or star-connected).

If the number of voltage levels is increased, the output voltage waveform can be closer to a sinusoidal voltage waveform. Under this condition, the converter does not need a large output filter and the output voltage and current harmonics can be minimized significantly. To highlight the main advantage of the multilevel converter compared to a conventional two-level converter, their output voltage waveforms in the time and frequency domain have been analyzed. As shown in Figure 7.25, the harmonic contents of the output voltage in the multilevel converter of Figure 7.25b are significantly less than that of the two-level converter of Figure 7.25a. During each switching transient, the voltage stress (dv/dt) across the load is reduced significantly in the multilevel converter. At each switching transient, the voltage level change is defined as $V_{dc}/(n-1)$, where n is the number of voltage levels in the converter. For example, in a conventional converter, there are two voltage levels, i.e. $n = 2$. Therefore, the voltage stress during each switching transient is V_{dc}. For a three-level inverter where $n = 3$, the voltage stress at the switching transients is reduced to half of V_{dc}. These two main advantages can reduce low- and high-frequency harmonics and EMI filter size and cost in different applications, such as in motor drive and grid-connected renewable energy systems.

The other advantage of a multilevel converter is its flexibility to have low switching losses. Utilizing fast and low-voltage semiconductor devices in a multilevel converter compared to slow and high-voltage semiconductors in a two-level converter can make them operate much faster during each switching transient (on and off) and at a lower switching frequency. Thus, a multilevel converter can be designed with a reduced switching loss suitable for different applications. Note that it is obvious that the number of the semiconductor switches in a multilevel converter is higher than a two-level converter but the DC voltage across each switch in a multilevel converter can be lower than the same voltage across a conventional converter. As the switching loss is proportional to the DC voltage level across the switch, the number of switches, the number of voltage levels, and the DC link voltage should be considered to calculate the total switching loss in a multilevel converter.

In addition to these advantages, the multilevel converter can be modulated in such a way that it has a reduced common-mode voltage in a three-phase system

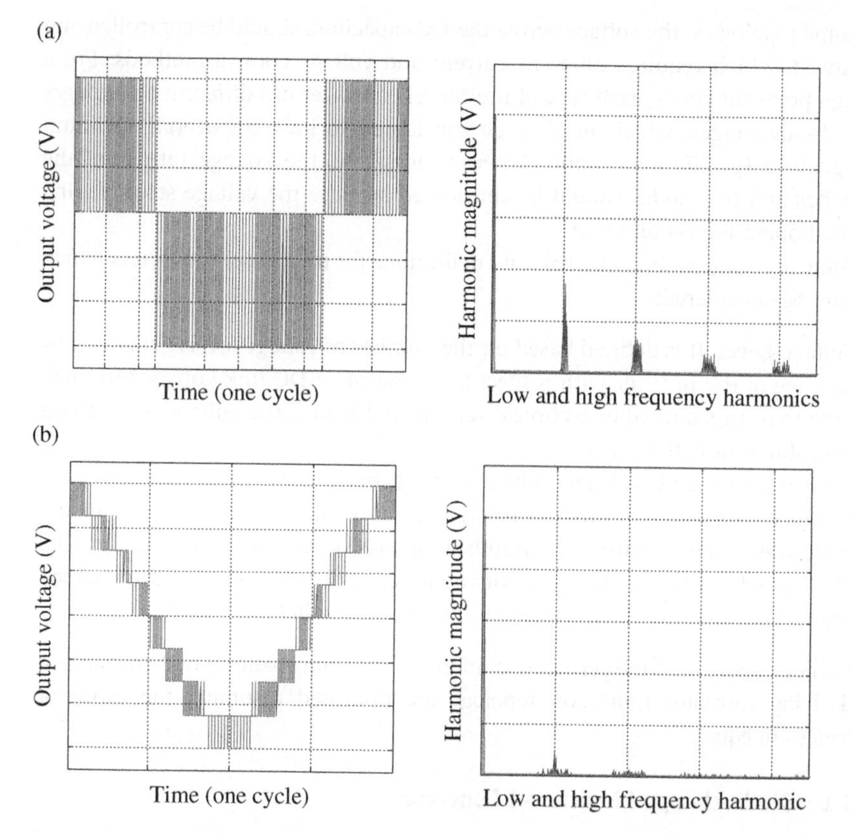

Figure 7.25 Output voltage waveform and harmonic spectrum of (a) two-level converter and (b) multilevel converter.

based on a proper modulation technique through the utilization of different zero-voltage vectors. This feature reduces the common mode and shaft voltage in motor drive systems with less voltage stress on a bearing system, thereby increasing the lifetime of the motor under control.

However, the main drawbacks of the multilevel converters are the need for a high number of passive and switching components and a complexity on the converter topology, control, and overall system design. Multilevel converters have different topologies and configurations based on different structures of a DC link voltage. The main multilevel converter topologies are:

- diode-clamped
- flying capacitor
- cascaded converters.

In some topologies, the voltage across the DC capacitors should be controlled and balanced, which requires different current and voltage control methods. From design points of views, each type of multilevel converter has different advantages and disadvantages, which need to be considered to have an overall optimum design in a specific application. For each topology, the voltage ratings of the switches and the diodes should be considered because the voltage stress during normal operations is different.

There are some technical terms and definitions for multilevel converters which should be considered:

- Voltage Level: It is defined based on the number of voltage levels generated by each leg of the inverter with respect to the negative DC link voltage terminal. Note that, in some other technical reports and books, the voltage level might have different definitions.
- Capacitor Voltage Level: The voltage across capacitors can be equal or unequal. The main advantage of the unequal voltage across the capacitors is to generate more voltage levels with a few numbers of passive and switching devices. The main drawback here is the capacitor voltage control during the operation which depends on the multilevel topology, control, and switching patterns.

In Sections 7.5.1–7.5.3, the general operation and switching patterns of diode-lamped and flying capacitor multilevel topology are discussed assuming the capacitor voltages are equal.

7.5.1 Diode-clamped Multilevel Converter

A diode-clamped (or neutral-point-clamped) multilevel converter is one of the multilevel converter topologies which have been widely utilized in different products such as variable speed motor drive, renewable energy, and power quality compensation systems. Figure 7.26a shows one leg of a three-level diode-clamped converter where the total DC link voltage is V_{dc} and $V_{c1} = V_{c2} = V_{dc}/2$.

The three-level converter, at each leg, has three voltage levels (0, $V_{dc}/2$, and V_{dc}) with respect to the negative DC link terminal. The DC link voltage is split into two voltage levels using two capacitors, C_1 and C_2, which are charged equally at $V_{dc}/2$. This structure consists of pairs of switches (S_{a1}, S_{a3} and S_{a2}, S_{a4}) and two clamped diodes (D_{ca1} and D_{ca2}) which connect the common terminals of the capacitors to the switches in each leg. For example, when switch S_{a1} is on, switch S_{a3} is off, and vice versa, and one of the diodes might be turned on depending on the converter switching state. The switching state of each leg is defined based on the switching states of the top switches, where 1 and 0 represent the on and off states respectively of the top switches. For example, the switching state of 01 in leg-a means that the

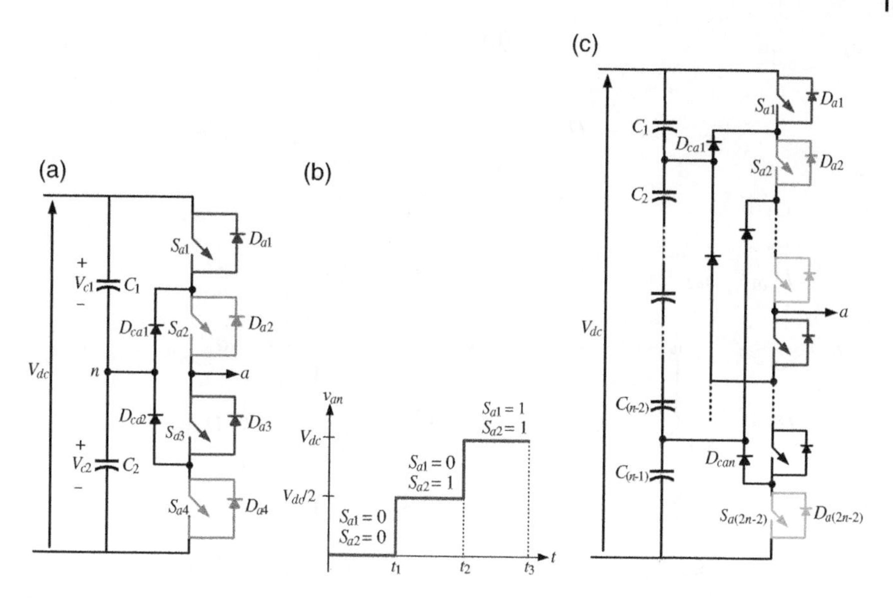

Figure 7.26 (a) One leg of a three-level diode-clamped converter, (b) its output voltage waveform, and (c) one leg of an n-level converter.

first top switch, S_{a1}, is off and the second top switch, S_{a2}, is on. Three voltage levels are generated based on different switching states, as shown in Figure 7.26b and explained in the following paragraphs.

To develop a single-phase or three-phase converter, more legs (switches in series) are required to be in parallel while the DC link configuration is not changed. However, more clamped diodes are required to connect the DC link to each leg of the converter. To increase the output voltage amplitude and levels, more DC capacitors and pairs of switches are required as well as the clamped diodes, according to the topology shown in Figure 7.26c.

7.5.2 Switching States of Diode-clamped Multilevel Converters

In this section, the different switching states of the diode-clamped converter are presented in detail.

- Case 1: When $S_{a1} = 0$ *and* $S_{a2} = 0$: In this switching state, switches S_{a1} and S_{a2} are off and the complement switches, S_{a3} and S_{a4}, are on, as shown in Figure 7.27. When the load current is positive, the antiparallel diodes D_{a3} and D_{a4} conduct, as shown in Figure 7.27a, while Figure 7.27b depicts the loop when

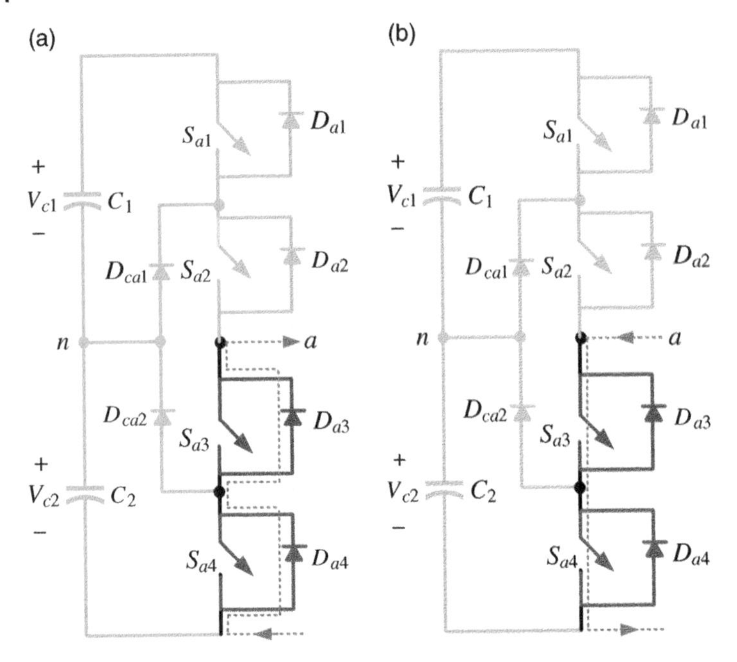

Figure 7.27 The current loops when S_{a1} = 0 and S_{a2} = 0 for (a) positive load current and (b) negative load current.

the load current is negative and S_{a3} and S_{a4} conduct. Thus, terminal a is connected to terminal n of the DC link and the output voltage is zero, i.e. $v_a = 0$ during the time interval $0 < t \leq t_1$ (see Figure 7.26b). According to the current loops shown in Figure 7.27, this switching state cannot affect the charging state of the DC link capacitors, as the capacitors C_1 and C_2 are not charged or discharged through the load current.

- Case 2: When $S_{a1} = 0$ *and* $S_{a2} = 1$: In this switching state, the switch S_{a1} is off and its complementary switch S_{a3} is on, while the switches S_{a2} and S_{a4} are on and off respectively. Figure 7.28a depicts the current flow path when the load current is positive. During the interval $t_1 < t \leq t_2$, switch S_{a2} and diode D_{ca1} conduct due to the polarity of the voltage across the diode and the load current direction. In this case, when the load current passes through the DC link capacitor, C_2 will discharge it. Figure 7.28b shows the current flow direction when the load current is negative, where the current flow direction will cause the capacitor C_2 to charge. Assuming that the voltage ripple is negligible, and the capacitors are charged equally, we have $V_{c1}(t) = V_{c2}(t) = V_{dc}/2$. During this interval of $t_1 < t \leq t_2$, the output voltage is $v_a = V_{dc}/2$.

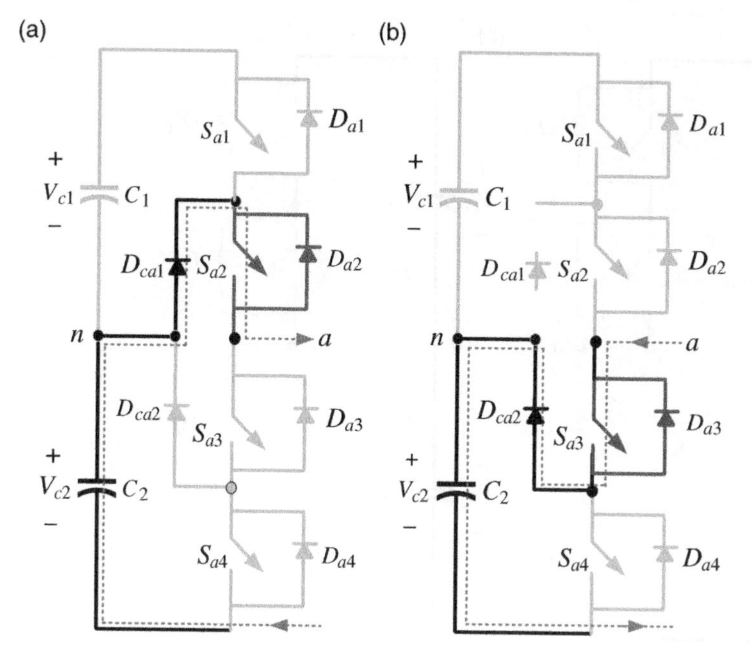

Figure 7.28 The current loops when $S_{a1} = 0$ and $S_{a2} = 1$ for (a) positive load current and (b) negative load current.

- Case 3: When $S_{a1} = 1$ *and* $S_{a2} = 1$: When the top switches S_{a1} and S_{a2} are on, the output voltage is V_{dc}. Assuming the load current is positive, S_{a1} and S_{a2} conduct and the load current passes through the DC link, as shown in Figure 7.29a. Usually, the DC link voltage is connected to a common DC source, and the total DC voltage does not change significantly during each switching state. Since the switching state $S_{a1} = 0$ and $S_{a2} = 0$ cannot affect the charging states of the DC link capacitors, as the capacitors C_1 and C_2 are assumed to be connected to a voltage source and the total DC link, the voltage across the capacitors is not changed.
- Case 4: When $S_{a1} = 1$ *and* $S_{a2} = 0$: This is a forbidden state for the diode-clamped topology. Figure 7.30 shows the two different output voltage levels based on positive and negative load currents. When the load current is positive, the diodes D_{a3} and D_{a4} conduct, and the output voltage is zero (see Figure 7.30a). On the other hand, when the load current is negative, the diodes D_{a1} and D_{a3} conduct, for which the output voltage is V_{dc}, as shown in Figure 7.30b. We cannot achieve an output voltage of $V_{dc}/2$ and its amplitude

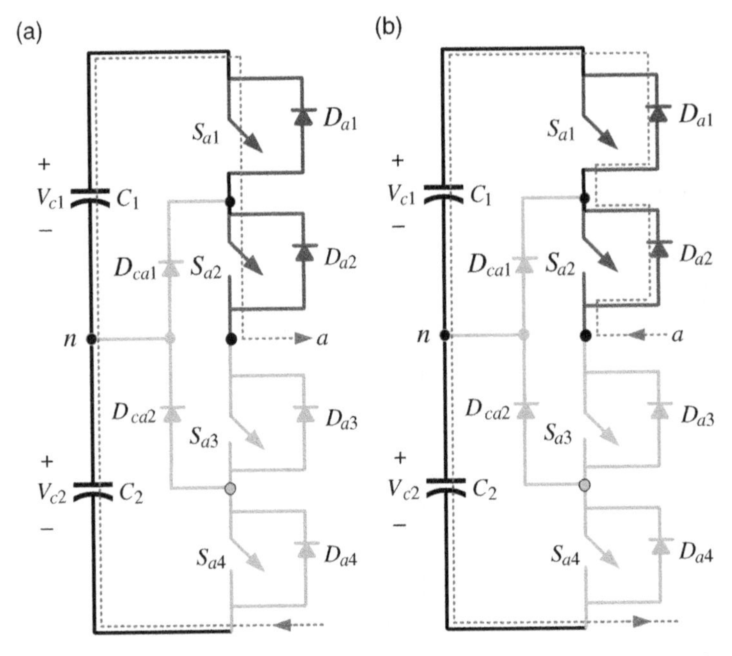

Figure 7.29 The current loops when $S_{a1} = 1$ and $S_{a2} = 1$ for (a) positive load current and (b) negative load current.

is changed from 0 to V_{dc}, depending on the load current. Thus, this switching state is not used in a diode-clamped topology as the output voltage depends on the load current.

So far only one leg of a diode-clamped converter has been analyzed and explained. A single-phase full-bridge or a three-phase converter configuration consists of two or three legs of the above-mentioned topology in parallel respectively. The common terminals of the capacitors are connected to the switches in each leg by the clamped diodes. These are shown in Figure 7.31. The DC link configuration is not changed, while the parallel legs are added to develop single-phase or three-phase converters.

Table 7.4 lists all switching states of a single-phase, diode-clamped three-level converter, while Table 7.5 lists the switching states of a three-level, three-phase converter. The output voltage (line-to-line) is measured based on two leg voltages. However, in a three-phase system, the phase voltage across the load should be extracted based on the leg and the common mode voltage. In Table 7.4, when the switching state of each leg is 10, the output voltage cannot be determined,

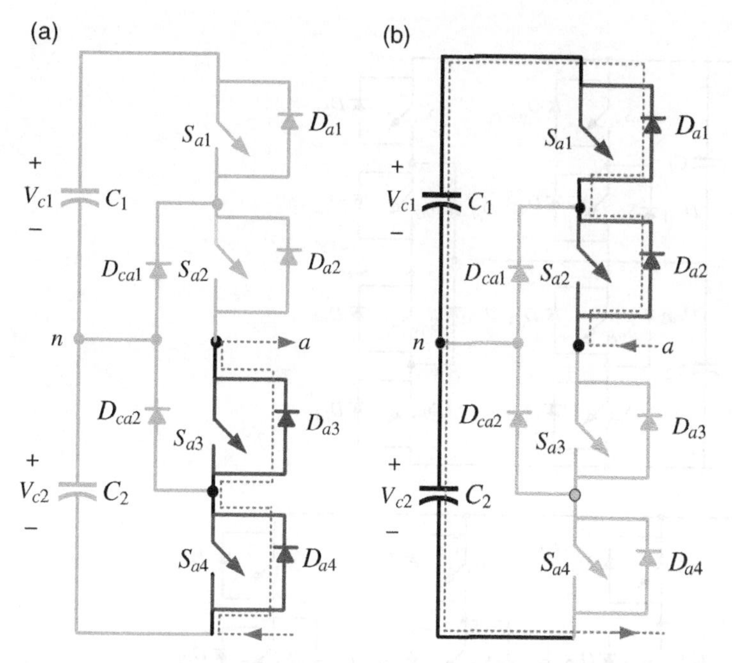

Figure 7.30 The current loops when $S_{a1} = 1$ and $S_{a2} = 0$ for (a) positive load current and (b) negative load current.

as the output voltage depends on the load current. This is indicated by a "?" where this occurs, and the rows are shaded in the table. Thus, these switching states are not used in the inverter operating modes.

7.5.3 Flying Capacitor Multilevel Converter

This multilevel converter topology comprises several capacitors that are connected in parallel with each leg. One leg of a three-level converter is shown in Figure 7.32a. In this, the switch pairs (S_{a1}, S_{a4}) and (S_{a2}, S_{a3}) work in complementary fashion, i.e. when one of the pairs (e.g. S_{a1}) is on, the other switch S_{a4} is off and/or vice versa. The capacitors in this structure are charged to different voltage levels. Different output voltage levels can be achieved by connecting the capacitors in parallel with the DC link voltage based on switching states of the semiconductor switches. Assuming that the voltage across the DC capacitors is $V_{dc}/2$ with no ripple, the leg voltage for different switching states is shown in Figure 7.32b. One leg of a general n-level flying capacitor converter is shown in Figure 7.32c.

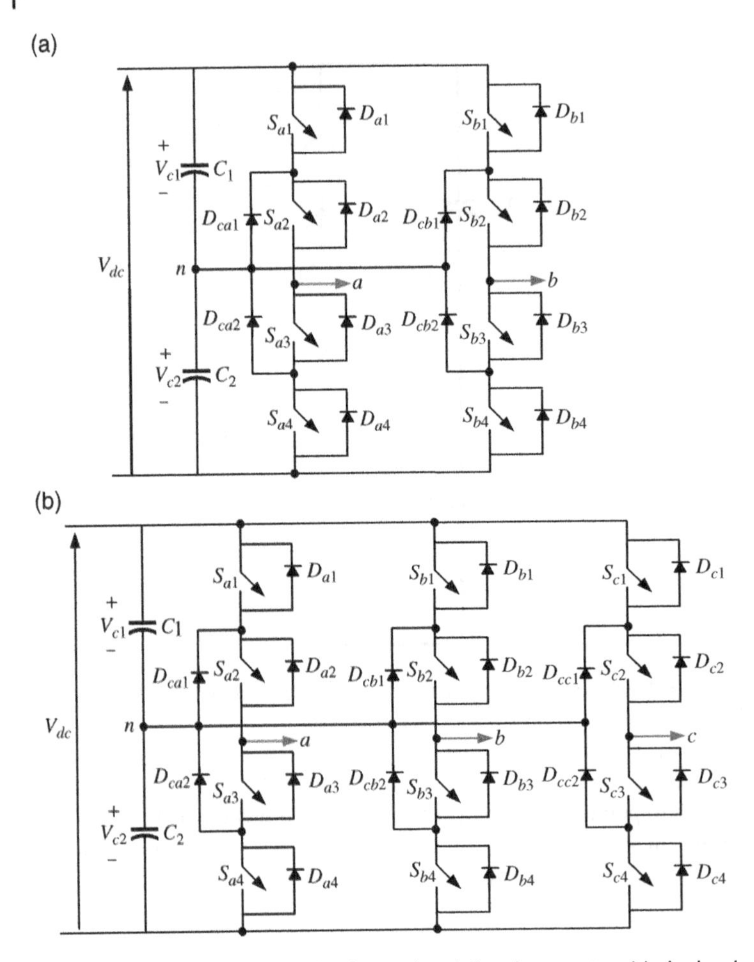

Figure 7.31 Three-level diode-clamped multilevel converter: (a) single-phase and (b) three-phase.

As in the case of a diode-clamped converter, the switching behaviors of the flying capacitor converter are dependent on the switching states. Three different cases are discussed here:

- Case 1: When $S_{a1} = 0$ *and* $S_{a2} = 0$: In this switching state, the complement switches S_{a3} and S_{a4} are on, as shown in Figure 7.33. The output current will flow through either the switches or the diodes depending on the direction of the current. The output voltage, however, will remain zero in either case.
- Case 2: When ($S_{a1} = 0$ *and* $S_{a2} = 1$) *or* ($S_{a1} = 1$ *and* $S_{a2} = 0$): There is no switching restriction in the flying capacitor topology and the top switches can have two

Table 7.4 Switching states for a three-level, single-phase diode-clamped converter.

Switching States $S_{a1}\,S_{a2}\,S_{b1}\,S_{b2}$	$v_a(t)$	$v_b(t)$	$v_{ab}(t)$
0000	0	0	0
0001	0	$V_{dc}/2$	$-V_{dc}/2$
0010	0	?	?
0011	0	$V_{dc}/2$	$-V_{dc}/2$
0100	$V_{dc}/2$	0	$V_{dc}/2$
0101	$V_{dc}/2$	$V_{dc}/2$	0
0110	$V_{dc}/2$	?	?
0111	$V_{dc}/2$	V_{dc}	$-V_{dc}/2$
1000	?	0	?
1001	?	$V_{dc}/2$	?
1010	?	?	?
1011	?	V_{dc}	?
1100	V_{dc}	0	V_{dc}
1101	V_{dc}	$V_{dc}/2$	$V_{dc}/2$
1110	V_{dc}	?	?
1111	V_{dc}	V_{dc}	0

different combinations to generate $V_{dc}/2$, as shown in Figure 7.34. Assuming that the voltage across the flying capacitor C_{a1} is $V_{dc}/2$, the output voltage for both switching states is $V_{dc}/2$ since:

- when $S_{a1} = 0$ and $S_{a2} = 1$, $v_a = V_{ca1} = V_{dc}/2$
- when $S_{a1} = 1$ and $S_{a2} = 0$, $v_a = V_{dc} - V_{ca1} = V_{dc}/2$

- Case 3: When $S_{a1} = 1$ *and* $S_{a2} = 1$: In this case both switches S_{a1} and S_{a2} are on and the complementary switches S_{a3} and S_{a4} are off. As shown in Figure 7.35, the output voltage is V_{dc} and the flying capacitor C_{a1} are not connected across the load.

As in the case of the diode-clamped topology, a single-phase and a three-phase flying capacitor topology can be developed by paralleling two or three legs, as shown in Figure 7.36. Since there is no restriction for the top switches, and there are four top switches (two in leg-a and two in leg-b), a total of $2^4 = 16$ different switching combinations for a single-phase, three-level converter is possible. These are listed in Table 7.6.

Table 7.5 Only possible switching states for a three-level, three-phase diode-clamped converter.

Switching States $S_{a1}\,S_{a2}\,S_{b1}\,S_{b2}\,S_{c1}\,S_{c2}$	$v_a(t)$	$v_b(t)$	$v_c(t)$	$v_{ab}(t)$	$v_{bc}(t)$	$v_{ca}(t)$
000000	0	0	0	0	0	0
010000	$V_{dc}/2$	0	0	$V_{dc}/2$	0	$-V_{dc}/2$
110 000	V_{dc}	0	0	V_{dc}	0	$-V_{dc}$
000100	0	$V_{dc}/2$	0	$-V_{dc}/2$	$V_{dc}/2$	0
010100	$V_{dc}/2$	$V_{dc}/2$	0	0	$V_{dc}/2$	$-V_{dc}/2$
110 100	V_{dc}	$V_{dc}/2$	0	$V_{dc}/2$	$V_{dc}/2$	$-V_{dc}$
001100	0	$V_{dc}/2$	0	$-V_{dc}$	V_{dc}	0
011100	$V_{dc}/2$	V_{dc}	0	$-V_{dc}/2$	V_{dc}	$-V_{dc}/2$
111 100	V_{dc}	V_{dc}	0	0	V_{dc}	$-V_{dc}$
000001	0	0	$V_{dc}/2$	0	$-V_{dc}/2$	$V_{dc}/2$
010001	$V_{dc}/2$	0	$V_{dc}/2$	$V_{dc}/2$	$-V_{dc}/2$	0
110 001	V_{dc}	0	$V_{dc}/2$	V_{dc}	$-V_{dc}/2$	$-V_{dc}/2$
000101	0	$V_{dc}/2$	$V_{dc}/2$	$-V_{dc}/2$	0	$V_{dc}/2$
010101	$V_{dc}/2$	$V_{dc}/2$	$V_{dc}/2$	0	0	0
110 101	V_{dc}	$V_{dc}/2$	$V_{dc}/2$	$V_{dc}/2$	0	$-V_{dc}/2$
001101	0	V_{dc}	$V_{dc}/2$	$-V_{dc}$	$V_{dc}/2$	$V_{dc}/2$
011101	$V_{dc}/2$	V_{dc}	$V_{dc}/2$	$-V_{dc}/2$	$V_{dc}/2$	0
111 101	V_{dc}	V_{dc}	$V_{dc}/2$	0	$V_{dc}/2$	$-V_{dc}/2$
000011	0	0	V_{dc}	0	$-V_{dc}$	V_{dc}
010011	$V_{dc}/2$	0	V_{dc}	$V_{dc}/2$	$-V_{dc}$	$V_{dc}/2$
110 011	V_{dc}	0	V_{dc}	V_{dc}	$-V_{dc}$	0
000111	0	$V_{dc}/2$	V_{dc}	$-V_{dc}/2$	$-V_{dc}/2$	V_{dc}
010111	$V_{dc}/2$	$V_{dc}/2$	V_{dc}	0	$-V_{dc}/2$	$V_{dc}/2$
110 111	V_{dc}	$V_{dc}/2$	V_{dc}	$V_{dc}/2$	$-V_{dc}/2$	0
001111	0	V_{dc}	V_{dc}	$-V_{dc}$	0	V_{dc}
011111	$V_{dc}/2$	V_{dc}	V_{dc}	$-V_{dc}/2$	0	$V_{dc}/2$
111 111	V_{dc}	V_{dc}	V_{dc}	0	0	0

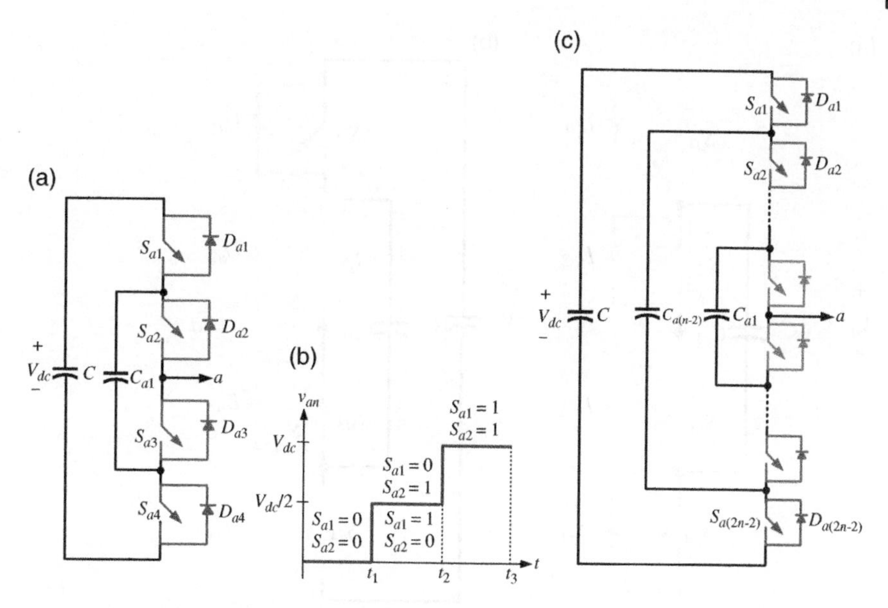

Figure 7.32 (a) One leg of a three-level flying capacitor converter, (b) its output voltage waveform, (c) and one leg of an n-level converter.

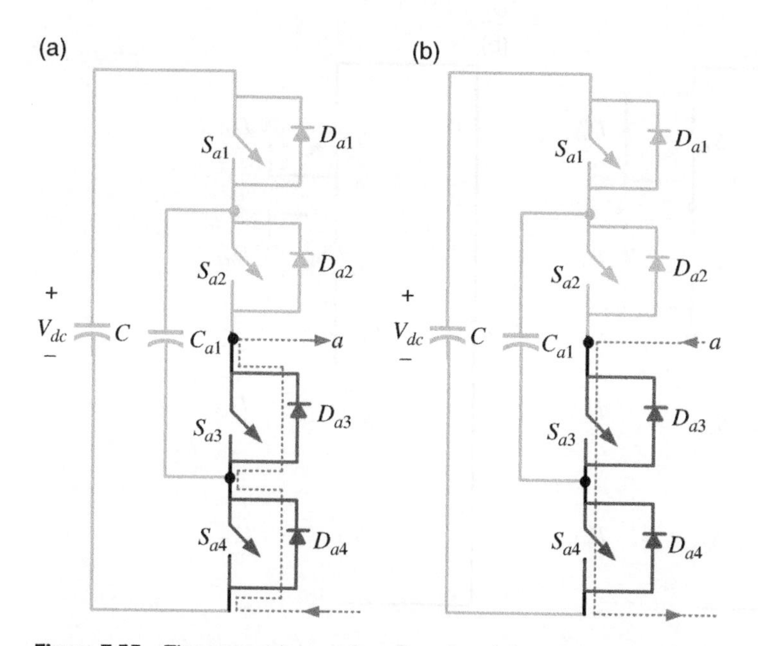

Figure 7.33 The current loops when $S_{a1} = 0$ and $S_{a2} = 0$ for (a) positive load current and (b) negative load current.

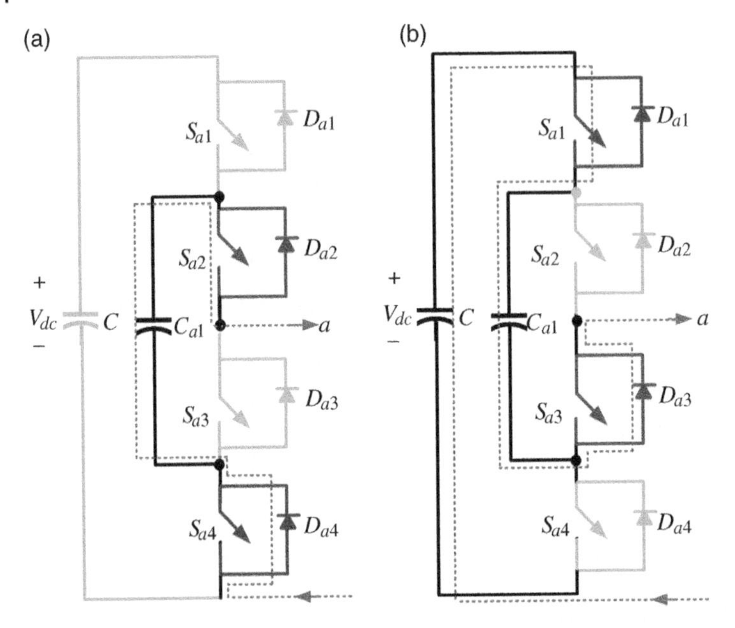

Figure 7.34 The current loops when (a) $S_{a1} = 0$ and $S_{a2} = 1$ and (b) $S_{a1} = 1$ and $S_{a2} = 0$.

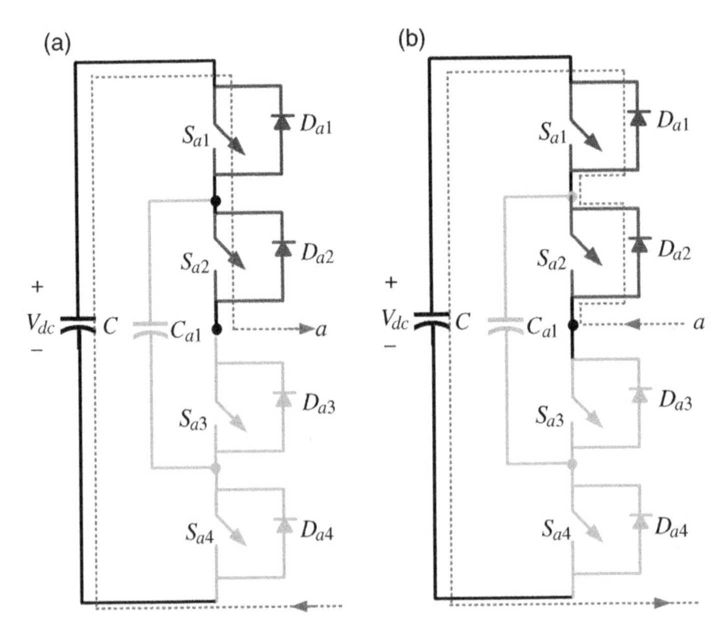

Figure 7.35 The current loops when $S_{a1} = 1$ and $S_{a2} = 1$ for (a) positive load current and (b) negative load current.

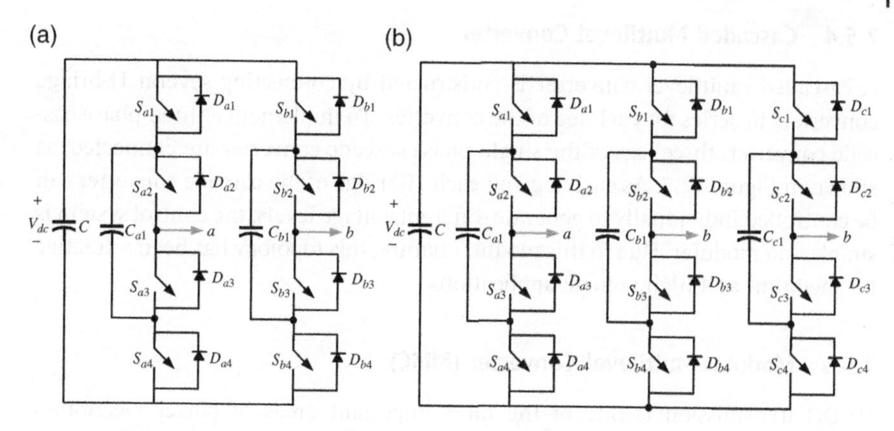

Figure 7.36 Three-level flying capacitor converter: (a) single-phase and (b) three-phase.

Table 7.6 Switching states for a three-level, single-phase flying capacitor converter.

Switching States $S_{a1}\,S_{a2}\,S_{b1}\,S_{b2}$	$v_a(t)$	$v_b(t)$	$v_{ab}(t)$
0000	0	0	0
0001	0	$V_{dc}/2$	$-V_{dc}/2$
0010	0	$V_{dc}/2$	$-V_{dc}/2$
0011	0	$V_{dc}/2$	$-V_{dc}/2$
0100	$V_{dc}/2$	0	$V_{dc}/2$
0101	$V_{dc}/2$	$V_{dc}/2$	0
0110	$V_{dc}/2$	$V_{dc}/2$	0
0111	$V_{dc}/2$	V_{dc}	$-V_{dc}/2$
1000	$V_{dc}/2$	0	$V_{dc}/2$
1001	$V_{dc}/2$	$V_{dc}/2$	0
1010	$V_{dc}/2$	$V_{dc}/2$	0
1011	$V_{dc}/2$	V_{dc}	$-V_{dc}/2$
1100	V_{dc}	0	V_{dc}
1101	V_{dc}	$V_{dc}/2$	$V_{dc}/2$
1110	V_{dc}	$V_{dc}/2$	$V_{dc}/2$
1111	V_{dc}	V_{dc}	0

7.5.4 Cascaded Multilevel Converter

A cascaded multilevel converter is constructed by connecting several H-bridge converters in series for each leg of the converter. To implement a three-phase cascade converter, three legs of the single-phase cascade converter are connected, as shown in Figure 7.37. As each leg and each H-bridge of the cascade converter can be controlled individually to generate different voltage levels, the control system is simple and modular. Due to this modular nature, this topology has been attractive for medium- and high-voltage applications.

7.5.5 Modular Multilevel Converter (MMC)

HVDC transmission is one of the most important areas of power electronics applications to power systems. Due to the large uptake of offshore windfarms, modular multilevel converters are gaining popularity these days for VSC-HVDC transmission. The MMCs are modular and therefore can be scalable to any desired voltage level. They have higher efficiency and lower harmonic components. The schematic diagram of one phase of an MMC is shown in Figure 7.38a. Each phase can have several identical submodules (SMs). In Figure 7.38a, there are n SMs on the positive half of the leg and another n SMs on the negative half of the leg. The SMs

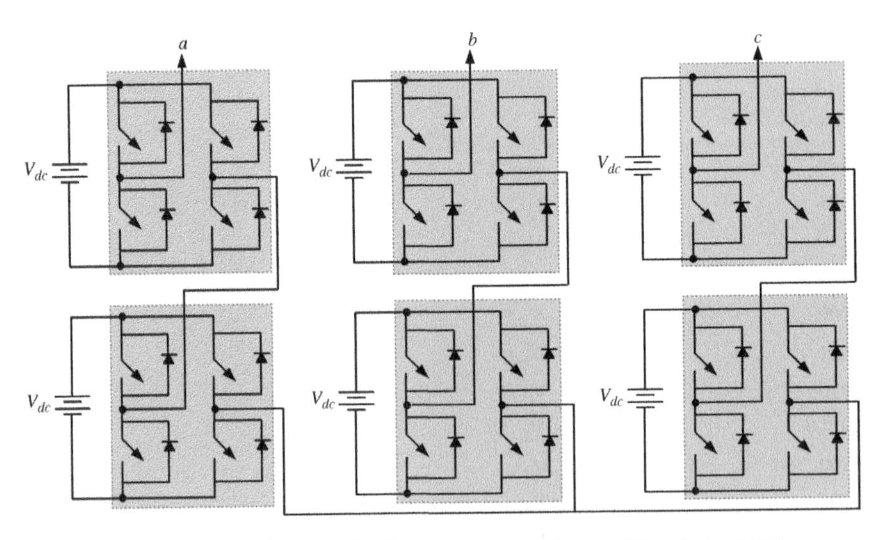

Figure 7.37 Schematic diagram of a three-phase cascaded multilevel converter.

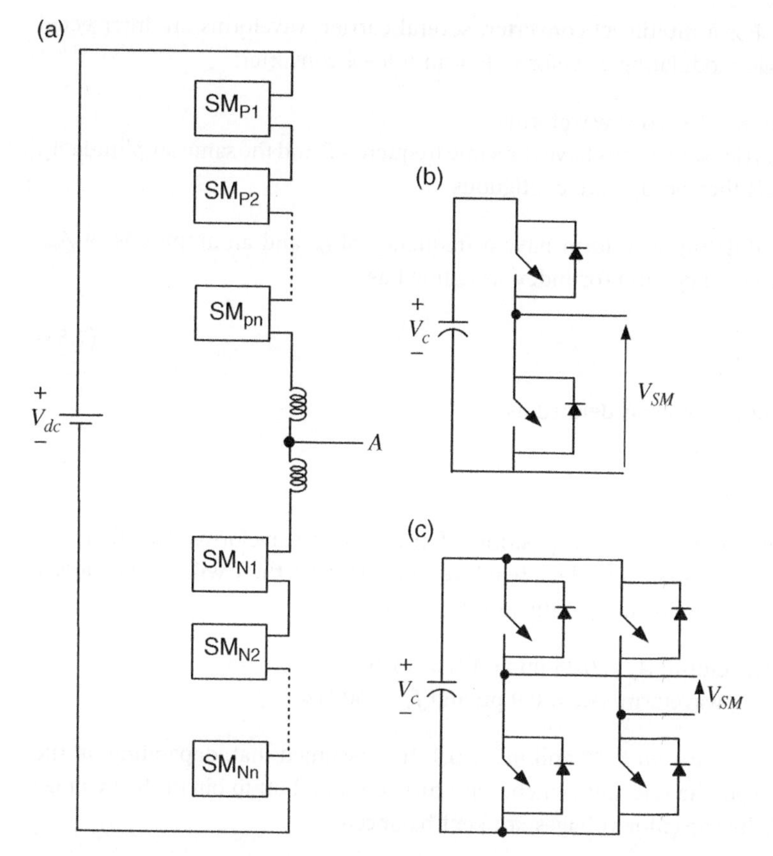

Figure 7.38 (a) Schematic diagram of an MMC and (b and c) two possible SM structures.

can be a half bridge or full bridge, as shown in Figure 7.38b,c. Other configurations are also possible. It is obvious that this converter can produce an almost sinusoidal voltage by phase shifting the firing pulses.

The SPWM of MMC is somewhat similar to that of multilevel converters. In Section 7.5.6, we briefly present the different possible PWM techniques for multilevel converters.

7.5.6 PWM of Multilevel Converters

We see in this chapter that a PWM voltage waveform is generated through the intersection of a sinusoidal modulating waveform with a triangular carrier waveform. For a multilevel converter, several carrier waveforms are intersected with a single modulating waveform. For an n-level converter:

- There are $n-1$ carrier waveforms.
- All the carrier waveforms have the same frequency f_c and the same amplitude A_c.
- The bands they occupy are contiguous.

Let the modulating waveform have a frequency of f_m and an amplitude of A_m. Then the frequency ratio (or index) is defined as

$$m_f = \frac{f_c}{f_m} \tag{7.53}$$

The modulation index is defined as

$$m_a = \frac{A_m}{(n-1)A_c} \tag{7.54}$$

Three different types of deposition of the carrier waveforms are discussed here. For this, we shall consider a five-level converter, i.e. there will be four carrier waveforms. The following parameters are chosen:

- Carrier waveform: $A_c = 0.25$ pu and $f_c = 1350$ Hz
- Modulating waveform: $A_m = 0.9$ pu and $f_m = 50$ Hz

Therefore, we have $m_f = 27$ and $m_a = 0.9$. It is assumed that, depending on the converter type, the relevant switches are turned on and off to obtain five voltage levels and the capacitor voltages are kept balanced.

- Phase Disposition (PD): In this method, even though the carriers are disposed, all of them are in phase, as shown in Figure 7.39a. The output voltage is shown in Figure 7.39b. In this method, the harmonics are centered around the carrier frequency f_c, as can be seen from Figure 7.39c.
- Alternative Phase Opposition Disposition (APOD): In this method, the contiguous carrier waveforms are phase displaced by 180° from each other. This is shown in Figure 7.40a, where the topmost and the third waveform from the top are in phase, while the other two waveforms are in phase opposition. The output voltage waveform is shown in Figure 7.40b. The significant harmonics appear as side bands of the carrier frequency f_c, as shown in Figure 7.40c.
- Phase Opposition Disposition (POD): There are two sets of carrier waveforms: those above and those below the reference line. The waveforms of each set are in phase. However, the sets are phase displaced by 180° from each other, as shown in Figure 7.41a. The output voltage waveform is shown in

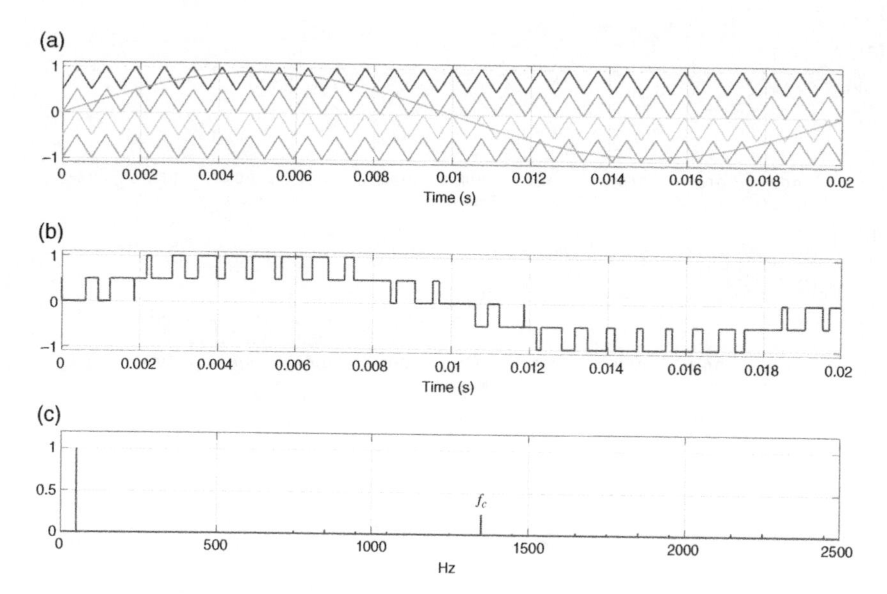

Figure 7.39 PD-PWM of a five-level converter: (a) carrier and modulating waveforms, (b) output voltage, and (c) harmonic spectrum.

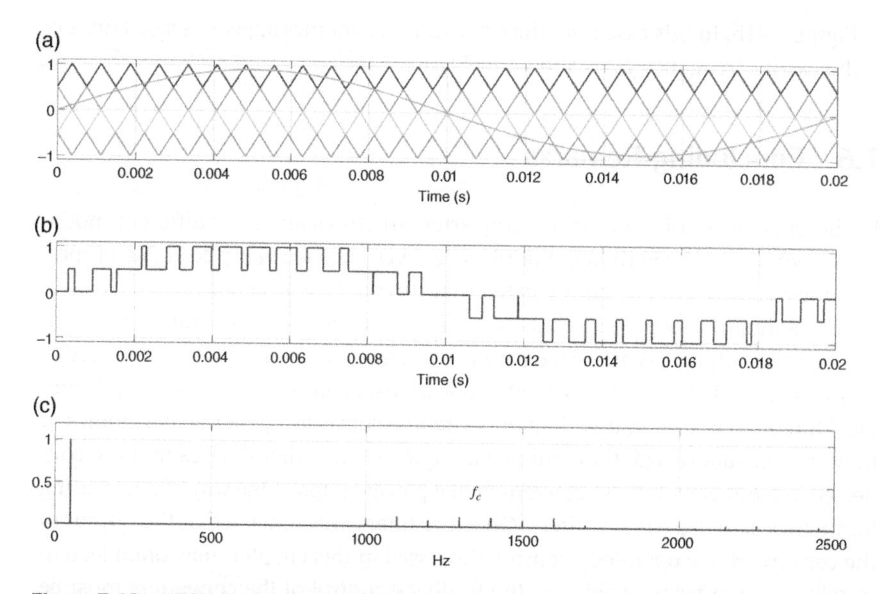

Figure 7.40 APOD-PWM of a five-level converter: (a) carrier and modulating waveforms, (b) output voltage, and (c) harmonic spectrum.

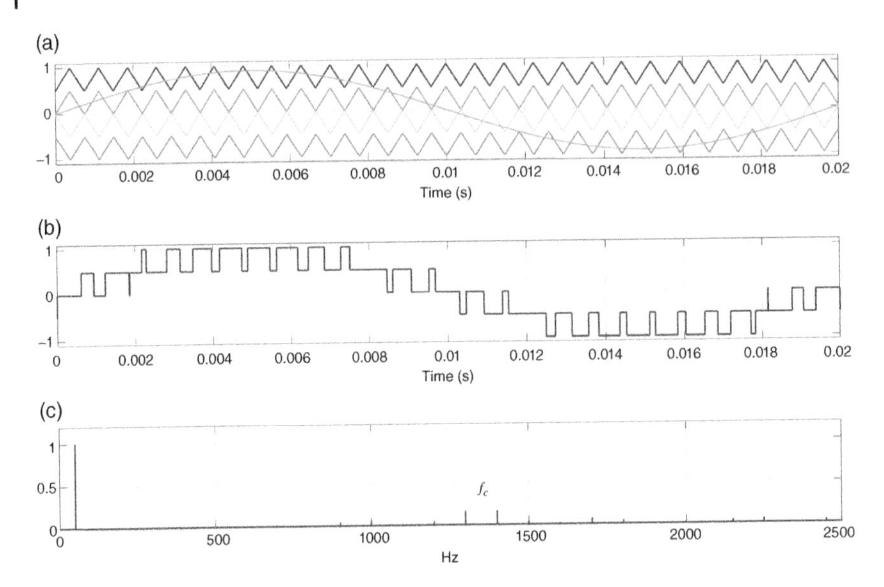

Figure 7.41 POD-PWM of a five-level converter: (a) carrier and modulating waveforms, (b) output voltage, and (c) harmonic spectrum.

Figure 7.41b. In this case also, the significant harmonics appear as side bands of the carrier frequency f_c, as shown in Figure 7.41c.

7.6 Concluding Remarks

In this chapter we discuss various converter structures and their different modulation techniques. Even though multilevel converters have many desirable properties, their control is more complicated than that of two-level converters.

The modulation techniques that are discussed in this chapter operate in the open loop, i.e. the output is synthesized based on a reference signal, which is assumed to be sinusoidal. However, in practical applications, open-loop control is not desirable. Moreover, as seen in this chapter, switching converters generate harmonics. If they are not suppressed, they can be damaging to the other devices that are connected in parallel with these converters in a power system. One way of eliminating harmonics is to use passive filters. Once the filters are connected at the output of the converters, the open-loop control, discussed in this chapter, may often lead to instability. In order to avoid this, the feedback control of the converters must be designed by including the filter dynamics along with that of the converter. This is discussed in Chapter 8.

Problems

7.1 A three-phase VSC is supplied by 2 kV at the DC side. It has to produce a line-to-line fundamental voltage with a peak magnitude of 1 kV at 60 Hz using bipolar SPWM. Determine the modulation index m_a.

7.2 The modulation index found in Problem 7.1 is used in bipolar SPWM of a single-phase VSC, which is supplied by a DC voltage of 2 kV. Determine the peak of the fundamental voltage.

7.3 A set of balanced, 50 Hz three-phase voltages is given in phasor domain by

$$V_{a0} = 400\angle 30^\circ \text{ V}, \quad V_{b0} = 400\angle -90^\circ \text{ V}, \quad V_{c0} = 400\angle 150^\circ \text{ V}$$

Find the space vector v_P when the time is 0.11 seconds from (7.13). Verify the result obtained from (7.14).

7.4 Find the quantities T_a, T_b, and T_0 in terms of T_s using the method presented in Section 7.3.3 when the space vector is

(a) $v_P = -78.8011 + j61.5661$ V, DC voltage $V_{dc} = 250$ V

(b) $v_P = 286.79 - j409.58$ V, DC voltage $V_{dc} = 1000$ V

(c) $v_P = 1000 \angle 210^\circ$ V, DC voltage $V_{dc} = 2500$ V

7.5 Verify the results obtained in Problem 7.4 using the timing calculation method of Section 7.3.4.

7.6 A VSC is controlled by SVPWM, where the modulating voltages are given by

$$v_{a0} = 300 \sin(100\pi t + 10^\circ) \text{ V}$$
$$v_{b0} = 300 \sin(100\pi t - 110^\circ) \text{ V}$$
$$v_{c0} = 300 \sin(100\pi t + 130^\circ) \text{ V}$$

Assuming the frequency of the carrier waveform as 10 kHz and the DC side voltage of 600 V, sketch the switching pattern when $t = 0.1502$ seconds.

Notes and References

An excellent tutorial on space vector modulation can be found in [16]. Summaries of different multilevel converter topologies are discussed in [17–19]. A comprehensive review of a multilevel converter, including PWM and SVPWM, is presented in [20]. Simulation of a modular multilevel converter using an electromagnetic transient program is discussed in [21]. Theoretical analysis of PWM for a multilevel converter is presented in [22], while several methods of PWM applications in multilevel converters are discussed in [23]. Multilevel converters with unbalanced voltage are discussed in [14–15].

1 Mohan, N., Undeland, T.M., and Robbins, W.P. (2002). *Power Electronics: Converters, Applications and Design*, 3e. New York: Wiley.
2 Rashid, M.H. (2013). *Power Electronics: Circuits, Devices and Applications*, 4e. Lincoln: Pearson Education Limited.
3 Zare, F., Zabihi, S., and Ledwich, G., (2007). An adaptive hysteresis current control for a multilevel inverter used in an active power filter. European Conference on Power Electronics and Applications, Aalborg, Denmark.
4 Zare, F. and Ledwich, G. (2007). A new hysteresis current control for three-phase inverters based on adjacent voltage vectors and time error. In: *IEEE Power Electronics Specialists Conference*, 431–436. Orlando, Florida: IEEE.
5 Zare, F. and Ledwich, G. (2002). A hysteresis current control for single-phase multilevel voltage source inverters: PLD implementation. *IEEE Trans. Power Electron.* 17 (5): 731–738.
6 Ghosh, A. and Ledwich, G. (2002). *Power Quality Enhancement Using Custom Power Devices*. New York: Springer Science+Business Media.
7 Quinn, C.A. and Mohan, N. (1992). Active filtering of harmonic currents in three-phase, four-wire systems with three-phase and single-phase nonlinear loads. In: *Applied Power Electronic Conference (APEC)*, 829–836. IEEE.
8 Quinn, C.A., Mohan, N., and Mehta, H. (1993). A four-wire, current-controlled converter provides harmonic neutralization in three-phase, four-wire systems. In: *Applied Power Electronic Conference (APEC)*, 841–846. IEEE.
9 Aredes, M., Hafner, J., and Heumann, K. (1997). Three-phase four-wire shunt active filter control strategies. *IEEE Trans. Power Electron.* 12 (2): 311–318.
10 Iyer, S.V., Ghosh, A., and Joshi, A. (2005). Inverter topologies for DSTATCOM applications: a simulation study. *Electr. Power Syst. Res.* 75 (2): 161–170.
11 Broeck, H.W.V.D., Skudelny, H.-C., and Stanke, G.V. (1988). Analysis and realization of a pulse width modulator based on voltage space vectors. *IEEE Trans. Ind. Appl.* 24 (1): 142–150.
12 Tran, P.H. (2012). MATLAB®/SIMULINK® implementation and analysis of three pulse-width-modulation (PWM) techniques, Master of science thesis, Boise State University.
13 Houldsworth, J.A. and Grant, D.A. (1984). The use of harmonic distortion to increase the output voltage of a three-phase PWM inverter. *IEEE Trans. Ind. Appl.* IA-20 (5): 1224–1228.

14 Nami, A., Zare, F., Ghosh, A., and Blaabjerg, F. (2011). A hybrid cascade converter topology with series-connected symmetrical and asymmetrical diode-clamped H-bridge cells. *IEEE Trans. Power Electron.* 26 (1): 51–65.

15 Nami, A., Zare, F., Ghosh, A., and Blaabjerg, F. (2010). Multi-output DC-DC converters based on diode-clamped converters configuration: topology and control strategy. *IET Power Electron.* 3 (2): 197–208.

16 Neacsu, D.O. (2001) Space vector modulation: an introduction. *27th IEEE Industrial Electronics Society Conference (IECON).* Denver, CO.

17 Lai, J.-S. and Peng, F.Z. (1996). Multilevel converters: a new breed of power converters. *IEEE Trans. Ind. Appl.* 32 (3): 509–517.

18 Rodríguez, J., Lai, J.-S., and Peng, F.Z. (2002). Multilevel inverters: a survey of topologies, controls, and applications. *IEEE Trans. Indust. Electron.* 49 (4): 724–738.

19 Rodríguez, J., Franquelo, L.G., Kouro, S. et al. (2009). Multilevel converters: an enabling technology for high-power applications. *Proc. IEEE* 97 (11): 1786–1817.

20 Debnath, S., Qin, J., Bahrani, B. et al. (2015). Operation, control, and applications of the modular multilevel converter: a review. *IEEE Trans. Power Electron.* 30 (1): 37–53.

21 Gnanarathna, U.N., Gole, A.M., and Jayasinghe, R.P. (2011). Efficient modeling of modular multilevel HVDC converters (MMC) on electromagnetic transient simulation programs. *IEEE Trans. Power Delivery* 26 (1): 316–324.

22 Carrara, G., Gardella, S., Marchesoni, M. et al. (1992). A new multilevel PWM method: a theoretical analysis. *IEEE Trans. Power Electron.* 7 (3): 497–505.

23 Agelidis, V.G. and Calais, M. (1998). Application specific harmonic performance evaluation of multicarrier PWM techniques. *Proc. IEEE Power Electron. Special. Con.* 172–178.

8

Control of DC-AC Converters

In this chapter, various techniques that can be employed for controlling the output voltage or current of a voltage source converter (VSC) are presented. The main aim of the closed-loop control is to control the switching functions of the VSC in such a way that a desired output is attained. Chapter 7 shows how a converter output voltage can be synthesized using pulse width modulation (PWM) techniques. These are, however, open-loop techniques where the output voltage is produced by comparing a modulating wave with a carrier wave. Such modulation can introduce harmonics into AC systems in which the VSCs are connected. In order to suppress the harmonics, passive filters are connected at the output of the VSC's output. The feedback controllers then will have to consider the presence of these filters. This implies that the dynamics of these filters must be considered for feedback control design. The output filters must be chosen carefully considering: (i) the harmonic spectrum on the converter side, (ii) possible resonance due to the interaction of the filter with the rest of the system, and (iii) variation in the harmonic emission with the operating point [1]. This chapter starts with a discussion about the structure and design of the output filters.

8.1 Filter Structure and Design

A DC-AC converter is a switching device, which generates $\pm V_{dc}$ voltage at the output, where V_{dc} is the DC bus voltage. Therefore, when connected to a power system, these converters inject harmonics. To prevent such harmonics from entering the system, passive filters are used at the output of the converters. Usually, three types of filters are used: L-type, LC-type, and LCL-type. These are shown in Figure 8.1.

Control of Power Electronic Converters with Microgrid Applications, First Edition.
Arindam Ghosh and Firuz Zare.

Published 2023 by John Wiley & Sons, Inc.

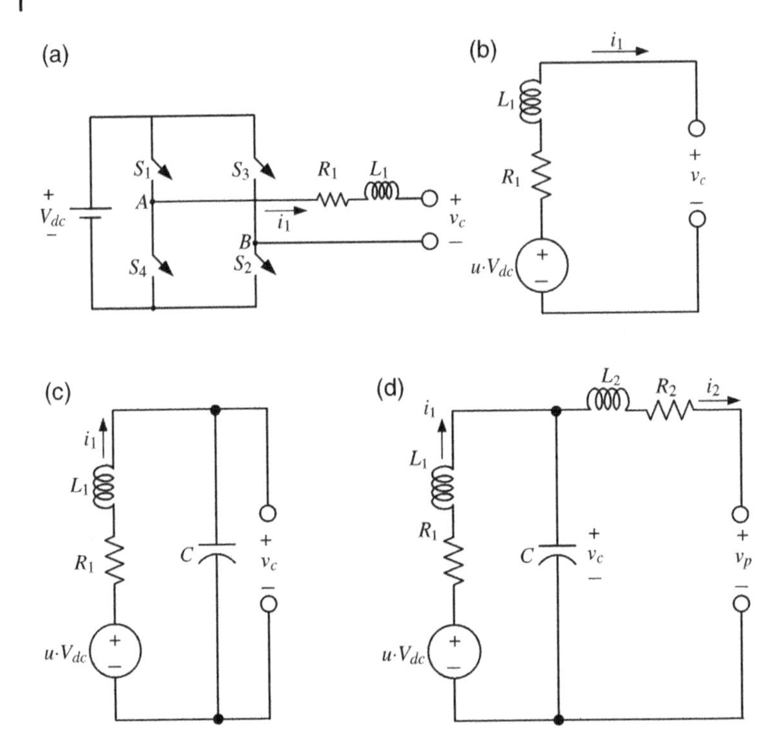

Figure 8.1 (a) Single-phase voltage source converter with L filter; converter equivalent representation with (b) L filter, (c) LC filter, and (d) LCL filter.

Consider the single-phase converter shown in Figure 8.1a, which has an output L-type filter with a converter side inductance of L_1. Note that the resistance R_1 is added due to the quality factor of the coil. This can also include the converter losses. Its equivalent circuit is shown in Figure 8.1b, in which the converter is represented by a voltage source $u \cdot V_{dc}$, where $u = \pm 1$. A converter with an LC filter is shown in Figure 8.1c. In addition to the inductor, this also has a capacitor C that is connected in shunt. The capacitor provides a low-impedance path to the harmonic currents – the higher the harmonics, the lower the impedance. Therefore, the harmonic currents are bypassed and do not appear at the output. The LCL-type filter is shown in Figure 8.1d. In addition to the LC filter, it has an additional inductor L_2 (and its associated resistor R_2), often called the outer or grid side inductor.

The main aim of a converter control is to generate the switching signals $u = \pm 1$. In the example of hysteresis current control in Chapter 7 (Section 7.1.1), current is controlled through a hysteresis band when the converter output voltage varies between $\pm V_{dc}$ (Figures 7.2 and 7.3). This is the simplest form of control, which is suitable for current control with only L-type filters. In the presence of LC or LCL filters, the closed-loop control system needs to consider the filter dynamics for system stability and tracking performance. However, these filters must be designed carefully. The design of the LCL filter is discussed in Section 8.1.1, and the design of the LC filter can be considered a subset of this design.

8.1.1 Filter Design

For the filter design, we start with the converter side inductance L_1. The size of this inductor will depend on the maximum current ripple ($\Delta i_{1\max}$). Note that, to reduce the current ripple, a larger inductor needs to be chosen, which will increase the size and core losses. On the other hand, a lower ripple will cause lower switching losses. Therefore, the choice of this inductor will be a compromise between the inductor size and the losses. Typically, the current ripple is chosen between 5 and 25% of the rated current [1]. Then, denoting the converter switching frequency as f_{SW}, the empirical formula for the inductor is given by [2]

$$L_1 = \frac{V_{dc}}{8 f_{SW}\, \Delta i_{1\max}} \tag{8.1}$$

Next, the value of the capacitance (C) will be selected. If this value is high, more reactive power will flow into the capacitor, resulting in larger current demand. This value cannot be too small either; otherwise, the inductor value must be increased to meet the attenuation requirement. In general, the capacitance value is chosen as a percentage (λ) of the reactive power with respect to the real power at rated conditions, given by [1]

$$C = \lambda \frac{S}{2\pi f\, V_{LL}^2} \tag{8.2}$$

where V_{LL} is the line-line voltage, f is the line frequency of 50 or 60 Hz, and S is the apparent power of the VSC.

According to [1], λ should be limited to 5%; otherwise, the power factor may decrease. The total inductance (i.e. $L_1 + L_2$) should be limited within a certain percentage to limit the voltage drop across the inductors [1]. A detailed and optimum LCL filter design approach is proposed in [3]. A mathematical approach is utilized to obtain a relation between the inverter side inductor and the ripple of the inverter output current.

In this chapter, voltage and current controllers for a VSC are developed. The function of a voltage controller is to produce a set reference voltage across the converter output. For this purpose, usually LC filters are used so that the voltage v_c

across the filter capacitor C can be directly impressed across a load or a power system. However, depending on prevailing situations, LCL filters may also be used, but the voltage across the filter capacitor still needs to be controlled. The L-type filter is the most basic filter that is used for current control. For more sophisticated current control, LCL filters are used. This is also discussed in this chapter.

Example 8.1 In this example, we shall design an LC filter based on (8.1) and (8.2), for a 400 V (L-L), 50 kVA converter. In a three-phase system, the DC link voltage should be kept as low as possible to minimize switching losses. However, this DC voltage must be higher than the maximum instantaneous line to neutral grid voltage to push the required current to the grid. In this example, the DC voltage is assumed to be $V_{dc} = 600$ V. The maximum per phase current is given by

$$i_{1\max} = \frac{S}{3V_{LN}} = \frac{S}{\sqrt{3}V_{LL}} = \frac{50\times 10^3}{\sqrt{3}\times 400} = 72.17\ \text{A}$$

Then assuming that the maximum current ripple is 5% of $i_{1\max}$, we have $\Delta i_{1\max} = 3.6$ A. Then, for a switching frequency of 15 kHz, the converter side inductor is

$$L_1 = \frac{600}{8\times 15000\times 3.6} = 1.39\ \text{mH} \approx 2\ \text{mH}$$

From (8.2), by choosing $\lambda = 5\%$, the filter capacitor is calculated as

$$C = 0.05\times\frac{50\times 10^3}{100\pi(400)^2} = 4.97\times 10^{-5} \approx 50\ \mu\text{F}$$

Consider the equivalent circuit of the LC filter shown in Figure 8.1c. Defining a state vector as $x = [v_c\ i_1]^T$, the state space equation of the system can be written as

$$\dot{\mathbf{x}} = \mathbf{A}\mathbf{x} + \mathbf{B}V_{dc}u_c \tag{8.3}$$

where u_c is the feedback control law, based on which the converter switching signal $u = \pm 1$ is generated and the matrices A and B are

$$\mathbf{A} = \begin{bmatrix} 0 & 1/C \\ -1/L_1 & -R_1/L_1 \end{bmatrix},\quad \mathbf{B} = \begin{bmatrix} 0 \\ 1/L_1 \end{bmatrix}$$

For the L and C values calculated above and choosing $R_1 = 0.1\ \Omega$, the Bode plot of the filter is shown in Figure 8.2, where the -3 dB point is around 800 Hz. This means that the filter will attenuate signals below the 16th harmonic. Thus, with proper control, the converter can cancel the lower-order harmonics below the 16th harmonic. This is an important property for active filtering applications, where the converter is required to cancel harmonics created by loads.

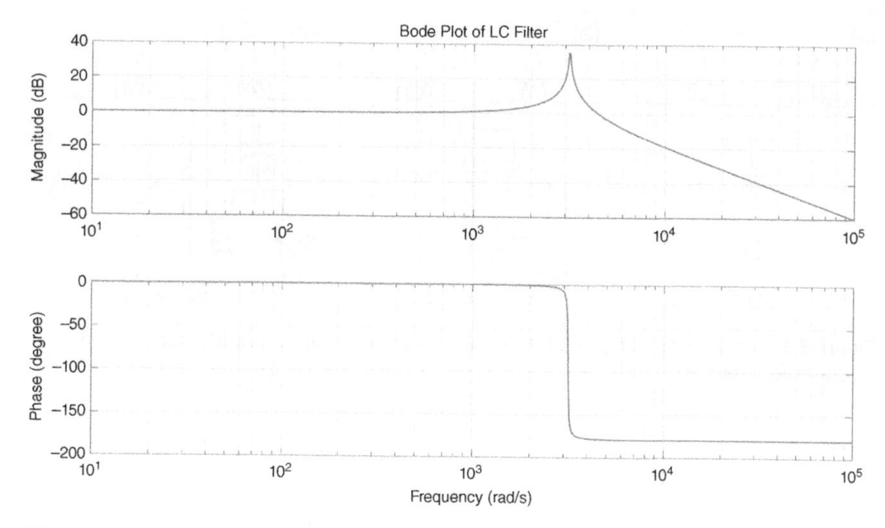

Figure 8.2 Bode plot of the LC filter.

8.1.2 Filter with Passive Damping

The filters discussed in Section 8.1.1 may have problems of resonance with the systems to which the converter is connected. To compensate for the resonance peak, passive damping circuits may be required [1, 2]. Obviously, these damping circuits will require resistors that may cause power loss. Some typical passive damping filters are shown in Figure 8.3. An LCL filter with a series resistor, as shown in Figure 8.3a, is the simplest of the passive dampers. An improved version of the series damper is the shunt RC damper, as shown in Figure 8.3b. The high-frequency attenuation of the LCL filter is retained in this structure. Good high-frequency attenuation that is achieved without causing much power loss through shunt RLC damper, is shown in Figure 8.3c. Here, the fundamental current in the damping resistor is bypassed by the damping inductance L_d. There are several other passive damping circuits that are discussed and analyzed in [1, 2]. We shall, however, not discuss the passive damping circuits further but will proceed to design controllers that will provide significant damping through active damping control.

8.2 State Feedback Based PWM Voltage Control

The purpose of a converter controller is to track a set of voltages or currents faithfully. In this section, we assume that the reference signals are instantaneous quantities in their abc coordinates. Consider the system of Figure 8.4a, which

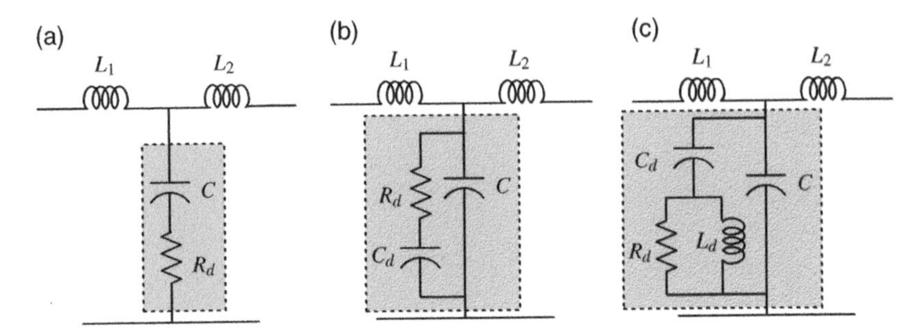

Figure 8.3 Passive damping LCL filters: (a) series damper, (b) shunt RC damper, and (c) shunt RLC damper.

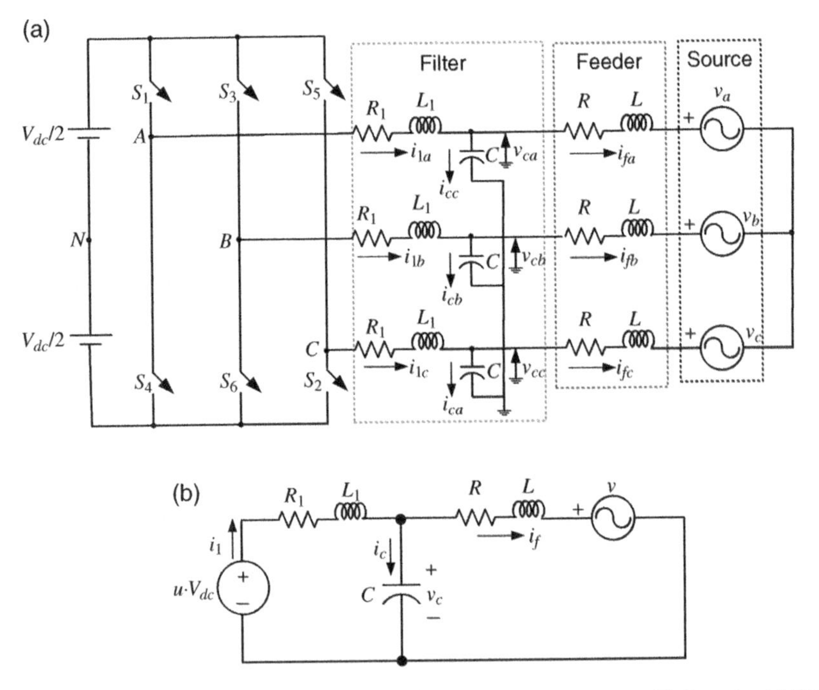

Figure 8.4 Schematic diagram of (a) a voltage source converter with its output LC filter connected to a source through a feeder and (b) its single-line diagram.

depicts a bridge converter that is connected to a source (back emf) through a feeder. The LC filters that are connected to the converter are also shown in the figure. The purpose of the converter control is to make the converter output voltages v_{ck}, $k = a, b, c$ to track a set of prespecified reference voltages. In this section, the closed-loop PWM based state feedback control is discussed.

The output filters of the converter, shown in Figure 8.4, are all grounded. Therefore, the three legs of the converter can be treated separately, and each leg can be controlled independent of the other two legs. Moreover, the DC bus is split into two, with a center point N. Nevertheless, the DC bus voltage will be equal to V_{dc}, while it contains two DC sources, each equal to $V_{dc}/2$. In a three-phase, four-wire system, when the center point N and the load neutral are connected together, a path is provided for the unbalanced (zero-sequence) component of the current to flow. The converter controller will be designed for just one phase with the understanding that a similar control law can also be derived for the other two phases.

As mentioned elsewhere, three identical controllers will be designed that will be applied to each of the three phases separately. The single-line diagram for one phase of the circuit is shown in Figure 8.4b, where the feeder current is denoted by i_f and the capacitor current is denoted by i_c. This contains an LC filter, the state space description of which is given in (8.3), where the state vector of $\mathbf{x} = [v_c \quad i_1]^T$ has been chosen. For the circuit of Figure 8.4b, this results in the following state space equation

$$\dot{\mathbf{x}} = \mathbf{Ax} + \mathbf{B}V_{dc}u_c + \begin{bmatrix} -\dfrac{1}{C} \\ 0 \end{bmatrix} i_f \tag{8.4}$$

The feeder current i_f is considered as a disturbance input and is not included in the control design. Therefore, (8.3) is used for control law computation. Assuming that the references for the states are available and are denoted by $\mathbf{x}_{ref} = [v_{cref} \quad i_{1ref}]^T$, the state feedback control law is given as

$$u_c = \mathbf{K}\left(\mathbf{x}_{ref} - \mathbf{x}\right) \tag{8.5}$$

where $\mathbf{K} = [k_1 \quad k_2]^T$ is the feedback gain matrix.

In the PWM control law, a triangular carrier waveform (v_{tri}) is generated that varies from -1 to $+1$ with a duty ratio of 0.5. The control output u_c is compared with the carrier waveform to generate u as per the following formula

$$\begin{aligned} &\text{if } u_c > v_{tri} \text{ then } u = +1 \\ &\quad \text{elseif } u_c < v_{tri} \text{ then } u = -1 \end{aligned} \tag{8.6}$$

The schematic diagram of the control law is given in Figure 8.5a.

For tracking the output voltage, the voltage reference (v_{cref}) for the output capacitor voltage can be prespecified. However, it is rather difficult to find a reference (i_{1ref}) for the converter output current i_1. One approach can be to set this reference to zero. This will, however, lead to an incorrect control action. To avoid this problem, a state transformation is used in [4]. This, however, is feasible only when the

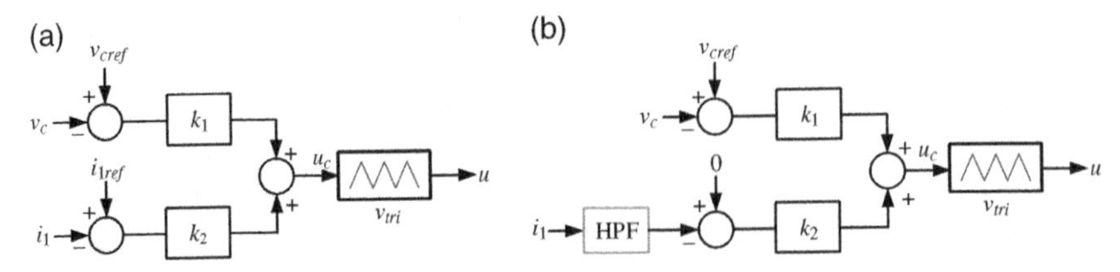

Figure 8.5 Two different feedback control structures: (a) full state feedback and (b) partial state feedback with HPF.

overall system structure and rough estimates of the system parameters are known a priori. Therefore, this cannot be stated as a general solution. There are two possible approaches to this problem: (i) to use a highpass filter (HPF) or (ii) to use an observer to estimate the current. These are discussed in Sections 8.1.1 and 8.1.2.

8.2.1 HPF-based Control Design

In this method, we note that the current i_1 should only contain lower-frequency components, while its high-frequency components should be zero. Therefore, if the current is passed through a HPF, then it is expected that the output (i_{1HPF}) of the filter is zero [5]. The modified control structure is shown in Figure 8.5b. Note that the control gains remain the same irrespective of the HPF. The HPF transfer function is given by

$$\frac{i_{1HPF}(s)}{i_1(s)} = \frac{s}{s+\alpha} \tag{8.7}$$

Equation (8.7) can be written as

$$i_{1HPF}(s) = \left(\frac{s}{s+\alpha}\right) i_1(s) = \left(1 - \frac{\alpha}{s+\alpha}\right) i_1(s) = i_1(s) - i_{1LPF}(s)$$

where $i_{1LPF}(s)$ is a lowpass filter, given by

$$i_{1LPF}(s) = \left(\frac{\alpha}{s+\alpha}\right) i_1(s)$$

This can be expressed in differential equation form as

$$\frac{d}{dt} i_{1LPF} = -\alpha\, i_{1LPF} + \alpha\, i_1 \tag{8.8}$$

Let us now define an extended state vector as $\mathbf{x}_e = \begin{bmatrix} v_c & i_1 & i_{1LPF} \end{bmatrix}^T$. Then, combining (8.8) with (8.3), we have

$$\dot{\mathbf{x}}_e = \mathbf{A}_e \mathbf{x}_e + \mathbf{B}_e V_{dc} u_c \tag{8.9}$$

where

$$\mathbf{A}_e = \begin{bmatrix} \mathbf{A} & 0 \\ [0 \quad \alpha] & -\alpha \end{bmatrix}, \quad \mathbf{B}_e = \begin{bmatrix} \mathbf{B} \\ 0 \end{bmatrix}$$

From (8.5) and Figure 8.5b, the following feedback control law is obtained

$$\begin{aligned} u_c &= k_1\left(v_{cref} - v_c\right) + k_2(0 - i_{1HPF}) = k_1\left(v_{cref} - v_c\right) - k_2(i_1 - i_{1LPF}) \\ &= [\,-k_1 \quad -k_2 \quad k_2\,]x_e + k_1 v_{cref} \end{aligned} \tag{8.10}$$

Substituting (8.10) in the state equation of (8.9), we have

$$\dot{\mathbf{x}}_e = (\mathbf{A}_e + \mathbf{B}_e[\,-k_1 \quad -k_2 \quad k_2\,])x_e + k_1\mathbf{B}_e v_{cref} \tag{8.11}$$

Example 8.2 Consider the system of Figure 8.4, where the LC filter designed in Example 8.1 is used. The feeder parameters are $R = 5\ \Omega$ and $L = 11.6$ mH. The system frequency is 50 Hz (i.e. $\omega = 100\pi$ rad/s) and source voltage and the chosen reference voltages are

$$\begin{aligned} v_a &= 326.6\sin(\omega t - 20^\circ) \qquad & v_{crefa} &= 326.6\sin(\omega t) \\ v_b &= 326.6\sin(\omega t - 140^\circ) \qquad & v_{crefb} &= 326.6\sin(\omega t - 120^\circ) \\ v_c &= 326.6\sin(\omega t + 100^\circ) \qquad & v_{crefc} &= 326.6\sin(\omega t + 120^\circ) \end{aligned}$$

A linear quadratic regulator is designed for the system of (8.3) with the LC parameters of Example 8.1, with the following weighting matrices

$$\mathbf{Q} = \begin{bmatrix} 150 & 0 \\ 0 & 1 \end{bmatrix}, \qquad r = 0.01$$

The resultant gain matrix is $\mathbf{K} = [\,122.47 \quad 10.59\,]$. The HPF is given by

$$\text{HPF}(s) = \frac{s}{s + 1000}$$

The Bode plot of the HPF is shown in Figure 8.6, where the cutoff frequency is about 1000 rad/s, i.e. 159 Hz. This means that the filter will pass the harmonics components above the third harmonic. The closed-loop eigenvalues of the system, computed from (8.11), are located at -4×10^6, -2.44×10^5, and -1×10^3. They are all on the left half of the s-plane.

A triangular carrier wave varying between ± 1 is chosen with a frequency of 15 kHz. The DC voltage (V_{dc}) is chosen as 800 V. The system response is shown in Figure 8.7, where the reference voltages and the actual voltages are shown. It can be seen that the tracking is almost perfect. The control inputs are shown in Figure 8.8. Barring the initial transients, these are restricted to ± 1, indicating that these signals remain in the undermodulated region. These signals will be in the overmodulated region for higher values of the gain matrix $\mathbf{K}$ and this will result

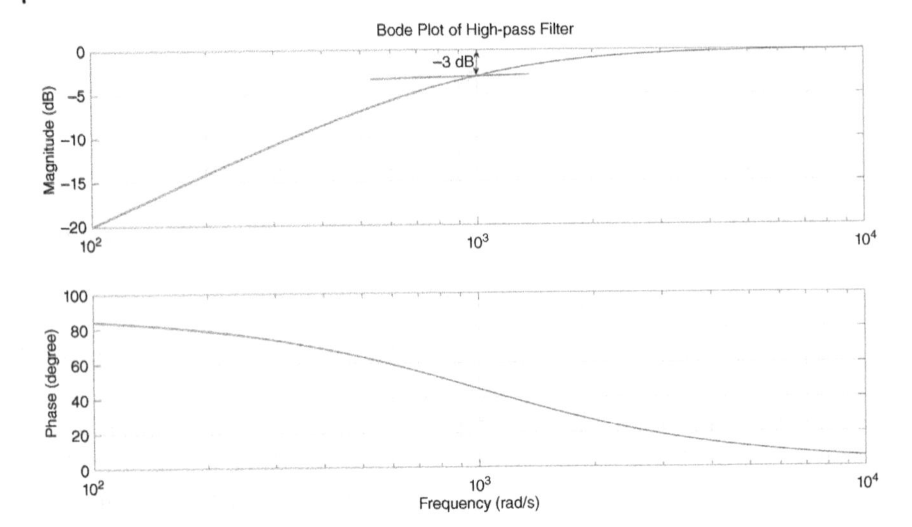

Figure 8.6 Bode plot of the HPF.

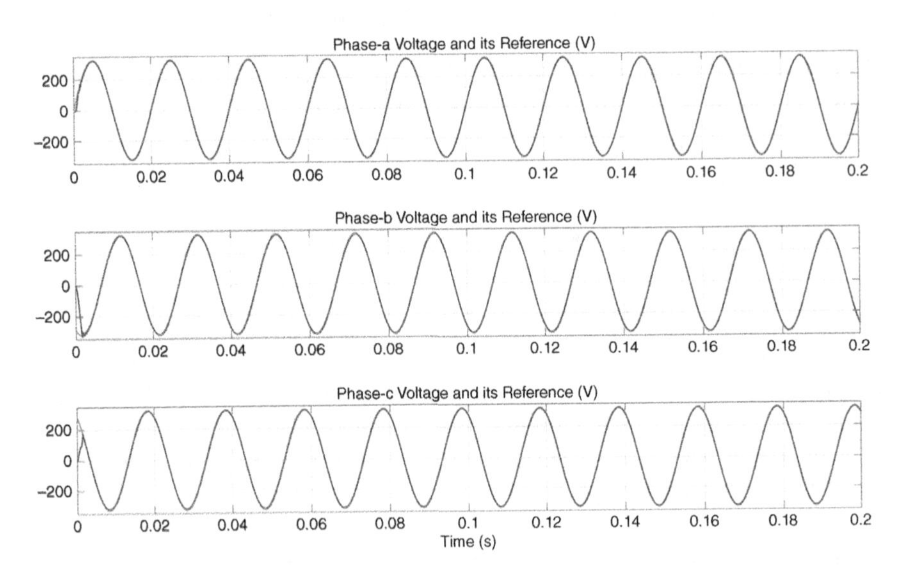

Figure 8.7 Tracking performance of the closed-loop system of Example 8.2.

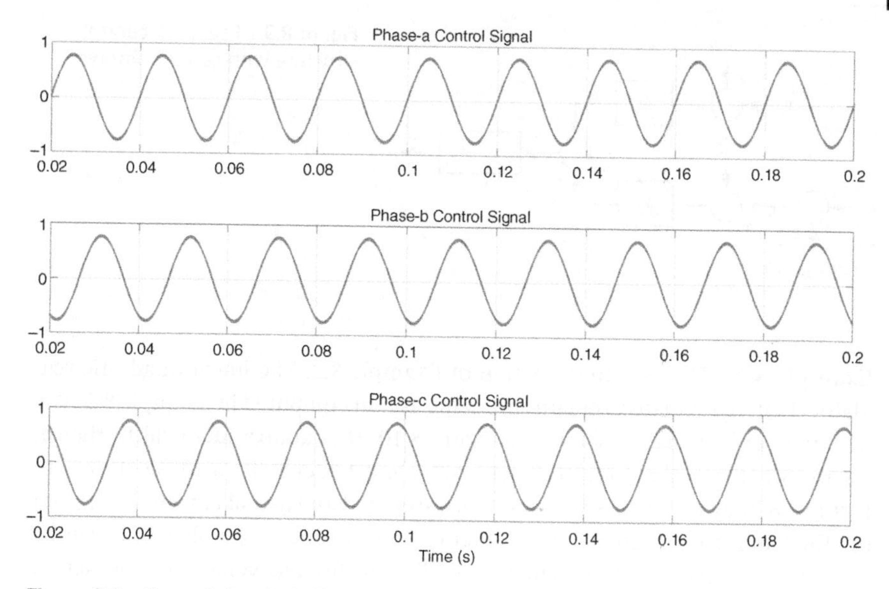

Figure 8.8 Control signals in Example 8.2.

in unnecessary ripples in the output voltages. In particular, the gain matrix **K** is very sensitive of the choice of parameters r, as the penalty on control reduces with the reduction on the value of r.

8.2.2 Observer-based Current Estimation

Consider again the system of Figure 8.4b. Applying KCL at the point of common coupling (PCC) of the feeder and converter, we get

$$i_1 = i_c + i_f \tag{8.12}$$

Let the reference capacitor voltage be assumed as

$$v_{cref} = V_m \sin(\omega t + \varphi)$$

Then, the reference for the capacitor current is written as

$$i_{cref} = C\frac{dv_{cref}}{dt} = CV_m \frac{d\sin(\omega t + \varphi)}{dt} = \omega C V_m \cos(\omega t + \varphi) \tag{8.13}$$

Now, for each phase, i_{cref} is computed using (8.13) and the feeder current i_f is measured. The reference for current i_1 is then formed from (8.12), which is termed as its estimate $\hat{i}_1$. The feedback control structure is shown in Figure 8.9.

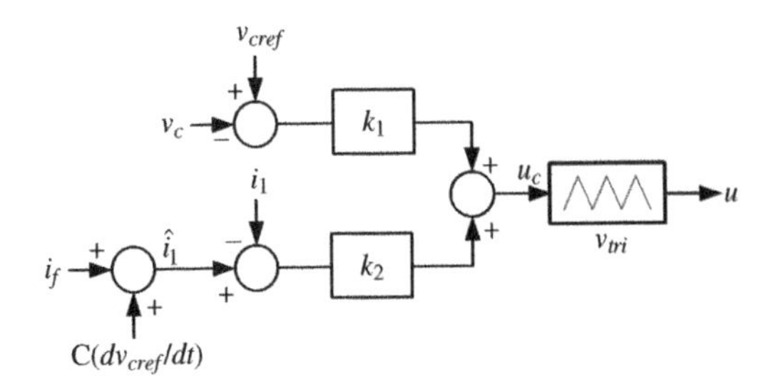

Figure 8.9 Feedback control structure with current observer.

Example 8.3 This is a continuation of Example 8.2. The linear quadratic regulator (LQR) gain matrix remains the same as that computed in Example 8.2. The tracking performance is shown in Figure 8.10. It is almost identical to that of Figure 8.7. Since the control signals are almost identical to those shown in Figure 8.8, they are not shown here. The estimated and actual current i_1 through the filter inductor L_1 are shown in Figure 8.11 for one of the phases. It can be seen that the estimated current is actually the filtered version of the actual current.

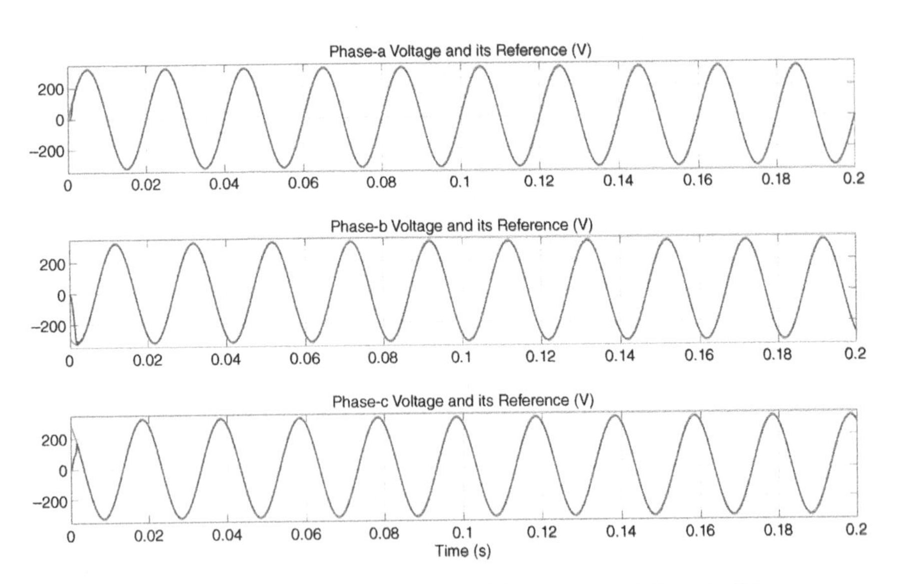

Figure 8.10 Tracking performance of the closed-loop system of Example 8.3.

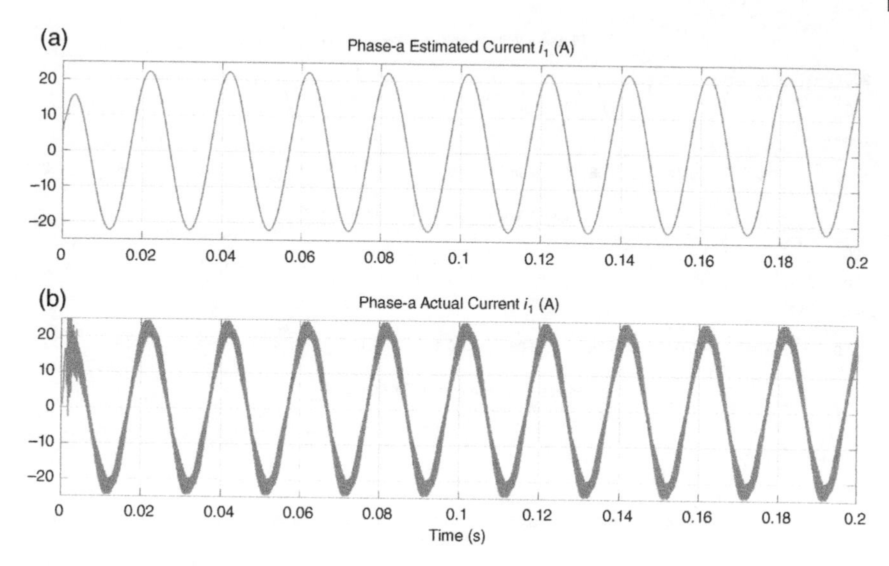

Figure 8.11 (a) Estimated and (b) actual phase-a current i_{1a} of Example 8.3.

8.3 State Feedback Based SVPWM Voltage Control

The concept of space vector pulse width modulation (SVPWM) closed-loop control is almost identical to that of the PWM control discussed in Section 8.2. The closed-loop switching computation is shown in Figure 8.12. First the control input u_c is computed for all the three phases using either the HPF based or observed based current estimation discussed in Section 8.2. Then follow the steps outlined in Section 7.3 of Chapter 7. First, using u_{ca}, u_{cb}, and u_{cc}, find the state vector V_P and its angle ϕ. Then, based on the angle, determine the sector to which the vector belongs. Thereafter, compute the timings T_a, T_b, and T_0. Then, sequence the

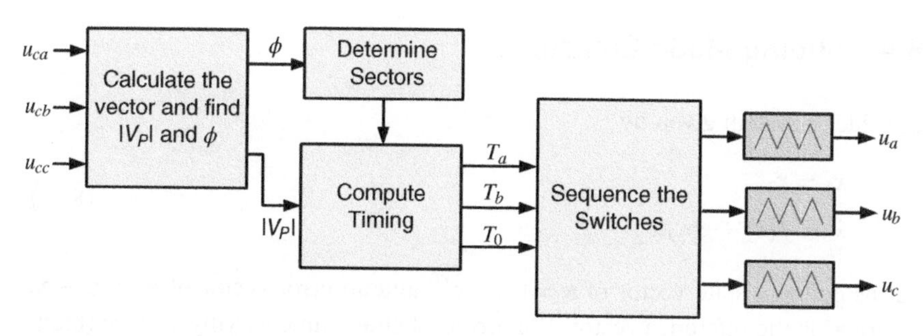

Figure 8.12 SVPWM feedback control structure.

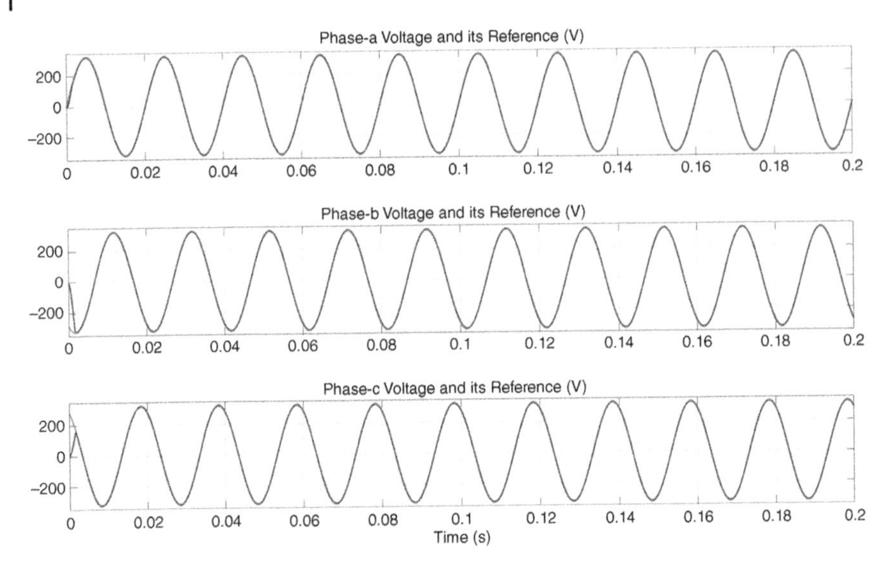

Figure 8.13 Tracking performance of the closed-loop system of Example 8.4.

switches and generate the switching pulses u_a, u_c, and u_c by comparing the output with a triangular carrier waveform.

Example 8.4 The same system as in Example 8.2, with the same LQR gain matrix, is used here. The switching frequency is chosen as 15 kHz. The tracking performance is shown in Figure 8.13. It can be seen that the performance of the SVPWM closed-loop controller is similar to those of PWM closed-loop controllers of Examples 8.2 and 8.3. The switching signals are shown in Figure 8.14. They are in the undermodulated region, as expected.

8.4 Sliding Mode Control

Consider a system given by

$$\begin{aligned}\dot{x}_1 &= x_2\\ \dot{x}_2 &= f(x) + g(x)u\end{aligned} \tag{8.14}$$

Let us define a state vector of $\mathbf{x} = [x_1 \quad x]^T$ and an error vector of $\mathbf{x}_e = \mathbf{x}^* - \mathbf{x}$, where $\mathbf{x}^*$ is the reference vector. We further define a time varying surface (manifold) as [6]

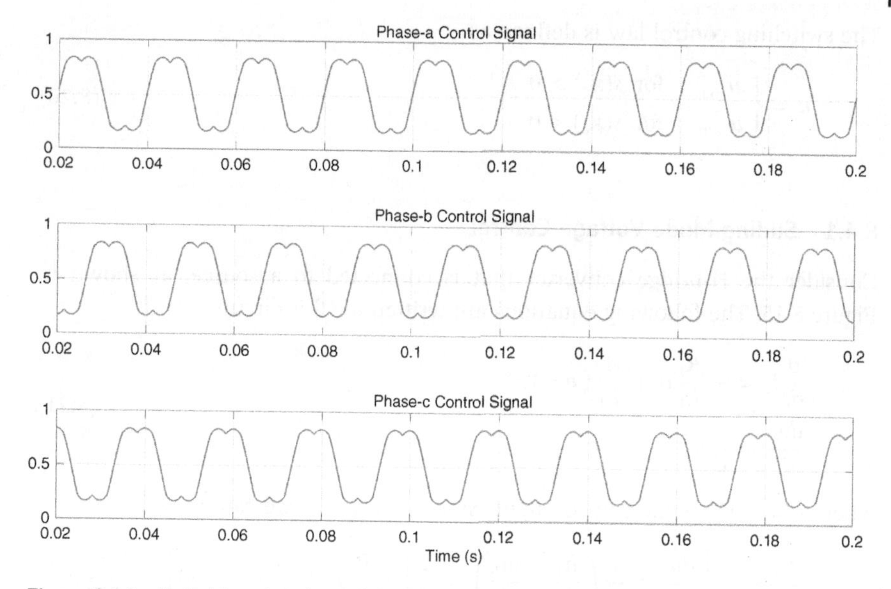

Figure 8.14 Switching signals of SVPWM control of Example 8.4.

$$s(\mathbf{x}_e) = \left(\frac{d}{dt} + \lambda\right)^{n-1} \mathbf{x}_e \tag{8.15}$$

where n is the order of the system. For the system of (8.14), the following can be written

$$s(\mathbf{x}_e) = \dot{\mathbf{x}}_e + \lambda \mathbf{x}_e \tag{8.16}$$

Then the tracking problem is converted into an equivalent form of remaining on the surface $s = 0$ for all $t > 0$.

The first-order problem of keeping $s(\mathbf{x}_e) = 0$ can be achieved by choosing the following control law [6]

$$\frac{1}{2}\frac{ds^2}{dt} \leq -\eta|s|, \quad n > 0 \tag{8.17}$$

The existence condition for the sliding mode is then written as

$$\begin{aligned} &\dot{s} > 0 \quad \text{when } s < 0 \\ &\dot{s} < 0 \quad \text{when } s > 0 \end{aligned} \tag{8.18}$$

This can be written in a compact form as

$$s\dot{s} < 0 \tag{8.19}$$

The switching control law is defined as

$$u = \begin{cases} u_{\max} & \text{for } s(\mathbf{x}_e) > 0 \\ u_{\min} & \text{for } s(\mathbf{x}_e) < 0 \end{cases} \tag{8.20}$$

8.4.1 Sliding Mode Voltage Control

Consider the H-bridge converter that is connected to a source, as shown in Figure 8.15. The following equations are written for this circuit

$$\begin{aligned} \frac{d}{dt} i_1 &= -\frac{R_1}{L_1} i_1 + \frac{1}{L_1}(u - v_C) \\ \frac{dv_C}{dt} &= \frac{1}{C} i_C \end{aligned} \tag{8.21}$$

where $u = \pm V_{dc}$ is the control input. Since $i_C = i_1 - i_L$, we can write

$$\begin{aligned} \frac{d^2 v_C}{dt^2} &= \frac{1}{C}\frac{di_C}{dt} = \frac{1}{C}\left(\frac{di_1}{dt} - \frac{di_L}{dt}\right) = \frac{1}{C}\left[-\frac{R_1}{L_1} i_1 + \frac{1}{L_1}(u - v_C)\right] - \frac{1}{C}\frac{di_L}{dt} \\ &= \frac{1}{C}\left[-\frac{R_1}{L_1}(i_C + i_L) + \frac{1}{L_1}(u - v_C)\right] - \frac{1}{C}\frac{di_L}{dt} \end{aligned}$$

The above equation is then written as

$$\frac{d^2 v_C}{dt^2} = -\frac{R_1}{L_1 C} i_C + \frac{1}{L_1 C}(u - v_C) - \frac{1}{C}\frac{di_L}{dt} - \frac{R_1}{L_1 C} i_L \tag{8.22}$$

Defining the following disturbance input

$$\text{dist} = -\frac{1}{C}\frac{di_L}{dt} - \frac{R_1}{L_1 C} i_L \tag{8.23}$$

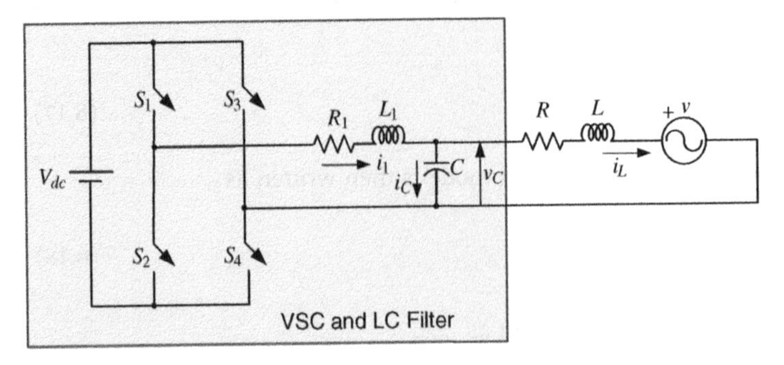

Figure 8.15 H-bridge VSC with LC filter connected to a back emf.

(8.21) is written as

$$\frac{d^2 v_C}{dt^2} = -\frac{R_1}{L_1 C} i_C + \frac{1}{L_1 C}(u - v_C) + \text{dist} \tag{8.24}$$

Let the state vector be chosen as

$$\mathbf{x} = \begin{bmatrix} \dot{v}_C \\ v_C \end{bmatrix}$$

Then the state space description of the system is written as

$$\begin{aligned} \dot{\mathbf{x}} &= \mathbf{A}\mathbf{x} + \mathbf{B}u + \mathbf{H}\text{dist} \\ y &= \mathbf{C}\mathbf{x} \end{aligned} \tag{8.25}$$

where

$$\mathbf{A} = \begin{bmatrix} -\frac{R_1}{L_1} & -\frac{1}{L_1 C} \\ 1 & 0 \end{bmatrix}, \quad \mathbf{B} = \begin{bmatrix} \frac{1}{L_1 C} \\ 0 \end{bmatrix}, \quad \mathbf{H} = \begin{bmatrix} 1 \\ 0 \end{bmatrix}, \quad \mathbf{C} = [0 \quad 1]$$

We now define the following error vector

$$\mathbf{x}_e = \begin{bmatrix} \dot{v}_C^* - \dot{v}_C \\ v_C^* - v_C \end{bmatrix}$$

The switching surface is then given from (8.16) as

$$s(\mathbf{x}_e) = \dot{\mathbf{x}}_e + \lambda \mathbf{x}_e = \left(\dot{v}_C^* - \dot{v}_C\right) + \lambda\left(v_C^* - v_C\right) \tag{8.26}$$

Assume that the converter has to track a reference voltage given by

$$v_C^* = V_m \sin(\omega t) \tag{8.27}$$

Then

$$\dot{v}_C^* = \omega V_m \cos(\omega t) \tag{8.28}$$

The following reference equations are now defined for an ideal oscillator [7]

$$\begin{aligned} \dot{\mathbf{x}}_m &= \mathbf{A}_m \mathbf{x}_m + \mathbf{B}_m r \\ y_m &= \mathbf{C}_m \mathbf{x}_m \end{aligned} \tag{8.29}$$

where

$$\mathbf{A}_m = \begin{bmatrix} 0 & -\omega^2 \\ 1 & 0 \end{bmatrix}, \quad \mathbf{B}_m = \begin{bmatrix} 1 \\ 0 \end{bmatrix}, \quad \mathbf{C}_m = [0 \quad 1]$$

Then

$$\frac{Y_m(s)}{R(s)} = \mathbf{C}_m(s\mathbf{I} - \mathbf{A}_m)^{-1}\mathbf{B}_m$$
$$= \frac{1}{(s^2 + \omega^2)}[0 \quad 1]\begin{bmatrix} s & -\omega^2 \\ 1 & s \end{bmatrix}\begin{bmatrix} 1 \\ 0 \end{bmatrix} = \frac{1}{(s^2 + \omega^2)} \tag{8.30}$$

Choosing

$$R(s) = V_m\omega \tag{8.31}$$

(8.30) is rewritten as

$$Y_m(s) = \frac{V_m\omega}{(s^2 + \omega^2)}$$

The inverse Laplace transform of the above equation is

$$y_m = V_m \sin(\omega t) \tag{8.32}$$

which is the desired output voltage.

Since the switching manifold $s(\mathbf{x}_e) = 0$, we can write

$$\dot{s}(\mathbf{x}_e) = \mathbf{K}(\dot{\mathbf{x}}_m^* - \dot{\mathbf{x}}) = 0 \tag{8.33}$$

where $\mathbf{K} = [1 \quad \lambda]$. Then, combining (8.25) and (8.29) with (8.33), we have

$$\dot{s}(\mathbf{x}_e) = \mathbf{K}(\mathbf{A}_m\mathbf{x}_m + \mathbf{B}_m r - \mathbf{A}\mathbf{x} - \mathbf{B}u - \mathbf{H}\text{dist}) = 0 \tag{8.34}$$

From (8.34), an equivalent control law is obtained, which is given as

$$u_{eq} = (\mathbf{KB})^{-1}\mathbf{K}(\mathbf{A}_m\mathbf{x}_m + \mathbf{B}_m r - \mathbf{A}\mathbf{x} - \mathbf{H}\text{dist}) \tag{8.35}$$

Note that, if infinitely large $u_{\max}$ and $u_{\min}$ were available, (8.35) would always be possible. This will guarantee that the sliding mode will exist. However, this is not possible, and the sliding mode will chatter between

$$u_{\min} \leq u_{eq} \leq u_{\max} \tag{8.36}$$

Now, from (8.29) and (8.25), we have

$$\dot{\mathbf{x}}_e = \dot{\mathbf{x}}_m - \dot{\mathbf{x}} = \mathbf{A}_m\mathbf{x}_m + \mathbf{B}_m r - \mathbf{A}\mathbf{x} - \mathbf{B}u - \mathbf{H}\text{dist}$$

Replacing u by u_{eq} from (8.35), the following equation is obtained [7]

$$\dot{\mathbf{x}}_e = \mathbf{A}_m\mathbf{x}_e, \quad \mathbf{A}_m = \begin{bmatrix} -\lambda & 0 \\ 1 & 0 \end{bmatrix} \tag{8.37}$$

The eigenvalues of the matrix $\mathbf{A}_m$ are located at 0 and $-\lambda$ and the system time constant is $1/\lambda$. Thus, by choosing the value of λ, the convergence on the switching line can be forced.

Example 8.5 Consider the system of Figure 8.15, where $L_1 = 2$ mH, $C = 50$ μF, $R = 5\ \Omega$, $V_{dc} = 500$ V, and $L = 11.6$ mH. The back emf and the reference voltage are given by

$$v = 326.6 \sin(\omega t - 20°), \quad v_c^* = 326.6 \sin(\omega t)$$

From (8.12), we get

$$i_C^* = \omega C V_m \cos(\omega t) = 5.13 \cos(\omega t)$$

Since $\dot{v}_C = i_C/C$, the error vector is written as

$$\mathbf{x}_e = \begin{bmatrix} \dot{v}_C^* - \dot{v}_C \\ v_C^* - v_C \end{bmatrix} = \begin{bmatrix} \frac{1}{C}(i_C^* - i_C) \\ v_C^* - v_C \end{bmatrix}$$

A hysteretic control law is then designed such that $\mathbf{x}_e$ is maintained within the upper bound $u_{\max}$ and lower bound $u_{\min}$.

The system response for two values of λ is shown in Figure 8.16. It can be seen that the system response is much faster when $\lambda = 200$ rather than when $\lambda = 20$ due to a smaller time constant.

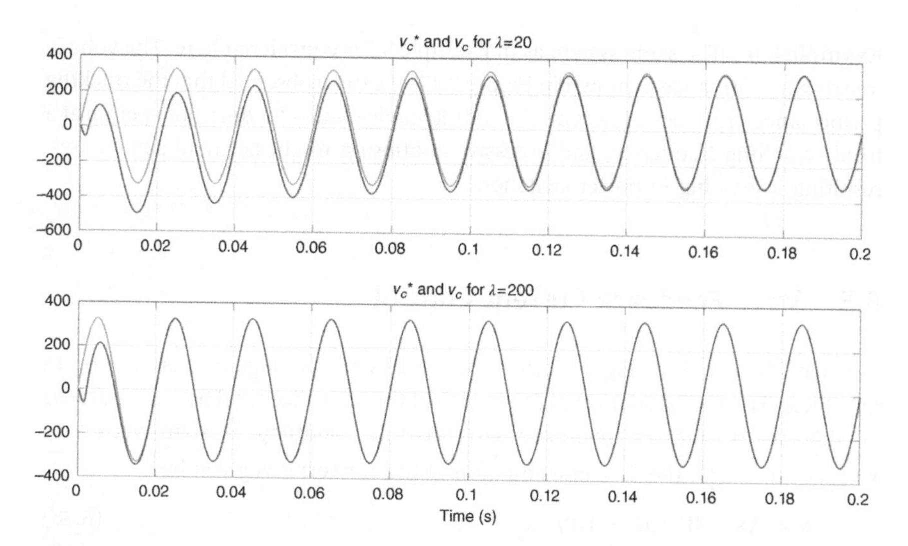

Figure 8.16 Tracking performance with sliding mode control H-bridge converter of Example 8.5.

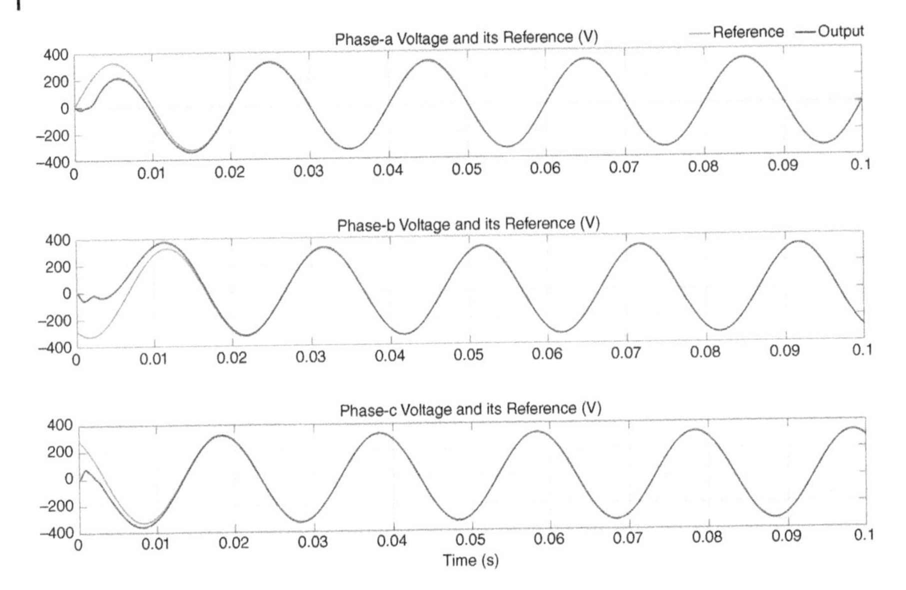

Figure 8.17 Tracking performance with sliding mode control of three-phase converter of Example 8.7.

Example 8.6 The same system as in Example 8.2 is considered here. The voltage tracking performance is shown in Figure 8.17. It can be observed that the tracking performance is perfect. However, the sliding mode controller does not operate at a fixed switching frequency, and excessive chattering might occur in some cases, resulting in heating in power switches.

8.5 State Feedback Current Control

Consider the circuit of Figure 8.18, which shows an H-bridge converter with an LCL filter. The VSC is connected to a back emf through a feeder. The main purpose of this is to track a reference current i_{2ref}. Defining a state vector as $\mathbf{x} = [v_C \quad i_1 \quad i_2]^T$, the dynamic equation of the converter is given by

$$\dot{\mathbf{x}} = \mathbf{A}\mathbf{x} + \mathbf{B}V_{dc}u_c + \mathbf{H}v_T \tag{8.38}$$

where v_T is the voltage at the terminal of the converter (PCC voltage) and

$$\mathbf{A} = \begin{bmatrix} 0 & 1/C & -1/C \\ -1/L_1 & -R_1/L_1 & 0 \\ 1/L_2 & 0 & -R_2/L_2 \end{bmatrix}, \quad \mathbf{B} = \begin{bmatrix} 0 \\ 1/L_1 \\ 0 \end{bmatrix}, \quad \mathbf{H} = \begin{bmatrix} 0 \\ 0 \\ -1/L_2 \end{bmatrix}$$

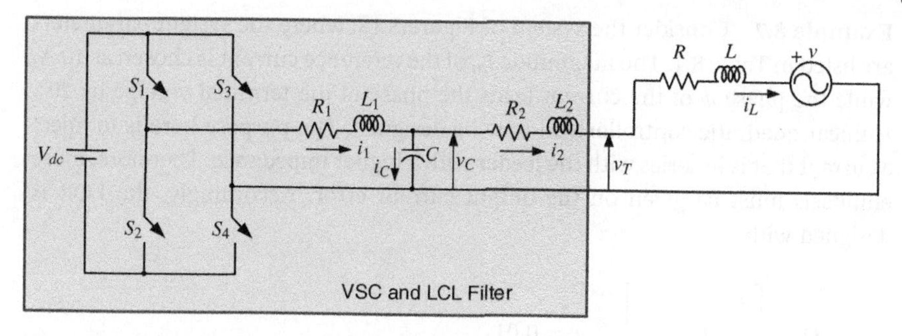

Figure 8.18 VSC in current control mode through an LCL filter.

Let the desired output current be

$$i_{2ref} = I_m \sin(\omega t + \phi)$$

where I_m is the desired magnitude and ϕ is the phase of the current with respect to the terminal voltage v_T. Since it is not possible to set references for the capacitor voltage v_C and the inner inductor current i_1, two HPFs are designed that are given by

$$\text{HPF-1} : i_{1HPF} = \left(\frac{s}{s+\alpha_1}\right) i_1, \quad \text{HPF-2} : v_{CHPF} = \left(\frac{s}{s+\alpha_2}\right) v_C$$

The feedback control structure is shown in Figure 8.19. First, a phase locked loop (PLL) is used to extract the phase θ of the terminal voltage v_T. The phase ϕ is then added with it to derive the reference i_{2ref} which the converter must track. A state feedback controller is then designed, and the resultant control signal is used for switching signal generation through PWM [8].

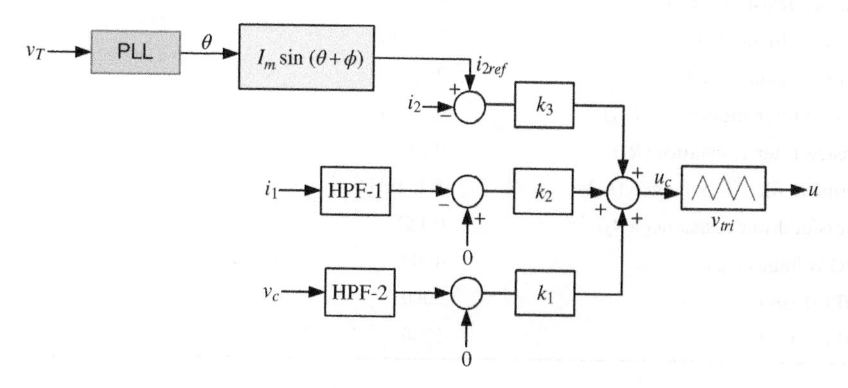

Figure 8.19 Current tracking state feedback control structure.

Example 8.7 Consider the system of Figure 8.18, where the system parameters are listed in Table 8.1. The magnitude I_m of the reference current is chosen as 10 A, while the phase ϕ of the current leads the phase of the terminal voltage by 20°. A linear quadratic controller will now be designed. The purpose here is to inject a current that is in series with the feeder with a higher impedance. Therefore, more emphasis must be given on the output current error. Accordingly, the LQR is designed with

$$\mathbf{Q} = \begin{bmatrix} 1 & & \\ & 1 & \\ & & 10^5 \end{bmatrix}, \quad r = 0.01$$

The resulting gain matrix is

$$\mathbf{K} = \begin{bmatrix} 10.25 & 16.15 & 3.15 \times 10^3 \end{bmatrix}$$

The results are shown in Figure 8.20. The current tracking is fairly accurate as the peak of the tracking error is 0.2 A. The total harmonic distortion (THD) of the output current is below 0.05% in the steady state. This implies that the LCL filter has eliminated the lower-order harmonics effectively.

Table 8.1 System parameters for Example 8.7.

System quantities	Parameter values
System frequency	50 Hz
Back emf	230 V (rms), with phase of 0°
Feeder resistance (R)	5 Ω
Load inductance (L)	11.6 mH
Filter capacitance (C)	25 μF
Inside filter inductance (L_1)	0.2 mH
Inside filter resistance (R_1)	0.1 Ω
Outside filter inductance (L_2)	1.25 mH
Outside filter resistance (R_2)	0.1 Ω
DC voltage (V_{dc})	450 V
HPF-1 (α_1)	5000
HPF-2 (α_2)	5000

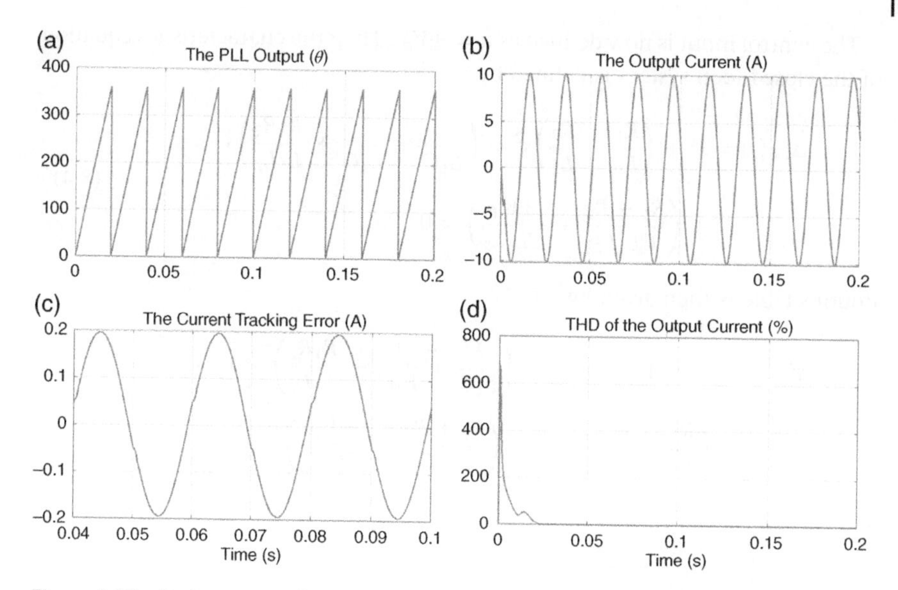

Figure 8.20 Performance of the state feedback based current controller. (a) The PLL Output (θ), (b) The Output Current (A), (c) The Current Tracking Error (A), (d) THD of the Output Current (%).

8.6 Output Feedback Current Control

A state feedback controller will require the measurements of the state variables for feedback. As is illustrated in Section 8.5, it is not possible to form or estimate the reference values for some of these variables and therefore two HPFs are used in Section 8.5. An output feedback control, on the other hand, will only require the measurements of the output variable, which is i_2 in this case. Hysteresis current control is very effective for current tracking, as demonstrated in Section 7.1.1 when the output of a VSC is connected with an L-type filter. To evaluate the behavior of hysteresis controller when the VSC has an output LCL filter, consider the state equation of (8.38). Neglecting the $\mathbf{H}v_T$ term, the transfer function between the input and the output is given by

$$\frac{i_2(s)}{u(s)} = [0 \quad 0 \quad 1](s\mathbf{I} - \mathbf{A})^{-1}\mathbf{B} \tag{8.39}$$

Solving (8.39), we have

$$\frac{i_2(s)}{u(s)} = \frac{\dfrac{1}{CL_1L_2}}{s^3 + \left(\dfrac{R_1}{L_1} + \dfrac{R_2}{L_2}\right)s^2 + \left(\dfrac{1}{L_1C} + \dfrac{1}{L_2C} + \dfrac{R_1R_2}{L_1L_2}\right)s + \dfrac{R_1 + R_2}{CL_1L_2}} \tag{8.40}$$

The control input is now defined as $u = \pm V_{dc}$. Then the characteristics equation of the closed-loop system is defined by

$$\Delta_{CL}(s) = s^3 + \left(\frac{R_1}{L_1} + \frac{R_2}{L_2}\right)s^2 + \left(\frac{1}{L_1C} + \frac{1}{L_2C} + \frac{R_1R_2}{L_1L_2}\right)s + \left(\frac{R_1 + R_2}{CL_1L_2} + \frac{\pm V_{dc}}{CL_1L_2}\right) = 0 \tag{8.41}$$

Routh's table is then given by

$$\begin{array}{lll} s^3 & 1 & \left(\dfrac{1}{L_1C} + \dfrac{1}{L_2C} + \dfrac{R_1R_2}{L_1L_2}\right) \\ s^2 & \left(\dfrac{R_1}{L_1} + \dfrac{R_2}{L_2}\right) & \left(\dfrac{R_1 + R_2}{CL_1L_2} + \dfrac{\pm V_{dc}}{CL_1L_2}\right) \\ s^1 & \alpha & \\ s^0 & \left(\dfrac{R_1 + R_2}{CL_1L_2} + \dfrac{\pm V_{dc}}{CL_1L_2}\right) & \end{array}$$

Then, for the system to be stable, $\alpha > 0$. For $u = V_{dc}$, this is satisfied when

$$V_{dc} < \frac{R_1L_2}{L_1} + \frac{R_2L_1}{L_2} + CR_1R_2\left(\frac{R_1}{L_1} + \frac{R_2}{L_2}\right)$$

Since the inductances are in mH and the capacitance is in μF, this is not possible. Furthermore, there will be two sign changes in Routh's table, and therefore there will be two roots of the closed-loop system in the right-half s-plane. On the other hand, when $u = -V_{dc}$, the term in the s^0 will be negative, even if $\alpha > 0$ is true. This implies that there will be one sign change in Routh's table, indicating one unstable closed-loop pole. Therefore, the hysteresis current control will result in an unstable system response. For the system of Example 8.7, the reference and output currents are shown in Figure 8.21, which shows that large, high-frequency current is flowing through the VSC, indicating unwanted behavior.

In order to avoid this, a discrete-time pole placement controller with a hysteresis band controller is designed instead. The control signal u_c is obtained using a pole placement controller, and the switching signal is generated in a hysteresis band, given by

$$\begin{aligned} &\text{If } u_c > h \text{ then } u = +1 \\ &\quad \text{elseif } u_c < -h \text{ then } u = -1 \end{aligned} \tag{8.42}$$

where $h > 0$is a small number. The control structure is shown in Figure 8.22. The following example demonstrates the effectiveness of the control design.

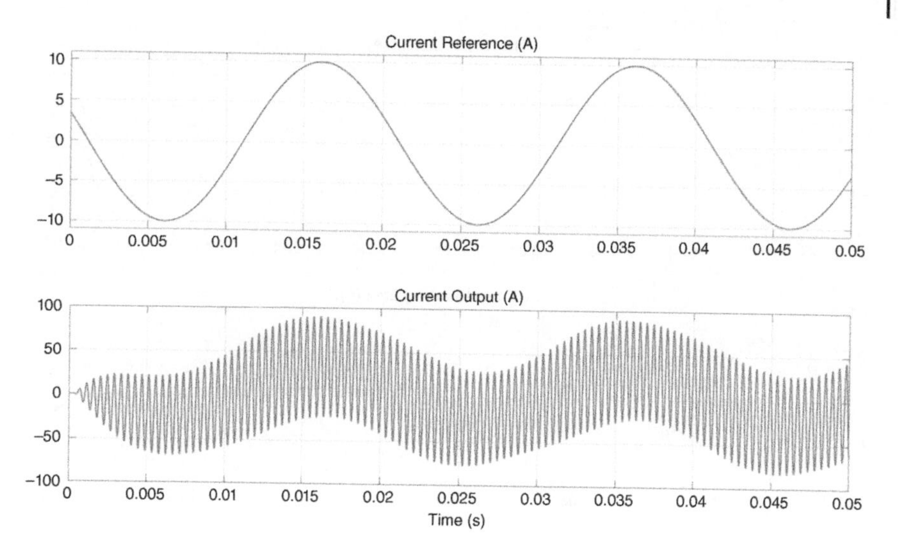

Figure 8.21 Current tracking failure due to hysteresis current control for VSC with LCL output filter.

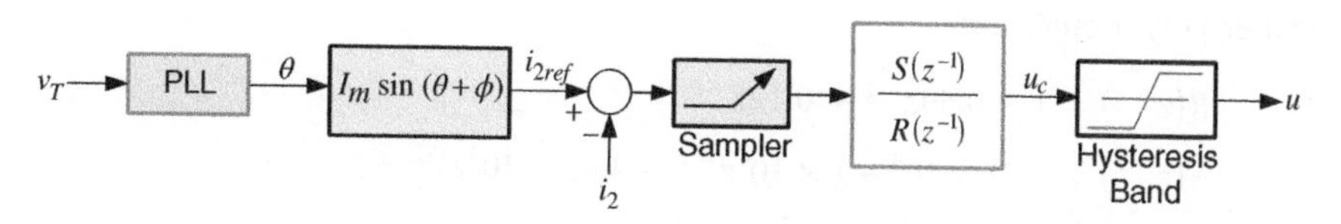

Figure 8.22 Current tracking output feedback control structure.

Example 8.8 Consider the system of Example 8.7. The first step in this process is to choose a sampling time, which is taken as 10 μs for this example. Then the input–output description of the system is given in difference equation form as

$$\begin{aligned} y(k) = &-a_1 y(k-1) - a_2 y(k-2) - a_3 y(k-3) \\ &+ [b_0 u_c(k-1) + b_1 u_c(k-2) + b_2 u_c(k-3)] + C(z^{-1})e(k) \end{aligned}$$

where $y(k) = i_2(k)$ is the output of the converter and $e(k) = v_T(k)$ is considered a disturbance input. For the parameters given in Table 8.1, the following coefficients of the difference equation are obtained

$$\begin{aligned} &a_1 = -2.98,\ a_2 = 2.97,\ a_3 = -0.995,\ b_0 = 3.33 \times 10^{-7}, \\ &b_1 = 13.27 \times 10^{-7},\ b_2 = 3.32 \times 10^{-7} \end{aligned}$$

For this system, one of the open-loop zeros is outside the unit circle. Therefore, a minimum variance controller cannot be designed for this system. Instead, we shall

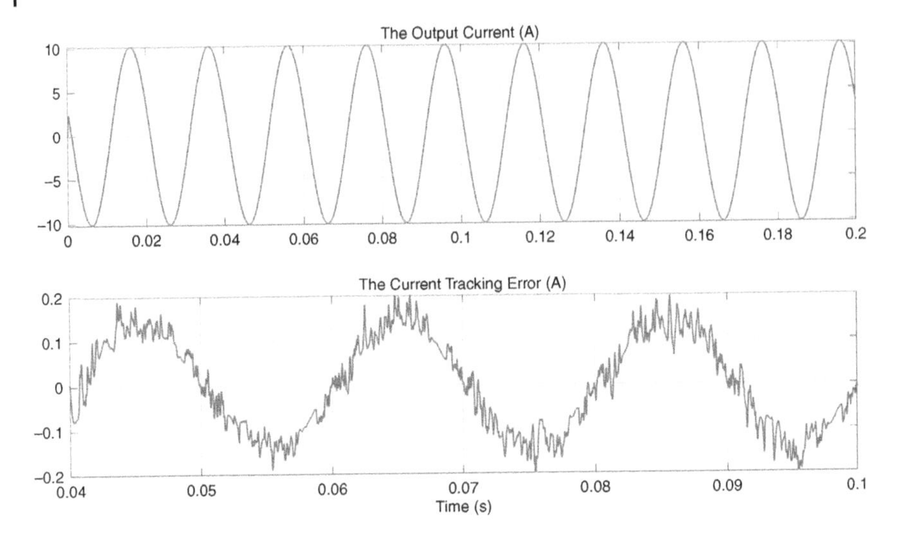

Figure 8.23 Performance of the output feedback based current controller.

design a pole shift controller with a pole shift factor of $\lambda = 0.75$. The resultant controller polynomials are

$$R(z^{-1}) = 1 + 0.59z^{-1} + 0.12z^{-2}$$
$$S(z^{-1}) = 5.71 \times 10^3 - 1 \times 10^4 z^{-1} + 4.62 \times 10^3 z^{-2}$$

The same reference as in Example 8.7 is chosen here as well. The hysteretic band chosen here is $h = 0.05$. The system output is shown in Figure 8.23a, while the tracking error is shown in Figure 8.23b. Since this is essentially a hysteretic controller, the chattering can be observed in the output current. The tracking performance, however, is acceptable. The THD in this case is 0.21%, which is higher than the state feedback control, even though it is much lower than the acceptable standard.

8.7 Concluding Remarks

In this chapter, we present the basic voltage and current control concepts using voltage source converters. These are developed further depending on the applications discussed in subsequent chapters. There are two main control aspects of power converters: fixed frequency and variable frequency. Fixed frequency controllers, using a PWM technique, are often desirable when looking to restrict the switching losses. The chapter shows that hysteretic current control will be unstable when the output of the VSC is connected to a higher-order filter. A hysteresis band controller that is designed based on a suitable discrete-time control technique is very stable. Even though these controllers operate at variable frequencies, they are easier to design, and usually have faster convergence and excellent tracking properties, albeit at the expense of higher switching losses. In subsequent chapters, both these techniques are studied depending on the problems being addressed.

Problems

8.1 Design an LC filter with the following specifications

- Voltage: 11 kV (L-L), frequency 50 Hz
- VA rating: 1.0 MVA
- DC voltage V_{dc}: 1.5 time the L-N voltage
- Switching frequency: 15 kHz
- Maximum current ripple: 2.5%
- λ: 5%

Choose $R_1 = 0.1\ \Omega$. Draw the Bode plot of the designed filter and find its bandwidth.

8.2 The filter designed in Problem 8.1 is used for the PWM voltage control of a single-phase VSC. Design an LQR state feedback controller with the following specifications

$$Q = \begin{bmatrix} 1000 & \\ & 10 \end{bmatrix}, \quad r = 0.1$$

Find the eigenvalues of the closed-loop system.

8.3 Assume that the filter designed in Problem 8.1 is used in the voltage control of a single-phase VSC, where the high-frequency components of the inner inductor current are suppressed by the HPF given in (8.7), where the parameter α is chosen as 5000. Then, choosing an extended state vector as $\mathbf{x}_e = \begin{bmatrix} v_c & i_1 & i_{1LPF} \end{bmatrix}^T$, the state space equation is given by (8.9). The control law is given by

$$u_c = \begin{bmatrix} k_1 & k_2 & k_3 \end{bmatrix} \times \begin{bmatrix} v_{cref} - v_c \\ 0 - i_{1HPF} \\ i_1 - i_{1LPF} \end{bmatrix}$$

Design an LQR state feedback controller with the following specifications

$$Q = \begin{bmatrix} 1000 & & \\ & 1 & \\ & & 10 \end{bmatrix}, \quad r = 0.1$$

Compute the eigenvalues of the closed-loop system and compare them with those obtained in Problem 8.2.

8.4 Assume that three identical LC filters are connected at the output of a three-phase VSC that is used for voltage control. The filter parameters are obtained from Problem 8.1. The VSC and load are connected so that each phase can be treated separately. Now assume that the state space equation in the dq domain is given by $\dot{\mathbf{x}}_{dq} = \mathbf{A}_{dq}\mathbf{x}_{dq} + V_{dc}\mathbf{B}_{dq}\mathbf{u}_{dq}$, where

$$\mathbf{x}_{dq} = \begin{bmatrix} v_{cd} & v_{cq} & i_{1d} & i_{1q} \end{bmatrix}^T, \quad \mathbf{u}_{dq} = \begin{bmatrix} u_d & u_q \end{bmatrix}^T$$

Then design an LQR controller with the following parameters

$$\mathbf{Q} = \begin{bmatrix} 1000 & & & \\ & 1000 & & \\ & & 10 & \\ & & & 10 \end{bmatrix}, \quad \mathbf{R} = \begin{bmatrix} 0.1 & \\ & 0.1 \end{bmatrix}$$

Compare the result with that obtained in Problem 8.2.

8.5 With the filter designed in Problem 8.1, add an outer inductor with $L_2 = 2$ mH and $R_2 = 0.5\ \Omega$ to form an LCL filter. Assuming the output is the current i_2, draw the Bode plot of the filter and find the bandwidth of the filter. What are the gain and phase margins?

8.6 For the filter designed in Problem 8.5, design a pole shift controller with pole shift factor (λ) of 0.7 and sampling time of 20 μs.

Notes and References

Discussion on sliding mode control can also be found in [9]. Fixed switching frequency application of sliding mode control is discussed in [10, 11], where its frequency domain characterization for a DSTATCOM application it also presented. Predictive current control using VSC has attracted much attention. Even though this is not presented in this chapter, this control method can be implemented on grid feeding converters. Those interested can refer to [12–15].

1 Beres, R.N., Wang, X., Liserre, M. et al. (2016). A review of passive power filters for three-phase grid-connected voltage-source converters. *IEEE J. Emerg. Selected Topics Power Electron.* 4 (1): 54–69.

2 Wang, T.C.Y., Ye, Z., Sinha, G., and Yuan, X. (2003) Output filter design for a grid-interconnected three-phase inverter. *IEEE 34th Annual Conference on Power Electronics Specialist Conference (PESC)*, Acapulco, Mexico.

3 Solatialkaran, D., Zare, F., Saha, T.K., and Sharma, R. (2020). A novel approach in filter design for grid-connected inverters used in renewable energy systems. *IEEE Trans. Sustain. Energy* 11 (1): 154–164.

4 Ghosh, A. and Ledwich, G. (2003). Load compensating DSTATCOM is weak ac systems. *IEEE Trans. Power Delivery* 18 (4): 1302–1309.

5 John, B., Ghosh, A., and Zare, F. (2018). Load sharing in medium voltage islanded microgrids with advanced angle droop control. *IEEE Trans. Smart Grid* 9 (6): 6461–6469.

6 Slotine, J.-J.E. and Li, W. (1991). *Applied Nonlinear Control.* Englewood Cliffs, NJ: Prentice Hall.

7 Carpita, M. and Marchesoni, M. (1996). Experimental study of a power conditioning system using sliding mode control. *IEEE Trans. Power Electron.* 11 (5): 731–742.

8 Ghosh, A. and Dewadasa, M. *CSIRO Intelligent Grid Research Cluster-Project 7: Operation Control and Energy Management of Distributed Generation.* iGrid http://igrid.net.au/resources/downloads/project7/Operation%20Control%20and%20Energy%20Management%20of%20Grid%20Connected%20DG%20Final%20Report.pdf, accessed 24 May 2022.

9 Khalil, H.K. (2002). *Nonlinear Systems*, 3e. Upper Saddle River, NJ: Prentice Hall.

10 Abrishamifar, A., Ahmad, A.A., and Mohamadian, M. (2012). Fixed switching frequency sliding mode control for single-phase unipolar inverters. *IEEE Trans. Power Electron.* 27 (5): 2507–2514.

11 Gupta, A. and Ghosh, A. (2006). Frequency-domain characterization of sliding mode control of an inverter used in DSTATCOM application. *IEEE Trans. Circuits Syst. 1: Regul. Pap.* 53 (3): 662–676.

12 Rodríguez, J. and Cortes, P. (2012). *Predictive Control of Power Converters and Electric Drives.* New York: Wiley.

13 Rodríguez, J., Pontt, J., Silva, C.A. et al. (2007). Predictive current control of a voltage source inverter. *IEEE Trans. Ind. Electron.* 54 (1): 495–503.

14 Kouro, S., Cortés, P., Vargas, R. et al. (2009). Model predictive control: a simple and powerful method to control power converters. *IEEE Trans. Ind. Electron.* 56 (6): 1826–1838.

15 Judewicz, M.G., González, S.A., Fischer, J.R. et al. (2018). Inverter-side current control of grid-connected voltage source inverters with LCL filter based on generalized predictive control. *IEEE J. Emerg. Selected Topics Power Electron.* 6 (4): 1732–1743.

9

VSC Applications in Custom Power

In this chapter, various techniques are presented which can be employed for controlling the output voltage or current of a voltage source converter (VSC) for power quality improvement in power distribution systems. The main aim is to control the switching functions of the VSC in such a way that a desired output is attained. One of the main applications of closed-loop converter control is in custom power, through which utilities can supply value-added power to customers [1]. The custom power technology uses power converters to alleviate various power quality-related issues, such as harmonics, voltage or load unbalance, poor power factor, etc. [2].

There are several custom power devices, such as:

- solid state current limiter (SSCL)
- solid state circuit breaker (SSCB)
- solid state transfer switch (SSTS)
- distribution static compensator (DSTATCOM)
- dynamic voltage restorer (DVR)
- unified power quality conditioner (UPQC).

Of these, the first three are network reconfiguring types, which can limit current during faults or can transfer loads from one feeder to the other. The last three are network compensating devices as they can compensate various power quality problems.

All the network compensating custom power devices are realized by VSCs, along with their output passive filters. DSTATCOM is connected in shunt at the point of common coupling (PCC) of the load and the feeder. The DVR is a series connected device that is usually connected at the load terminals such that it can tightly maintain the required load bus voltage in the face of voltage sag or swell in the upstream network. A UPQC has a shunt VSC and a series VSC that are connected together

Control of Power Electronic Converters with Microgrid Applications, First Edition.
Arindam Ghosh and Firuz Zare.

Published 2023 by John Wiley & Sons, Inc.

through a DC bus capacitor. In this chapter, the control of a DSTATCOM is discussed in detail. For DVR or UPQC control, refer to [2].

A DSTATCOM can be controlled in both voltage and current control modes. In the voltage control mode, the DSTATCOM can tightly regulate the PCC voltage such that it is balanced and sinusoidal, irrespective of unbalance and/or distortion in the load currents such that the upstream voltage and current are balanced sinusoidally. In the current control mode, the DSTATCOM can work as an active filter where it cancels out the load current harmonics, thereby preventing them from flowing into the upstream network. A significantly better option is complete load compensation in which the DSTATCOM can inject currents that can simultaneously perform power factor correction, harmonic filtering, and load balancing. These aspects are also discussed in this chapter.

9.1 DSTATCOM in Voltage Control Mode

A DSTATCOM is a versatile device that can be used in both voltage and current control modes. In this section, the voltage control application of DSTATCOM is discussed. Let us begin our discussion with an example.

Example 9.1 Consider the radial distribution system shown in Figure 9.1, where the source supplies two loads through two feeders. Of the two loads, Load Z_2 contains an unbalanced RL circuit and a diode rectifier load, while Load Z_1 is a balanced RL load. The system parameters are given in Table 9.1. The load currents are shown in Figure 9.2. Load Z_2 currents are unbalanced and distorted, causing Load Z_1 currents to be unbalanced and distorted as well.

This example shows that, when a distribution bus voltage is distorted, it affects the rest of the network. A DSTATCOM that is working in voltage control mode will be used to alleviate this problem. The possible DSTATCOM structures that are used in voltage control mode are shown in Figures 9.3–9.5. All of them have a DC storage capacitor C_{dc} supplying the DC bus. Furthermore, all of them have output LC filters. The voltages (v_{ca}, v_{cb}, and v_{cc}) across the filter capacitors (C) are to be controlled. Figure 9.3 shows a three-leg VSC, while Figure 9.4 shows a VSC that has been constructed through three H-bridges and three single-phase transformers. In these, the capacitors are connected together at point N, which needs to be connected with the load neutral, so that the zero-sequence current circulates in this path. Figure 9.5 shows a four-leg VSC, where the fourth leg is used to cancel the zero-sequence current.

Now consider the same system as in Figure 9.1 that is redrawn with a DSTATCOM, as shown in Figure 9.6. Here the DSTATCOM is connected to the point of common coupling (PCC) of Load Z_2 and Feeder-2. If the DSTATCOM maintains

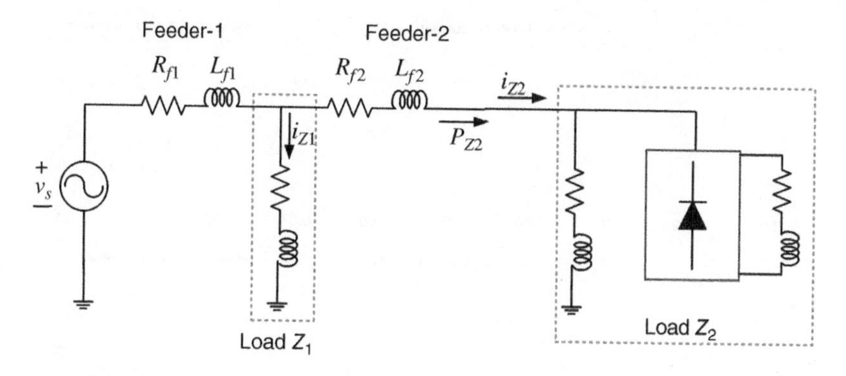

Figure 9.1 Single-line diagram of a radial system supplying two loads.

Table 9.1 Parameters of the system of Figure 9.1.

Components	**Parameters**
Source	Voltage: 11 kV (L-L) rms, angle 0°
	Frequency: 50 Hz
Feeders	Feeder-1: $0.4033 + j4.0527\ \Omega$
	Feeder-2: $0.2017 + j2.042\ \Omega$
Load Z_1	Balanced: $Z_1 = 121.0 + j39.77\ \Omega$
Load Z_2	Unbalanced RL:
	$Z_{2a} = 121.0 + j39.77\ \Omega$ $Z_{2b} = 91.0 + j31.42\ \Omega$ $Z_{2c} = 221.0 + j71.19\ \Omega$
	Diode rectifier with $R = 100\ \Omega$ and $L = 100$ mH

the voltage at the PCC (v_c) as a balanced sinusoid, then the current flowing through the upstream Feeder-1 and Feeder-2 will be balanced. Consequently, the Load Z_1 bus voltage will also be balanced, and an undistorted balanced current will be drawn by this load. The question here is how to make the PCC voltage balanced.

The first step in this process is to choose a set of reference voltages for the PCC. Let us set the references as

$$\begin{aligned} v_{ca}^* &= |V_m| \sin(\omega t + \delta) \\ v_{cb}^* &= |V_m| \sin(\omega t + \delta - 120^\circ) \\ v_{cc}^* &= |V_m| \sin(\omega t + \delta + 120^\circ) \end{aligned} \tag{9.1}$$

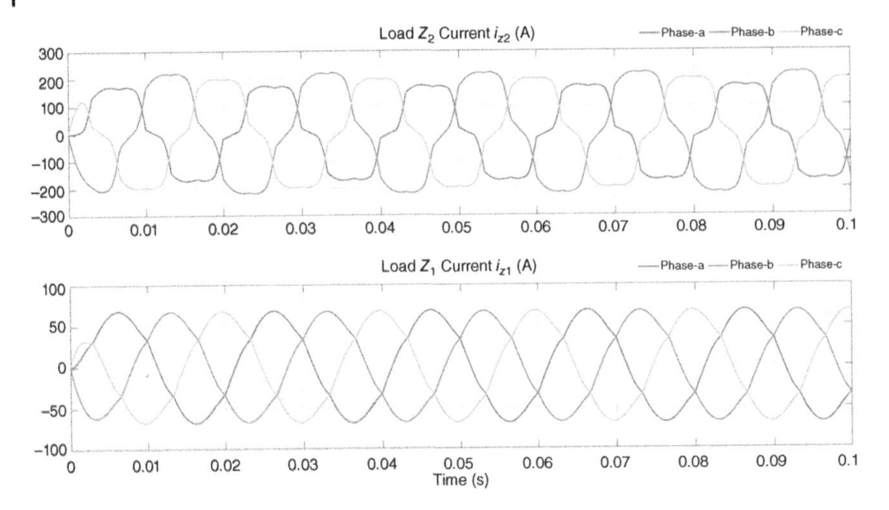

Figure 9.2 The loads currents in Example 9.1.

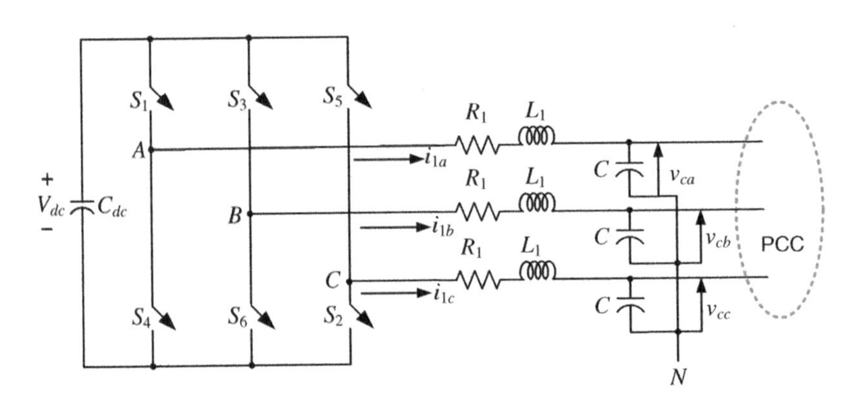

Figure 9.3 Three-leg DSTATCOM structure.

where $|V_m|$ is a prespecified voltage magnitude. In any VSC, the DC link voltage should roughly be around 1.5 times the line-to-line AC voltage to be synthesized. Once the DC capacitor is charged, the DC link voltage will remain constant if no power is drawn from it. However, a VSC will have its switching and other losses (such as those due to quality factor of inductor or transformer core losses, etc.). This implies that, to hold the DC voltage constant, the converter losses must be supplied by the grid. Failing this, the DC capacitor will get discharged and its voltage will continue to fall.

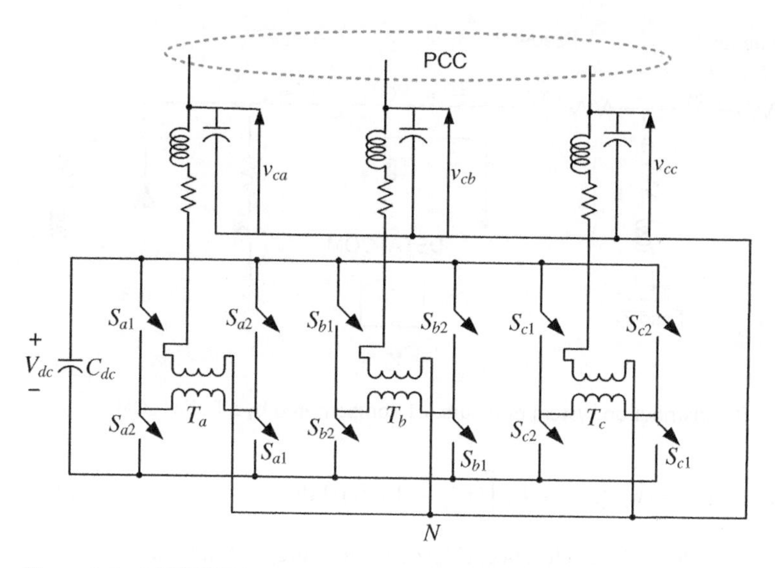

Figure 9.4 DSTATCOM structure with four H-bridges.

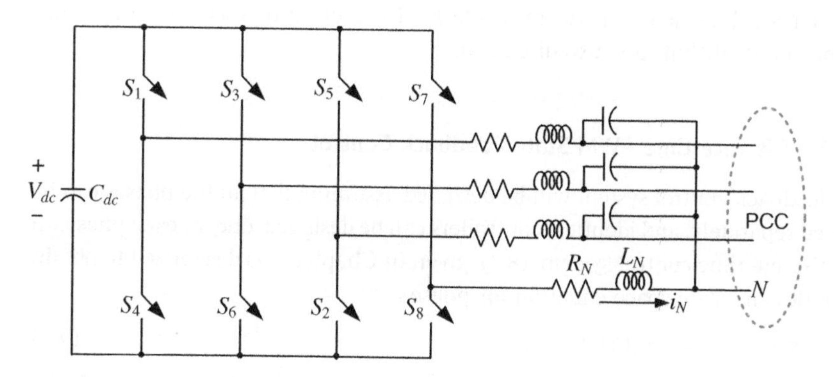

Figure 9.5 Four-leg DSTATCOM structure.

It is well known that the power flow over a line connecting two AC sources mainly depends on the angle difference between the sources. Therefore, taking the source voltage angle as the reference, the angle δ in (9.1) should be so adjusted that the sum total of the power required by Load Z_2 and converter losses should flow from the source to the PCC. Once the required amount of power flows to the PCC, the DC capacitor voltage will remain constant. Therefore, the angle δ is adjusted to hold the DC capacitor voltage constant through the following proportional plus integral (PI) controller.

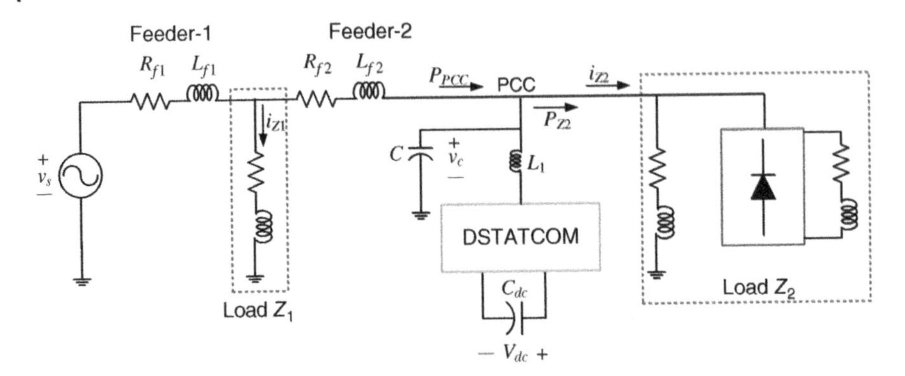

Figure 9.6 The distribution system of Figure 9.1 compensated by a DSTATCOM.

$$\delta = K_P\left(V_{dc}^* - \langle V_{dc}(t)\rangle\right) + K_I\int\left(V_{dc}^* - \langle V_{dc}(t)\rangle\right)dt \tag{9.2}$$

where V_{dc}^* is the DC voltage reference, $\langle V_{dc}(t)\rangle$ is the average of the DC voltage V_{dc}, and K_P and K_I are the PI gains. The VSC of the DSTATCOM will then track the voltage to balance the PCC voltage [3]. Any of the controllers discussed in Chapter 8 will be able to perform this task. However, in this section, other types of control algorithms are also discussed.

9.1.1 Discrete-time PWM State Feedback Control

The feedback control system will be designed assuming that all the phases can be treated separately, and identical controllers can be designed, one for each phase. In the discrete-time control system, (8.3), given in Chapter 8, is discretized to obtain the following state space equation for phase-a

$$\mathbf{x}_a(k+1) = \mathbf{F}\mathbf{x}_a(k) + \mathbf{G}u_{ca}(k) \tag{9.3}$$

where the state vector is $\mathbf{x}_a = [v_{ca}\ i_{1a}]^T$. For a sampling frequency of T, the matrices $\mathbf{F}$ and $\mathbf{G}$ are given by

$$\mathbf{F} = e^{\mathbf{A}T}, \quad \mathbf{G} = \left(\int_0^T e^{\mathbf{A}\tau}d\tau\right)\mathbf{B}$$

The control law is then given as

$$u_{ca}(k) = \mathbf{K}\left[\mathbf{x}_{refa}(k) - \mathbf{x}_a(k)\right] \tag{9.4}$$

where $\mathbf{K}$ is the feedback gain matrix and $\mathbf{x}_{ref} = [v_{ca}^* \quad i_{1a}^*]^T$ is the reference vector. It is mentioned in Chapter 8 that the reference of the current i_{1a} is unknown.

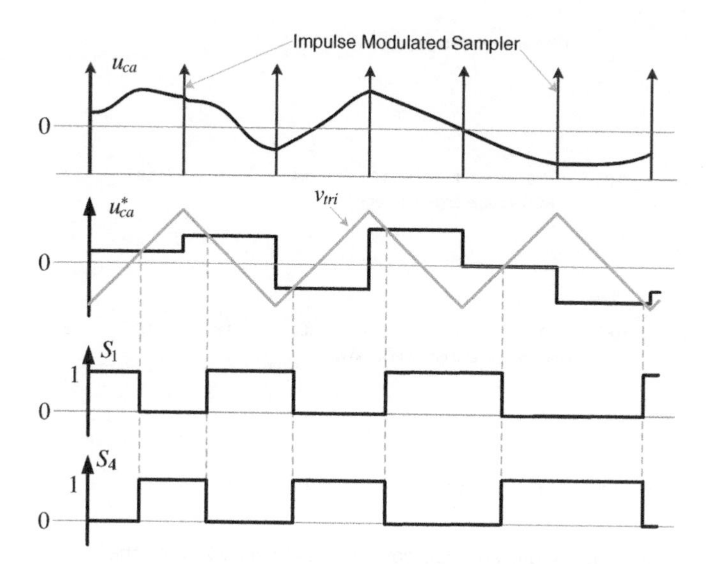

Figure 9.7 Firing pulse generation for discrete-time control.

Thus, the highpass filter (HPF) as shown in Figure 8.5b, can be used to remove the low-frequency components of i_{1a}. Alternately, the observer-based method shown in Figure 8.9 can also be used.

The switching scheme for phase-a is shown in Figure 9.7 for the VSC of Figure 9.3. The control output u_c is sampled twice in each cycle, one at the negative peak of the carrier waveform and the other at the positive peak. Assuming an impulse modulated sampling, the output of the sampler is held by a zero-order hold (ZOH) circuit to obtain u^*_c. This is then compared with a triangular carrier waveform (v_{tri}). The switching signals are generated from the comparison of the carrier waveform and the sampled output as

$$\begin{aligned} &\text{if } u^*_{ca}(k) > v_{tri} \text{ then } u = +1, \text{i.e. switch } S_1 \text{ is on and } S_4 \text{ is off} \\ &\text{elseif } u^*_{ca}(k) < v_{tri} \text{ then } u = -1, \text{i.e. switch } S_4 \text{ is on and } S_1 \text{ is off} \end{aligned} \tag{9.5}$$

A similar algorithm is also employed to control phase-b through switches S_3 and S_6 and to control phase-c through switches S_5 and S_2.

Example 9.2 This is a continuation of Example 9.1, where the DSTATCOM is connected to the PCC of the system, as in Figure 9.6. The following DSTATCOM parameters are chosen

$$\begin{aligned} &R_1 = 1\,\Omega, L_1 = 3.3\text{mH}, C = 50\mu\text{F, and } C_{dc} = 5000\,\mu\text{F} \\ &|V_m| = 9.0\,\text{kV}, \omega = 100\pi\text{ rad/s (i.e.50 Hz)} \\ &V^*_{dc} = 16\,\text{kV}, K_P = -0.05 \text{ and } K_I = -0.1 \end{aligned}$$

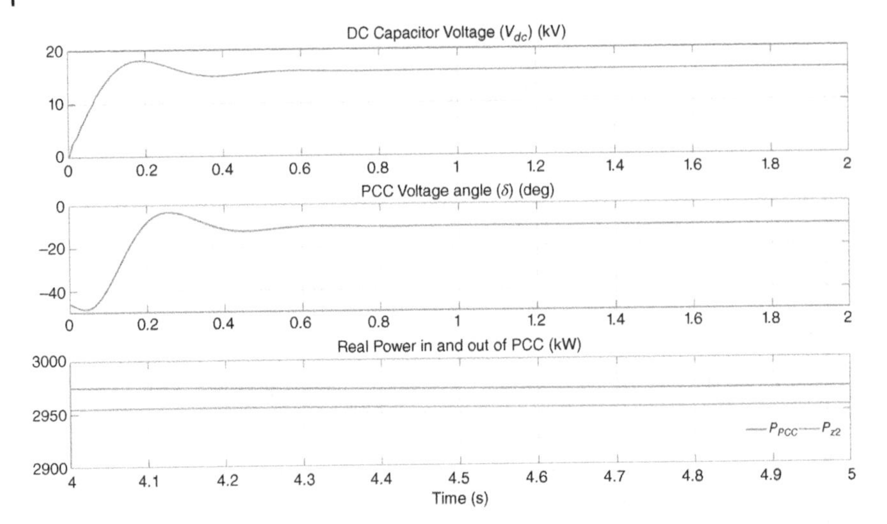

Figure 9.8 The DC voltage, load angle, and PCC powers in Example 9.2 when the DSTATCOM is connected.

Note that the negative signs in the PI gains indicates that the PCC voltage should lag the source voltage to facilitate power flow from the source to the PCC. A discrete-time LQR is designed for the system using the principle mentioned in this section, The carrier wave frequency is chosen as 15 kHz, and therefore the sampling frequency ($1/T$) is 30 kHz.

The results are shown in Figures 9.8 and 9.9. From Figure 9.8, it can be seen that the DC voltage (V_{dc}) settles to 16 kV and the load angle (δ) settles to $-10.5°$ in about 0.6 seconds. The steady state power entering and leaving the PCC is shown in Figure 9.8c. While the power entering the PCC (P_{PCC}) is about 2.972 MW, the power supplied to the load P_{L2} is about 2.955 MW. This means that DSTATCOM losses are about 17 kW. Figure 9.9 shows the PCC voltages and Load Z_1 currents. It can be seen that both are balanced and free of harmonics.

9.1.2 Discrete-time Output Feedback PWM Control

The state feedback control needs the references for both the states, which is not readily available. That is why the low-frequency components of the inductor current are eliminated from the feedback such that its high-frequency components can be equated to zero or an observer needs to be designed to estimate this current.

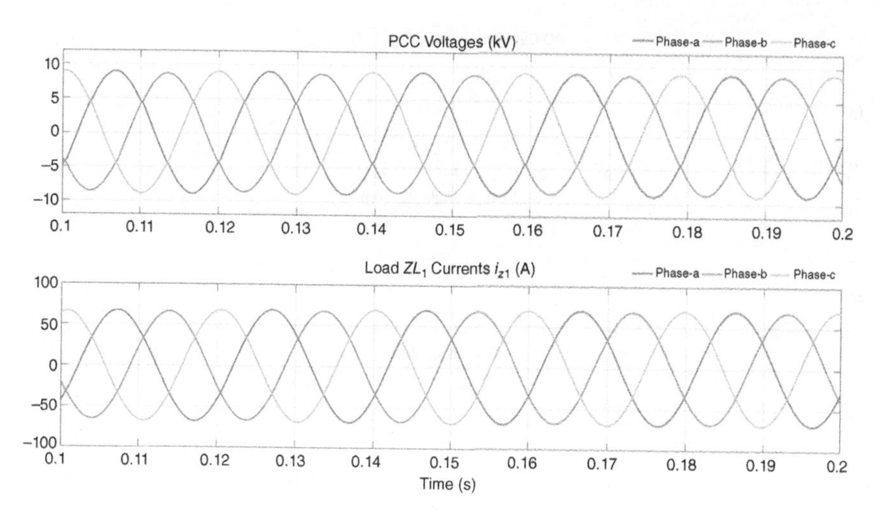

Figure 9.9 PCC voltages and load Z_1 currents in Example 9.9 when the DSTATCOM is connected.

In an output feedback control, only the reference for the capacitor voltage is required. Therefore, discrete-time output feedback control will be designed in this section for the DSTATCOM. Examples 9.3 and 9.4 discuss two examples of the DSTATCOM implementation for the same system of Example 9.2.

Example 9.3 (MV Control)

The application of minimum variance control for the DSTATCOM is illustrated in this example. Assuming the system output for phase-a is v_{ca}, the discrete-time state space equation of (9.3) is converted into the following transfer function

$$\frac{Y_a(z)}{U_{ca}(z)} = [1 \quad 0](s\mathbf{I} - \mathbf{F})^{-1}\mathbf{G}$$

where $y_a = v_{ca}$. Then the above equation is converted into input–output form as

$$y(k) + a_1 y(k-1) + a_2 y(k-2) = b_0 u(k-1) + b_1 u(k-2)$$

The frequency of the triangular carrier waveform is chosen as 10 kHz, such that the sampling frequency is 20 kHz. With this sampling frequency and the parameters of Example 9.2, the polynomial coefficients are

$$a_1 = -1.9699, a_2 = 0.985, b_0 = 7.5281 \times 10^{-3}, \text{and } b_1 = 7.4902 \times 10^{-3}$$

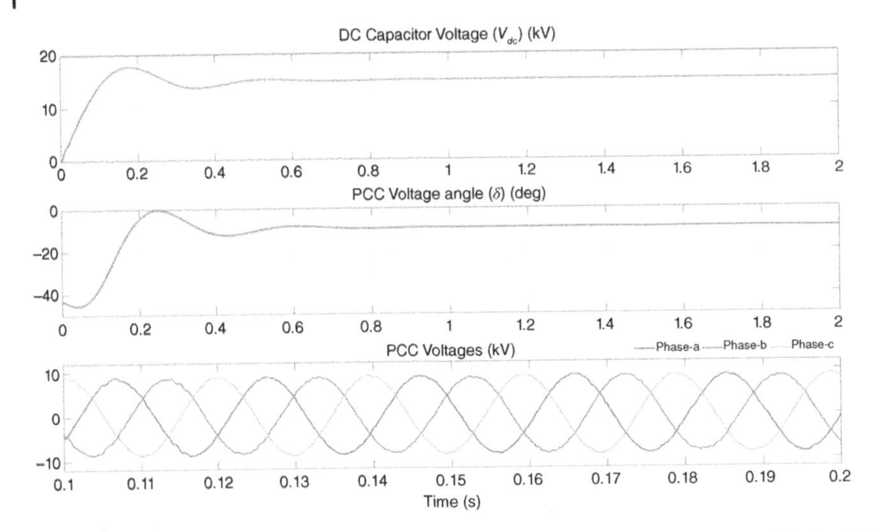

Figure 9.10 DC capacitor voltage and PCC voltage angle and PCC voltages with the MV control of Example 9.3.

The MV control law is then given by

$$u(k) = \frac{1}{b_0}\left[y_r(k+1) + a_1 y(k) + a_2 y(k-1) - b_1 u(k-1)\right]$$

The control output is then compared with a triangular carrier waveform to obtain the firing signals, as shown in Figure 9.7.

The system response with the MV controller is shown in Figure 9.10, where the DC capacitor voltage and its angle and the PCC voltages are also shown. The system response in this case is almost similar to that of Example 9.2, except that the PCC voltage waveforms with a state feedback controller are smoother.

Example 9.4 PS Control

In Example 9.3, the open loop zero is at −0.995, i.e. almost at the edge of the stability boundary that can cause instability in MV controllers. Instead, a pole-shift controller is designed in this example. For the input–output difference equation given in Example 9.3, the following controller polynomials are defined

$$S(z^{-1}) = s_0 + s_1 z^{-1}$$

$$R(z^{-1}) = 1 + r_1 z^{-1}$$

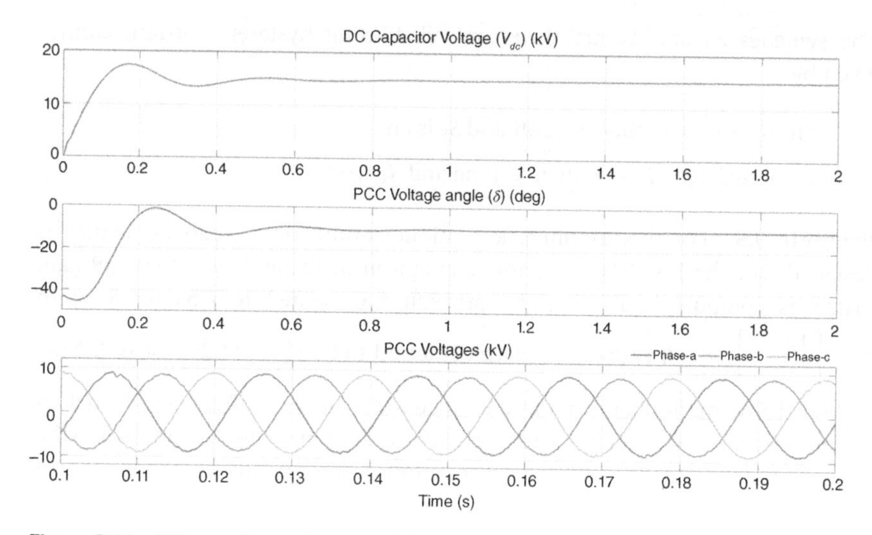

Figure 9.11 DC capacitor voltage and PCC voltage angle and PCC voltages with the PS control of Example 9.4.

The coefficients of these polynomials are then synthesized from

$$\begin{bmatrix} 1 & b_0 & 0 \\ a_1 & b_1 & b_0 \\ a_2 & 0 & b_1 \end{bmatrix} \begin{bmatrix} r_1 \\ s_0 \\ s_1 \end{bmatrix} = \begin{bmatrix} a_1(\lambda - 1) \\ a_2(\lambda^2 - 1) \\ 0 \end{bmatrix}$$

For a sampling frequency of 20 kHz and with $\lambda = 0.25$, the controller polynomials are given by

$$S(z^{-1}) = 115.87 - 79.58z^{-1}, R(z^{-1}) = 1 + 0.605z^{-1}$$

The system response with the PS controller is shown in Figure 9.11. It can be seen that the response is very similar to that with the MV controller.

9.1.3 Voltage Control Using Four-leg Converter

This converter structure is shown in Figure 9.5, in which the center point fourth leg is connected to the load neutral through a resistor and an inductor. One of the main aspects of the PCC voltage balancing is the cancelation of the zero-sequence current. Thus, the current through the neutral i_N is used to cancel the zero-sequence components of the load by controlling the switches S_7 and S_8. Therefore, the reference current for the fourth leg is the negative sum of the three load currents, i.e.

$$i_N^* = -(i_{La} + i_{Lb} + i_{Lc}) \tag{9.6}$$

The switches S_7 and S_8 are then controlled using hysteretic current control given by

$$\text{If } i_N \geq i_N^* + h \text{ then } S_7 \text{ is off and } S_8 \text{ is on}$$
$$\text{elseif } i_N \leq i_N^* - h \text{ then } S_7 \text{ is on and } S_8 \text{ is off}$$

Example 9.5 For this example, a continuous-time state feedback control is designed. For the DSTATCOM parameters given in Example 9.2, the LQR gain matrix is computed following the procedure presented in Chapter 8. With $\mathbf{Q} = \begin{bmatrix} 15 & \\ & 1 \end{bmatrix}$, $r = 0.1$, the gain matrix is found to be $\mathbf{K} = \begin{bmatrix} 11.29 & 36.63 \end{bmatrix}$. Note that Load Z_2 contains an unbalanced RL load and a diode rectifier load which does not produce a zero-sequence current. Therefore, the fourth leg is connected to the neutral of the unbalanced RL load. The output L-filter parameters of the fourth leg are

$$R_N = 0.1\,\Omega \text{ and } L_N = 1.0\,\text{mH}$$

The results are shown in Figures 9.12 and 9.13. The DC capacitor voltage, load angle, and PCC voltage responses are similar to those shown in Example 9.2. The neutral current reference is shown in Figure 9.13a and the actual neutral current is shown in Figure 9.13b. They are almost identical, as evident from the tracking error of Figure 9.13c.

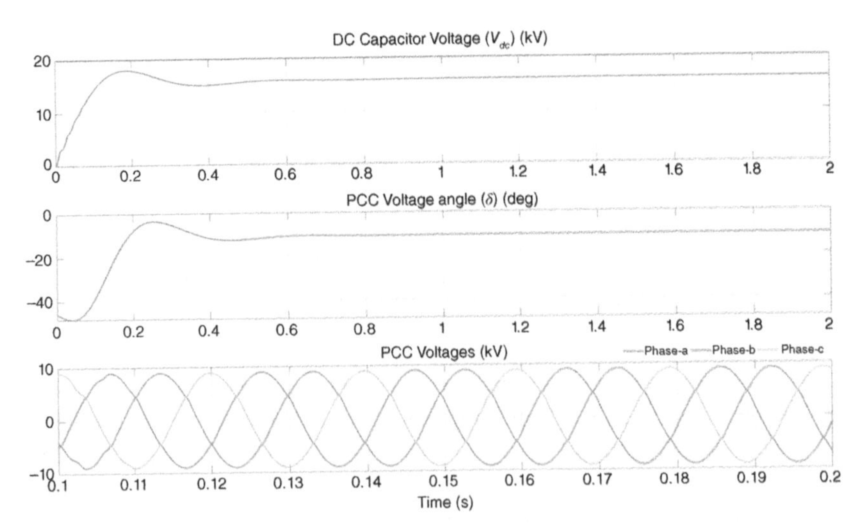

Figure 9.12 The DC voltage, load angle, and PCC voltage with a four-leg DSTATCOM.

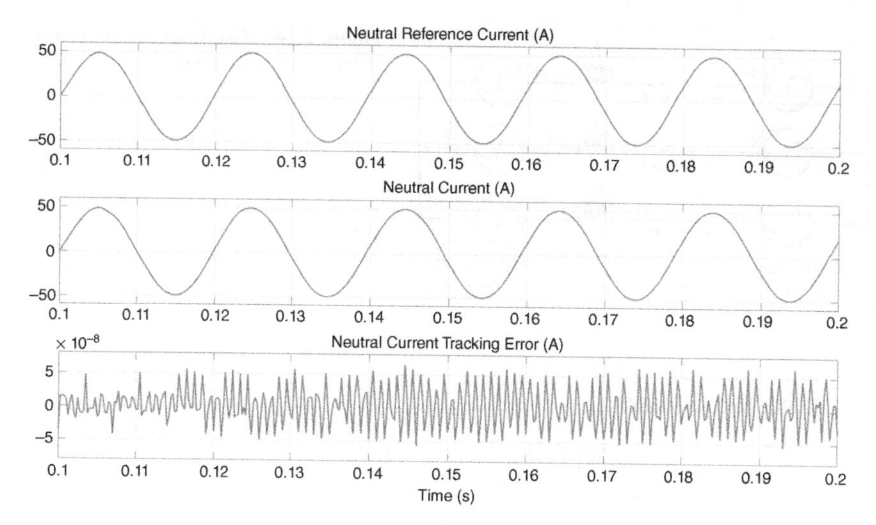

Figure 9.13 Neutral reference and actual current and current tracking error with a four-leg DSTATCOM.

9.1.4 The Effect of System Frequency

Consider Figure 9.14 in which the two sources are connected through an inductance. Assume that the sending end has a frequency of ω_1 and the receiving end operates at the synchronous frequency of ω. The sending and receiving end voltages are denoted by

$$\begin{aligned} v_{Sa} &= V_m \sin(\omega t + \Delta\omega t) & v_{Ra} &= V_m \sin(\omega t) \\ v_{Sb} &= V_m \sin(\omega t + \Delta\omega t - 120^\circ) & v_{Rb} &= V_m \sin(\omega t - 120^\circ) \\ v_{Sc} &= V_m \sin(\omega t + \Delta\omega t + 120^\circ) & v_{Rc} &= V_m \sin(\omega t + 120^\circ) \end{aligned}$$

where $\omega_1 = \omega + \Delta\omega$. Note that if $\Delta\omega = 0$ then the power transfer between the sending and receiving end will be zero, as there is no angle difference between the voltages.

Aligning the dq reference with the synchronous reference frame, i.e. $\theta = \omega t$, we use (2.37) to get

$$\begin{bmatrix} v_{Rd} \\ v_{Rq} \end{bmatrix} = \begin{bmatrix} V_m \\ 0 \end{bmatrix}$$

Again using (2.37), the sending end d-axis and q-axis voltages are given by

$$\begin{bmatrix} v_{Sd} \\ v_{Sq} \end{bmatrix} = \frac{2V_m}{3} \begin{bmatrix} \sin(\theta) & \sin(\theta - 120^\circ) & \sin(\theta + 120^\circ) \\ \cos(\theta) & \cos(\theta - 120^\circ) & \cos(\theta + 120^\circ) \end{bmatrix} \begin{bmatrix} \sin(\theta + \Delta\omega t) \\ \sin(\theta + \Delta\omega t - 120^\circ) \\ \sin(\theta + \Delta\omega t + 120^\circ) \end{bmatrix} \tag{9.7}$$

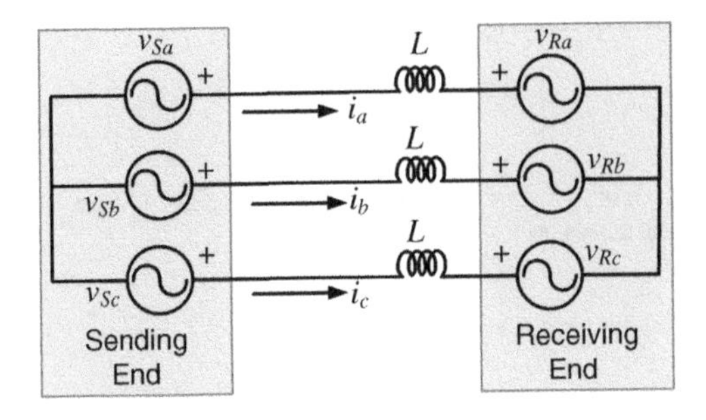

Figure 9.14 Two sources connected through an inductance.

Now using the identities

$$\sin(a)\sin(b) = \frac{\cos(a-b) - \cos(a+b)}{2},$$

$$\sin(a)\cos(b) = \frac{\sin(a+b) + \sin(a-b)}{2}$$

the terms in (9.7) are calculated as

$$\begin{aligned} v_{Sd} &= V_m \cos(\Delta\omega t) \\ v_{Sq} &= V_m \sin(\Delta\omega t) \end{aligned} \tag{9.8}$$

Using the similar approach in which (2.46) of Chapter 2 is derived

$$\frac{d}{dt}(i_{dq}) = \begin{bmatrix} 0 & \omega \\ -\omega & 0 \end{bmatrix} i_{dq} + \frac{V_m}{L}\begin{bmatrix} \cos(\Delta\omega t) - 1 \\ \sin(\Delta\omega t) \end{bmatrix} \tag{9.9}$$

The state transition matrix for the state equation of (9.9) is

$$\begin{aligned} \phi(t) &= \mathcal{L}^{-1}\left\{\begin{bmatrix} s & -\omega \\ \omega & s \end{bmatrix}^{-1}\right\} = \mathcal{L}^{-1}\left\{\frac{1}{(s^2+\omega^2)}\begin{bmatrix} s & \omega \\ -\omega & s \end{bmatrix}\right\} \\ &= \begin{bmatrix} \cos(\omega t) & \sin(\omega t) \\ -\sin(\omega t) & \cos(\omega t) \end{bmatrix} \end{aligned}$$

Therefore, the solution of (9.9), assuming that the initial conditions of the currents are zero, is

$$i_{dq} = \frac{V_m}{L}\int_0^t \begin{bmatrix} \cos(\omega(t-\tau)) & \sin(\omega(t-\tau)) \\ -\sin(\omega(t-\tau)) & \cos(\omega(t-\tau)) \end{bmatrix}\begin{bmatrix} \cos(\Delta\omega\tau) - 1 \\ \sin(\Delta\omega\tau) \end{bmatrix} d\tau \tag{9.10}$$

Now

$$\begin{bmatrix} \cos(\omega(t-\tau)) & \sin(\omega(t-\tau)) \\ -\sin(\omega(t-\tau)) & \cos(\omega(t-\tau)) \end{bmatrix} \begin{bmatrix} \cos(\Delta\omega\tau)-1 \\ \sin(\Delta\omega\tau) \end{bmatrix}$$
$$= \begin{bmatrix} \cos(\omega t-\omega\tau-\Delta\omega\tau) \\ -\sin(\omega t-\omega\tau-\Delta\omega\tau) \end{bmatrix} + \begin{bmatrix} -\cos(\omega(t-\tau)) \\ \sin(\omega(t-\tau)) \end{bmatrix}$$
$$= \begin{bmatrix} \cos(\omega t-\omega_1\tau) \\ -\sin(\omega t-\omega_1\tau) \end{bmatrix} + \begin{bmatrix} -\cos(\omega(t-\tau)) \\ \sin(\omega(t-\tau)) \end{bmatrix}$$

Substitution of the above equation in (9.11) results in

$$i_{dq} = \frac{V_m}{L}\int_0^t \begin{bmatrix} \cos(\omega t-\omega_1\tau) \\ -\sin(\omega t-\omega_1\tau) \end{bmatrix} d\tau + \frac{V_m}{L}\int_0^t \begin{bmatrix} -\cos(\omega(t-\tau)) \\ \sin(\omega(t-\tau)) \end{bmatrix} d\tau \tag{9.11}$$

The first integral of (9.11) is evaluated as

$$\begin{aligned} \int_0^t \begin{bmatrix} \cos(\omega t-\omega_1\tau) \\ -\sin(\omega t-\omega_1\tau) \end{bmatrix} d\tau &= \frac{V_m}{\omega_1 L}\begin{bmatrix} \sin(-\omega t+\omega_1\tau) \\ -\cos(-\omega t+\omega_1\tau) \end{bmatrix}_0^t \\ &= \frac{1}{\omega_1}\left\{\begin{bmatrix} \sin(\Delta\omega t) \\ -\cos(\Delta\omega\tau) \end{bmatrix} - \begin{bmatrix} \sin(\omega t) \\ -\cos(\omega t) \end{bmatrix}\right\} \end{aligned} \tag{9.12}$$

The solution of the second integral of (9.11) is

$$\int_0^t \left\{\begin{bmatrix} -\cos(\omega(t-\tau)) \\ \sin(\omega(t-\tau)) \end{bmatrix}\right\} d\tau = \frac{1}{\omega}\begin{bmatrix} \sin(\omega(t-\tau)) \\ \cos(\omega(t-\tau)) \end{bmatrix}_0^t = \frac{1}{\omega}\begin{bmatrix} \sin(\omega t) \\ 1-\cos(\omega t) \end{bmatrix} \tag{9.13}$$

Let us assume that $\Delta\omega$ is small and therefore $X = \omega L \approx \omega_1 L$. Then the substitution of (9.12) and (9.13) in (9.11) results in

$$\begin{aligned} i_{dq} &= \frac{V_m}{X}\left\{\begin{bmatrix} \sin(\Delta\omega t) \\ -\cos(\Delta\omega\tau) \end{bmatrix} - \begin{bmatrix} \sin(\omega t) \\ -\cos(\omega t) \end{bmatrix} + \begin{bmatrix} \sin(\omega t) \\ 1-\cos(\omega t) \end{bmatrix}\right\} \\ &= \frac{V_m}{X}\begin{bmatrix} \sin(\Delta\omega t) \\ -\cos(\Delta\omega\tau)+1 \end{bmatrix} \end{aligned} \tag{9.14}$$

Power delivered at the receiving end is

$$P = \frac{3}{2}\left(v_{Rd} i_d - v_{Rq} i_q\right) = \frac{3V_m^2}{2X}\sin(\Delta\omega t) \tag{9.15}$$

Example 9.6 Consider the system of Figure 9.14, where V_m = 11 kV (L-L), $\Delta\omega = 0.2$ Hz, and $L = 77$ mH. When both the source and receiving end voltages operate at 50 Hz, the line impedance is $L = 77 \times 10^{-3} \times 100\pi = 24.19\,\Omega$. Therefore, the maximum power that can be transferred over the line is $3 \times \left(V_m/\sqrt{3}\right)^2/X = 5$ MW. Figure 9.15 plots the receiving end power for two values of the sending frequency: 49.8 and 50.2 Hz. The power oscillates at 0.2 Hz in both these cases with peaks of 5 MW. However, the direction of power flow at the beginning is negative when $\Delta\omega$ is negative, while it is positive when $\Delta\omega$ is positive.

The consequence of the frequency mismatch for the four-leg DSTATCOM of Example 9.5 is shown in Figure 9.16 for two different values of the frequency. Figure 9.16a shows the DC voltage when the source frequency is 49.8 Hz. In this case, the power will flow from the receiving end to the sending at the beginning, and, as a result, the DC capacitor will start feeding power toward the source. Consequently, the capacitor voltage will start collapsing and the PCC voltage angle will diverge rapidly, as evident from Figures 9.16a,c respectively. This will cause tracking failure in the VSC, resulting in a system collapse. The opposite is true when the frequency is 50.2 Hz. A sudden rush of power from the sending end will cause the capacitor voltage to rise, as shown in Figure 9.16b, and the angle to run away, as shown in Figure 9.16d.

The protection circuit will prevent capacitor overvoltage and therefore the DSTATCOM will be taken off the system.

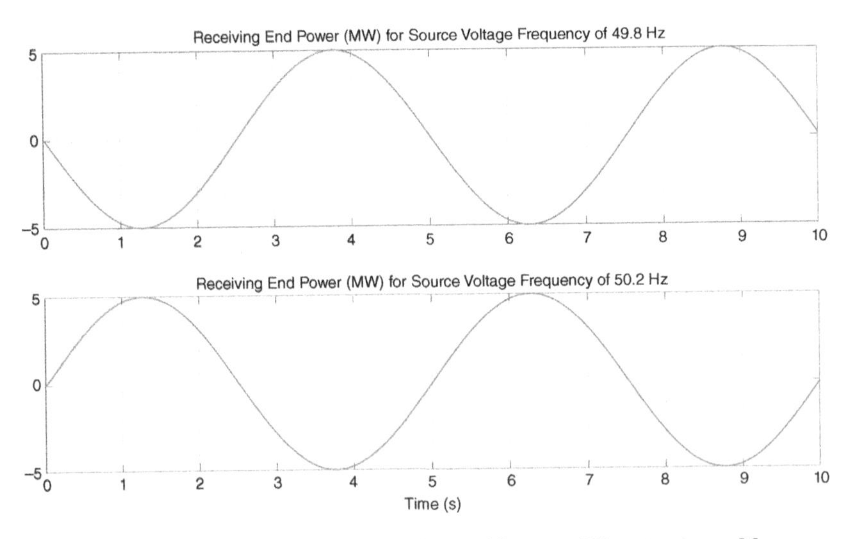

Figure 9.15 Power received at the receiving end for two different values of frequency.

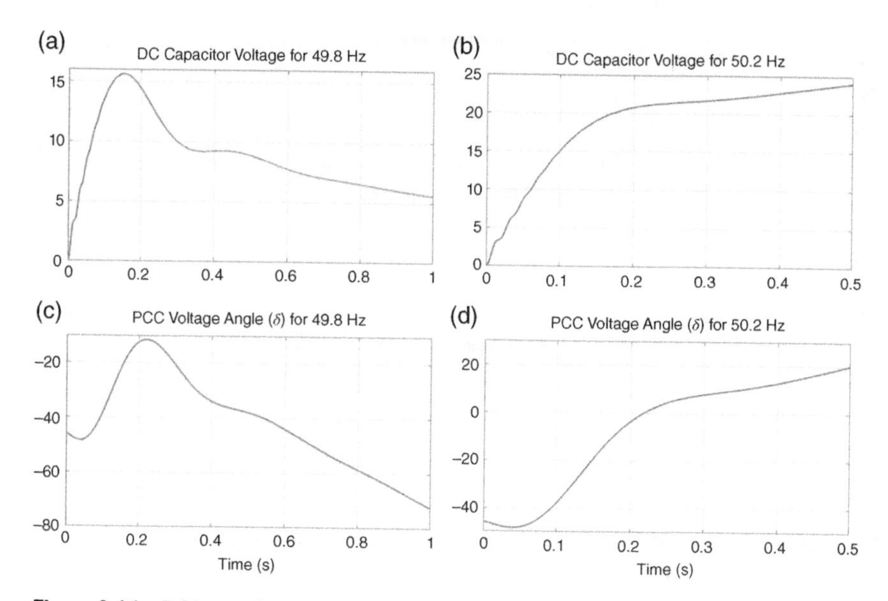

Figure 9.16 DC bus voltage and PCC voltage angle during frequency mismatch. (a) DC Capacitor Voltage for 49.8 Hz, (b) DC Capacitor Voltage for 50.2 Hz, (c) PCC Voltage Angle (δ) for 49.8 Hz, (d) PCC Voltage Angle (δ) for 50.2 Hz.

Example 9.7 We shall now use the frequency estimation algorithm discussed in Chapter 2 to estimate the system frequency. The system has been operating at the steady state at the beginning at 50 Hz, when the source frequency changes to 50.2 Hz at 1 seconds. The system response is shown in Figure 9.17. The estimated frequency is shown in Figure 9.17a, where the change in frequency is tracked within about 0.15 seconds. Consequently, the DC capacitor voltage settles to 16 kV after the initial transient following the frequency change, as shown in Figure 9.17b. The PCC voltage angle is shown in Figure 9.17c.

9.1.5 Power Factor Correction

The PCC reference voltage, given in (9.1), is assumed to have a fixed magnitude. This may cause, depending on the situation, the DSTATCOM to draw unnecessary excess reactive power from the source or even feed reactive power toward the upstream network. Let us assume that two voltage sources are connected through an impedance. The sending end voltage is given by $V_S = |V| \angle 0°$, the receiving end voltage is denoted by $V_R = |V_P| \angle -\delta$, while the impedance connecting them is given by $R + jX$. Therefore, the current I flowing between the sources is

$$I = \frac{|V| - |V_P|\angle{-\delta}}{R + jX} \tag{9.16}$$

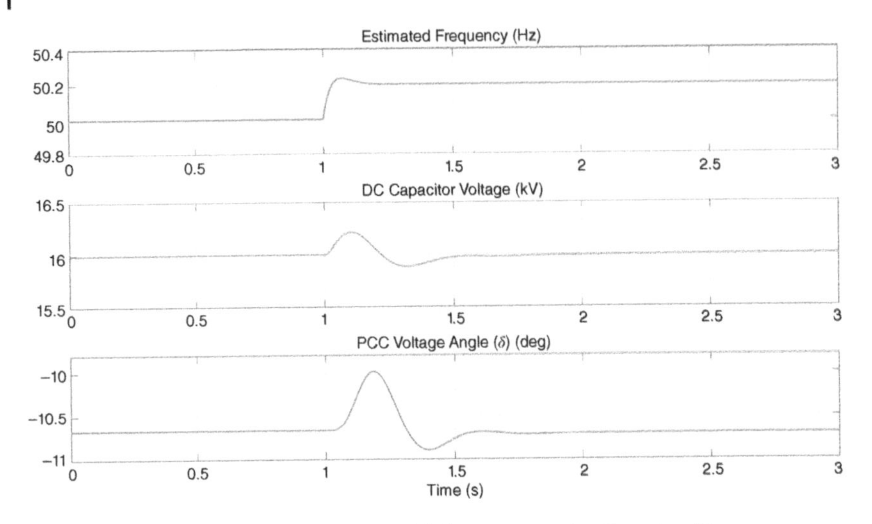

Figure 9.17 DSTATCOM performance with frequency estimation algorithm.

Therefore, the complex power injected to the receiving end (PCC) from the source is

$$P_R + jQ_R = (|V_P|\angle -\delta) \times I^* \tag{9.17}$$

Expanding (9.17), the real and reactive power injected at the receiving end are given by

$$P_s = \frac{1}{R^2 + X^2}\left[|V||V_P|R\cos\delta + |V||V_P|X\sin\delta - |V_P|^2 R\right] \tag{9.18}$$

$$Q_R = \frac{1}{R^2 + X^2}\left[|V||V_P|X\cos\delta - |V||V_P|R\sin\delta - |V_P|^2 X\right] \tag{9.19}$$

Assume that a DSTATCOM is connected at the receiving end (PCC) and Thévenin impedance of the upstream network is represented by $R + jX$. When the DSTATCOM operates in the voltage control mode, it adjusts the angle δ to draw power from the source in such a way that the load and the converter losses can be supplied from the source. However, it is evident from (9.19) that the reactive power injected at the PCC Q_R cannot be made equal to zero by only controlling δ. Moreover, from (9.18), it is evident that the real power injection P_R will depend on both $|V_P|$ and δ, given that $|V|$ and $R + jX$ are constant. Therefore, the reactive power control can be achieved without sacrificing real power control. It is desired that the DSTATCOM operates in such a way that the reactive power injected at the PCC is zero. From (9.19), it can be seen that $Q_R = 0$ when

$$|V_P| = |V|\left[\cos\delta - \frac{R}{X}\sin\delta\right] \tag{9.20}$$

Therefore, the reactive power can be forced to zero by controlling the PCC voltage magnitude $|V_P|$.

To control the PCC voltage magnitude, a PI controller is employed, which is of the form

$$\begin{aligned} e_Q &= 0 - Q_{PCC} \\ |V_m| &= |V_m^*| + K_P e_Q + K_I \int e_Q dt \end{aligned} \tag{9.21}$$

where Q_{PCC} is the measured reactive power entering the PCC, $|V_m|$ is the voltage magnitude used in (9.1) and $|V_m^*|$ is the reference voltage. Example 9.5 will now be repeated with the reactive power control. The four-leg DSTATCOM is used here as well, where $|V_m^*|$ is chosen as 9 kV, peak phase voltage for L-L voltage of 11 kV. The PI gains are $K_P = -0.05$ and $K_I = -1$. The results are shown in Figures 9.18 and 9.19. Figure 9.18 shows the DC capacitor voltage and the PCC voltage angle. It can be seen that they are the same as those shown in Figure 9.12. Figure 9.19 shows the controlled PCC voltage magnitude and the reactive power injected to the PCC from the source. The reactive power becomes zero after 3.5 seconds and the resultant PCC voltage magnitude becomes 8.61 kV, indicating a voltage drop of 4.33%, which is within the acceptable range.

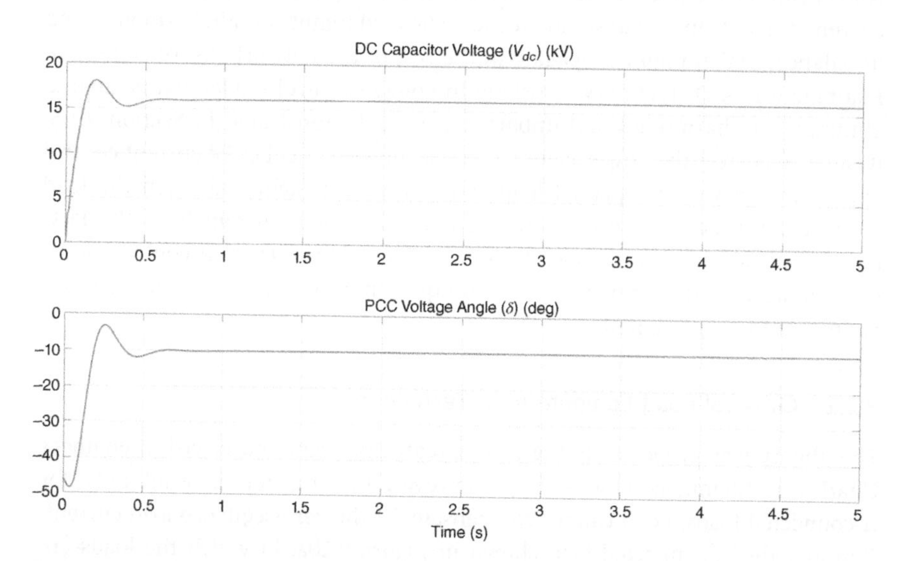

Figure 9.18 The DC voltage and load angle with the power factor correcting four-leg DSTATCOM.

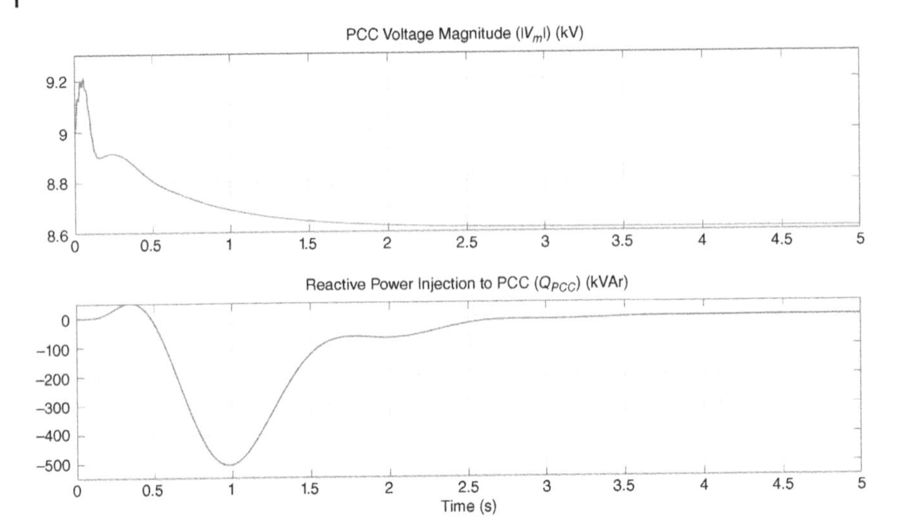

Figure 9.19 The PCC voltage magnitude and injected reactive power with the power factor correcting four-leg DSTATCOM.

9.2 Load Compensation

A load can have a poor power factor; this will result in more than the required current drawn from the upstream feeder. The load might contain harmonics and unbalance, which might propagate in the rest of the network, as we can see in Example 9.1. A DSTATCOM, working in voltage control mode, will be able to eliminate the harmonics and unbalance. As is demonstrated in Section 9.1.5, it can also correct the power factor. There is, however, a class of current control application in which a perfect load compensation is achieved when the load draws a unity power factor, balanced, pure sinusoidal current from the PCC bus. Before we discuss the operation of DSTATCOM in current control mode, the classical load compensation technique that was proposed by Charles Steinmetz [4] is presented.

9.2.1 Classical Load Compensation Technique

This theory was proposed by the genius mathematician and electrical engineer Charles P. Steinmetz (1865–1923). However, this method is valid only for Δ-connected loads, i.e. it cannot compensate for the zero-sequence load current. Consider the Δ-connected load shown in Figure 9.20a, in which the loads are denoted by Z_{Lab}, Z_{Lbc}, and Z_{Lca}, and they are connected between the phases a, b, and c.

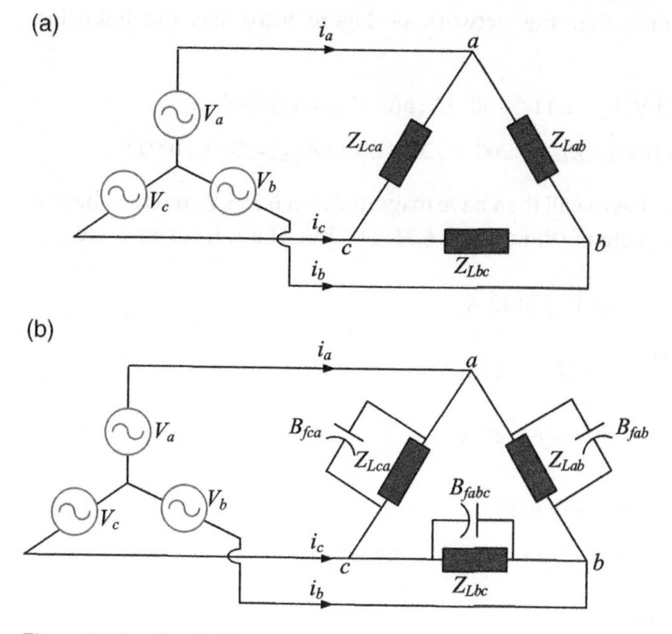

Figure 9.20 Classical load compensator: (a) uncompensated and (b) compensated network.

The load admittances are defined as

$$
\begin{aligned}
Y_{Lab} &= \frac{1}{Z_{Lab}} = G_{Lab} + jB_{Lab} \\
Y_{Lbc} &= \frac{1}{Z_{Lbc}} = G_{Lbc} + jB_{Lbc} \\
Y_{Lca} &= \frac{1}{Z_{Lca}} = G_{Lca} + jB_{Lca}
\end{aligned}
\tag{9.22}
$$

It is possible to have a pure reactive compensation in which three susceptances are connected, each one in parallel with a load in the Δ-connection. This is shown in Figure 9.20b. These susceptances are given by [4]

$$
\begin{aligned}
B_{fab} &= -B_{Lab} + \frac{G_{Lca} - G_{Lbc}}{\sqrt{3}} \\
B_{fbc} &= -B_{Lbc} + \frac{G_{Lab} - G_{Lca}}{\sqrt{3}} \\
B_{fca} &= -B_{Lca} + \frac{G_{Lbc} - G_{Lab}}{\sqrt{3}}
\end{aligned}
\tag{9.23}
$$

The remarkable property of this connection is that the line currents will be balanced and will have a unity power factor. Let us consider Example 9.8.

Example 9.8 Assume that the network of Figure 9.20a has the following parameters

$$V_{ab} = 11\angle 30^\circ \text{ kV}, V_{bc} = 11\angle -90^\circ \text{ kV}, \text{and } V_{ca} = 11\angle 150^\circ \text{ kV}$$

$$Z_{Lab} = 100 + j100\,\Omega, Z_{Lbc} = 200 + j200\,\Omega, \text{and } Z_{Lca} = 50 + j100\,\Omega$$

Note that the phase voltages will then have magnitudes of 6.35 kV, and the phase-a voltage will have an angle of 0°, i.e. $V_a = 6.35 \angle 0^\circ$ kV. The L-L currents are

$$I_{ab} = \frac{11\angle 30^\circ}{Z_{Lab}} = 75.13 - j20.13 \text{ A}$$

$$I_{bc} = \frac{11\angle -90^\circ}{Z_{Lbc}} = -27.5 - j27.5 \text{ A}$$

$$I_{ca} = \frac{11\angle 150^\circ}{Z_{Lca}} = 5.89 + j98.21 \text{ A}$$

The real power supplied to the load is

$$S_1 = \text{Re}\left(V_{ab}I_{ab}^* + V_{bc}I_{bc}^* + V_{ca}I_{ca}^*\right) = 1.3915 \text{ MW}$$

The load reactances are

$$Y_{Lab} = \frac{1}{Z_{Lab}} = G_{Lab} + jB_{Lab} = 0.005 - j0.005\,\mho$$

$$Y_{Lbc} = \frac{1}{Z_{Lbc}} = G_{Lbc} + jB_{Lbc} = 0.0025 - j0.0025\,\mho$$

$$Y_{Lca} = \frac{1}{Z_{Lca}} = G_{Lca} + jG_{Lca} = 0.004 - j0.008\,\mho$$

Then the compensator susceptances calculated from (9.23) are

$$B_{fab} = -B_{Lab} + \frac{G_{Lca} - G_{Lbc}}{\sqrt{3}} = 0.0059\,\mho$$

$$B_{fbc} = -B_{Lbc} + \frac{G_{Lab} - G_{Lca}}{\sqrt{3}} = 0.0031\,\mho$$

$$B_{fca} = -B_{Lca} + \frac{G_{Lbc} - G_{Lab}}{\sqrt{3}} = 0.0066\,\mho$$

Therefore, the compensated admittances are

$$Y_{cab} = Y_{Lab} + jB_{fab} = 0.005 + j0.0009\,\mho$$

$$Y_{cbc} = Y_{Lbc} + jB_{fbc} = 0.025 + j0.0006\,\mho$$

$$Y_{cca} = Y_{Lcca} + jB_{fca} = 0.004 - j0.0014\,\mho$$

Once the compensator is placed, the L-L currents will change. These are given by

$$I_{ab} = V_{ab}Y_{cab} = 42.67 + j35.75 \text{ A}$$
$$I_{bc} = V_{bc}Y_{cbc} = 6.35 - j27.5 \text{ A}$$
$$I_{ca} = V_{ca}Y_{cca} = -30.17 + j35.75 \text{ A}$$

From the L-L currents, the line currents are computed as

$$I_a = I_{ab} - I_{ca} = 73.0348 \text{ A}$$
$$I_b = I_{bc} - I_{ab} = -36.5174 - j63.25 = 73.0348\angle -120^\circ \text{ A}$$
$$I_c = I_{ca} - I_{bc} = -36.5174 + j63.25 = 73.0348\angle 120^\circ \text{ A}$$

It is obvious that the line currents are in phase with phase voltages, indicating unity power factor operation. The power supplied to the compensated load is

$$P = |V_aI_a| + |V_bI_b| + |V_cI_c| = 1.3915 \text{ MW}$$

This is the same as in the case of the uncompensated load since the compensators are pure susceptances and therefore do not consume any power.

This is a remarkable technique. However, the main drawback here is that it requires the accurate measurements of the load and can only be applied to Δ-connected loads. Since the load can change any time, the application of this technique is restrictive.

9.2.2 Load Compensation Using VSC

Consider the three-phase distribution system shown in Figure 9.21, which contains a DSTATCOM that is used as a current compensator. In this, the voltage source v_s supplies a load that may be unbalanced and nonlinear. The load current drawn is denoted by i_L. In the absence of the DSTATCOM, the load and source currents will be the same, i.e.

$$i_{sk} = i_{Lk}, \quad k = a, b, c \tag{9.24}$$

However, when the DSTATCOM is placed in the circuit, the KCL at PCC gives

$$i_{sk} + i_{fk} = i_{Lk} \Rightarrow i_{fk} = i_{Lk} - i_{sk}, \quad k = a, b, c \tag{9.25}$$

Let the load currents contain a fundamental and a harmonic component, given by

$$i_{Lk} = i_{Lkf} + i_{Lkh}, \quad k = a, b, c \tag{9.26}$$

where the subscripts f and h denote the fundamental and harmonic components respectively. The DSTATCOM should operate in such a way that it supplies the harmonic components of the load currents, i.e.

$$i_{fk} = i_{Lkh}, \quad k = a, b, c \tag{9.27}$$

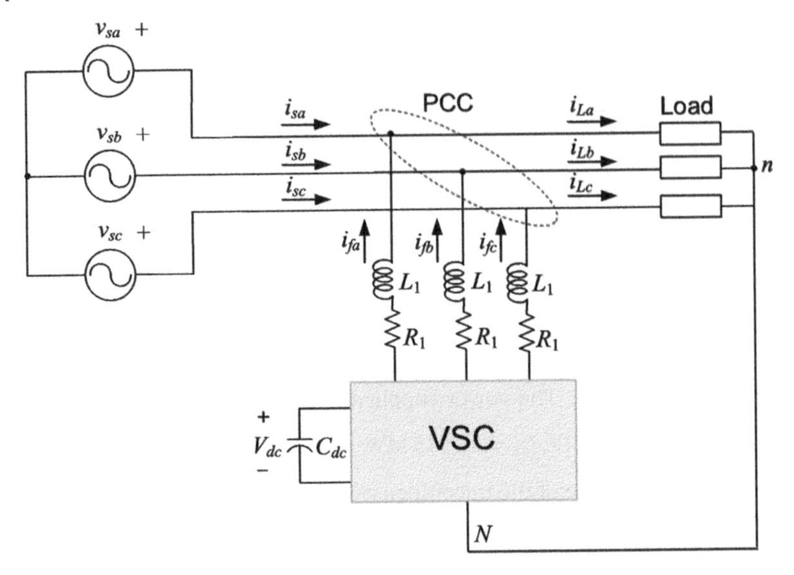

Figure 9.21 Schematic diagram of a load compensator circuit using a DSTATCOM.

Then the source currents will become

$$i_{sk} = i_{Lkf}, \quad k = a, b, c \tag{9.28}$$

This means that the source will supply only the fundamental components of the load currents. This procedure is called active filtering and the compensator here ensures that no harmonic current flows in the upstream network.

The active filtering method eliminates the harmonic current, but it does not guarantee that the power factor correction or load balancing. For example, in (9.28), if the fundamentals of load currents have unequal magnitude, the source currents will also have unequal magnitudes. Also, the power factor will be dominated by the fundamental load currents. Therefore, an alternate formulation for achieving power factor correction, load balancing, and harmonic filtering together is required. It is required that the compensating currents need to fulfill the following three objectives:

- To force the zero-sequence components of the source currents to zero.
- To make the source currents in phase with the source voltages.
- The DSTATCOM should neither supply nor absorb any real power. This implies that the source supplies the total real power requirement of the load.

To satisfy these three objectives, we get the following three equations

$$i_{sa} + i_{sb} + i_{sc} = 0 \tag{9.29}$$

$$\angle I_{sk} = \angle V_{sk}, \quad k = a, b, c \tag{9.30}$$

$$v_{sa}i_{sa} + v_{sb}i_{sb} + v_{sc}i_{sc} = P_{Lav} \tag{9.31}$$

Equation (9.29) stipulates that the instantaneous sum of the source current is zero and therefore the source currents will not contain any zero-sequence current. Equation (9.30) ensures a unity power factor operation. Now, if the source currents get balanced due to the DSTATCOM action, the power supplied to the load $v_{sa}i_{sa} + v_{sb}i_{sb} + v_{sc}i_{sc}$ will be a DC quantity and will be equal to the average of the real power (P_{Lav}) drawn by the load, as given in (9.31). The compensator currents are then given by [5]

$$\begin{aligned} i_{fa}^* &= i_{la} - v_{sa} \times P_{Lav}/\Delta \\ i_{fb}^* &= i_{lb} - v_{sb} \times P_{Lav}/\Delta \\ i_{fc}^* &= i_{lc} - v_{sc} \times P_{Lav}/\Delta \end{aligned} \tag{9.32}$$

where

$$\Delta = v_{sa}^2 + v_{sb}^2 + v_{sc}^2$$

Since the DSTATCOM is supplied by a DC capacitor, as mentioned in Section 9.1, the capacitor voltage will be discharged unless the converter losses are replenished by drawing power from the source. Therefore, a PI controller of the same form as (9.2) will be used to draw power from the source. This is given by

$$P_{loss} = K_P\left(V_{dc}^* - \langle V_{dc}(t)\rangle\right) + K_I \int \left(V_{dc}^* - \langle V_{dc}(t)\rangle\right) dt \tag{9.33}$$

Consequently, (9.32) is modified as

$$\begin{aligned} i_{fa}^* &= i_{la} - v_{sa} \times (P_{Lav} + P_{loss})/\Delta \\ i_{fb}^* &= i_{lb} - v_{sb} \times (P_{Lav} + P_{loss})/\Delta \\ i_{fc}^* &= i_{lc} - v_{sc} \times (P_{Lav} + P_{loss})/\Delta \end{aligned} \tag{9.34}$$

Example 9.9 The network of Figure 9.21 has the following parameters

- $V_s = 11$ kV, frequency = 50 Hz, $L_1 = 19.3$ mH, $R_1 = 0.1\ \Omega$
- Unbalanced load: $Z_{La} = 24.2 + j60.5\ \Omega$, $Z_{Lb} = 12.2 + j31.4\ \Omega$, and $Z_{LC} = 48.2 + j94.2\ \Omega$
- Nonlinear load: Diode bridge rectifier supplying 100 Ω load
- DC capacitor: $C_{dc} = 1000\ \mu$F, $V_{dc}^* = 12$ kV, $K_P = 1$, and $K_I = 10$

Any of the DSTATCOM configurations presented in Section 9.1 can be used. The only aspect that needs to be considered is that its neutral point N must be connected to the load neutral *n*. The DSTATCOM in this example is operated under a hysteretic current control.

The results are shown in Figures 9.22 and 9.23. Figure 9.22a shows the load currents, which are unbalanced and harmonically contaminated. Figure 9.22b shows the compensated line currents, which are balanced and are almost free of harmonics. However, some notches are visible in the source currents. These notches are caused when the currents in each phase of the diode rectifier commutates. There are sharp, almost discontinuous, changes in the load currents at these points. Due to the presence of the inductance L_1, the DSTATCOM cannot make its output current change rapidly enough to counter and smoothen the notches. Figure 9.22c shows the phase-a of the source voltage and source current, which are in phase, indicating a unity power factor operation. The DC capacitor voltage, shown in Figure 9.23a, has a steady state value of 12 kV, as per the reference voltage chosen. The average load power and the power loss are shown in Figure 9.23b, where the power loss is almost negligible.

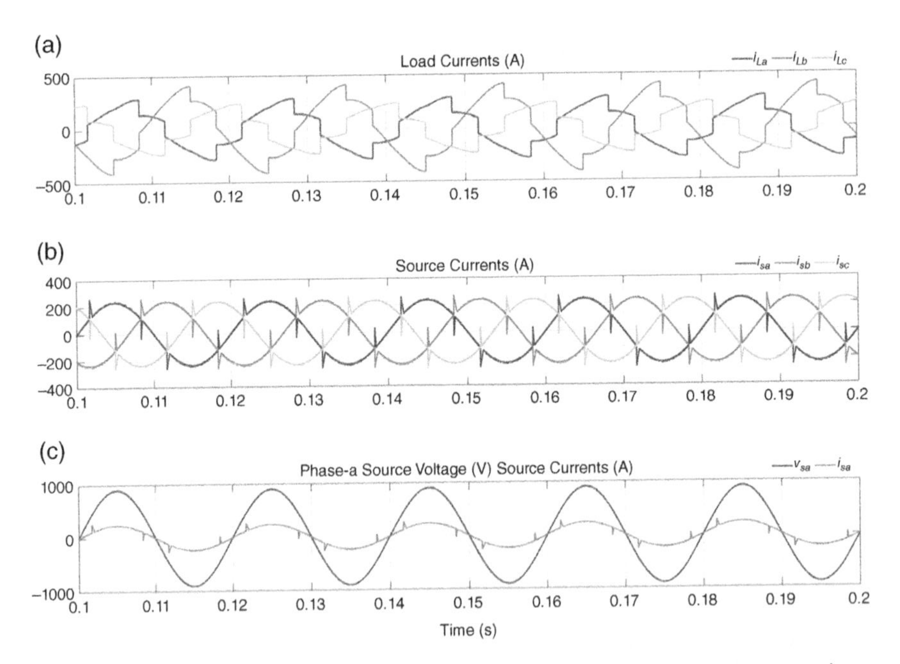

Figure 9.22 (a) Load currents, (b) load currents, and (c) source voltage and currents in Example 9.9.

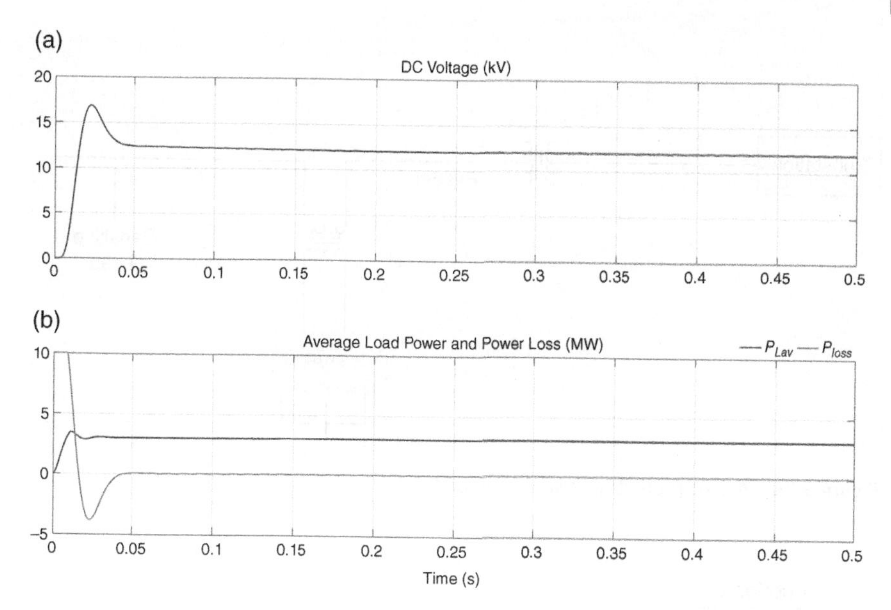

Figure 9.23 (a) Average DC capacitor voltage and (b) average load power and power loss in Example 9.9.

9.3 Other Custom Power Devices

There are two more types of compensating custom power devices. One of them is called a dynamic voltage restorer (DVR). This is essentially a VSC that is connected in series to a distribution feeder, as shown in Figure 9.24. The main idea is to protect a sensitive or critical load from voltage disturbance in the upstream feeder. As shown in the figure, if a voltage sag occurs at the substation end, the DVR inserts the amount of voltage needed such that the voltage supplied to the sensitive load remains balanced and sinusoidal.

The DVR structure, shown in Figure 9.25, is a VSC, with its associated transformer (with leakage inductance L_f) and a filter capacitor C_d. It injects a voltage v_d such that the sensitive load voltage $v_{L1} = v_P + v_d$ remains balanced and sinusoidal. In the network of Figure 9.25, it is assumed that the load current i_{L1} is unbalanced, as shown in Figure 9.26a. This causes the PCC voltage v_P to become unbalanced as well (see Figure 9.26b). However, due to the DVR action, the critical load bus voltage v_{L1} remains balanced and sinusoidal, as can be seen from Figure 9.26c. With the system operating in the steady state, a voltage sag takes place at 0.05 seconds, and lasts till 0.125 seconds. Due to the action of the DVR, the voltage v_{L1} across the sensitive load remains balanced and supplies the load with a specified voltage magnitude of 9 kV (peak).

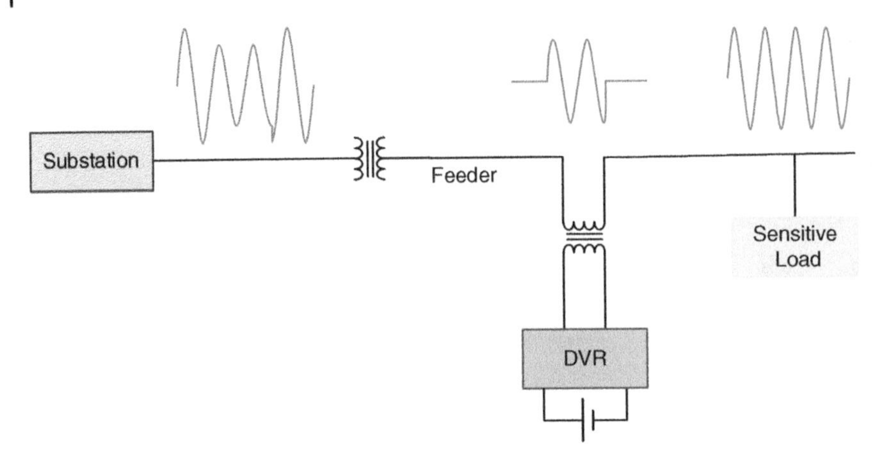

Figure 9.24 A DVR protecting a sensitive load.

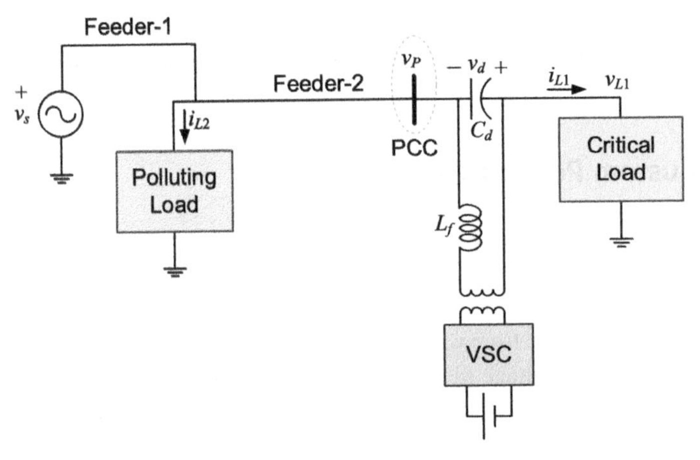

Figure 9.25 Schematic diagram of a distribution system where a DVR protecting a sensitive load from a polluting load and voltage sag/swell.

One more custom power device that has attracted some attention is the unified power quality conditioner (UPQC), the schematic diagram of which is shown in Figure 9.27. The UPQC contains a converter connected in shunt with the feeder and another converter that is connected in series with the feeder. These two converters are connected together by a DC capacitor C_{dc} on the DC side. The series converter injects the voltage v_d in series to protect a sensitive load, while the shunt converter injects current i_f for current compensation. Furthermore, the shunt

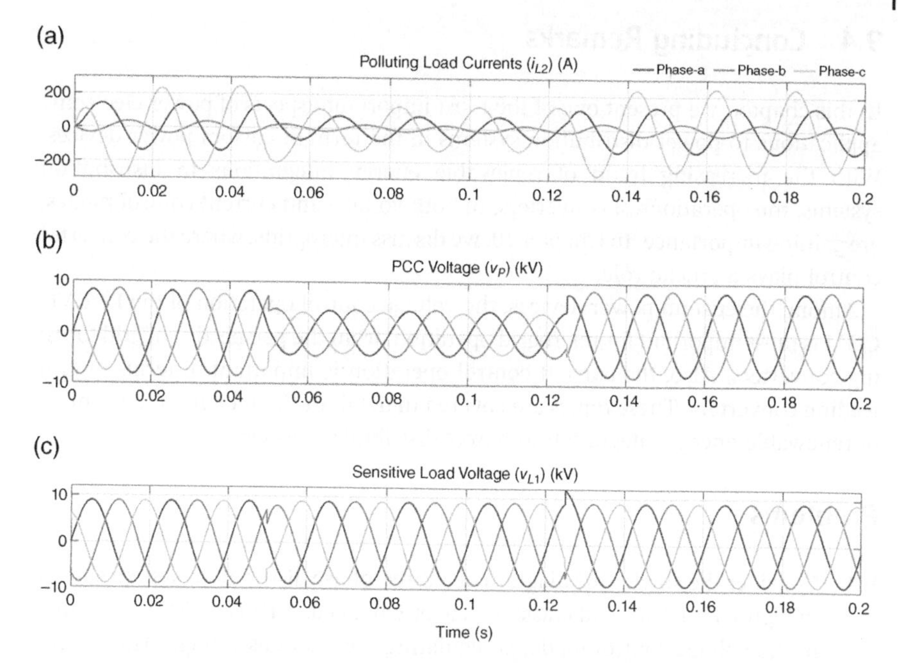

Figure 9.26 (a) Polluting load current, (b) PCC voltage, and (c) sensitive load voltage for the distribution system of Figure 9.25.

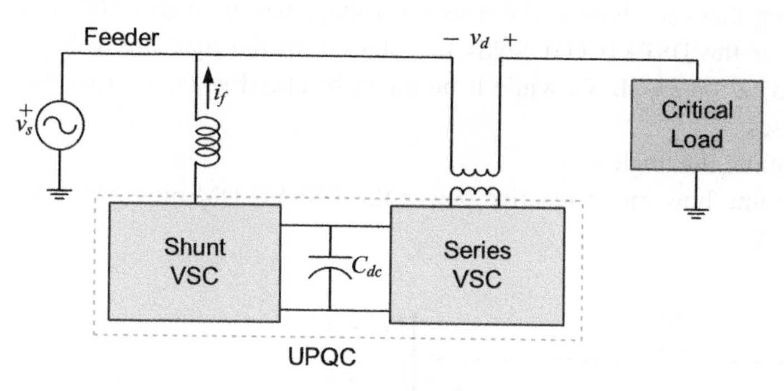

Figure 9.27 Schematic diagram of a UPQC.

compensator draws the required current from the source to maintain the change on the capacitor C_{dc}. Note that the placement of these converters can also be reversed (i.e. shunt on the right and series on the left) depending on applications. These placement arrangements are sometimes referred to as left UPQC or right UPQC, depending on the relative positions of the shunt and series compensators.

9.4 Concluding Remarks

In this chapter, we present one of the most important aspects of power electronic applications to power distribution systems in the form of custom power devices. With the increasing trend of renewable energy integrations in distribution systems, the operations of converters, in both voltage and current control modes, are gaining importance. In Chapter 10, we discuss microgrids, where the converter control plays a crucial role.

Among the custom power devices, the voltage control operation of the DSTATCOM is most important for microgrid operation, from the perspective of grid forming converters. Also, the current control operation is important in terms of grid feeding converters. These topics are covered in detail in Chapter 10 in the context of renewable energy integration to power distribution systems.

Problems

9.1 Consider the single-line diagram of a three-phase distribution system shown in Figure P9.1. The load is assumed to be balanced. A DSTATCOM, operating in the voltage control mode, is regulating the load bus voltage. The system parameters are

$$V_S = 11.0\,\text{kV (L} - \text{L)}, X = 18.18\,\Omega, \text{and } Z_L = 120 + j12\,\Omega$$

Assuming that the phase-a of the source voltage has an angle of 0°, it is desired that the DSTATCOM holds the phase-a of the load bus voltage (V_L) to 6.35 ∠ $-\delta$ kV (L-N), while it produces balanced voltage across the three phases.

(a) Determine the angle δ_L.

(b) Determine how much reactive power the DSTATCOM must inject to the PCC.

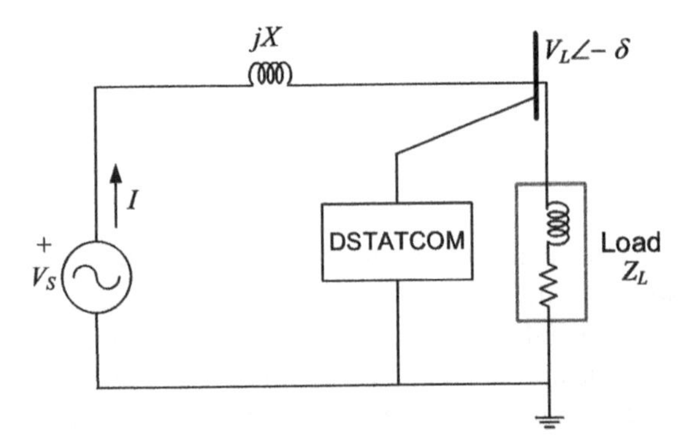

Figure P9.1 DSTATCOM compensated distribution system of Problems 9.1 and 9.2.

9.2 Consider the single-line diagram of a three-phase distribution system shown in Figure P9.1 again. In this case the load is assumed to be unbalanced. The system parameters are as follows

- Base voltage: 11 kV (L-L, rms)
- Source voltage (V_S): 1.1 per unit
- Feeder reactance (X): 24.2 Ω
- Unbalanced load impedances (Z_L):
- Phase-a: 100 + j61 Ω
- Phase-b: 100 + j50 Ω
- Phase-c: 90 + j25 Ω
- DSTATCOM losses: 10 kW

Assuming that the phase-a of the source voltage angle 0°, it is desired that the DSTATCOM holds the load bus voltage (V_L) to $1\angle-\delta$ per unit.

(a) Determine δ, and hence
(b) Compute the reactive power (Q_S) injected by the source.

9.3 Consider the single-line diagram of a wye-connected three-phase system shown in Figure P9.3, where the relevant system data are also shown. A PV is connected to all three phases of the load bus and is injecting 600 kW power at unity power factor. Together, the PV and the utility supply a constant power load of 6.0 MVA at a lagging power factor of 0.95. The DSTATCOM has to regulate the load bus voltage in such a way that the reactive power injected from the utility to the PCC is zero.

(a) Determine the angle δ of the load bus voltage.
(b) Determine the magnitude of the load bus voltage.
(c) Determine the reactive power injected by the source.

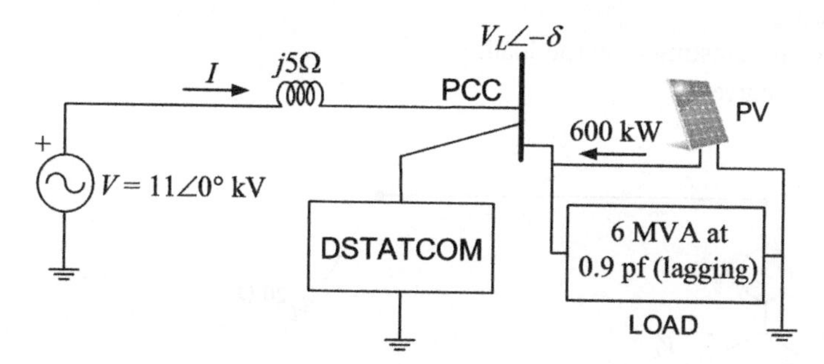

Figure P9.3 DSTATCOM compensated distribution system of Problem 9.3.

9.4 Consider the three-phase circuit, the single-line diagram of which is shown in Figure P9.4. In this circuit, the magnitude of the load bus line-to-neutral voltage V_L has to be regulated at $|V| \angle 0°$ V by adjusting the series compensator that injects the voltage V_F. The sending end line-to-neutral voltage is $V_S = |V| \angle \delta$ V.

It is stipulated that the series compensator does neither supply nor absorb any real power from the AC system, i.e. $\text{Re}\left(V_F I_S^*\right) = 0$. Then determine V_F, assuming $Z_L = 20 + j20\ \Omega$ and $Z_f = 0.5 + j2\ \Omega$.

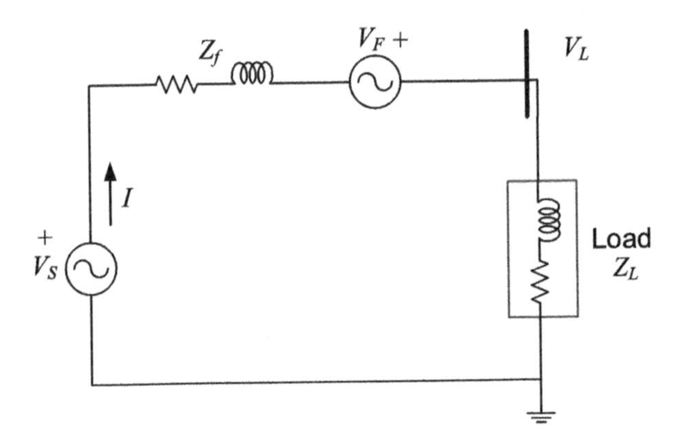

Figure P9.4 Series compensated distribution system of Problem 9.4.

9.5 The load in Figure P9.5 is supplied by an 11 kV (L-L), 50 Hz supply. Connect a parallel delta-connected fixed purely reactive compensator such that the load gets balanced and draws currents at unity power factor. Calculate the following:

(a) Power consumed by the load.

(b) Line currents.

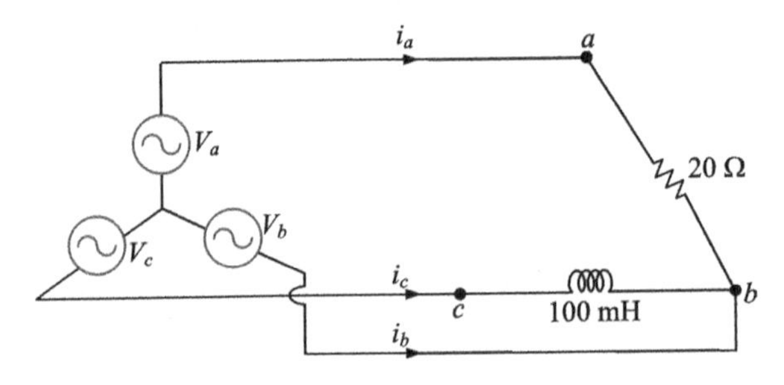

Figure P9.5 Open-delta connection for Problem 9.4.

9.6 Consider the compensator circuit shown in Figure 9.21, where the load is supplied by a 50 Hz balanced supply, given in per unit as

$$v_{sa} = \sqrt{2}\sin \omega t, \quad v_{sb} = \sqrt{2}\sin(\omega t - 120^\circ), \quad v_{sc} = \sqrt{2}\sin(\omega t + 120^\circ)$$

It is assumed that the load is unbalanced and contain fifth and seventh harmonics. The instantaneous load currents are given in per unit as

$$i_{La} = 0.21\sin(\omega t - 16.5^\circ) + 0.05\sin(5\omega t) + 0.3\sin(7\omega t)$$
$$i_{Lb} = 0.42\sin(\omega t - 110.7^\circ) + 0.05\sin[5(\omega t - 120^\circ)] + 0.3\sin[7(\omega t - 120^\circ)]$$
$$i_{Lc} = 0.31\sin(\omega t + 118.9^\circ) + 0.05\sin[5(\omega t + 120^\circ)] + 0.3\sin[7(\omega t + 120^\circ)]$$

It is assumed that the shunt compensator that does not supply any real power to the load, such that the entire amount of real power must then come from the supply. Furthermore, the compensator balances the load, and the compensated load draws power at unity power factor. Then find the instantaneous compensator and source currents.

Notes and References

The instantaneous PQ theory, which was first reported in [6] has gained a worldwide attention, due to the remarkable way in which instantaneous reactive power is defined. Based on this, reference generation for load compensation is reported in [7]. This method has been widely used since then for distribution system applications of shunt active filters, series active filters, and even for transmission system applications. Various other interpretations of this method are given in [8–11].

The theory of load compensation using instantaneous symmetrical components is proposed in [5]. In this chapter, we present a gist of this method with its application for a distribution system having a stiff source. However, in general, a polluting load can be placed further down a distribution feeder, where the switching frequency harmonics can pollute the PCC voltage. To avoid this, an LC filter and its associated feedback controller will be required, as is reported in [12].

The early installations of DVRs are reported in [13, 14] to protect process plants such as automated rug manufacturing, milk food processing, etc. The DC side of a DVR can be supported by a battery storage unit [15] or rectifier [16]. The operation of DVRs that are supported by a DC storage capacitor was first reported in [17], where it was ensured that the DVR does not supply any real power to maintain its charge. This technique has also been extended to systems containing unbalanced and distorted loads, where the limits of performance are also evaluated [18]. A comprehensive review of UPQC can be found in [19]. An analysis of the applications of DSTATCOM is a power distribution system containing several DERs is presented in [20].

1 Hingorani, N. (1995). Introducing custom power. *IEEE Spectr.* 32 (6): 41–48.
2 Ghosh, A. and Ledwich, G. (2002). *Power Quality Improvement Using Custom Power Devices*. New York: Springer Science+Business Media.
3 Mishra, M.K., Ghosh, A., and Joshi, A. (2003). Operation of a DSTATCOM in voltage control mode. *IEEE Trans. Power Deliv.* 18 (1): 258–264.
4 Miller, T.J.E. (1982). *Reactive Power Control in Electric Systems*. New York: Wiley.
5 Ghosh, A. and Joshi, A. (2000). A new approach to load balancing and power factor correction in power distribution system. *IEEE Trans. Power Deliv.* 15 (1): 417–422.
6 Akagi, H., Kanazawa, Y., Fujita, K., and Nabae, A. (1983). Generalized theory of the instantaneous reactive power and its application. *Electr. Eng. Jpn.* 103 (4): 58–65.
7 Akagi, H., Kanazawa, Y., and Nabae, A. (1984). Instantaneous reactive power compensators comprising switching devices without energy storage components. *IEEE Trans. Indus. App.* IA-20 (3): 625–630.
8 Akagi, H., Nabae, A., and Atoh, S. (1986). Control strategy of active power filters using multiple voltage-source PWM converters. *IEEE Trans. Indus. App.* IA-22 (3): 460–465.
9 Furuhashi, T., Okuma, S., and Uchikawa, Y. (1990). A study on the theory of instantaneous reactive power. *IEEE Trans. Indus. Electron.* 37 (1): 86–90.
10 Willems, W. (1992). A new interpretation of the Akagi-Nabae power components for nonsinusoidal three phase situations. *IEEE Trans. Instrum. Meas.* 41 (4): 523–527.
11 Watanabe, E.D., Stephan, R.M., and Aredes, M. (1993). New concepts of instantaneous active and reactive powers in electrical systems with generic load. *IEEE Trans. Power Deliv.* 8 (2): 697–703.
12 Ghosh, A. and Ledwich, G. (2003). Load compensating DSTATCOM in weak AC systems. *IEEE Trans. Power Deliv.* 18 (4): 1302–1309.
13 Woodley, N.H., Morgan, L., and Sundaram, A. (1999). Experience with an inverter-base dynamic voltage restorer. *IEEE Trans. Power Deliv.* 14 (3): 1181–1185.
14 Woodley, N.H., Sundaram, A., Coulter, B., and Morris, D. (1998) Dynamic voltage restorer demonstration project experience. *12th Conf. Electric Power Supply Industry (CEPSI)*, Pattaya, Thailand.
15 Ramachandaramurthy, V.K., Fitzer, C., Arulampalam, A. et al. (2002). Control of a battery supported dynamic voltage restorer. *Proc. IEE Gener. Transm. Distrib.* 149 (5): 533–542.
16 Huang, C.J., Huang, S.J., and Pai, F.S. (2003). Design of dynamic voltage restorer with disturbance-filtering enhancement. *IEEE Trans. Power Electron.* 18 (5): 1202–1210.
17 Ghosh, A. and Ledwich, G. (2002). Compensation of distribution system voltage using DVR. *IEEE Trans. Power Deliv.* 17 (4): 1030–1036.

18 Ghosh, A., Jindal, A.K., and Joshi, A. (2004). Design of a capacitor supported dynamic voltage restorer (DVR) for unbalanced and distorted loads. *IEEE Trans. Power Deliv.* 19 (1): 405–413.

19 Khadkikar, V. (2012). Enhancing electric power quality using UPQC: a comprehensive overview. *IEEE Trans. Power Electron.* 27 (5): 2284–2297.

20 Ghosh, A. and Shahnia, F. (2015). Applications of power electronic devices in distribution systems. In: *Transient Analysis of Power Systems: Solution Techniques, Tools and Applications* (ed. J.A. Martinez-Velasco). IEEE.

10

Microgrids

A microgrid, essentially, is a small power distribution grid where the generations and loads are placed in closed proximity. The term microgrid was proposed by the Consortium for Electric Reliability Technology Solutions (CERTS) [1]. There are several definitions of a microgrid. However, most of them have some common themes. These are that a microgrid [2]:

- Incorporates multiple loads.
- Incorporates multiple distributed energy resources (DERs).
- Is able to island from the utility grid.
- Can act as a single controllable entity.
- Can operate nominally in grid-connected mode.
- Has clear boundaries.

Summarizing the above points, a microgrid can be defined as an aggregation of electrical loads and generation. To the utility, a microgrid is an electrical load that can be controlled in magnitude. This load can (i) be constant, (ii) increase at night when the price of electricity is cheaper, (iii) be zero during time of grid system stress, and (iv) even supply power to the grid when the microgrid generation is more than its internal load demand.

A thematic diagram of a microgrid is shown in Figure 10.1. A microgrid may contain distributed generators (DGs), like photovoltaic (PV), wind turbine, microturbine, fuel cell, and diesel (maybe biodiesel) generators. It can also contain storages like battery energy storage systems (BESS) and flywheels. Together, the DGs and storages are called DERs. These two terms will be used interchangeably in the chapter. To control such diverse types of DERs, a microgrid has a central controller, as shown in Figure 10.2. However, a microgrid may have different hierarchies of control. This is discussed in Section 10.8.

Control of Power Electronic Converters with Microgrid Applications, First Edition.
Arindam Ghosh and Firuz Zare.

Published 2023 by John Wiley & Sons, Inc.

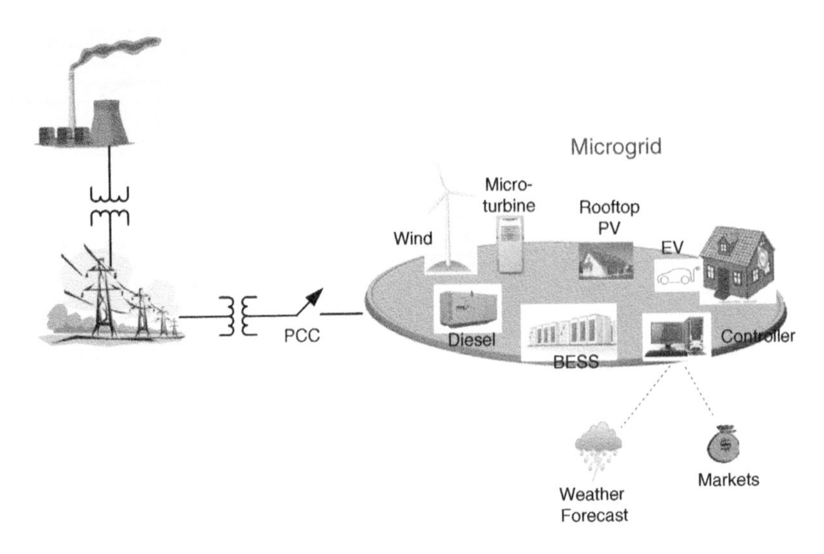

Figure 10.1 Thematic diagram of a microgrid.

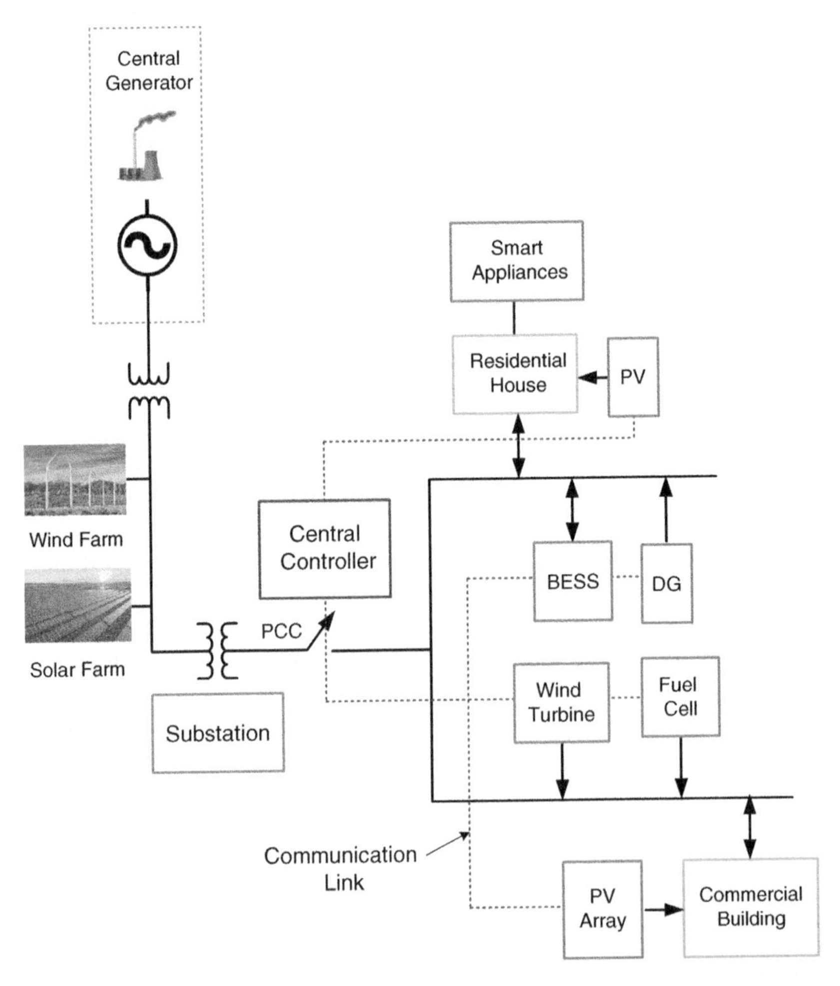

Figure 10.2 Schematic diagram of a typical microgrid.

As can be seen from Figures 10.1 and 10.2, a microgrid is connected with the utility grid at a point of common coupling (PCC). The utility grid may contain central generation and large-scale renewable generators like wind farms or solar farms. The operation of a microgrid is not influenced by the generators on the utility side. A microgrid can operate in a grid-connected mode or be islanded (also known as autonomous mode). In the latter mode of operation, the DERs connected to a microgrid will have to cater to all the local loads in the microgrid. Therefore, a microgrid will also be required to install demand response and load shedding protocols.

An AC microgrid can contain different types of DERs with various characteristics:

- Inertial: These have a rotating mass that can store kinetic energy. These include, for example, diesel or natural gas turbines employing (usually) synchronous generators. The advantage of these is that they can hold the frequency for a longer period of time by releasing the stored kinetic energy temporarily. Disadvantages include the use of polluting resources and the lack of availability of fuel, especially in remote areas. These usually have slower response times.
- Noninertial: These usually are interfaced through converters that can change their outputs almost instantaneously (very fast response time). These include, for example, batteries, fuel cells, microturbines, and solar photovoltaic cells (PV). These DERs cannot store rotational energy, and frequency/voltage collapse can result if immediate remedial action is not taken with respect to faults or load increase.

Additionally, DERs are also classified according to their ability to control their output powers:

- Dispatchable: The power output of these DERs can be controlled depending on the load demand. These include diesel or natural gas gen-sets, batteries, fuel cells, microturbines, etc.
- Nondispatchable: The maximum available power is harnessed in maximum power point tracking mode in renewable energy sources, such as solar PV and wind. The power output from these sources is intermittent since the PV output can get affected by a passing cloud or the wind speed can change suddenly.

If a microgrid contains such a mix, the maximum power is obtained from nondispatchable sources, while the dispatchable sources supply the rest. The problem with such a scheme can occur when the power available from the renewable resources is more than the demand or when it is not sufficient to meet the load demand. In both cases, storage devices will be required to absorb excess generated power or to supply a power shortfall.

Dispatchable DGs, such as diesel generators and microturbines, can be easily controlled by the associated controllers to follow the desired generation patterns. However, the nondispatchable sources, which are mainly the renewable DGs such as wind and solar, cannot be properly controlled due to the fluctuation of their input power. Intermittency and volatility are the main characteristics of this type of DG, and these can deteriorate the power quality indices if their penetration level is high. Usually, these side effects can be avoided by installing a proper capacity of electric energy storage (ESS) units. The ESS can smooth fluctuations of the output power of the DGs by absorbing or injecting the appropriate amount of power. Therefore, the combination of ESS and the nondispatchable DGs can yield dispatchable energy sources that, beside other dispatchable DGs, can be employed to control both the frequency and the voltage in a microgrid.

The discussions in this chapter start with the different operating modes of converters for grid integration of renewable sources – both in microgrids and in power distribution systems.

10.1 Operating Modes of a Converter

A voltage source converter (VSC) can operate in three different modes: grid forming, grid feeding, and grid supporting. The dynamic operation of a converter depends on its operating principles [3–4].

- Grid forming: The main objective of this type of converter operation is to regulate network voltage and frequency. A converter operates in this mode when supplying small standalone systems like uninterrupted power supply (UPS) or islanded microgrids. A grid forming converter works as a controlled voltage source, with fixed voltage amplitude and frequency. Here, the real power and the reactive power are not directly controlled, but these are determined by the interaction of the converter with the network. It is to be noted that since grid forming converters operate at a fixed frequency, it can operate in parallel with other converters only when all of them operate at the same frequency.
- Grid feeding: In grid feeding converters, the active and the reactive are directly controlled, while the voltage magnitude and frequency are determined by the interaction of the converter with the grid. These converters are used for delivering power from renewable energy sources to the grid. For example, in a solar PV, the active power is harnessed by a maximum power point (MPPT), while the reactive power is usually set as zero. It is important to note that, since these converters are designed to operate in parallel, they cannot operate in isolation.
- Grid supporting: These converters operate like synchronous generators by delivering power to the grid. Here, frequency and voltage magnitude are controlled by a droop mechanism, thereby imposing a fair power sharing regime.

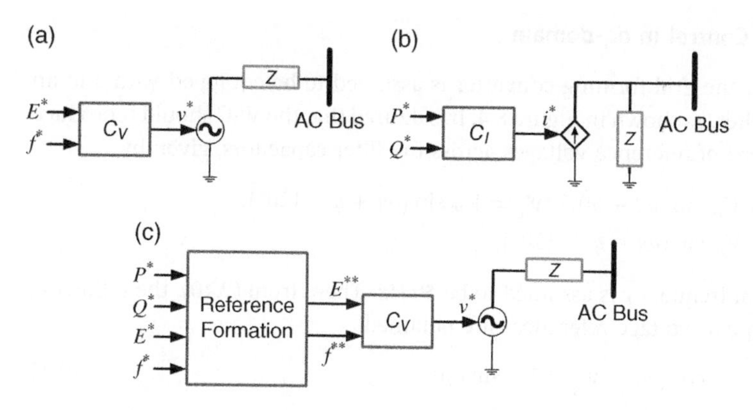

Figure 10.3 Grid-connected VSC operation modes: (a) grid forming, (b) grid feeding, and (c) grid supporting.

These converters can be connected in parallel and can also work in isolation. As is reported in [3], these converters can either be current-source based or voltage-source based. Only the voltage-source based converters are discussed here since they have a direct relation with droop equations.

The schematic diagrams of these three operation modes are shown in Figure 10.3, where C_V defines the voltage reference generation and C_I defines the current reference generation. In grid forming mode, shown in Figure 10.3a, the converter output voltage is connected to the AC grid through an output impedance (Z). The converter reference voltage depends on the prespecified voltage amplitude and frequency. In the grid feeding mode, shown in Figure 10.3b, the converter behaves like a current source in parallel with a high impedance. In this method, it delivers specified active and reactive power to the grid. The current injected should be synchronized with the grid frequency. Finally, in the grid supporting mode of Figure 10.3c, the converter behaves like an ideal voltage source that is connected to the grid through an impedance. The voltage source reference depends on active power, reactive power, voltage magnitude, and frequency. Often, the link impedance is emulated by an internal control loop [3].

10.2 Grid Forming Converters

The feedback control principle of a grid forming converter is discussed in Chapter 8 (Section 8.2) in the abc-frame. In this section, we design the control based on the dq-domain. Two different types of control of such converters – (i) using multiple proportional plus integral (PI) controllers and (ii) using state feedback control – are discussed. For this, the same system as shown in Figure 8.4 will be considered.

10.2.1 PI Control in dq-domain

The VSC for the grid forming converter is assumed to be equipped with and an output LC filter, as shown in Figure 8.4. It is desired that the VSC should reproduce a balanced set of reference voltages across the filter capacitors, given by

$$v_{ca}^{*} = V_m \sin(\omega t + \varphi), \quad v_{cb}^{*} = V_m \sin(\omega t + \varphi - 120°),$$
$$v_{cc}^{*} = V_m \sin(\omega t + \varphi + 120°)$$

The system frequency is assumed to be 50 Hz. Then, from (2.40), the following d-axis and q-axis voltage references are obtained

$$v_{cd}^{*} = V_m \cos(\varphi), \quad v_{cq}^{*} = V_m \sin(\varphi) \tag{10.1}$$

Two PI controllers will now be used to regulate these voltages to their desired values. The closed-loop control system is shown in Figure 10.4. The advantage of this scheme, as opposed to those discussed in Chapter 8, is that the filter dynamics need not be considered here.

Example 10.1 Consider the same system as in Example 8.2 with the same network, converter, and filter parameters, where the magnitude of the voltage is chosen as $V_m = 326.6$ V. The AC source and the converter reference voltages are written by choosing $\varphi = 20°$ as

$$v_a = V_m \sin(\omega t), \quad v_b = V_m \sin(\omega t - 120°), \quad v_c = V_m \sin(\omega t + 120°)$$
$$v_{ca}^{*} = V_m \sin(\omega t + 20°), \quad v_{cb}^{*} = V_m \sin(\omega t - 100°), \quad v_{cc}^{*} = V_m \sin(\omega t + 140°)$$

Then, as per (10.1), their reference voltages are given by $v_{cd}^{*} = 306.9\,\text{V}$ and $v_{cq}^{*} = 111.7\,\text{V}$. The gains of both the PI controllers are chosen as $K_P = 1$ and $K_I = 100$.

The system tracking performance is shown in Figures 10.5 and 10.6. From these figures, it is obvious that the steady state tracking performance is very good. However, the transient performance needs improvement as significant ripples can be observed at the beginning.

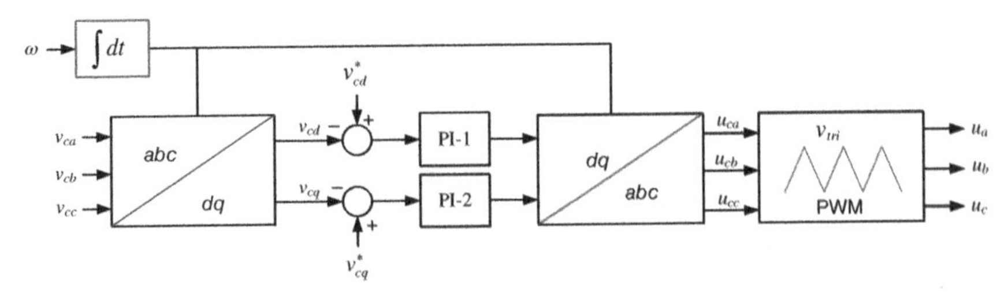

Figure 10.4 Block diagram of closed-loop voltage control for grid forming converter.

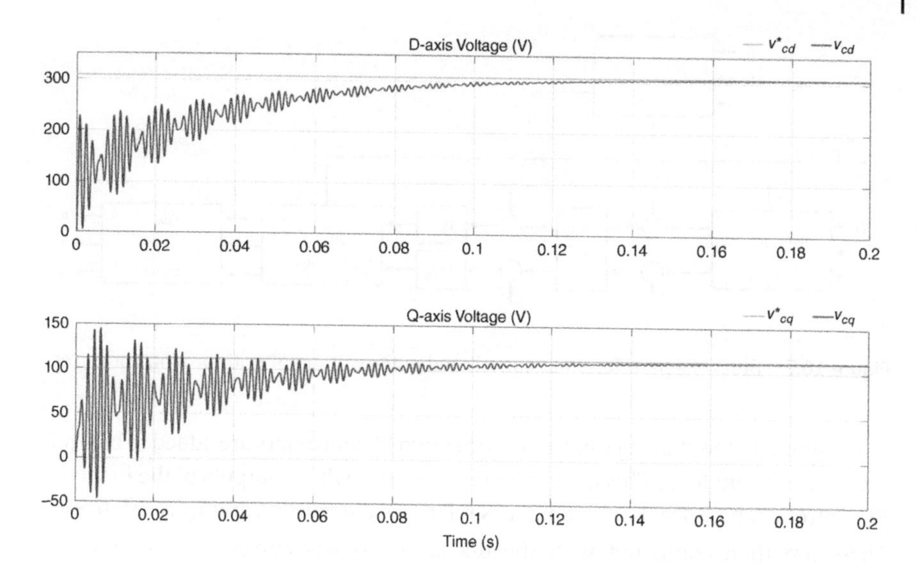

Figure 10.5 d- and q-axis voltages for grid forming converter with voltage feedback.

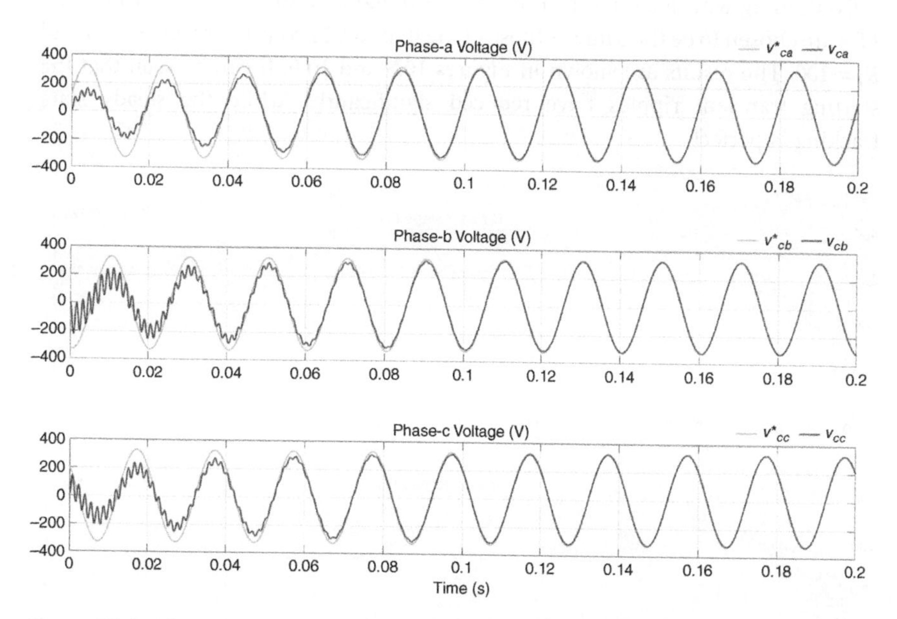

Figure 10.6 Phase voltages for grid forming converter with voltage feedback.

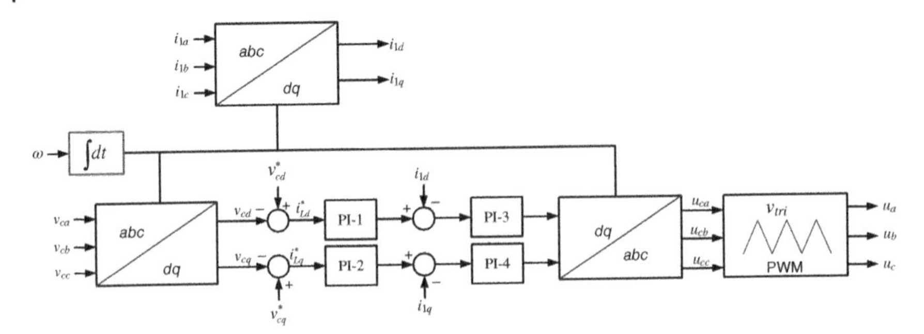

Figure 10.7 Block diagram for inner current and outer voltage of grid forming converter.

To eliminate the starting ripples, a further two PI controllers are added to control the inner current loop. This is shown in Figure 10.7, where outputs of the first two PI controllers produce d-axis and q-axis reference currents i_{1d}^* and i_{1q}^* respectively. These are then compared with the d-axis and q-axis currents of i_{1d} and i_{1q} respectively and passed through the second set of PI controllers to produce the switching signals.

Continuing with Example 10.1, the proportional and integral gains of PI-3 and PI-4 are chosen to be the same as those of the first two PI controllers, i.e. $K_P = 1$ and $K_I = 100$. The results are shown in Figures 10.8 and 10.9. It can be seen that the starting transient ripples have reduced significantly, while the steady state tracking is perfect.

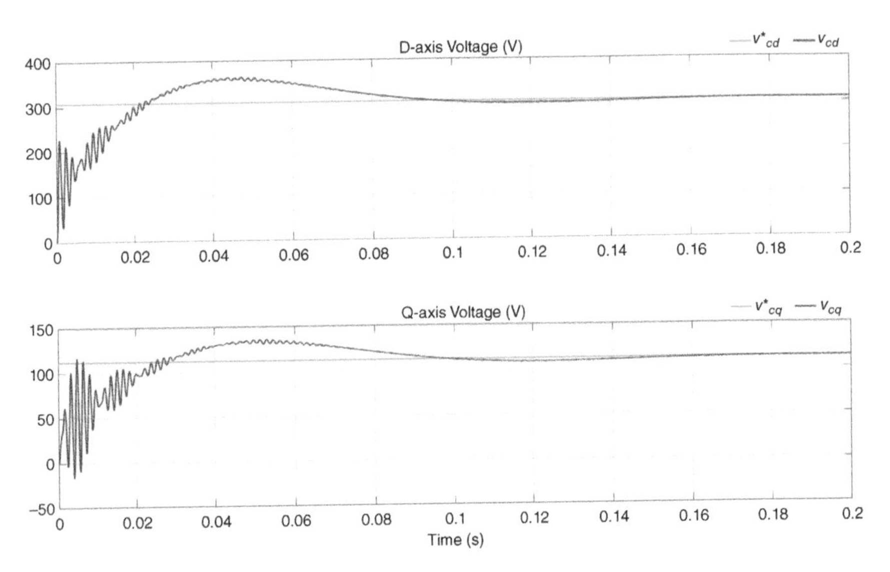

Figure 10.8 d- and q-axis voltages for grid forming converter with two loop controls.

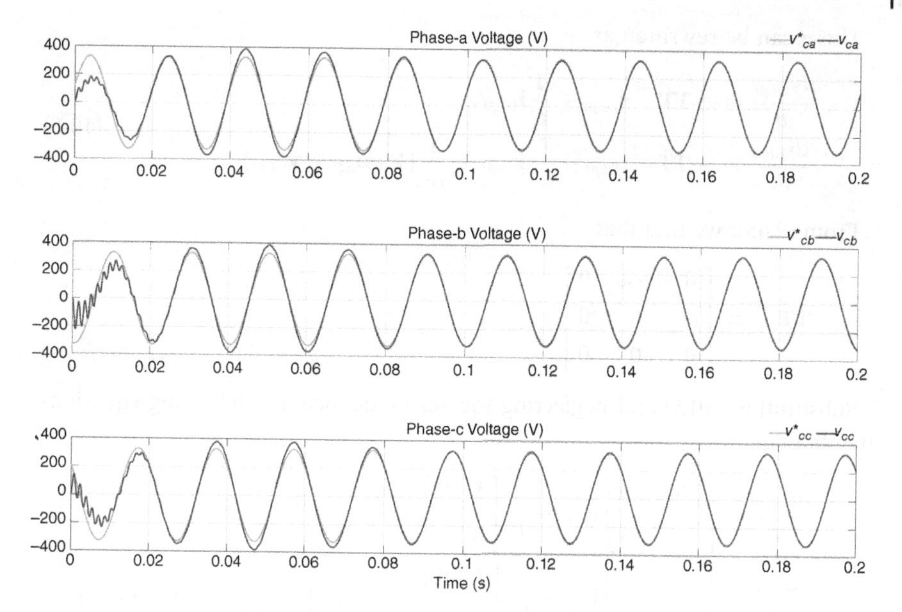

Figure 10.9 Phase voltages for grid forming converter with two loop controls.

10.2.2 State Feedback Control in dq-domain

Consider the single-line diagram of the LC filter structure shown in Figure 8.1c. The differential equations for the capacitor voltage v_c and the inductor current i_1 are given by

$$\begin{aligned}\frac{dv_c}{dt} &= \frac{1}{C}i_1\\ \frac{di_1}{dt} &= -\frac{R_1}{L_1}i_1 + \frac{1}{L_1}(V_{dc}u - v_c)\end{aligned} \tag{10.2}$$

These can be rewritten in a compact form for all the three phases as

$$\begin{aligned}\frac{d\mathbf{v}_{cabc}}{dt} &= \frac{1}{C}\mathbf{i}_{1abc}\\ \frac{d\mathbf{i}_{1abc}}{dt} &= -\frac{R_1}{L_1}\mathbf{i}_{1abc} + \frac{1}{L_1}(V_{dc}\mathbf{u}_{abc} - \mathbf{v}_{cabc})\end{aligned} \tag{10.3}$$

From (2.38), the dq-axis transformations of the above two equations in (10.3) are

$$\begin{aligned}\mathbf{T}^{-1}\frac{d\mathbf{v}_{cdq0}}{dt} + \dot{\mathbf{T}}^{-1}v_{cdq0} &= \frac{1}{C}\mathbf{T}^{-1}\mathbf{i}_{1dq0}\\ \mathbf{T}^{-1}\frac{d\mathbf{i}_{1dq0}}{dt} + \dot{\mathbf{T}}^{-1}\mathbf{i}_{1dq0} &= -\frac{R_1}{L_1}T^{-1}\mathbf{i}_{1dq0} + \frac{1}{L_1}\mathbf{T}^{-1}\left(V_{dc}\mathbf{u}_{dq0} - \mathbf{v}_{cdq0}\right)\end{aligned} \tag{10.4}$$

These can be rewritten as

$$\begin{aligned}\frac{d\mathbf{v}_{cdq0}}{dt} &= -\mathbf{T}\dot{\mathbf{T}}^{-1}\mathbf{v}_{cdq0} + \frac{1}{C}\mathbf{i}_{1dq0}\\ \frac{d\mathbf{i}_{1dq0}}{dt} &= -\mathbf{T}\dot{\mathbf{T}}^{-1}\mathbf{i}_{1dq0} - \frac{R_1}{L_1}\mathbf{i}_{1dq0} + \frac{1}{L_1}\left(V_{dc}\mathbf{u}_{dq0} - \mathbf{v}_{cdq0}\right)\end{aligned} \tag{10.5}$$

From (2.45), we find that

$$\mathbf{T}\dot{\mathbf{T}}^{-1} = \begin{bmatrix} 0 & -\omega & 0 \\ \omega & 0 & 0 \\ 0 & 0 & 0 \end{bmatrix}$$

Substituting (10.5) and neglecting the zero-sequence, the following equations are obtained

$$\begin{aligned}\frac{d\mathbf{v}_{cdq}}{dt} &= \begin{bmatrix} 0 & \omega \\ -\omega & 0 \end{bmatrix}\mathbf{v}_{cdq} + \frac{1}{C}\begin{bmatrix} 1 & 0 \\ 0 & 1 \end{bmatrix}\mathbf{i}_{1dq}\\ \frac{d\mathbf{i}_{1dq}}{dt} &= \begin{bmatrix} 0 & \omega \\ -\omega & 0 \end{bmatrix}\mathbf{i}_{1dq} - \frac{R_1}{L_1}\begin{bmatrix} 1 & 0 \\ 0 & 1 \end{bmatrix}\mathbf{i}_{1dq} + \frac{1}{L_1}\begin{bmatrix} 1 & 0 \\ 0 & 1 \end{bmatrix}\left(V_{dc}\mathbf{u}_{dq} - \mathbf{v}_{cdq}\right)\end{aligned} \tag{10.6}$$

Let the state and control vectors be defined by

$$\mathbf{x}_{dq} = \begin{bmatrix} v_{cd} \\ v_{cq} \\ i_{1d} \\ i_{1q} \end{bmatrix}, \quad \mathbf{u}_{dq} = \begin{bmatrix} u_d \\ u_q \end{bmatrix}$$

Then (10.6) is rewritten as

$$\dot{\mathbf{x}}_{dq} = \mathbf{A}\mathbf{x}_{dq} + \mathbf{B}V_{dc}\mathbf{u}_{dq} \tag{10.7}$$

where

$$\mathbf{A} = \begin{bmatrix} 0 & \omega & 1/C & 0 \\ -\omega & 0 & 0 & 1/C \\ -1/L_1 & 0 & -R_1/L_1 & \omega \\ 0 & -1/L_1 & -\omega & -R_1/L_1 \end{bmatrix}, \quad \mathbf{B} = \begin{bmatrix} 0 & 0 \\ 0 & 0 \\ 1/L_1 & 0 \\ 0 & 1/L_1 \end{bmatrix}$$

The state feedback control law is given by

$$\mathbf{u}_{dq} = \mathbf{K}\left(\mathbf{x}_{dq}^{*} - \mathbf{x}_{dq}\right) \tag{10.8}$$

The feedback control structure is shown in Figure 10.10, where the reference vector $\mathbf{x}_{dq}^{*}$ is computed as per Example 10.2.

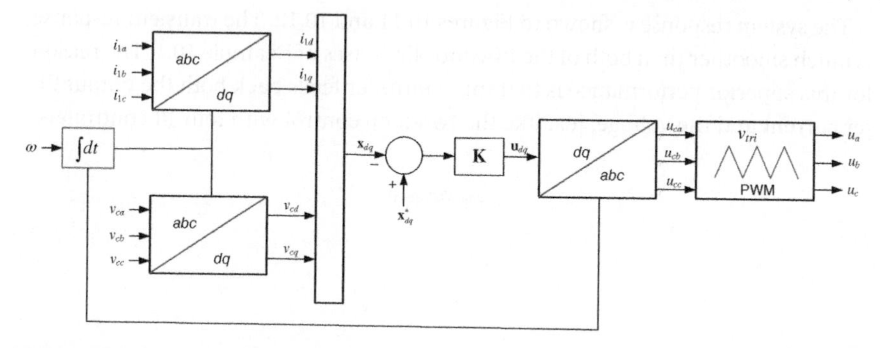

Figure 10.10 Block diagram of the state feedback control structure of grid forming converter.

Example 10.2 Consider the same system as in Example 10.1, where the reference voltages and therefore their dq components are the same as those computed. The reference capacitor currents are computed using (8.13), i.e.

$$i_{cref} = \omega C V_m \cos(\omega t + \varphi)$$

For $C = 50$ μF and the system frequency of 50 Hz, the reference capacitor currents are computed as

$$I_m = C\omega V_m = 5.13\text{ A}$$

$$i_{ca}^* = I_m \cos(\omega t + 20°),\quad i_{cb}^* = I_m \cos(\omega t - 100°),\quad i_{cc}^* = I_m \cos(\omega t + 140°)$$

The d-axis and q-axis components of the capacitor currents are then computed as

$$i_{cd}^* = -1.75,\quad i_{cq}^* = 4.82$$

The references for the inductor L_1 are computed from (8.12) as

$$i_{1d}^* = i_{cd}^* + i_{fd},\quad i_{1q}^* = i_{cq}^* + i_{fq}$$

A linear quadratic controller is now designed with the following gain matrices (the same as those shown in Examples 8.2 and 8.3)

$$Q = \begin{bmatrix} 150 & & & \\ & 150 & & \\ & & 1 & \\ & & & 1 \end{bmatrix},\quad R = \begin{bmatrix} 0.1 & \\ & 0.1 \end{bmatrix}$$

The resulting gain matrix is

$$K = \begin{bmatrix} 122.47 & 0 & 10.06 & 0 \\ 0 & 122.47 & 0 & 10.06 \end{bmatrix}$$

The system response is shown in Figures 10.11 and 10.12. The transient response is much smoother than both of the PI-controlled cases of Example 10.1. The reason for this superior performance is that this controller feeds back both the output filter current and the voltage, just like the two-loop control with four PI controllers.

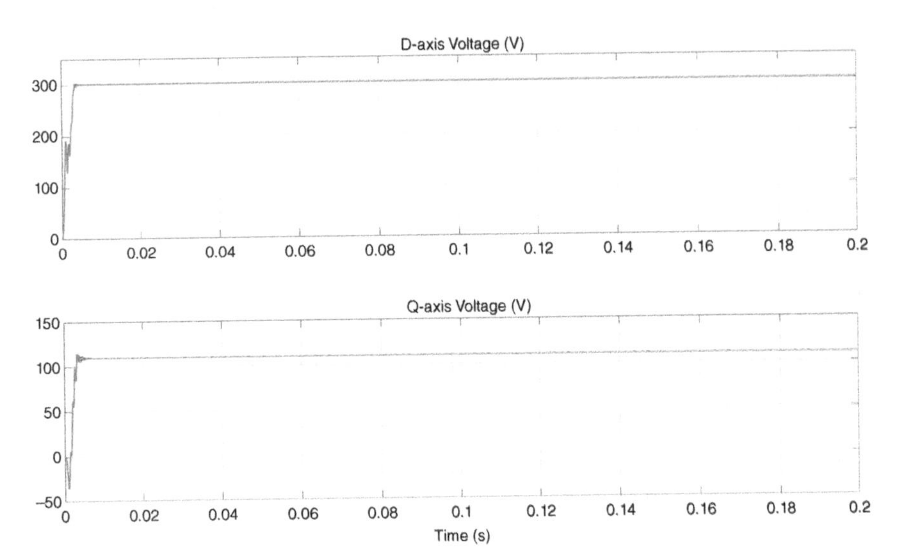

Figure 10.11 d- and q-axis voltages for grid forming converter with state feedback control.

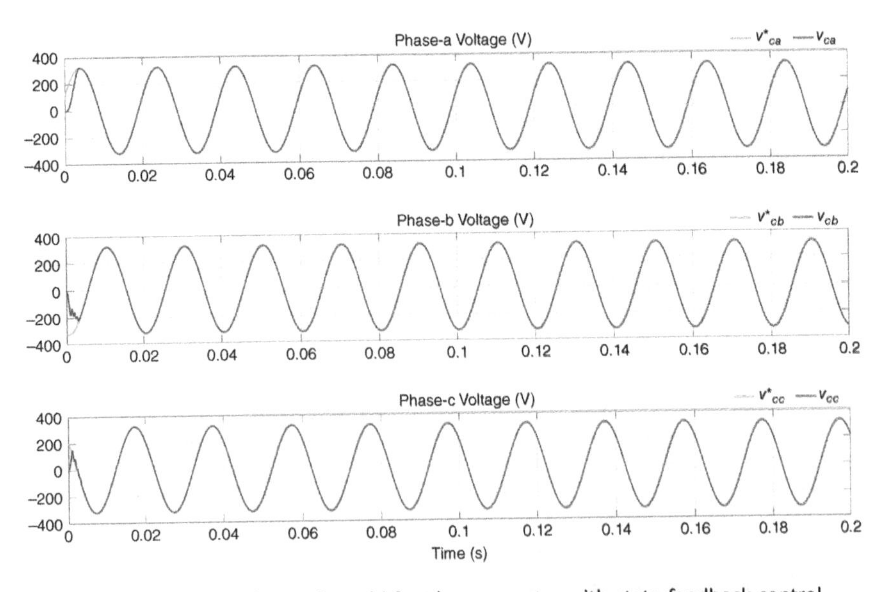

Figure 10.12 Phase voltages for grid forming converter with state feedback control.

However, the controller in this case is designed based on the optimal control law. Such controllers are easier to design than PI controllers. Moreover, optimal controllers are very robust to parameter variations.

10.3 Grid Feeding Converters

A grid feeding converter should be synchronized with the grid frequency and is required to inject prespecified amounts of active and reactive power to the grid. One of the simplest approaches is to use a hysteretic current control, where the reference is generated in αβ-frame.

Assume that the converter output is connected with the utility bus through an external reactance only (L-type filter). We now denote the following:

- Utility bus voltages: v_a, v_b and v_c
- Converter output currents: i_a, i_b and i_c
- Desired active power: P^*
- Desired reactive power: Q^*.

The measured utility bus voltages (v_a, v_b, and v_c) are converted to their equivalent αβ components to obtain v_α and v_β. The desired real and reactive powers can be defined as

$$\begin{aligned} P^* &= \frac{3}{2}\left(v_\alpha i_\alpha^* + v_\beta i_\beta^*\right) \\ Q^* &= \frac{3}{2}\left(v_\alpha i_\beta^* - v_\beta i_\alpha^*\right) \end{aligned} \tag{10.9}$$

where i_α^* and i_β^* are the desired output currents. The solution of (10.9) can be written in the following form

$$\begin{bmatrix} i_\alpha^* \\ i_\alpha^* \end{bmatrix} = \frac{2}{3\left(v_\alpha^2 + v_\beta^2\right)} \begin{bmatrix} v_\alpha & -v_\beta \\ v_\beta & v_\alpha \end{bmatrix} \begin{bmatrix} P^* \\ Q^* \end{bmatrix} \tag{10.10}$$

Once i_α^* and i_β^* are computed, the desired output currents i_a^*, i_b^*, and i_c^* are obtained using inverse αβ transform. These are then compared with i_a, i_b, and i_c to generate the switching pulses using hysteretic current control.

One of the main advantages of using αβ transform is that the signal is automatically synchronized with the grid frequency. The grid feeding inductor is now connected with a stiff utility bus of voltage 400 V (L-L) through a 1 mH inductor. The converter DC side voltage is chosen as 600 V. The converter operates in a hysteresis current control with a hysteresis band of 5 A. It is required that the converter injects $P^* = 50$ kW and $Q^* = 25$ kVAr power to the utility bus, the frequency of which is 50 Hz at the beginning. It is changed to 49.75 Hz at 1.1 seconds. The results are shown in Figure 10.13. It can be seen that the frequency change has

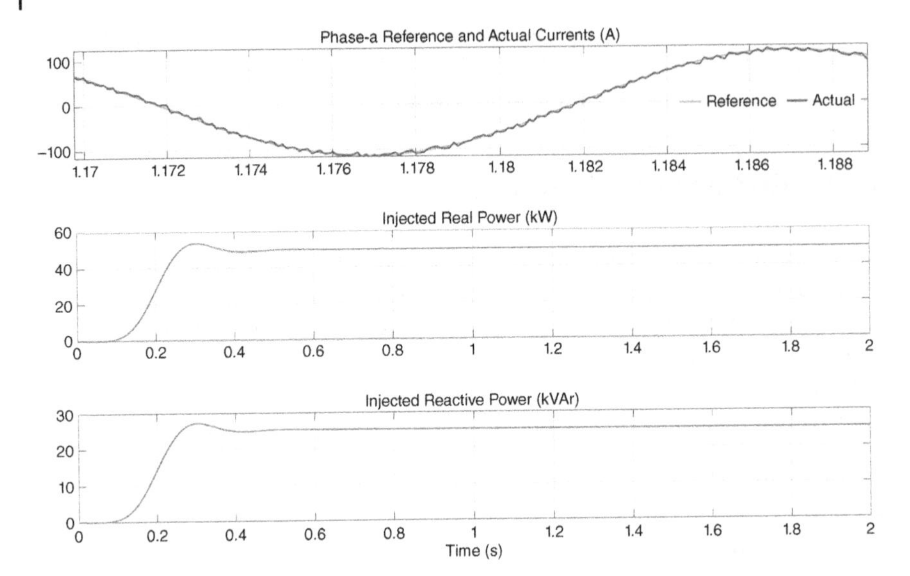

Figure 10.13 Performance of the grid feeding converter.

no impact on the operation of the system and that the desired active and reactive power are supplied by the converter.

Consider the system shown in Figure 10.14, in which a utility substation supplies a load through Feeder-1. There is a battery ESS connected to the system through

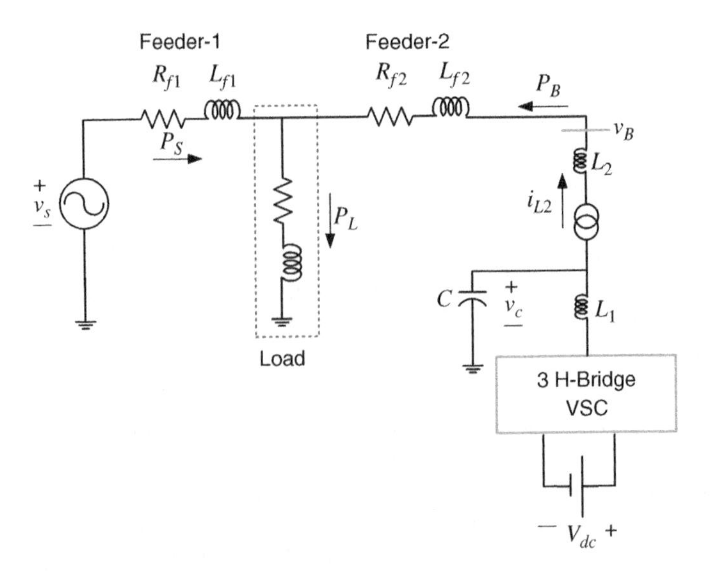

Figure 10.14 Battery-connected grid feeding converter in a distribution network.

another feeder (Feeder-2). The battery system is interfaced with the AC system through a three H-bridge configuration of VSC shown in Figure 7.11a. The VSC has an LCL output filter, where L_2 is the leakage inductance of the transformer. The system parameters chosen for the study are:

- System voltage (v_s): 11 kV (L-L), 50 Hz
- Feeder-1 and -2: $L_{f1} = L_{f2} = 57.8$ mH, $R_{f1} = R_{f2} = 3.025\ \Omega$
- RL load: $L_L = 372$ mH, $R_L = 242\ \Omega$
- LC filter: $L_1 = 38.5$ mH, $C = 3.76\ \mu$F
- Transformer: 1 MVA, 440 V: 11 kV, leakage inductance of 10%.

An output feedback current control is designed using the pole shift control in the abc-frame as in Example 8.7 with a sampling time of 10 μs. The magnitude and the phase angle of the battery bus voltage v_B is first extracted using a fast Fourier transformer (FFT) block. Let this be denoted by $|V_{Bm}| \angle \delta_B$. It is assumed that the battery injects power at a unity power factor. If the power reference is denoted by P_B^*, the reference for the current that will be injected by the battery is given by

$$I_m = \frac{\sqrt{2}}{3} \times \frac{P_B^*}{|V_{BM}|}$$

$$i_{L2a}^* = I_m \sin(\omega t + \delta_B), \quad i_{L2b}^* = I_m \sin(\omega t + \delta_B - 120°), \quad i_{L2c}^* = I_m \sin(\omega t + \delta_B + 120°)$$

The results are shown in Figure 10.15. At the beginning, the source frequency is 50.1 Hz and the battery neither absorbs nor supplies any real power. Then, at 0.5, the battery starts absorbing 135 kW of power at unity power factor. Subsequently,

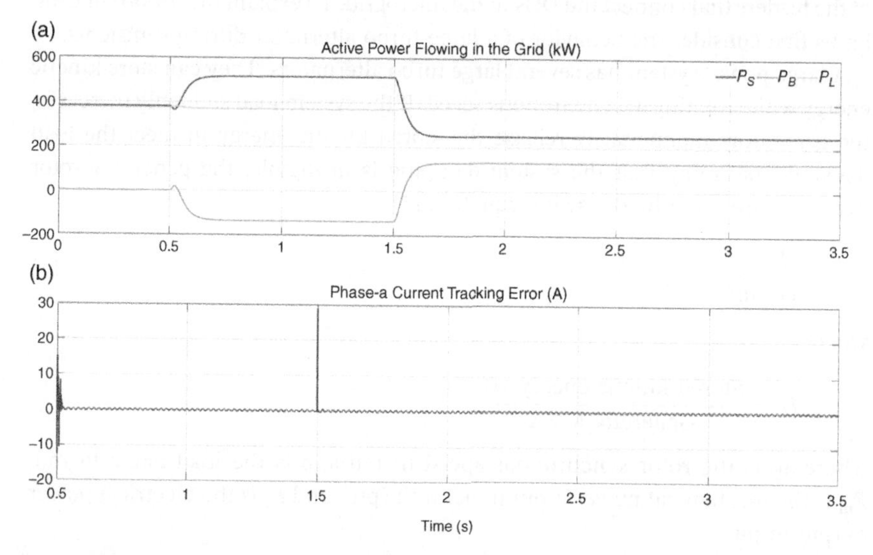

Figure 10.15 (a) Power flow in the network and (b) current tracking error when the battery operates in grid feeding mode.

at 1.5 seconds, the battery supplies 130 kW of power at unity power factor. The source frequency changes to 49.8 Hz at 2.5 seconds. It can be seen from Figure 10.15a that the power supplied by the source increases at 0.5 seconds to supply the battery and reduces at 1.5 seconds since the battery is now supplying power. The load power remains constant throughout. The change in frequency has no impact on the power flow since the battery is operating in the grid feeding mode. The current tracking error for phase-a is shown in Figure 10.15b. It can be seen that there is a jump discontinuity when the battery power reference changes. However, the tracking error is negligible. Also, the change in frequency does not have any impact on this tracking error.

10.4 Grid Supporting Converters for Islanded Operation of Microgrids

In this section, the islanded (or autonomous) operation of a microgrid is discussed, where the converters operate in grid supporting mode, as shown in Figure 10.3c. The inputs to these converters are DGs (DERs) on the DC side. The converters are required to produce output voltages depending on active and reactive power, voltage magnitude, and frequency. The DERs that are connected to the microgrid share power according to their ratings. A step-by-step approach for designing droop gains such that an accurate real/reactive power sharing can occur is presented here. However, the power sharing usually is dependent on the R/X ratio of the feeders that connect the DGs in the microgrid. To explain the droop concept, let us first consider the behavior of a large turbo alternator during transients.

A large power system has several large turbo alternators. They can store kinetic energy while rotating at synchronous speed. If the system load suddenly increases, the generators momentarily release the stored kinetic energy to meet the load demand. Assuming that the system damping is negligible, the generator rotor dynamics are given by the swing equation [5]

$$\frac{2H}{\omega_s}\frac{d\delta^2}{dt^2} = P_m - P_e \tag{10.11}$$

where

$$H = \frac{\text{Stored kinetic energy MJ}}{\text{Generator MVA}}$$

where ω_s is the rotor synchronous speed in rad/s, δ is the load angle in rad, P_m is the mechanical power input in per unit (pu), and P_e is the electrical power output in pu.

The load angle is related to the generator speed ω by the relation

$$\frac{d\delta}{dt} = \omega - \omega_s \tag{10.12}$$

For a large turbo alternator, the mechanical input speed cannot be changed instantaneously due to its large rotational inertia. Thus, when a sudden load increase occurs, the term $P_m - P_e$ becomes negative. Equation (10.12) indicates that the generator speed ω drops below the synchronous speed ω_s, i.e. the generator slows down. It tries to maintain the balance between the mechanical and electrical power through the release of the stored rotational energy such that the angle does not diverge, causing the generator to lose synchronism. This, however, is a temporary process and can last only a few cycles. In order to supply the increased load, the generator turbine-governor action is required. This is achieved through a steady state frequency–power relation, called the droop equation [5]. In Section 10.4.1, the droop equations for different R/X ratios of distribution feeders are discussed.

10.4.1 Active and Reactive Over a Feeder

Consider the simple system shown in Figure 10.16, in which two sources are connected together through a feeder with an impedance of $R + jX$. Assuming $V_S = V_1 \angle \delta_1$ and $V_R = V_2 \angle \delta_2$, the current through the line is given by

$$I = \frac{V_1\angle\delta_1 - V_2\angle\delta_2}{R + jX} \tag{10.13}$$

Therefore, the complex power flowing from the source is

$$P_1 + jQ_1 = V_1 \times I^* = V_1\angle\delta_1 \times \frac{V_1\angle-\delta_1 - V_2\angle-\delta_2}{R - jX} = \frac{V_1^2 - V_1V_2\angle(\delta_1 - \delta_2)}{R - jX} \tag{10.14}$$

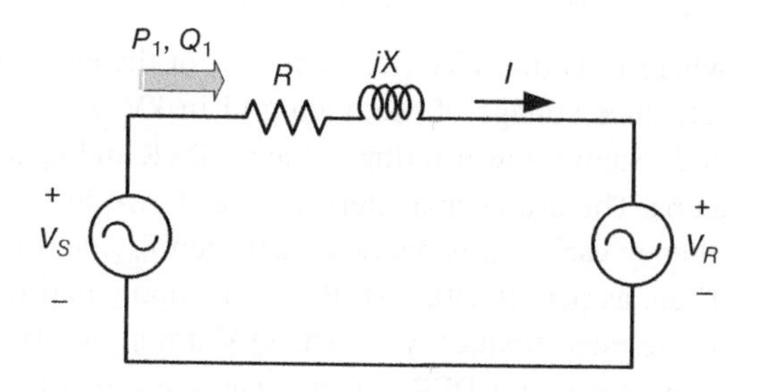

Figure 10.16 Two AC source connected together through a feeder.

Defining $\delta = \delta_1 - \delta_2$ and expanding (10.14), the real and imaginary parts are given by

$$P_1 = \frac{V_1}{R^2 + X^2}[R(V_1 - V_2 \cos\delta) + XV_2 \sin\delta] \tag{10.15}$$

$$Q_1 = \frac{V_1}{R^2 + X^2}[-RV_2 \sin\delta + X(V_1 - V_2 \cos\delta)] \tag{10.16}$$

10.4.2 Inductive Grid

Consider the case when the grid is predominantly inductive, i.e. $X > > R$, such that we can assume $R \approx 0$. Furthermore, assume that the load angle δ is very small, such that $\sin(\delta) \approx \delta$ and $\cos(\delta) \approx 1$. Therefore, (10.15) and (10.16) can be rewritten as

$$P_1 \approx \frac{V_1 V_2}{X}\delta \Rightarrow \delta \approx \frac{XP_1}{V_1 V_2} \tag{10.17}$$

$$Q_1 \approx \frac{V_1^2 - V_1 V_2}{X} \Rightarrow V_1 - V_2 \approx \frac{XQ_1}{V_1} \tag{10.18}$$

These two relations define the direct relationship:

- Between the load angle δ and the active power P_1.
- Between the voltage difference $V_1 - V_2$ and the reactive power Q_1.

The first relation has resulted in a P-δ droop, as is reported in [6–7]. However, note from (10.12) that the derivative of angle δ is proportional to the frequency difference. Therefore, a relation between the active power and the frequency can be established. Also, the second relation results in a voltage droop that varies with reactive power.

Assume that a microgrid has a total N number of DERs. Then the P-f and Q-V droop equations for each DER are given by

$$f_i = f^* + n_i \times (0.5P_i^* - P_i), \quad i = 1, 2, ..., N \tag{10.19}$$

$$V_i = V^* + m_i \times (0.5Q_i^* - Q_i), \quad i = 1, 2, ..., N \tag{10.20}$$

where f^* is the reference frequency of the entire microgrid in Hz and V^* is the reference voltage of the microgrid in kV; P_i^* and Q_i^* respectively are the active and reactive power rating of the ith DER and n_i and m_i are their respective droop gains. The droop characteristics are shown in Figure 10.17. Assume that the frequency variation is restricted between f_{max} and f_{min}, as shown in Figure 10.17a. Then, as per (10.19), a DER should supply half its rated power when it operates at the rated frequency f^*. The Q-V droop line is shown in Figure 10.17b, which stipulates that a DER voltage magnitude should be 1.0 pu when it supplies half its rated reactive power.

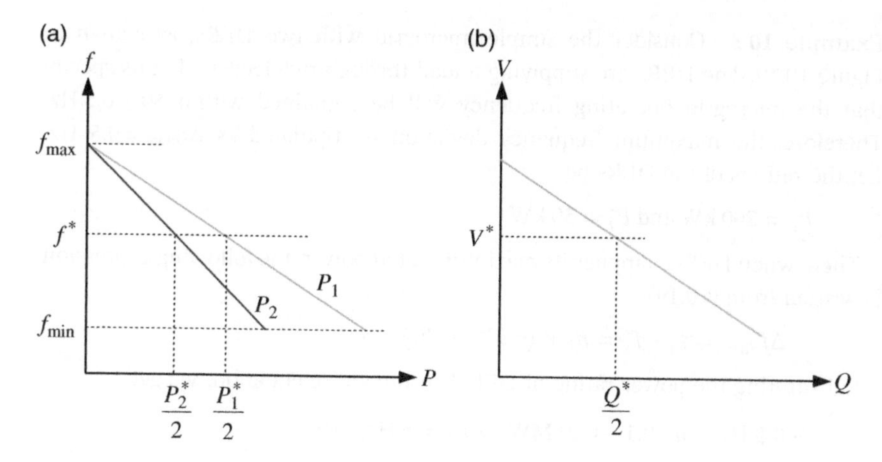

Figure 10.17 Droop characteristics for inductive grid: (a) P-f droop and (b) Q-V droop.

Consider a microgrid where only two DERs are present. The droop lines are shown in Figure 10.18. Assume that the DERs are operating at the rated frequency of 50 Hz, supplying half their rated power, when the load increases suddenly. It is shown in Chapter 9 (Section 9.1.4) that all generators in a grid must operate at the same frequency, failing which a large amount of current will flow through the network. Therefore, DER-1 should supply $P_1^*/2 + \Delta P_1$ amount of power, while DER-2 should supply $P_2^*/2 + \Delta P_2$ such that the operating frequency of both the DERs becomes $50 - \Delta f$ Hz. This will be the basis of a droop gain selection. Note that the droop line of DER-1 is shallower than that of DER-2. The inequality $P_1^* > P_2^*$ implies that the angle of the droop line with the y-axis is larger for DERs with higher power ratings than those with lower power ratings.

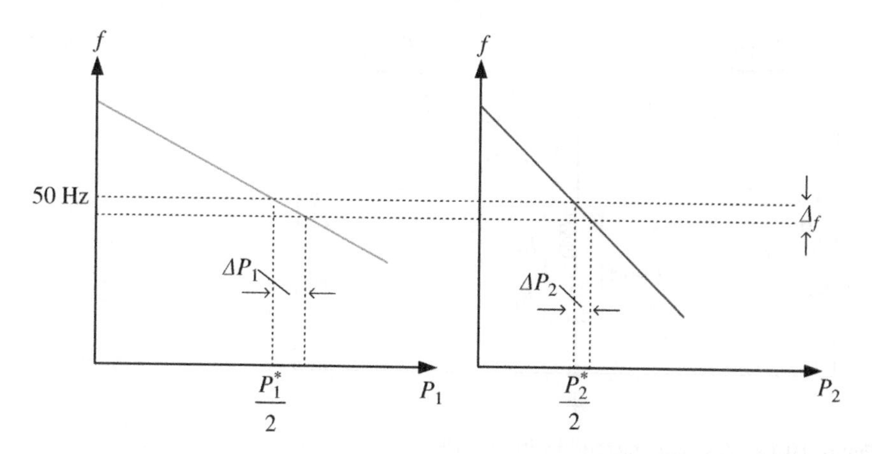

Figure 10.18 P-f droop lines of a two-DG microgrid.

Example 10.3 Consider the simple microgrid with two DERs, as shown in Figure 10.19. The DERs are supplying a load through two feeders. Let us specify that the microgrid operating frequency will be contained within 50 ± 0.5 Hz. Therefore, the maximum frequency deviation is stipulated as $\Delta f_{max} = 0.5$ Hz. Let the ratings of the DERs be

$$P_1^* = 200\ \text{kW and}\ P_2^* = 50\ \text{kW}$$

Then, when DER-1 supplies its maximum rated power, the following expression is written from (10.19)

$$-\Delta f_{\ max} = f_1 - f^* = n_1 \times (0.5P_1{}^* - P_1)$$

Substituting the power rating of DER-1 in the above equation, we get

$$-0.5\ \text{Hz} = n_1(0.1 - 0.2)\ \text{MW} \Rightarrow n_1 = 5\ \text{Hz/MW}$$

In a similar way, the droop gain for DER-2 is calculated as

$$n_2 = \frac{0.5}{0.025} = 20\ \text{Hz/MW}$$

Note from these values that

$$\frac{P_1^*}{P_2^*} = 4 = \frac{n_2}{n_1}$$

In general, a microgrid containing N number of DERs will obey the following expression if the frequency of all of them are to be restricted between $\pm\Delta f_{max}$ Hz

$$P_1{}^* n_1 = P_2{}^* n_2 = \cdots = P_N^* n_N = \alpha \tag{10.21}$$

Thus, the droop gains for all the DERs should be chosen such that α remains constant.

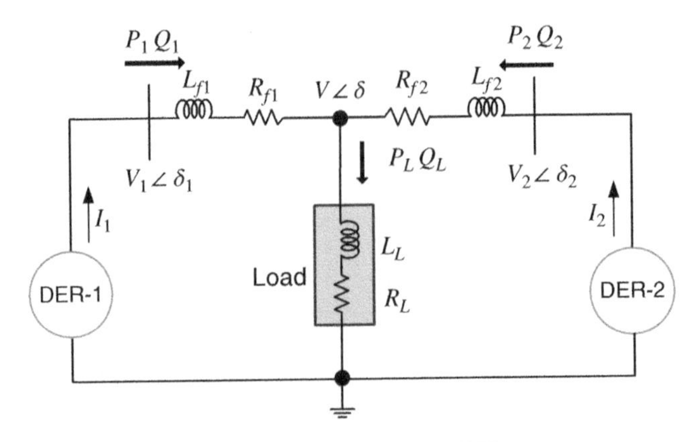

Figure 10.19 A simple microgrid with two DERs.

Once the required voltage magnitude and the frequency are obtained, the reference voltages for each DER are obtained as

$$\begin{cases} v_{ai}^* = \sqrt{2}V_i \sin(2\pi f_i t) \\ v_{bi}^* = \sqrt{2}V_i \sin(2\pi f_i t - 120^\circ) \quad i = 1, 2, ..., N \\ v_{ci}^* = \sqrt{2}V_i \sin(2\pi f_i t + 120^\circ) \end{cases} \tag{10.22}$$

Each DER is realized by a VSC, connected to the microgrid through an output LC filter. The reference voltages given in Example 8.2 are then reproduced across the filter capacitor in an abc frame using a linear quadratic regulator (LQR) state feedback controller discussed in Chapter 8. It is assumed that the DERs are supplied by fixed DC sources.

For the system of Example 10.3, the reactive power limits are chosen as half the real power limits, i.e. $Q_1^* = 100$ kVAr and $Q_2^* = 25$ kVAr. The voltage droop gains are chosen as $m_1 = 10$ V/MVAr and $m_2 = 40$ V/MVAr. The feeder impedances are $L_{f1} = 51.4$ mH, $R_{f1} = 3\ \Omega$, $L_{f2} = 77$ mH, and $R_{f2} = 2.42\ \Omega$. A balanced RL load is chosen as $R_L = 672\ \Omega$ and $L_L = 0.4265$ H.

The results are shown in Figure 10.20. Figure 10.20a shows the real powers supplied by the DERs, which, in the steady state, are $P_1 = 142$ kW and $P_2 = 35.5$ kW. Thus, the power sharing ratio is 4 : 1, which is as expected from the droop relations. From (10.19), and with the droop gain of $n_1 = 5$ Hz/MW, the steady frequency of DER-1 should be

$$f_1 = 50 + 5 \times (0.1 - 0.142) = 49.79 \text{ Hz}$$

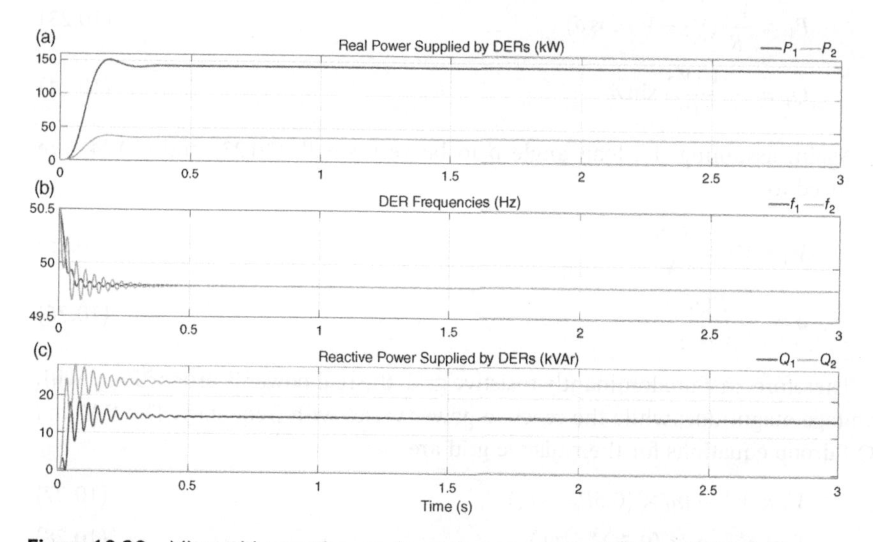

Figure 10.20 Microgrid operation results while operating with P-f and Q-V droop control. (a) Real power supplied by DERs (kW), (b) DER frequencies (Hz), (c) Reactive power supplied by DERs (kVAr).

which is as shown in Figure 10.20b. As the droop gains chosen such that the frequency remains the same throughout the microgrid, both the DERs supply power at this frequency. The reactive powers supplied are $Q_1 = 14.4$ kVAr and $Q_2 = 23.6$ kVAr, shown in Figure 10.20c. They do not obey the desired sharing ratio. There is a potential problem with this discrepancy in the reactive power sharing.

Let us assume that the DER output voltages are roughly equal to 6.35 kV (L-N). Then the currents supplied by the DERs are

$$I_1 = \frac{\sqrt{142^2 + 14.4^2}}{3 \times 6.35} = 7.49\text{ A and } I_2 = \frac{\sqrt{35.5^2 + 23.6^2}}{3 \times 6.35} = 2.24\text{ A}$$

Thus, the current sharing ratio is 3.34 : 1, i.e. the converter with a lower rating has more share of the current. It might so happen that a DER with lower rating may have to supply more reactive power than other DERs with higher ratings, so much so that it reaches its maximum current limit. The DER protection system may then take it offline, thereby endangering the system stability. Therefore, proper care must be taken to avoid this action.

10.4.3 Resistive Grid

Power transmission systems usually have a high a X/R ratio. On the contrary, distribution feeders are mostly resistive and therefore the reactance X in (10.15) and (10.16) can be neglected. These two equations are then rewritten as

$$P_1 = \frac{V_1}{R}(V_1 - V_2 \cos\delta) \tag{10.23}$$

$$Q_1 = -\frac{V_1 V_2}{R}\sin\delta \tag{10.24}$$

Again, assuming the load angle δ to be very small, (10.23) and (10.24) are reduced to

$$V_1 - V_2 = \frac{RP_1}{X} \tag{10.25}$$

$$\delta = -\frac{RQ_1}{V_1 V_2} \tag{10.26}$$

Therefore, in a predominantly resistive grid, the real power is proportional with voltage magnitude, while the reactive power varies with frequency. The P-V and Q-f droop equations for the resistive grid are

$$V_i = V^* + m_i \times (0.5P_i^* - P_i) \tag{10.27}$$

$$f_i = f^* - n_i \times (0.5Q_i^* - Q_i) \tag{10.28}$$

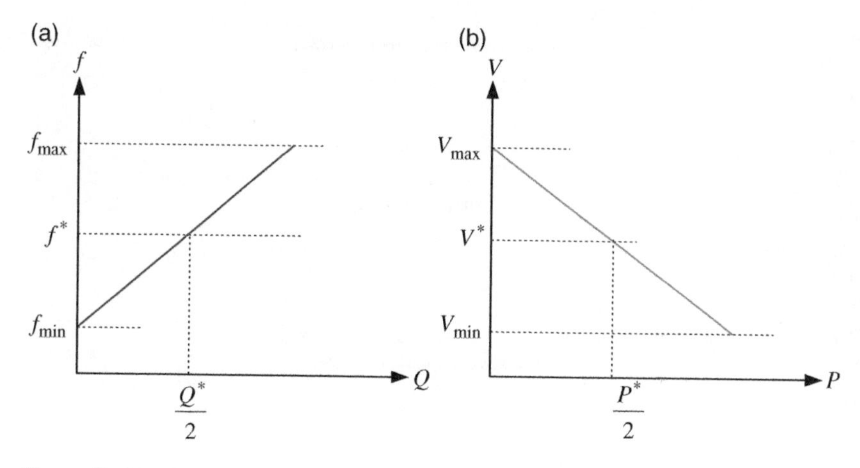

Figure 10.21 Droop characteristics for resistive grid: (a) Q-f droop and (b) P-V droop.

Note that the negative sign in (10.28) results from the negative sign in Q - δ relation in (10.26). The Q-f and P-V droop lines are shown in Figure 10.21.

Example 10.4 Consider the system described in Example 10.3. The DER output voltages have a specified value of 11 kV (L-L). Therefore, the L-N voltage will be 6.35 kV. Let the voltage variation be restricted between ±0.06 pu (6%), i.e. between 5.969 and 6.732 kV. Thus, when DER-1 is not supplying any power, from (10.27) we have

$$(1.06-1)\times 6.35\text{ kV} = m_1 \times 0.1\text{ MW}$$
$$\Rightarrow m_1 = \frac{0.381}{0.1} = 3.81\text{ kV/MW}$$

Since DER-2 rating is one-fourth of that of DER-1, its droop gain will be $m_2 =$ 15.42 kV/MW.

For the Q-f droop gains, the frequency deviation is restricted to be within ±0.5 Hz. In Example 10.3, the reactive power limits are chosen as $Q_1^* = 100$ kVAr and $Q_2^* = 25$kVAr. Using (10.27), when DER-1 is supplying its maximum rated reactive power, the following expression is obtained

$$0.5 = n_1 \times 0.05 \Rightarrow n_1 = 10\text{ Hz/MVAr}$$

Therefore, the droop gain for DER-2 is $n_2 = 40$ Hz/MVAr.

The feeder impedances are chosen as $L_{f1} = L_{f2} = 0.8$ mH and $R_{f1} = R_{f2} = 10\ \Omega$, i.e. the X/R ratio of both the feeders are 0.2513 : 10 (nearly 1 : 40) at 50 Hz frequency. A balanced RL load is chosen as $R_L = 672\ \Omega$ and $L_L = 0.8556$ H. The DERs are supplied through two DC sources, while the converters operate in the LQR state

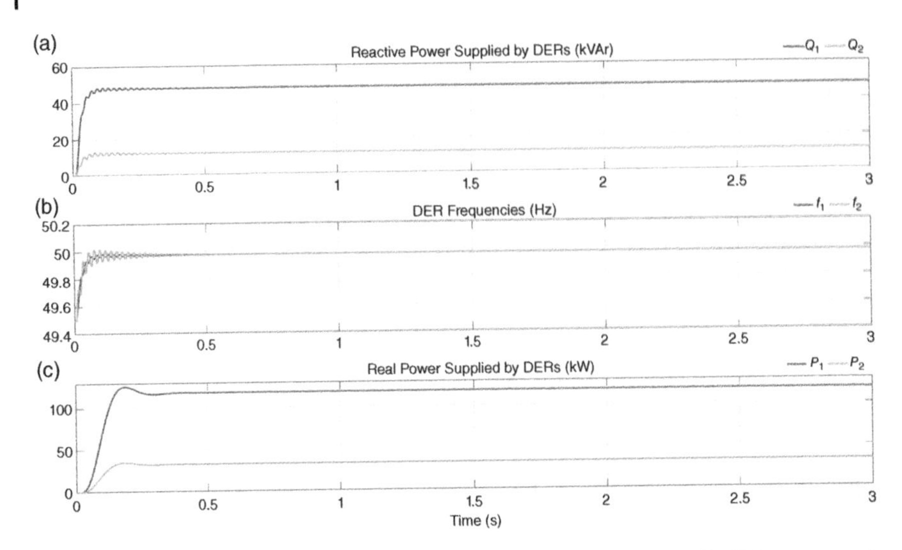

Figure 10.22 Microgrid operation results while operating with Q-f and P-V droop control. (a) Reactive power supplied by DERs (kVAr), (b) DER frequencies (Hz), (c) Real power supplied by DERs (kW).

Table 10.1 Steady state quantities obtained for the resistive grid.

DER-1	DER-2
$P_1 = 117.9$ kW	$P_2 = 32.6$ kW
$Q_1 = 47.68$ kVAr	$Q_2 = 11.92$ kVAr
$f_1 = 49.777$ Hz	$f_2 = 49.777$ Hz

feedback control mode. The results are shown in Figure 10.22. The steady state quantities are summarized in Table 10.1. Since the reactive power supplied by the DERs is nearly half their reactive power ratings, the operating frequency is near 50 Hz, exactly $f_1 = 50 - 10 \times (0.05 - 0.04768) = 49.777$ Hz. From this table, it is found that $Q_1 : Q_2 = 4$ and $P_1 : P_2 = 3.62$. Thus, while the reactive sharing is fairly accurate, the real power sharing does not meet the requirement.

10.4.4 Consideration of Line Impedances

In the two droop control methods discussed in Sections 10.4.2 and 10.4.3, the droop equations are derived by neglecting either the line resistance or the line reactance. This results in suboptimal performances, as evident from Examples 10.3 and 10.4.

One approach to include the line impedances in the droop equations is proposed in [8]. With respect to Figure 10.16, let us define

$$|Z|\angle\theta = R + jX \tag{10.29}$$

where $\theta = \tan^{-1}(X/R)$. An orthogonal rotational transformation matrix is defined as [8]

$$\begin{bmatrix} P' \\ Q' \end{bmatrix} = \mathbf{T}\begin{bmatrix} P \\ Q \end{bmatrix} \tag{10.30}$$

where

$$\mathbf{T} = \begin{bmatrix} \sin\theta & -\cos\theta \\ \cos\theta & \sin\theta \end{bmatrix} = \begin{bmatrix} \dfrac{X}{|Z|} & -\dfrac{R}{|Z|} \\ \dfrac{R}{|Z|} & \dfrac{X}{|Z|} \end{bmatrix}$$

The following observations can be made from (10.30):

- When $\dfrac{R}{X} = 0$ $(\theta = 90°)$, $P' = P$ and $Q' = Q$.
- When $\dfrac{R}{X} = 1$ $(\theta = 45°)$, $P' = \dfrac{P-Q}{\sqrt{2}}$ and $Q' = \dfrac{P+Q}{\sqrt{2}}$.
- When $\dfrac{R}{X} = \infty$ $(\theta = 0°)$, $P' = -Q$ and $Q' = P$.

The droop control equations are then [8]

$$\begin{aligned} f - f^* &= n \times (P'^* - P') \\ &= n\left[\frac{X}{|Z|}(P^* - P) - \frac{R}{|Z|}(Q^* - Q)\right] \end{aligned} \tag{10.31}$$

$$\begin{aligned} V - V^* &= m \times (Q'^* - Q') \\ &= m\left[\frac{R}{|Z|}(P^* - P) + \frac{X}{|Z|}(Q^* - Q)\right] \end{aligned} \tag{10.32}$$

Example 10.5 Consider the same system discussed in Example 10.4. The feeder impedances for this case are chosen as $L_{f1} = L_{f2} = L = 0.8$ mH and $R_{f1} = R_{f2} = R = 1\ \Omega$, i.e. the R/X ratio is 3.99 at 50 Hz frequency. Then, $X = 100\pi \times 0.0008 = 0.2513\ \Omega$ and $\theta = \tan^{-1}(0.2513) = 14.1°$. Furthermore, $|Z| = \sqrt{0.2513^2 + 1} = 1.031\ \Omega$, and therefore $R/|Z| = 0.97$ and $X/|Z| = 0.244$. The droop gains are chosen as

$$n_1 = 0.1, \quad n_2 = 0.4, \quad m_1 = 4, \quad m_2 = 16$$

The results are shown in Figure 10.23. The DER-1 supplies 135.41 kW of real power and 54.18 kVAr of reactive power, while the real and reactive power

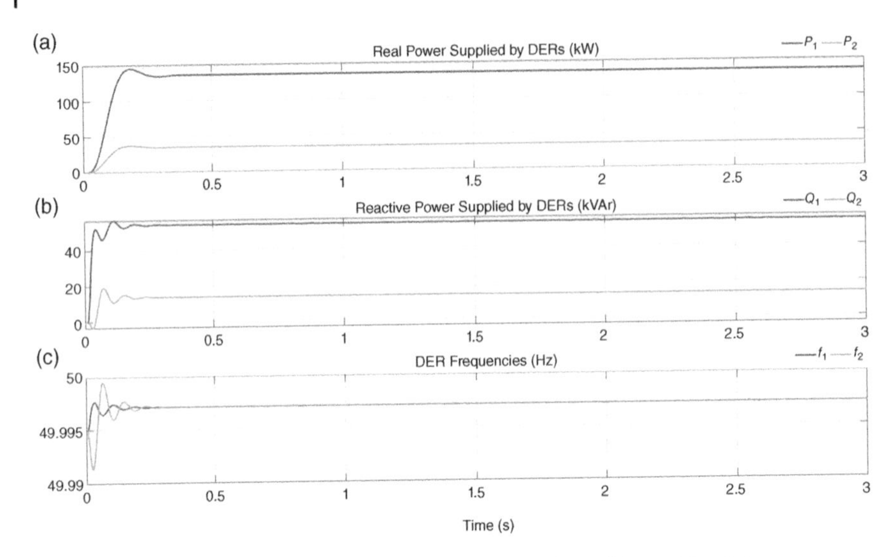

Figure 10.23 Microgrid operation results while operating with combined droop considering line parameters of Example 10.5. (a) Real power supplied by DERs (kW), (b) Reactive power supplied by DERs (kVAr), (c) DER frequencies (Hz).

supplied by DER-4 respectively are 34.44 kW and 13.71 kVAr. Both the real and reactive power sharing ratios are (nearly) 4 : 1. The frequency of the DERs are 49.997 Hz.

As demonstrated in Example 10.5, this method accommodates the effects of line impedance and, because of this, both real and reactive power sharing ratios are fairly accurate. In this example, it is assumed that the feeder impedances are known a priori. However, this assumption is not valid in real-life situations since the load will not be lumped and placed in a single location but will be distributed throughout the network. Therefore, a change in some of the loads will alter the Thévenin equivalence of the feeder impedance. Moreover, in the case of plug-and-play DERs, a DER coming online or going offline will change the Thévenin impedance. Thus, for the method to be successfully implemented, a parameter estimation algorithm and communication network will be required. In [9], a method is proposed with angle droop which can facilitate accurate power sharing in highly resistive lines, however, using a communication network.

10.4.5 Virtual Impedance

It is evident from the discussions in this chapter that the power sharing ratio is heavily dependent on the R/X ratio of the distribution feeders. It is not possible to change the feeder impedance and therefore alternate means will be required to counter the influence of the X/R ratios. One way of overcoming the effect of higher resistance is to use a large output inductor in an LCL filter. This, however, is not desirable, since it may cause unnecessary voltage drops and degrade power factors. An alternative to this is the use of a virtual impedance, which was first proposed in [10] and subsequently appeared in several other publications. For a detailed explanation of this method, see [10–12].

The main idea here is to use the droop equation discussed in Sections 10.4.2 and 10.4.3. However, the voltage reference is then modified by subtracting the output current multiplied by the virtual impedance. Consider, for example, the microgrid containing two DERs of Figure 10.19. The voltage references generated through the droop equation for DERs are given in (10.22). These are then modified through negative feedback of the output currents. For DER-1, the modified voltage references will be

$$v_{refk1} = v_{k1}^* - Z_V i_{k1}, \quad k = a, b, c \tag{10.33}$$

A similar expression can also be written for DER-2. Therefore, without adding any physical hardware that may cause the power or voltage to drop, the effect of an output impedance is introduced in the feedback loop. Let us now consider Examples 10.6 and 10.7.

Example 10.6 In this example, the effect of choosing virtual resistors in predominantly resistive lines will be demonstrated. For this, the same P-f droop gains, as in Example 10.3 (i.e. $n_1 = 5$ Hz/MW and $n_2 = 20$ Hz/MW), are selected, with the same DER power ratings. The feeder impedances are $L_{f1} = 1.9$ mH, $R_{f1} = 3\ \Omega$, $L_{f2} = 6.4$ mH, and $R_{f2} = 10\ \Omega$. The R/X ratios of both feeders at 50 Hz are about 5. It is obvious that these are highly resistive feeders, where a P-f droop will not work properly. A balanced RL load is chosen with $R_L = 672\ \Omega$ and $L_L = 0.4265$ H. The Q-V droop gains are chosen as $m_1 = 10$ V/MVAr and $m_2 = 40$ V/MVAr. In order to counter the highly resistive line, the effect of the feeder resistances is nullified using two virtual resistors, which are chosen as $Z_{V1} = R_{V1} = 5\ \Omega$ and $Z_{V2} = R_{V2} = 20\ \Omega$.

The results are shown in Figure 10.24. The steady state DER output powers are $P_1 = 137.5$ kW and $P_2 = 34.38$ kW, which are in the ratio of 4 : 1. The corresponding frequency in the steady state is 49.81 Hz, which obeys the droop Eq. (10.19). The reactive powers, however, do not obey the droop relations. In fact, Q_1 is less than Q_2 in this case since the Feeder-2 impedance is higher than that of Feeder-1.

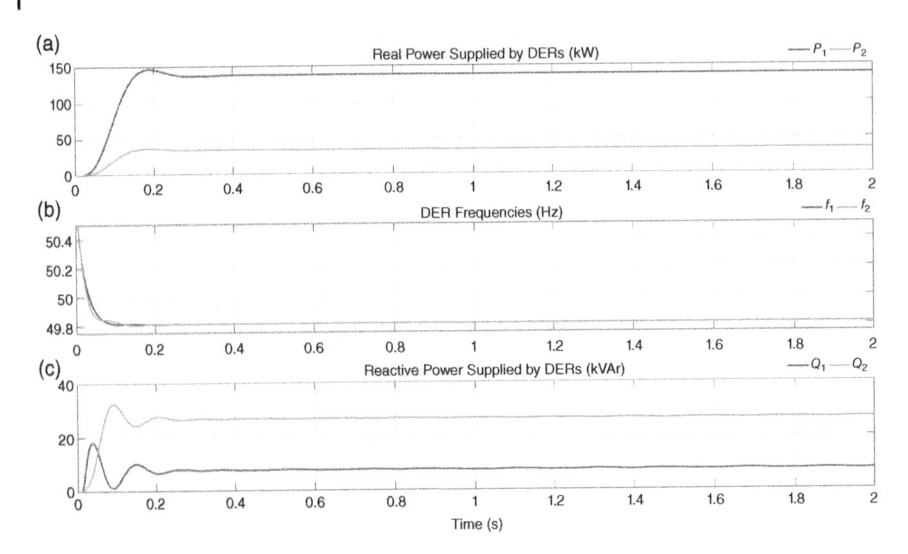

Figure 10.24 Microgrid operation results while operating with P-f and Q-V droop control with virtual resistance. (a) Real power supplied by DERs (kW), (b) DER frequencies (Hz), (c) Reactive power supplied by DERs (kVAr).

Example 10.7 For this example, the predominantly resistive circuit of Example 10.4 is chosen with the same line and load parameters. The Q-f and P-V droop parameters remain the same as those calculated in Example 10.4, which shows that even though the reactive power sharing was accurate the real power sharing was not acceptable. To counter the effect of the line resistance, a virtual inductor is now chosen such that (10.33) is rewritten as

$$v_{refi} = v_i^* - L_{Vi}\frac{di_i}{dt}, \quad i = 1, 2$$

Often the derivative part is not implemented in the form shown in the above equation. An ideal derivative action can generate spikes every time the set point changes, as is pointed out in (3.32) of Chapter 3. Instead, the derivative action in the above equation is modified as

$$L_{Vi}\frac{di_i}{dt} = L_{Vi}\frac{Ns}{s+N}i_i, \quad i = 1, 2$$

The virtual inductor parameters are chosen as $L_{V1} = 50$ mH, $L_{V2} = 550$ mH, and $N = 100$. The results are shown in Figure 10.25. The values of the reactive powers and the frequencies are identical to those listed in Table 10.1. However, the real power sharing has improved considerably. These are given by $P_1 = 119.2$ kW and $P_2 = 29.5$ kW. They have a power sharing ratio of 4 : 1, as desired.

These two examples, which present the merits of including the virtual impedance, can be very effective in improving power sharing in an islanded microgrid.

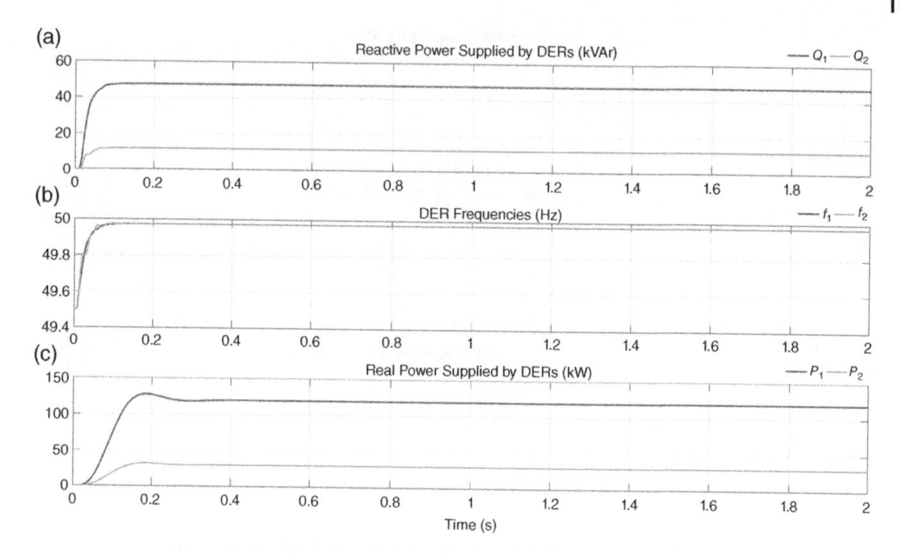

Figure 10.25 Microgrid operation results while operating with Q-f and P-V droop control with virtual inductance. (a) Reactive power supplied by DERs (kVAr), (b) DER frequencies (Hz), (c) Real power supplied by DERs (kW).

These impedances, however, can lead to system instability. The design aspects are discussed in [10–12]. A systematic study of an interconnected microgrid will be required, which may provide guidelines for the selection of these impedances. The eigenvalue analysis of an inverter-based microgrid is presented in [13]. This paper can be used as a guideline for checking the limits of these impedances for a stable operation of microgrids.

10.4.6 Inclusion of Nondispatchable Sources

A microgrid, be it islanded or grid-connected, can have different generation types. Usually, nondispatchable sources like wind or solar PV operate in a maximum power tracking mode. A solar PV, when connected to an islanded microgrid, will inject active power in the grid feeding mode. This can be taken as a negative load and the dispatchable generators can supply the rest of the load in the droop control mode, as illustrated by Example 10.8.

Example 10.8 Let us consider the inductive grid of Example 10.3, except that a grid feeding converter is connected to the system through another feeder that is placed to the right of DER-2 in Figure 10.19. The impedance parameters of this feeder are 77 mH and 2.42 Ω. With the system operating in steady state, the grid feeding inverter is connected at 1 second. The power that the grid feeding converter injects varies at discrete intervals of time. The results are shown in Figure 10.26, and these are summarized in Table 10.2, where the power injected

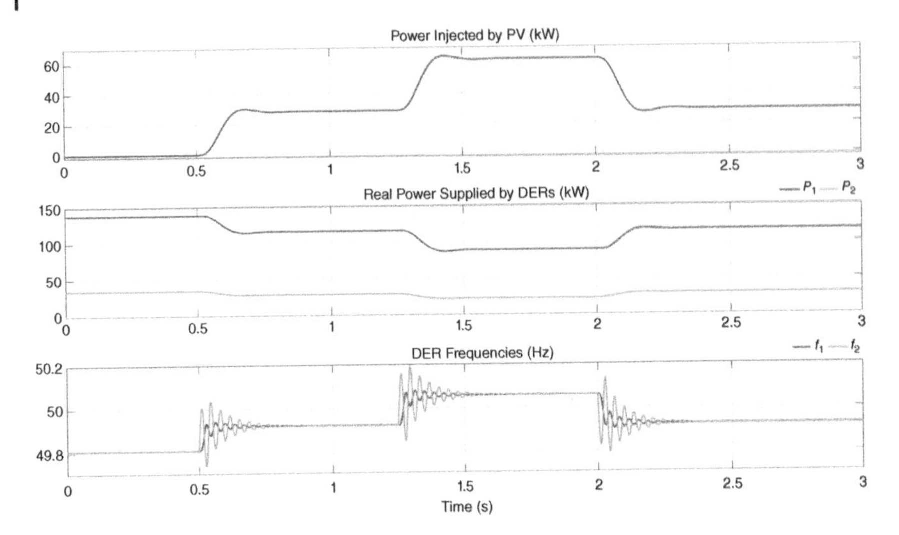

Figure 10.26 Operation of islanded microgrid with a grid feeding converter.

Table 10.2 Steady state quantities obtained in Example 10.8.

Time t (s)	P_{PV} (kW)	P_1 (kW)	P_2 (kW)	Frequency (Hz)	$P_1 : P_2$
$0 < t < 0.5$	0	138.5	34.62	49.807	4 : 1
$0.5 \leq t < 1.25$	28.4	115.65	28.91	49.92	4 : 1
$1.25 \leq t < 2$	62.1	88.6	22.15	50.057	4 : 1
$t \geq 2$	28.4	115.65	28.91	49.92	4 : 1

by the grid feeding converter is denoted by P_{PV}. From these results, it is obvious that the inclusion of the nondispatchable source does not affect the droop sharing of the dispatchable DERs.

10.4.7 Angle Droop Control

The angle droop control is suitable for a predominantly reactive grid, even though it is feasible to implement this on grids with high R/X ratios using communication networks [9]. Consider the two-DER microgrid system of Figure 10.19, which is redrawn in Figure 10.27. In this figure, it is assumed that both the DERs are equipped with output LCL filters. The inductances L_1 and L_2 represent the outer inductances of the LCL filters. However, the DERs still operate in voltage control

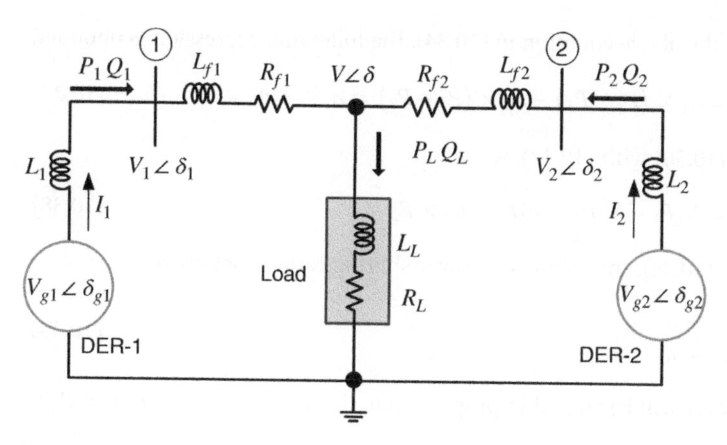

Figure 10.27 Microgrid structure for angle droop control.

mode in which they control the voltage across the filter capacitors. These capacitor voltages are denoted by $V_{g1}\angle\delta_{g1}$ and $V_{g2}\angle\delta_{g2}$ in Figure 10.27. The outer inductors L_1 and L_2 are used for real and reactive power flow control, as will be explained in this section. The load bus voltage is denoted by $V\angle\delta$.

It is shown in (10.17) that the real power is directly proportional to the angle difference between two AC sources. Therefore, the droop equations for DERs are given by

$$\delta_i = \delta^* + n_i \times (P_i^* - P_i), \quad i = 1, 2 \tag{10.34}$$

$$V_i = V^* + m_i \times (Q_i^* - Q_i), \quad i = 1, 2 \tag{10.35}$$

where δ^* is the reference angle and V_i and δ_i define the bus voltage magnitude and its angle, as indicated in Figure 10.27. Assuming that the feeder resistances are negligible and applying a DC load flow, the following expressions are obtained for the microgrid of Figure 10.27 [6–7].

$$\begin{aligned} \delta_1 - \delta &= X_1 P_1 \\ \delta_2 - \delta &= X_2 P_2 \end{aligned} \tag{10.36}$$

where $X_1 = \dfrac{\omega L_{f1}}{VV_1}$ and $X_2 = \dfrac{\omega L_{f2}}{VV_2}$ ω being the fundamental frequency in rad/s.

In an AC system, the power flow depends on the relative angle difference, and therefore the reference angle δ^* can be taken as 0. However, all the DERs in a microgrid must measure their angles with respect to this reference angle, and hence a global clock will be required for synchronizing all the units. The angle droop gains must be chosen as per (10.21), hence

$$P_1{}^* n_1 = P_2{}^* n_2$$

Substituting the above equation in (10.34), the following expression is obtained

$$\delta_1 - \delta_2 = n_1 \times (P_1^* - P_1) - n_2 \times (P_2^* - P_2) = n_1 P_1 - n_2 \times P_2 \tag{10.37}$$

Comparing (10.36) with (10.37), we get

$$\delta_1 - \delta_2 = X_1 P_1 - X_2 P_2 = n_1 P_1 - n_2 \times P_2 \tag{10.38}$$

Rearranging (10.38), the following power sharing ratio is obtained

$$\frac{P_1}{P_2} = \frac{X_1 + n_1}{X_2 + n_2} \tag{10.39}$$

Then the power will be shared in proportion to the droop gains, i.e. $P_1 n_1 = P_2 n_2$ provided that

$$n_1 >> X_1 \text{ and } n_2 >> X_2$$

The droop Eqs. (10.34) and (10.35) will produce the bus voltage magnitudes and their angles. From these quantities, the references for the filter capacitor voltages need to be calculated. Consider, for example, DER-1. From Figure 10.27, the current I_1 flowing from DER-1 to Bus-1 is given by

$$I_1 = \frac{V_{g1}\angle\delta_{g1} - V_1\angle\delta_1}{j\omega L_1}$$

Then the complex power injected by the DER to the microgrid bus is given by

$$P_1 + jQ_1 = V_1\angle\delta_1 \left[\frac{V_{g1}\angle-\delta_{g1} - V_1\angle-\delta_1}{-j\omega L_1}\right] = \frac{V_{g1}V_1\angle(\delta_1 - \delta_{g1}) - V_1^2}{-j\omega L_1} \tag{10.40}$$

Separating the real and imaginary components, the real and reactive powers injected in Bus-1 are given by

$$P_1 = \frac{V_{g1}V_1 \sin(\delta_1 - \delta_{g1})}{\omega L_1} \tag{10.41}$$

$$Q_1 = \frac{V_{g1}V_1 \cos(\delta_1 - \delta_{g1}) - V_1^2}{\omega L_1} \tag{10.42}$$

Equations (10.41) and (10.42) can be rewritten as

$$\sin(\delta_1 - \delta_{g1}) = \frac{\omega L_1 P_1}{V_{g1}V_1}$$

$$\cos(\delta_1 - \delta_{g1}) = \frac{\omega L_1 Q_1 + V_1^2}{V_{g1}V_1}$$

From these expressions, the angle of the DER voltage is computed as

$$\delta_{g1} = \delta_1 - \tan^{-1}\left(\frac{\omega L_1 P_1}{\omega L_1 Q_1 + V_1^2}\right) \tag{10.43}$$

In (10.43), the value of ωL_1 is known a priori. The Bus-1 voltage V_1 and its angle δ_1 are computed from the droop equation. Furthermore, P_1 and Q_1 are measured. Therefore, knowing these quantities, the angle δ_{g1} is calculated from (10.43). Once this is obtained, the DER reference voltages for the three phases are decided as

$$\begin{aligned} v_{g1a} &= V_{m1} \sin\left(\omega t + \delta_{g1}\right) \\ v_{g1b} &= V_{m1} \sin\left(\omega t + \delta_{g1} - 120^\circ\right) \\ v_{g1c} &= V_{m1} \sin\left(\omega t + \delta_{g1} + 120^\circ\right) \end{aligned} \tag{10.44}$$

where V_{m1} is computed from (10.41) as

$$V_{m1} = \sqrt{2}\frac{\omega L_1 P_1}{V_1 \sin\left(\delta_1 - \delta_{g1}\right)}$$

In a similar way, the DER-2 voltage and angle can also be derived.

Example 10.9 Consider the predominantly inductive system of Example 10.3, with the same DER real and reactive power ratings. The other system parameters chosen are

$$V^* = 6.35\,\text{kV}, \delta^* = 0^\circ, L_1 = L_2 = 10\,\text{mH}$$

$$n_1 = 1^\circ/\text{MW}, n_4 = 4^\circ/\text{MW}, m_1 = 0.05\text{kV/MVAr, and } m_2 = 0.2\text{kV/MVAr}$$

The results are shown in Figures 10.28 and 10.29. Figure 10.28 shows the real and reactive powers supplied by the DERs, which are $P_1 = 137.5$ kW, $P_2 =$ 39.4 kW, $Q_1 = 50$ kVAr, and $Q_2 = 38.8$ kVAr. The active power sharing ratio 3.5 : 1 is not very accurate. The accuracy can be improved by increasing the droop gains. This might, however, make the system unstable and auxiliary controllers may be required for stabilization [6]. The bus voltages and their angles are shown in Figure 10.29.

In this section, the islanded operation of microgrids is discussed, where it is assumed that all the DERs operate in dispatchable mode and all of them are interfaced with the microgrid through VSCs. However, even rotary generators like biofuel-based generators can easily be integrated to a microgrid through a frequency droop equation. Hydrogen fuel cells and microturbines are dispatchable. Solar PVs and wind turbines can also be dispatchable if they are not operated in a maximum power point tracking mode. Furthermore, renewable generators with battery ESSs can also act like a dispatchable generator. For example, consider

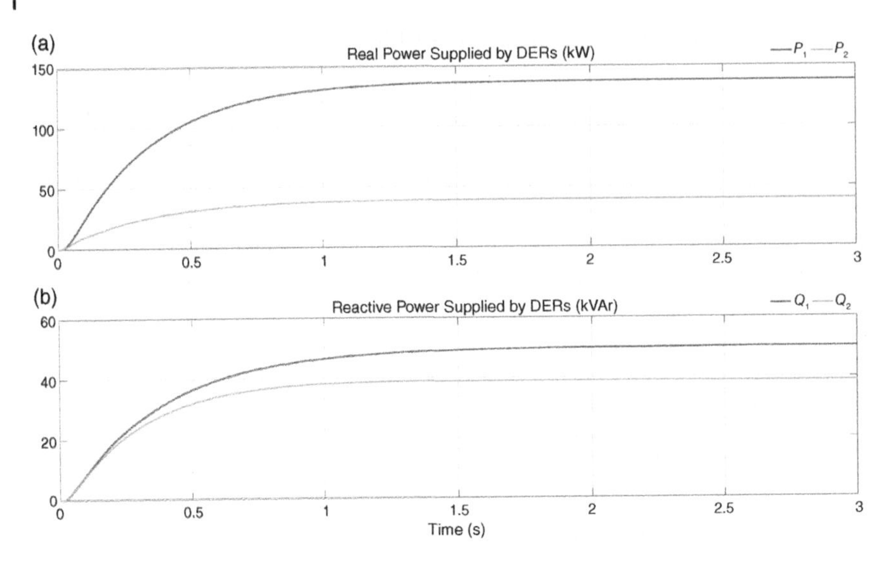

Figure 10.28 (a) Real and (b) reactive power supplied by the DERs, when operated under angle droop control.

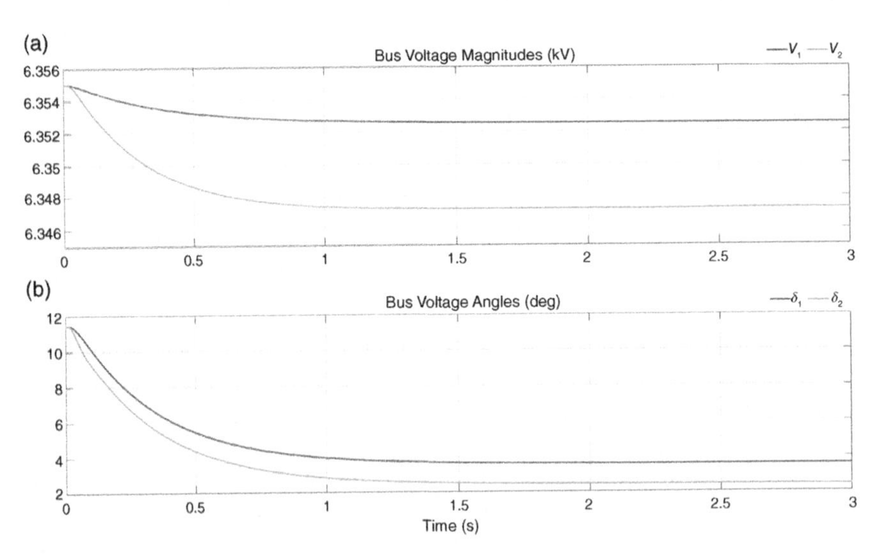

Figure 10.29 Bus voltage (a) magnitude and (b) angle when the microgrid operates under angle droop control.

the case in which a solar PV and battery are connected to the same bus, and they work in tandem. Then, during the sunlight hours, the PV–battery combination can supply a fixed amount of power in most of the time. If the PV generation is high, it can charge the battery while maintaining the rated output power constant as well. On the other hand, when the PV generation is low, the battery can discharge, thereby maintaining the rated output power constant. The battery storages can also act like a dispatchable source if they are fully charged, especially in the evenings. To take into account these variabilities, microgrid planning becomes an important aspect that needs to be considered for the successful implementation and operation of a microgrid.

10.5 Grid-connected Operation of Microgrid

Consider the system shown in Figure 10.30, in which a microgrid is connected to a utility substation through a feeder of impedance R_S - L_S. There is a circuit breaker CB placed at the PCC that connects/disconnects the utility from the microgrid. There are two modes of operation: (i) grid-connected mode and (ii) islanded mode. The DERs are equipped with output LC filters and will operate in voltage control mode. The main aim is to compute the reference voltages depending on the mode of operation of these converters. These two modes are defined as:

- Islanded mode: The DERs will share power according to their ratings under a suitable droop control regime.

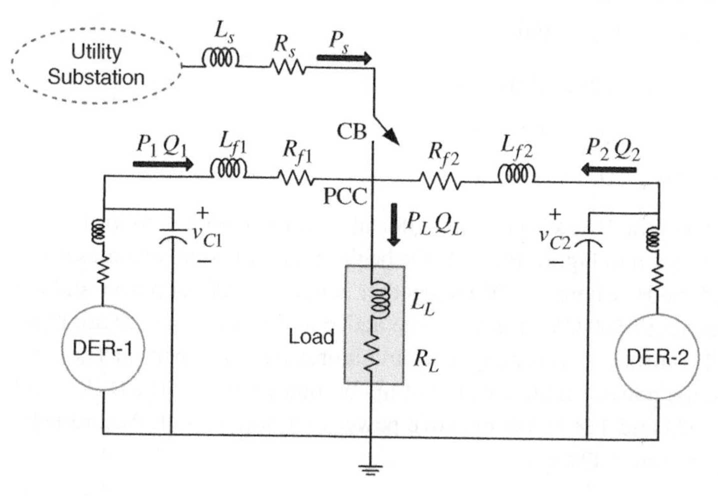

Figure 10.30 Schematic diagram of a utility connected microgrid.

- Grid-connected mode: The DERs operate as voltage-controlled grid feeding converters. They will supply fixed amounts of real and reactive power to the local load. The rest of the power requirement will come from the utility system.

Let us denote the real and reactive power references of DER-1 by P_1^* and Q_1^* respectively. Also, the DER-1 output voltage is defined as

$$v_{C1} = V_{m1} \sin(\omega t + \delta_1) \tag{10.45}$$

where ω is the frequency of the grid, synchronized through a PLL. The angle δ_1 should be so adjusted that the required amount of active power P_1^* flows out of the DER. Similarly, the voltage magnitude V_{m1} is adjusted through reactive power feedback. These are accomplished through two PI controllers, given by

$$\delta_1 = K_{P\delta 1}\left(P_1^* - P_1\right) + K_{I\delta 1}\int\left(P_1^* - P_1\right)dt \tag{10.46}$$

$$V_{m1} = K_{PV1}\left(Q_1^* - Q_1\right) + K_{IV1}\int\left(Q_1^* - Q_1\right)dt \tag{10.47}$$

In the same manner, the real and reactive powers of DER-2 are also controlled in the grid-connected mode.

Example 10.10 The microgrid of Figure 10.30 is operated with the parameters given in Table 10.3. In the islanded mode, the microgrid is operated in P-f and Q-V droop control mode. For the grid-connected mode, the PI controller parameters chosen for both the DERs are

$$K_{P\delta} = 0.001, \quad K_{I\delta} = 10$$
$$K_{PV} = 0.001, \quad K_{IV} = 100$$

Also, the power references chosen are

$$P_1^* = 300\text{ kW}, \quad Q_1^* = 150\text{ kVAr}$$
$$P_2^* = 100\text{ kW}, \quad Q_2^* = 50\text{ kVAr}$$

Both the VSCs of the DERs operate in LQR state feedback voltage control mode. The results are shown in Figure 10.31. At the beginning, the system operates in the grid-connected mode, where the DERs together supply 400 kW of power and the utility supplies about 500 kW of power. The active and reactive power are regulated by the PI controllers accurately. The circuit breaker (CB) opens at 1 second. Once it opens, the power supplied by the utility becomes zero and the DER-1 and DER-2 supply 675 and 168.75 kW of active power in a ratio of 4 : 1. Accordingly, the frequency becomes 49.83 Hz.

Table 10.3 System parameters chosen for Example 10.8.

Quantities	Parameters
Utility	
Voltage	11 kV (L-L)
Frequency	50 Hz
Feeder	0.605 Ω and 19.3 mH
DER-1	
Power references	$P_1^* = 1$ MW and $Q_1^* = 600$ kVAr
Droop gains	$n_1 = 1$ Hz/MW and $m_1 = 0.54$ kV/MVAr
Filter parameters	3.3 mH, 0.1 Ω and 50 μF
DER-2	
Power references	$P_2^* = 250$ kW and $Q_1^* = 150$ kVAr
Droop gains	$n_4 = 4$ Hz/MW and $m_2 = 2.16$ kV/MVAr
Filter parameters	3.3 mH, 0.1 Ω, and 50 μF
Microgrid feeder and load parameters	
Feeder-1	$R_{f1} = 3$ Ω and $L_{f1} = 57.8$ mH
Feeder-2	$R_{f2} = 6$ Ω and $L_{f2} = 115.6$ mH
Balanced RL load	120 Ω and 125.5 mH

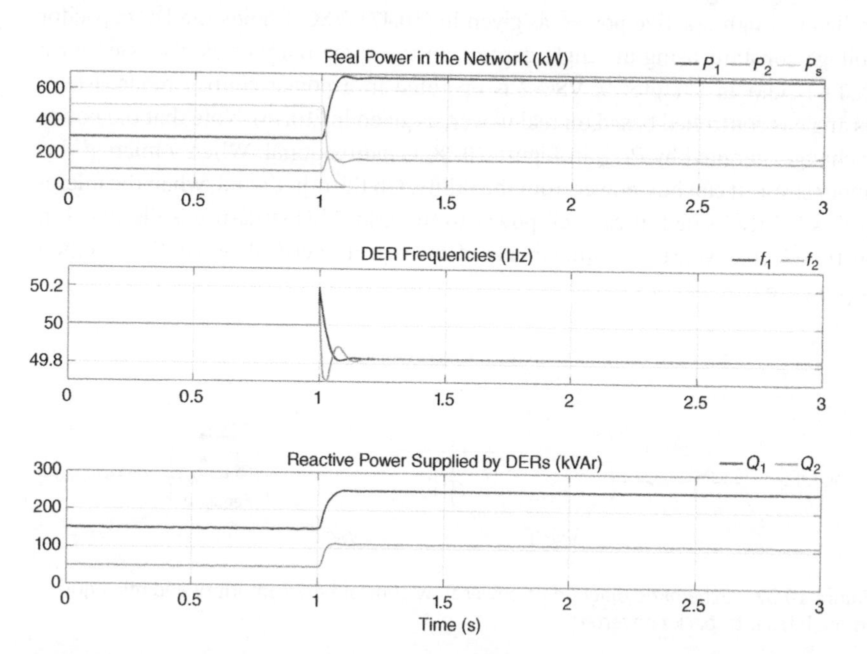

Figure 10.31 Operation of a microgrid, when connected to a utility and subsequently islanded.

Resynchronization of microgrids with the utility network is an active area of research. In general, an AC source can be synchronized with another AC system, when they have the same voltage, the same frequency, the same phase sequence, and the same phase angle. In undergraduate laboratories, this synchronization is performed by using synchroscopes and three lamps. However, this process must be automated for microgrid synchronization. In [14], the grid synchronization is performed through an intelligent connection agent (ICA), which consists of a switch and a grid-connected VSC. The VSC is fed from an external DC source that acts as an energy storage unit. The synchronization is performed through a PLL that obtains the phase angle of the grid voltage. There are other aspects that need to be considered before the microgrid can be synchronized with the utility grid. As shown in Example 10.10, the microgrid frequency becomes 49.83 Hz in the islanded mode. However, to bring the frequency back to the grid frequency before the synchronization can occur, an isochronous controller can be used. In [15], a wind energy integration with an islanded microgrid is presented, where the droop line is shifted using the isochronous action before the system can be integrated with the wind energy system. In [16], an interconnecting switch is used for coupling two microgrids when they operate in different droop control regimes.

A scheme for controlled power flow between a utility system and a microgrid is proposed in [7]. This employs a set of back-to-back VSCs, as shown in Figure 10.32. The voltage magnitudes of the VSCs are assumed to be specified, or can be controlled through reactive power, as given in (10.47). VSC-1 holds the DC capacitor voltage constant using an angle control, in the same manner as discussed for a DSTATCOM in Chapter 9. VSC-2 is operated in a power control mode, where its angle is controlled based on real power, as given in (10.46). Note that the power exchange, denoted by P_{Link} in Figure 10.28, is bidirectional. When a microgrid is under stress, it can buy power from the utility. On the other hand, when the microgrid is lightly loaded, it can sell power to the grid. This structure is discussed in Section 10.11, where a frequency-based microgrid overload prevention method is presented.

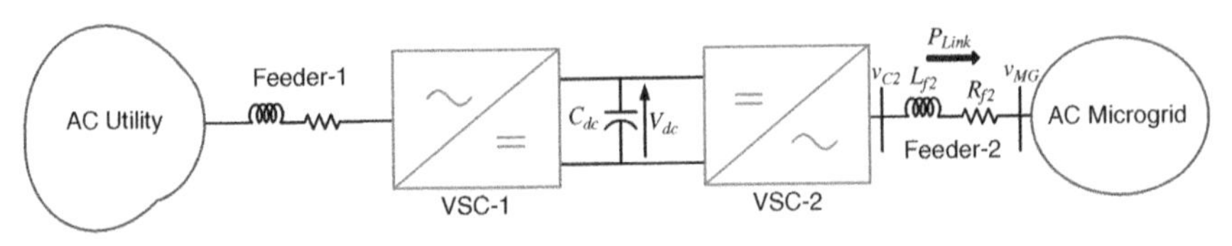

Figure 10.32 Schematic diagram of power flow control between utility and microgrid through back-to-back converters.

10.6 DC Microgrids

Many electrical loads, such as LED lighting, adjustable speed motors, electric vehicles, computing, and communication equipment need DC supply. At the same time, many DERs – like PV, batteries, and fuel cells – produce power at the DC voltage level, which is then converted into AC for grid connection. Moreover, there is potential to directly connect a wind turbine or microturbine to a DC grid. Therefore, there is renewed interest in DC grids at the distribution level as they reduce conversion losses significantly. Voltage transformation from one level to another in a DC grid can be achieved through DC-DC converters, which have more than a 95% efficiency.

The basic building block of a typical DC microgrid (DCMG) is shown in Figure 10.33. The DCMG is connected to the utility system through a transformer and an interlinking AC-DC converter. Renewable energy sources, like wind or solar PV, can be directly connected to the DC bus of the microgrid. The DCMG can also have BESS units. Usually, such units operate with constant output voltages. However, there may be variations in the output voltages depending on the battery chemistry, current, ambient temperature, and state of charge (SoC). The direct connections of a battery to the DC bus can result in fluctuations in

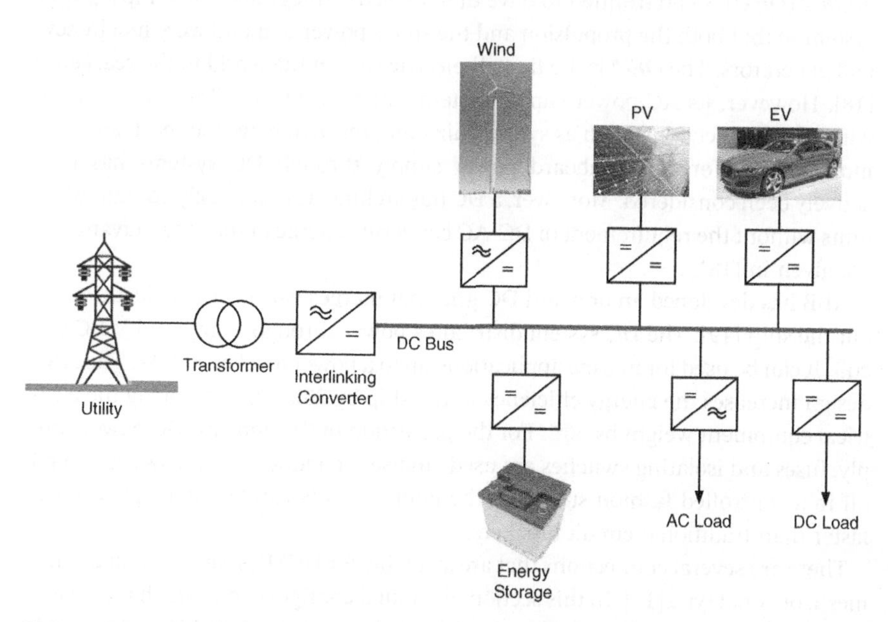

Figure 10.33 The basic building blocks of a DCMG system. *Source:* foxbat / Shutterstock.com and Tesla / flickr.

the bus voltage and inrush current, thereby shortening the lifetime of the battery [17]. Such fluctuations can affect the DC bus voltage as well. Thus, these battery storage units are connected through DC-DC converters, as shown in Figure 10.33.

There are advantages and disadvantages of DC systems. Some of the advantages are [17]:

- There is no skin effect in a DC system. Therefore, the current can flow through the entire cable and not just the outer edges. This reduces losses and makes it possible to use smaller cables for the same amount of current.
- The problem associated with the synchronization with renewable sources with AC grids does not exist in DC systems.
- There is no requirement for reactive power control.

Some of the disadvantages of the DC systems are [17]:

- There is no inherent current zero crossing in DC systems. Therefore, the protection of a DC system is more difficult than that of an AC system.
- DC systems will require new standards for products and voltage levels.
- DC system grounding and corrosion issues need to be investigated and resolved.

One of the interesting and important applications of DCMG that is emerging these days is in the maritime onboard power supply. The cruise ship *Queen Elizabeth 2* (*QE-2*) was retrofitted to have diesel-electric integrated with ship's power system so that both the propulsion and the ship's power demand were met by several generators. The *QE-2* is the first all-electric ship in the world in the real sense [18]. However, its AC power supply system suffers from the ailments associated with such connections, such as synchronization, reactive power support, and harmonics. Therefore, the onboard power supply through DC systems has now actively been considered. Moreover, a DC bus architecture can easily contain BESS units without the requirement of DC-AC converters. Some of the other advantages are given in [18].

ABB has developed an onboard DC grid that merges various DC links throughout the ship [19]. The DC system distributes power through a single 1 kV DC circuit. It can be used for marine applications up to a power level of 20 MW. The ABB design increases the energy efficiency of the ship by 20% while reducing the electrical equipment weight by 30%. For the protection of the onboard DC power supply, fuses and isolating switches are used, and semiconductor switches are turned off in a controlled fashion such that the fault currents can be interrupted much faster than traditional circuit breakers.

There are several connections that are available for DCMGs, such as radial, ring, mesh, or zonal type [17]. In this section, the radial configuration, which is the simplest structure, is considered. First, the islanded operation of DCMG is discussed, where two different droop control structures are presented.

10.6.1 P-V Droop Control

As in the case of an AC microgrid, the DCMG, when operating in the standalone (islanded) mode, must manage its own load. This implies that all the DGs in the microgrid must supply the local load demand. Consider the schematic diagram of a DCMG shown in Figure 10.34. It contains two DC-DC converters. Converter-1 and 2 are supplied by DGs, which are represented by voltage sources. These converters supply a load R_L. The output voltages of Conveters-1 and 2 are denoted by V_1 and V_2 respectively, while the load voltage is denoted by V_L. The feeder resistances are represented by R_{f1} and R_{f2}.

In the P-V droop control scheme, the desired output voltages of the DC-DC converters are generated from the power flowing out of them. This is given by

$$\begin{aligned} V_1^* &= V^* - n_1 P_1 \\ V_2^* &= V^* - n_2 P_2 \end{aligned} \tag{10.48}$$

where $P_1 = V_1 \times I_1$, $P_2 = V_2 \times I_2$, V^* is the reference voltage for the microgrid, and n_1 and n_2 are droop gains. The DC-DC converters, through their duty ratio control actions, need to follow the references generated such that $V_1 \approx V_1^*$ and $V_2 \approx V_2^*$. Equation (10.48) can be written in terms of currents as

$$\begin{aligned} V_1^* &= V^* - n_1 V_1 I_1 \\ V_2^* &= V^* - n_2 V_2 I_2 \end{aligned} \tag{10.49}$$

If the DC-DC converters can reproduce the desired reference voltage output accurately, then Kirchhoff's voltage law at the load bus in Figure 10.30 will produce

$$V_L = V_1 - R_{f1} I_1 = V_2 - R_{f2} I_2 \tag{10.50}$$

Substitution of (10.49) in (10.50) yields

$$V^* - n_1 V_1 I_1 - R_{f1} I_1 = V^* - n_2 V_2 I_2 - R_{f2} I_2$$

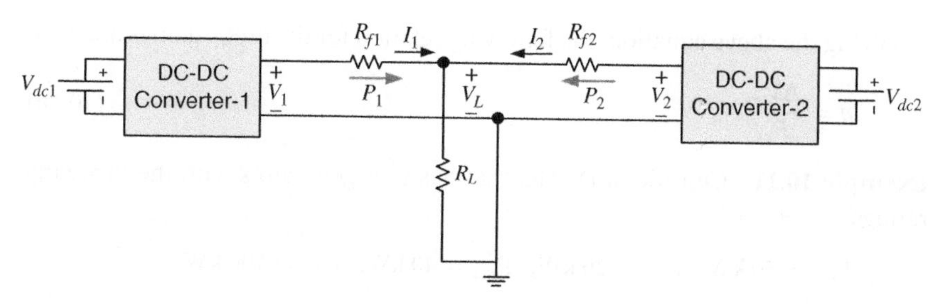

Figure 10.34 Schematic diagram of a simple DCMG.

This implies that

$$\frac{I_1}{I_2} = \frac{n_2V_2 + R_{f2}}{n_1V_1 + R_{f1}} \tag{10.51}$$

Under the assumptions

$$n_1V_1 >> R_{f1}, \quad n_2V_2 >> R_{f2} \tag{10.52}$$

(10.51) becomes

$$\frac{I_1}{I_2} \cong \frac{n_2V_2}{n_1V_1} \tag{10.53}$$

Therefore, the ratio of the output powers is given by

$$\frac{P_1}{P_2} = \frac{V_1I_1}{V_2I_2} = \frac{n_2}{n_1} \tag{10.54}$$

Note that the above relation is the same as that obtained for an AC microgrid. If a DCMG contains M DGs, with output powers of P_1, P_2, ..., P_M and with droop gains of n_1, n_2, ..., n_M, the droop sharing ratios will be given by

$$P_1 \times n_1 = P_2 \times n_2 = \cdots = P_M \times n_M = \alpha \tag{10.55}$$

Here we outline a procedure for the selection of droop gain selection. This procedure will produce rating-based power sharing so long as the assumptions of (10.52) are not violated.

The droop gain selection for DCMGs depends on the maximum available power from a DG and the maximum allowable voltage drop at the output of the converter that can be allowed. Let the maximum available power be denoted by P_m, while the maximum allowable voltage drops at the output of the converter be denoted by ΔV_m. Then, noting from (10.48) that the maximum voltage drop occurs when the converter is supplying the maximum power, the following expression can be written for each converter

$$V - V^* = -\Delta V_m = -nP_m$$

Solving the above equation, the following relation for the droop gain is obtained

$$n = \frac{\Delta V_m}{P_m} \text{ V/W} \tag{10.56}$$

Example 10.11 Consider a DCMG that has four generators with the following ratings

$$P_{m1} = 50\text{ kW}, \quad P_{m2} = 20\text{ kW}, \quad P_{m3} = 40\text{ kW}, \quad P_{m2} = 100\text{ kW}$$

Note that the standard for voltage level for DCMGs is currently emerging to be 380 V, and therefore we shall choose this as our reference voltage. Furthermore, the voltage drop will be restricted to be within 10% of the nominal voltage, i.e.

$$V^* = 380\,\text{V}, \quad \Delta V_m = 38\,\text{V}$$

Therefore, using (10.56), the following droop gains for the DGs are obtained

$$n_1 = \frac{38}{50} = 0.76\,\text{V/kW}, \quad n_2 = \frac{38}{20} = 1.9\,\text{V/kW},$$
$$n_3 = \frac{38}{40} = 0.95\,\text{V/kW}, \quad n_4 = \frac{38}{100} = 0.38\,\text{V/kW}$$

It can be easily verified that α in (10.55) is 38 V, i.e. equal to ΔV_m.

10.6.2 The Effect of Line Resistances

Assuming that the voltage tracking by the DC-DC converters is perfect, (10.57) is obtained from (10.49)

$$V^* = V_1 + n_1 V_1 I_1 = V_2 + n_2 V_2 I_2 \tag{10.57}$$

Now, (10.53) can be rearranged as

$$I_1 = I_2 \frac{n_2 V_2}{n_1 V_1}$$

Therefore, the above two equations are combined to get

$$V_1 + n_1 V_1 I_1 = V_1 + n_1 V_1 I_2 \frac{n_2 V_2}{n_1 V_1} = V_1 + n_2 V_2 I_2 \tag{10.58}$$

Comparing (10.57) with (10.58), it can be seen that $V_1 = V_2$, only if the assumptions of (10.55) are true, and for $R_{f1} = R_{f2}$ and $P_{m1} = P_{m2}$. Since these conditions are not easily satisfied, we can at best have $V_1 \approx V_2$. However, the power sharing can be influenced by the line resistances and by ΔV_m.

To determine the actual power flow in the network, a set of nonlinear equations needs to be solved. There are four unknown quantities in the network: V_1, I_1, V_2, and I_2. Therefore, four equations are needed to solve the power flow equations. The first two of these are obtained from (10.49) as

$$\begin{aligned} g_1 &= V_1(1 + n_1 I_1) - V^* = 0 \\ g_2 &= V_2(1 + n_2 I_2) - V^* = 0 \end{aligned} \tag{10.59}$$

Equation (10.50) is rewritten as

$$g_3 = V_1 - R_{f1}I_1 - V_2 + R_{f2}I_2 = 0 \tag{10.60}$$

Kirchhoff's current law at the load bus of Figure 10.34 gives

$$\frac{V_L}{R_L} = I_1 + I_2$$

The load bus voltage V_L can be written from (10.50) as $V_L = V_1 - R_{f1}I_1$. Therefore, substituting this voltage in the above equation, the final equation is obtained as

$$g_4 = I_1 + I_2 - \frac{V_1 - R_{f1}I_1}{R_L} = 0 \tag{10.61}$$

To solve these four equations, the following two vectors are formed – one for the four unknowns and the other for the four functions

$$\mathbf{x} = [V_1 \ \ V_2 \ \ I_1 \ \ I_2]^T \text{ and } \mathbf{g} = [g_1 \quad g_2 \quad g_3 \quad g_4]^T$$

Then, the Newton–Raphson method is used to solve (10.59) to (10.61). The first step is to form a Jacobian matrix, which is given by

$$\mathbf{J} = \begin{bmatrix} \frac{\partial g_1}{\partial V_1} & \frac{\partial g_1}{\partial V_2} & \frac{\partial g_1}{\partial I_1} & \frac{\partial g_1}{\partial I_2} \\ \frac{\partial g_2}{\partial V_1} & \frac{\partial g_2}{\partial V_2} & \frac{\partial g_2}{\partial I_1} & \frac{\partial g_2}{\partial I_2} \\ \frac{\partial g_3}{\partial V_1} & \frac{\partial g_3}{\partial V_2} & \frac{\partial g_3}{\partial I_1} & \frac{\partial g_3}{\partial I_2} \\ \frac{\partial g_4}{\partial V_1} & \frac{\partial g_4}{\partial V_2} & \frac{\partial g_4}{\partial I_1} & \frac{\partial g_4}{\partial I_2} \end{bmatrix} = \begin{bmatrix} 1 + n_1 I_1 & 0 & n_1 V_1 & 0 \\ 0 & 1 + n_2 I_2 & 0 & n_2 V_2 \\ 1 & -1 & -R_{f1} & R_{f2} \\ -\frac{1}{R_L} & 0 & 1 + \frac{R_{f1}}{R_L} & 1 \end{bmatrix} \tag{10.62}$$

Then, choosing an initial vector of $\mathbf{x}^{(0)}$, the first step in the iteration process is to determine the small perturbation of the states from

$$\Delta\mathbf{x}^{(k)} = \left(\mathbf{J}^{(k)}\right)^{-1}\Delta\mathbf{g}^{(k)} = \left(\mathbf{J}^{(k)}\right)^{-1}\left[0 - \mathbf{g}^{(k)}\right], \quad k = 0, 1, \cdots \tag{10.63}$$

where k is the iteration number. The states are then updated from

$$\mathbf{x}^{(k+1)} = \mathbf{x}^{(k)} + \Delta\mathbf{x}^{(k)}, \quad k = 0, 1, \cdots \tag{10.64}$$

The process terminates when minimum $|\Delta\mathbf{x}^{(k)}|$ is less than a small positive number. This is illustrated by Example 10.12.

Example 10.12 Consider a DCMG that has two generators with the following ratings

$$P_{m1} = 50 \text{ kW and } P_{m2} = 20 \text{ kW}$$

Then, for $V^* = 380$ V, $\Delta V_m = 38$ V, the droop gains are calculated from Example 10.11 as $n_1 = 0.76$ V/kW and $n_2 = 1.9$ V/kW. The load resistance is assumed to be $R_L = 4\ \Omega$. Then, assuming the maximum voltage drop across the line, the load bus voltage is $V_L = 0.9 \times 380 = 342$ V. Therefore, the DGs will deliver $V_L^2/R_L = 29.24$ kW power to the load. The Newton–Raphson method mentioned in Section 10.6.2 is now employed for three $R_{f1} : R_{f2}$ ratios. The results are shown in Figure 10.35. The power sharing accuracy increases as the line resistance decreases in all the three cases. The DG output voltages increase with line resistances to cater for an increased drop across the lines. Furthermore, the $R_{f1} : R_{f2}$ ratio plays an important role in the power sharing ratios.

10.6.3 I-V Droop Control

An alternate droop formulation, where the output voltages depend on currents, is discussed in this section. Consider the system of Figure 10.34 again. The droop equations are given by

$$\begin{aligned} V_1 &= V^* - n_1 I_1 \\ V_2 &= V^* - n_2 I_2 \end{aligned} \tag{10.65}$$

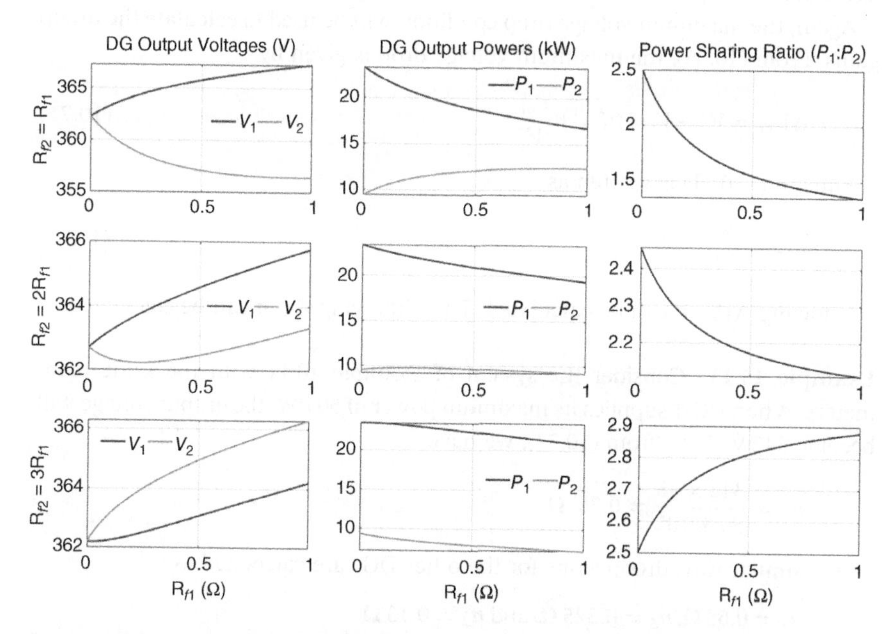

Figure 10.35 The effect of line resistance on the output voltage, power, and power sharing ratio.

where the droop gains are now defined in ohms (Ω). Substituting the relations in (10.50), we get

$$V_L = V^* - n_1 I_1 - R_{f1} I_1 = V^* - n_2 I_2 - R_{f2} I_2 \tag{10.66}$$

This implies that

$$\frac{I_1}{I_2} = \frac{n_2 + R_{f2}}{n_1 + R_{f1}} \tag{10.67}$$

Under the assumption

$$n_1 >> R_{f1}, \quad n_2 >> R_{f2} \tag{10.68}$$

(10.67) can be modified as

$$\frac{I_1}{I_2} \cong \frac{n_2}{n_1} \tag{10.69}$$

Furthermore, assuming $V_1 \approx V_2$, the power relation is given as

$$\frac{P_1}{P_2} \approx \frac{V_1 I_1}{V_2 I_2} = \frac{n_2}{n_1} \tag{10.70}$$

It can be seen that (10.70) is exactly the same as (10.54), and hence (10.55) also remains valid. However, the droop gains in this case are defined in terms of resistance, while in (10.53) they are defined in V/kW. Therefore, a different methodology must be adopted for the choice of the droop gains for the I-V droop control.

Again, the maximum voltage drop condition will be used to calculate the droop gains. From (10.65), the maximum voltage drop is given by

$$\Delta V_m = V^* - V = nI = n\frac{P_m}{V} \tag{10.71}$$

Equation (10.71) is written as

$$n = \frac{V \times \Delta V_m}{P_m} \tag{10.72}$$

Knowing ΔV_m, V can be calculated. Then knowing P_m, n can be calculated.

Example 10.13 Consider the system of Example 10.11 with the same parameters. When DG-1 supplies its maximum power of 50 kW, the output voltage will become 342 V. Then from (10.72), we have

$$n_1 = \frac{342 \times 38}{50 \times 10^3} = 0.26\ \Omega$$

In a similar way, droop gains for the other DGs are calculated as

$$n_2 = 0.65\,\Omega, n_3 = 0.325\,\Omega, \text{and } n_2 = 0.13\,\Omega$$

10.6.4 DCMG Operation with DC-DC Converters

The DCMG operation discussed in Section 10.6.3 assumes that the load is directly connected across the DC line with a maximum voltage of 380 V. The voltage supplied to the load can vary (drop) depending on the load resistance. In practice, however, there might be loads connected to the microgrid that have a much lower voltage rating. Consider, for example, the microgrid system shown in Figure 10.36. In this, the DG voltages are raised to microgrid voltage level by the two boost converters. This voltage might be higher than the voltage that can be tolerated by the load. Therefore, a buck converter is connected at the load terminal that regulates the voltage V_L across the load. Note that the voltage V_m is applied as the input to the buck converter. To eliminate the ripples in buck converter input voltage, a filter capacitor C_f is connected across this point. The working principle of this system is demonstrated with both the droop controllers mentioned in Sections 10.6.1 and 10.6.3.

Example 10.14 In the DCMG of Figure 10.36, the power ratings of the DGs are taken as

$$P_{m1} = 50\ \text{kW and } P_{m2} = 20\ \text{kW}$$

The line resistances are $R_{f1} = 0.01\ \Omega$ and $R_{f2} = 0.06\ \Omega$. The P-V droop gains for these are computed in Example 10.11 and the I-V droop gains are given in Example 10.13. All the three converters are assumed to have the following parameters

$$L = 4\ \text{mH}, C = 500\ \mu\text{F}, \text{and } f = 10\ \text{kHz}$$

In addition, $C_f = 500\ \mu$F and $V_{dc1} = V_{dc2} = 250$ V are also chosen and the reference voltage for the buck converter output voltage $V_L^* = 150$ V.

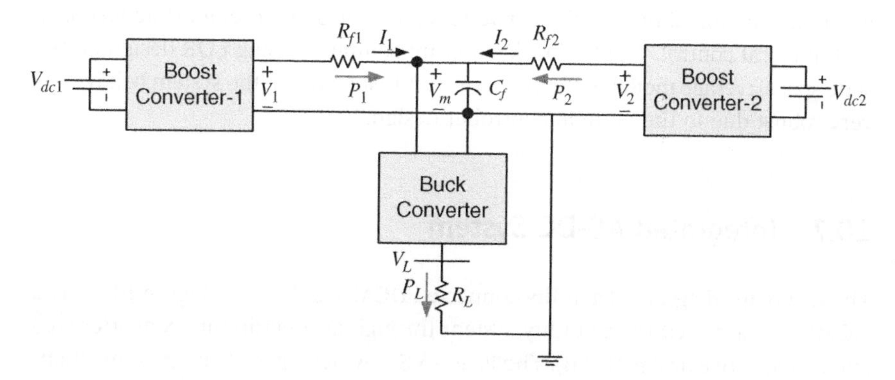

Figure 10.36 Schematic diagram of a DCMG containing buck and boost converters.

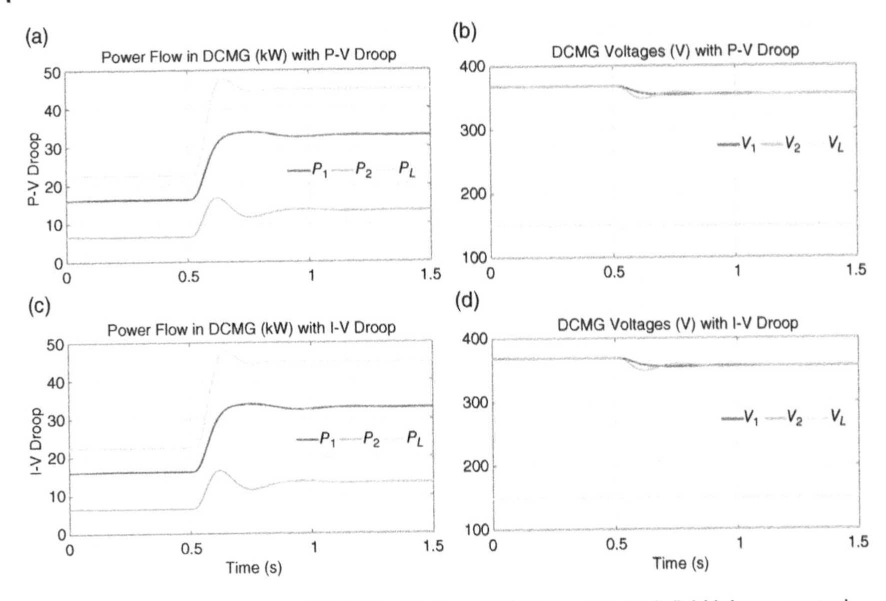

Figure 10.37 Performance of DCMG with (a and b) P-V and (c and d) I-V droop control.

All the three converters are operated in LQR state feedback with integral control using the state space averaging method.

First, the load resistance R_L is chosen as 1 Ω such that the load draws 22.5 kW of power. Then, at 0.5 seconds, the load resistance is decreased to 0.5 Ω such that the load now becomes 45 kW. The results with both types of droop control are shown in Figure 10.37. These are almost identical. The power sharing ratio $P_1 : P_2$ is 2.5 : 1. The buck converter holds the load voltage (V_L) constant at 150 V. The stability (eigenvalue) analysis for the DCMG circuit of Figure 10.36 is reported in [20], where it is assumed that all the converters are controlled through state feedback with integral control. The feedback gains are computed using LQR design on the state space average model of the converters. It is shown that the system behavior is very robust due to the optimal controller design.

10.7 Integrated AC-DC System

The schematic diagram of a utility connected DCMG is shown in Figure 10.38. The DCMG is connected to the utility system through an interlinking converter (IC) and a dual active bridge (DAB). The IC is a VSC, which operates in the same manner as a voltage controlled DSTATCOM. Therefore, it holds the DC voltage V_{dc1}

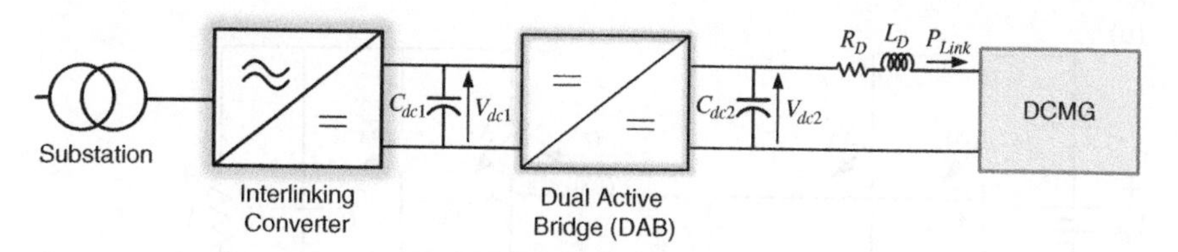

Figure 10.38 Schematic diagram of a utility connected DCMG.

across the capacitor C_{dc1} constant by drawing the power required by the DCMG from the utility. The DAB is a bidirectional DC-DC converter that also isolates the AC and DC systems. This device modulates the voltage V_{dc2} across the capacitor C_{dc2} to facilitate the bidirectional power flow (P_{Link}) between the utility and the DCMG. The operation and control of the DAB is briefly discussed in Section 10.7.1.

10.7.1 Dual Active Bridge (DAB)

A DAB is an isolated, bidirectional, buck and boost DC-DC converter topology that can be used in high-power applications as solid state transformers. The advantages of DAB converters include a lower number of passive components, high-power density, and high-power efficiency resulting from zero voltage switching (ZVS) [21–23]. The schematic diagram of a DAB is shown in Figure 10.39a. It contains two H-bridge converters that are connected together by a high-frequency transformer. A DAB can be controlled by:

- Phase shifting the switching of the two bridges, while they work at fixed duty ratio.
- Controlling the duty ratio of the bridges.
- Modulating the switching frequency.

We shall derive a controller using the first method.

The equivalent circuit of a DAB is shown in Figure 10.39b. In this, R represents converter and transformer losses and L is the leakage reactance of the transformer. All the quantities are referred to the secondary side of the transformer. The voltage in the primary side of the transformer is denoted by v_P, while that of the secondary side is denoted by v_s. Since the converter contains two H-bridge converters, there are four possible switching states, given by

$$
\begin{aligned}
v_p &= \begin{cases} V_{dc} & \text{when } S_1 \text{ is on and } S_2 \text{ is off} \\ -V_{dc} & \text{when } S_2 \text{ is on and } S_1 \text{ is off} \end{cases} \\
v_s &= \begin{cases} V_0 & \text{when } S_3 \text{ is on and } S_4 \text{ is off} \\ -V_0 & \text{when } S_4 \text{ is on and } S_3 \text{ is off} \end{cases}
\end{aligned}
\tag{10.73}
$$

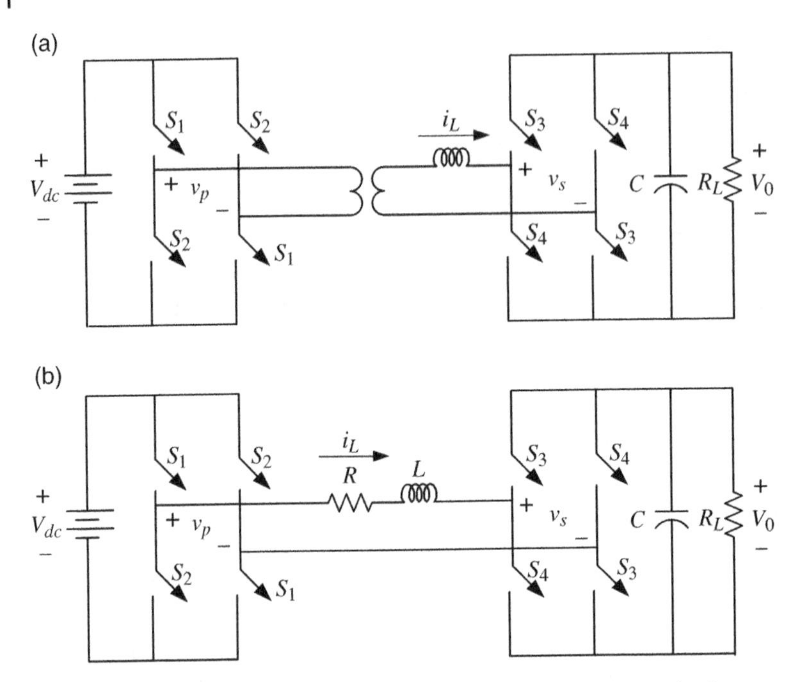

Figure 10.39 Schematic diagram of (a) DAB and (b) its equivalent circuit.

It is assumed that both the H-bridges are operated at a 50% duty ratio. However, their switching is phase shifted by an instant ϕ, as shown in Figure 10.40a.

There are four modes of operation of the DAB. These are defined as:

- Mode-1: When switches S_1 and S_4 are closed.
- Mode-2: When switches S_1 and S_3 are closed.
- Mode-3: When switches S_2 and S_3 are closed.
- Mode-4: When switches S_2 and S_4 are closed.

These modes of operations are shown in Figure 10.40b–e. Defining a state vector as $\mathbf{x} = \begin{bmatrix} V_0 & i_L \end{bmatrix}^T$, the state space equations for the different modes are given by

$$\text{Mode-1 } (t_0 \le t < t_1): \dot{\mathbf{x}} = \begin{bmatrix} -1/R_L C & -1/C \\ 1/L & -R/L \end{bmatrix} \mathbf{x} + \begin{bmatrix} 0 \\ 1/L \end{bmatrix} V_{dc} \tag{10.74}$$

$$\text{Mode-2 } (t_1 \le t < t_2): \dot{\mathbf{x}} = \begin{bmatrix} -1/R_L C & -1/C \\ -1/L & -R/L \end{bmatrix} \mathbf{x} + \begin{bmatrix} 0 \\ 1/L \end{bmatrix} V_{dc} \tag{10.75}$$

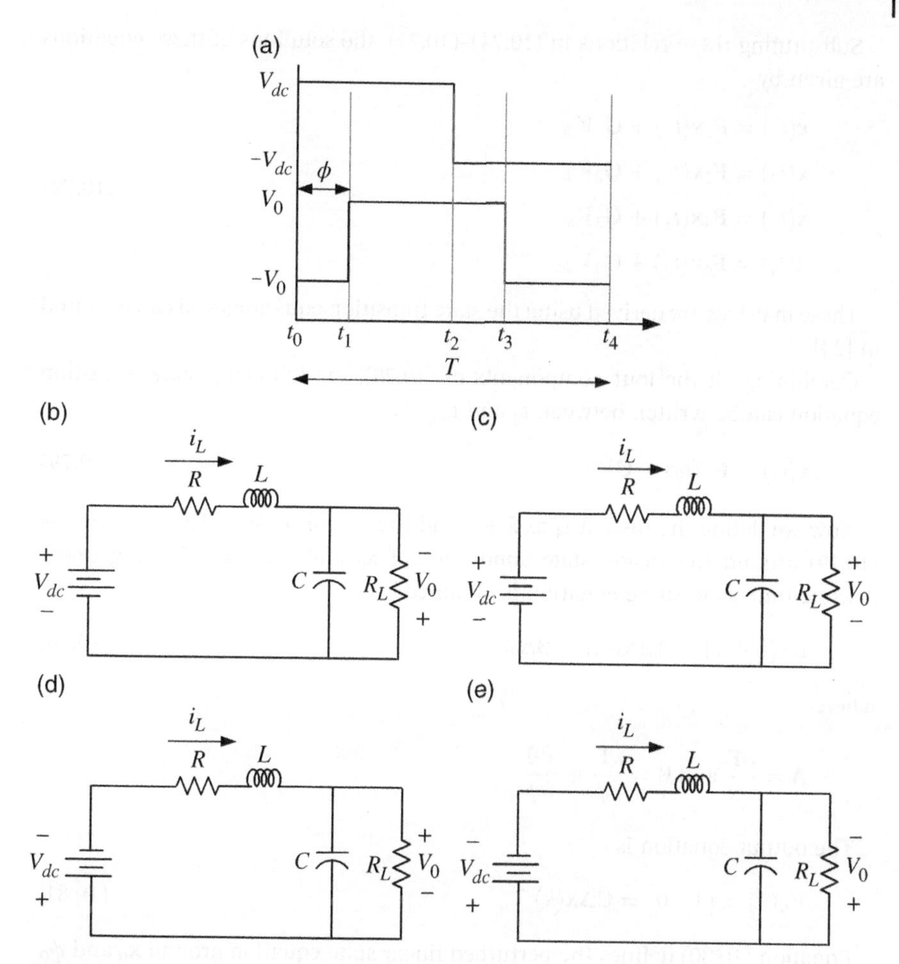

Figure 10.40 (a) Switching configuration of a DAB and its equivalent circuit in (b) Mode-1, (c) Mode-2, (d) Mode-3, and (e) Mode-4.

$$\text{Mode-3 } (t_2 \le t < t_3) : \dot{\mathbf{x}} = \begin{bmatrix} -1/R_L C & -1/C \\ 1/L & -R/L \end{bmatrix} \mathbf{x} + \begin{bmatrix} 0 \\ -1/L \end{bmatrix} V_{dc} \quad (10.76)$$

$$\text{Mode-3 } (t_3 \le t < t_4) : \dot{\mathbf{x}} = \begin{bmatrix} -1/R_L C & -1/C \\ 1/L & -R/L \end{bmatrix} \mathbf{x} + \begin{bmatrix} 0 \\ -1/L \end{bmatrix} V_{dc} \quad (10.77)$$

From Figure 10.40a, the following steady state conditions are defined

$$t_1 - t_0 = t_3 - t_2 = \phi, \quad t_2 - t_1 = t_4 - t_3 = \frac{T}{2} - \phi, \quad t_4 - t_0 = T$$

Substituting these relations in (10.74)–(10.77), the solutions of these equations are given by

$$\begin{aligned}\mathbf{x}(t_1) &= \mathbf{F}_1\mathbf{x}(t_0) + \mathbf{G}_1V_{dc}\\ \mathbf{x}(t_2) &= \mathbf{F}_2\mathbf{x}(t_1) + \mathbf{G}_2V_{dc}\\ \mathbf{x}(t_3) &= \mathbf{F}_3\mathbf{x}(t_1) + \mathbf{G}_3V_{dc}\\ \mathbf{x}(t_4) &= \mathbf{F}_4\mathbf{x}(t_3) + \mathbf{G}_4V_{dc}\end{aligned} \tag{10.78}$$

These matrices are derived using the state transition equations and can be found in [24].

Combining all the four components of (10.78), the following state transition equation can be written between t_0 and t_4

$$\mathbf{x}(t_4) = \mathbf{F}\mathbf{x}(t_0) + \mathbf{G}V_{dc} \tag{10.79}$$

Now we define the instant t_4 as $k+1$ and the instant t_0 as k. Then linearizing (10.79) around the steady state conditions of $\mathbf{x}_0$ and ϕ_0, the following linear discrete-time state space equation is obtained

$$\Delta\mathbf{x}(k+1) = \mathbf{A}\Delta\mathbf{x}(t_0) + \mathbf{B}\Delta\phi \tag{10.80}$$

where

$$\mathbf{A} = \frac{\partial \mathbf{F}}{\partial \mathbf{x}} \text{ and } \mathbf{B} = \frac{\partial \mathbf{F}}{\partial \phi} + \frac{\partial \mathbf{B}}{\partial \phi}$$

The output equation is

$$V_0(k) = [1 \quad 0] = \mathbf{C}\Delta\mathbf{x}(k) \tag{10.81}$$

Equation (10.80) defines the perturbed linear state equation around $\mathbf{x}_0$ and ϕ_0. The steady state quantities must be derived first before a control law can be computed. Assuming the converter to be lossless, the power transfer relationship is given by [22, 23]

$$P_1 = P_2 = \frac{V_{dc}V_0\phi(\pi-\phi)}{2\pi^2 fL} \tag{10.82}$$

where f is the switching frequency. From (10.82), it is evident that the power transfer is zero either when $\phi = 0°$ or when $\phi = \pi$ and the maximum power is transferred when $\phi = \pi/2$.

It can be seen from (10.82) that, for a constant V_{dc} and ϕ, the output voltage V_0 will be proportional to the output power P_2. The first step in the linearization process is to decide a nominal value of the output power P_{20} at the nominal voltage of V_{00}. Equation (10.82) can then be rewritten in terms of these quantities as

$$\phi_0^2 - \pi\phi_0 + \frac{2\pi^2 fL}{V_{dc}V_{00}}P_{20} = 0 \tag{10.83}$$

Solving the quadratic equation and choosing the lesser of the two values, the nominal value of the phase shift ϕ_0 is obtained.

From the nominal values of the output voltage V_{00} and output power P_{20}, the nominal value of load resistance is calculated as $R_{L0} = V_{00}^2/P_{20}$. With this value of R_{L0} and ϕ_0, Eqs. (10.74) to (10.79) are evaluated. The steady state vector is then calculated from (10.79) as

$$\mathbf{x}_{ss} = (\mathbf{I} - \mathbf{F})^{-1}\mathbf{G}V_{dc} \tag{10.84}$$

This vector and ϕ_0 are now used for the derivation of the linearized model of (10.80–10.81).

10.7.2 AC Utility Connected DCMG

In Figure 10.38, the interlinking VSC holds the DC voltage V_{dc1} across the capacitor C_{dc1}, while the DAB modulates the voltage V_{dc2} across the capacitor C_{dc2} to facilitate the bidirectional power flow. The reference voltage V_{dc2}^* is obtained by the link power flow through the following PI controller

$$V_{dc2}^* = V^* + K_P\left(P_{Link}^* - P_{Link}\right) + K_I\int\left(P_{Link}^* - P_{Link}\right)dt \tag{10.85}$$

where V^* is the DCMG reference voltage and P_{Link}^* is the desired link power.

Example 10.15 The DCMG is the same as that discussed in Example 10.14. The DAB transformer is chosen as 1 kV:1 kV while its leakage inductance is 0.4 mH. The DAB steady state model is derived for a nominal voltage of 380 V and power of 50 kW. The DAB then is controlled using LQR state feedback with integral control. The DAB PI controller gains are $K_P = 2.0$ and $K_I = 100$. The results are shown in Figure 10.41. At the beginning, the DCMG operates in the steady state, supplying a load of 45 kW, where the DGs share power the ratio of 2.5 : 1. During this time, the DAB holds its output voltage in such a way that no power exchange takes place between the utility and the DCMG, i.e. $P_{Link} = 0$. Then, at 0.25 seconds, 15 kW of power is drawn from the utility. From Figure 10.41a, it can be seen that the load power remains constant all through this process. However, since 15 kW of power is supplied by the utility after 0.25 seconds, the DGs supply $P_1 = 22.1$ kW and $P_2 = 8.85$ kW maintaining a ratio of 2.5 : 1. The DAB output voltage and its reference are shown in Figure 10.41b, while the DAB control output ϕ is shown in Figure 10.41c.

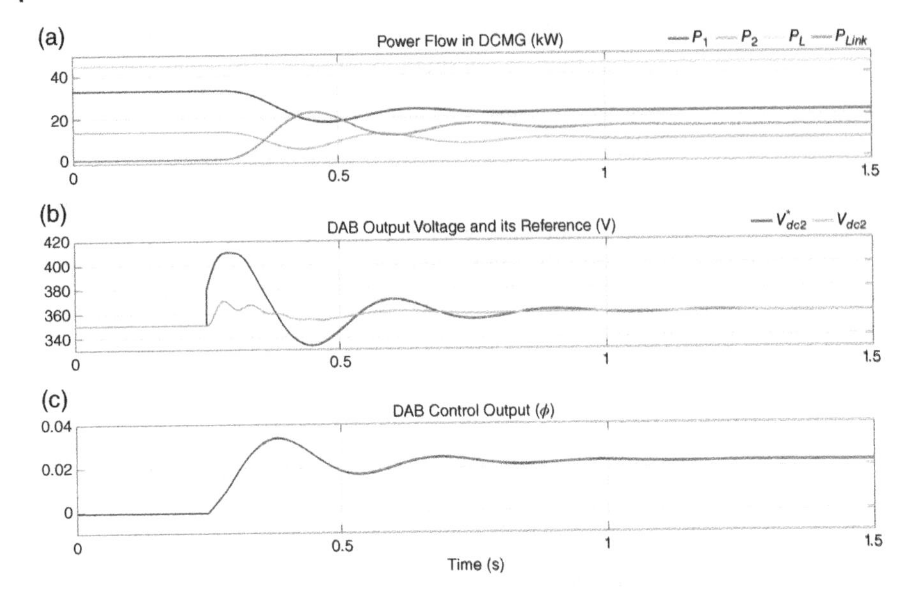

Figure 10.41 Performance of utility connected DCMG. (a) Power flow in DCMG (kW), (b) DAB output voltage and its reference (V), (c) DAB control output (ϕ).

10.8 Control Hierarchies of Microgrids

The droop control of AC microgrids is essentially a local control of DERs that can operate without communication networks. However, to operate a microgrid in the mix of several other microgrids and distribution networks, this local controller is not adequate. There are different levels of microgrid controllers summarized in [25]. Figure 10.42 shows the control hierarchy of a microgrid, which consists of three levels of control: primary, secondary, and tertiary. These are briefly discussed in this section.

10.8.1 Primary Control

The droop control discussed in Sections 10.4.2–10.4.5 and 10.4.7 is the main type of primary control. However, there are several drawbacks with this method, as is pointed out in [25]. One of the obvious ones is the dependence of droop gains on the X/R ratio. To overcome this problem, a virtual impedance can be used. However, the stability issues for the selection of virtual impedance need to be addressed through an eigenvalue analysis. For this, a simplified model of the system needs to be derived.

The droop control is basically a steady state concept in which the load dynamics are ignored. This can lead to instability during fast load changes. Furthermore, the

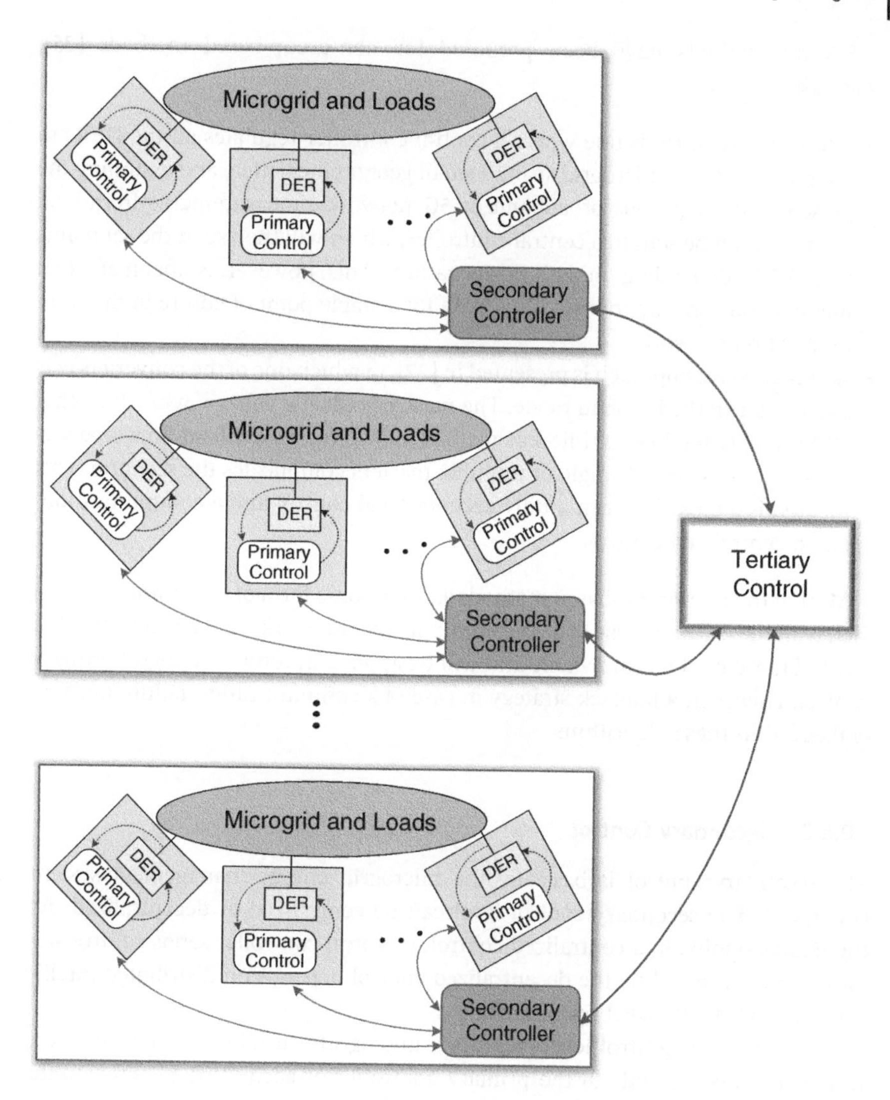

Figure 10.42 Three-level control structure of microgrids.

nonlinear and/or unbalanced loads can cause power quality issues in the system. One way to approach this issue is to stipulate that each DG must compensate for its polluting local load, while they share common load as per their ratings [26]. This, however, will require the placement of a compensating DG with each polluting load, which is very restrictive. There are other approaches that can be found from the references listed in [25].

Several methods have been proposed for non-droop-based methods [25], such as:

- Centralized control is one where a central controller regulates all the DERs in the microgrid. If a microgrid spans a small geographical area, and assuming the presence of ubiquitous broadband or 5G networks, measurements from load terminals can be sent to a central controller, which will then send the set points to all DERs depending on their power ratings. This, however, is not an efficient method and can lead to instability, even for a single point of failure in the communication network.
- A master–slave approach is presented in [27], in which one of the converters acts as a master in the islanded mode. The master holds the voltage, while the other DERs supply the load. While this method is inspired by the load flow type scenario, the presence of single or multiple masters complicates the scenario. This method is a cross between a fully decentralized control and a communication-based centralized control.

Most of the non-droop-based methods assume some form of communication. In fact, with the increased use of data communication networks, it is conceivable that most of these methods are viable and will be applied in future microgrid primary control. However, a fallback strategy in case of a communication failure must be embedded in these algorithms.

10.8.2 Secondary Control

The secondary control is basically the microgrid energy management system (MGEMS). The secondary control level can be centralized or decentralized. As the names signify, in a centralized control, a central controller sends control signals to the DERs, while the decentralized control depends on distributed intelligence to coordinate DER operations.

The centralized control level includes a microgrid central controller (MGCC), which sends commands to the primary controller of each DER to compensate for the deviation in their output voltage and frequency from their respective set points. Through this, the coordination amongst the local controllers of the microgrids can be realized at this level. A communication system is required to implement this level of control since the information and commands should be regularly transferred between the local regulators and the secondary controller. Usually, the operation timeframe of this control is slower than that of the primary level, and therefore a lower bandwidth communication system is sufficient. Apart from the data regarding the DERs and loads, the MGCC can consider the relevant weather forecasting data to determine the changes in the set points required [25].

The main problem of a centralized controller is that it can have a deleterious effect even for a single point of failure in the communication system. In a decentralized control scheme, each DER in a microgrid is provided with more autonomy than that given in the centralized approach. In [28], a distributed control architecture is discussed for power electronic based DERs such that each of them has its secondary controller. Each of these secondary controllers can produce appropriate control correction for their primary controllers by considering the measurements from the other DER units. In [29], a multiagent-based secondary controller is proposed. The objective of a multiagent system (MAS) is to segment a total system into several entities, each called an agent. These agents can then interact amongst themselves to solve a complex problem. A review of an MAS application in a smart grid can be found in [30]. The schematic diagram of the MAS-based secondary controller is shown in Figure 10.43, in which all the agents communicate amongst themselves and with the MGCC [29].

10.8.3 Tertiary Control

This level can be considered as a part of the main grid operation [25]. At this level, the interactions and power exchanges amongst the microgrids as well as between the microgrids and the main grid are controlled. Generally, these actions are organized to achieve various technical, economic, and environmental objectives within the whole system, while satisfying some prespecified constraints. One of the objectives could be the reactive power management through the coordination

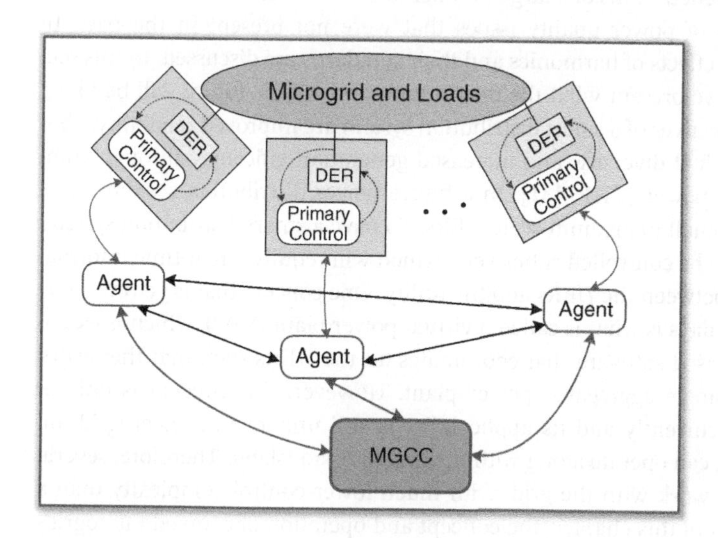

Figure 10.43 MAS-based distributed secondary control structure of a microgrid.

of different microgrids. In a coupled microgrid scenario, the tertiary controller should determine the amount of power to be exchanged amongst the subsystems along with the associated timings. This controller is placed at the highest level, and therefore it is responsible for coordinating the operation of the microgrids by sending appropriate commands to the secondary controllers. It is to be noted that this is also the slowest control level that should operate in minutes rather than in seconds. Some of the objectives of the tertiary level are:

- Microgrid overload relieving.
- Reliability improvement.
- Resilience enhancement.
- Loss minimization within the entire system.

10.9 Smart Distribution Networks: Networked Microgrids

In the past, power distribution systems were radial in nature, i.e. power would flow from a distribution substation down the feeder in a unidirectional manner. Currently, the situation is changing rapidly. Several rooftop PVs are getting connected to power distribution systems, causing voltage rise, reverse power flow, and power quality problems. Moreover, PVs are intermittent in nature: a sudden shading in PVs can cause a power drop or power fluctuations that need to be smoothened out. Moreover, the penetrations of a large number of power electronic converters have been the cause of power quality issues that were not present in the past. In Chapter 11, the effects of harmonics and their standards are discussed. In this section, however, we present what the distribution grids of the future will be like.

The main objectives of a smart distribution system are improved reliability, ability to self-heal, fuel diversity, and increased generation efficiency through combined heat and power (CHP) [31]. In a future power distribution system, there will be several small to medium-scale DERs. To have a smart distribution system, all these need to be controlled cohesively, which will require a real-time information exchange between the DERs and the utility. One concept that is getting some attention these days is what is called a virtual power plant (VPP), which is essentially a cloud-based software that coordinates all the DERs such that they work together as a single aggregated power plant. However, this concept is only at the trial stage currently and its applicability is still unknown. A microgrid, on the other hand, can operate along with the grid or as an island. Therefore, several microgrids can work with the grid, with much lower control complexity than a VPP. In the rest of this chapter, the concept and operation of coupled microgrids and microgrid clusters are discussed.

There are several advantages in forming distribution systems with a network of microgrids. It is pointed out in [32] that networked microgrids can be used for improving resilience in power systems. The main argument in favor of microgrid-based resilience enhancement is that a microgrid can island itself in case of an extreme event. It can supply its local load demand with a much lesser impact in the event of the grid failure. A microgrid, while interacting with a utility grid, can provide ancillary support, voltage support, black start support, etc. [32]. Oak Ridge National Laboratory in Tennessee, USA, has published a scoping study on networked microgrids in 2016 [33]. In this report, the authors discuss the interconnection layout, communication and control architectures, and the potential benefits of networking microgrids. Several research needs are pointed out in this article. Some of these are:

- Coordination of energy trading and ancillary service management through microgrids, addressing DER control and feeder reconfiguration.
- Multi-objective optimization considering peak shaving, loss minimization, and voltage regulation across several interconnected microgrids.
- Comprehensive energy management for each microgrid, considering the presence of networked microgrids and distribution systems.
- Distribution energy market and distribution ancillary service market, both managed by the distribution service operators.
- Demand response incentive signal design to allow utility scale energy management.
- Defining microgrid-to-microgrid communication framework.

A comprehensive review of networked microgrids is given in [34]. In Section 10.10, the interconnection layout and operational strategies of networked microgrids are discussed.

10.9.1 Interconnection of Networked Microgrids

It is possible to have different layout structures of networked microgrids. However, all of these are the subsets of the three general layouts shown in Figures 10.44–10.46. Figure 10.44 shows the configuration of microgrids connected in series to a single feeder, where the microgrid management system (MMS) controls the operation of microgrids and the connecting switches SW-1 and SW-2. Depending on the operating principles, the MMS can control four possible operating scenarios. These are listed in Table 10.4, where the microgrids are abbreviated as microgrids. The MMS will coordinate these actions and generate signals for the switches to open or close.

Figure 10.45 shows the configuration of two microgrids that are connected in parallel to a single feeder. The microgrids are connected by the switch SW-12,

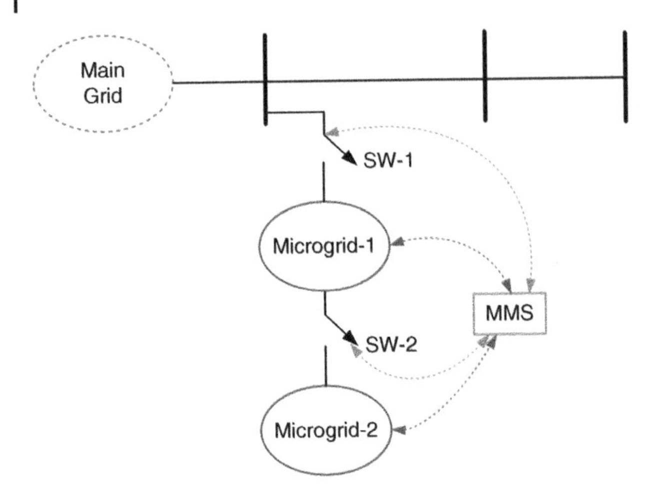

Figure 10.44 Serial microgrids on a single feeder.

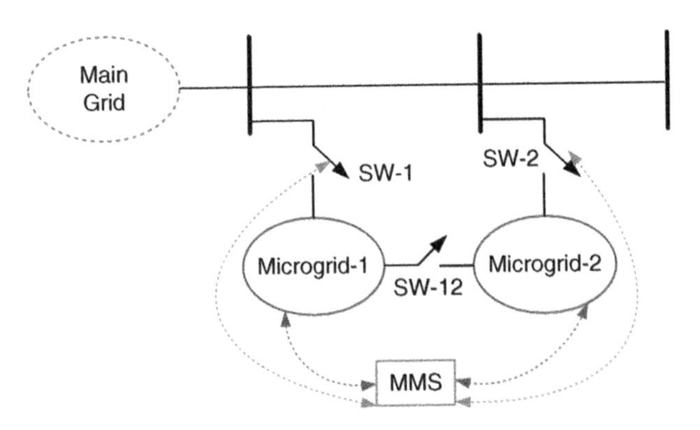

Figure 10.45 Parallel microgrids on a single feeder.

while switches SW-1 and SW-2 connect them with the utility feeder. Since there are three switches, $2^3 = 8$ operating conditions are possible, which are listed in Table 10.5.

Figure 10.46 shows the interconnections of microgrids on multiple feeders. Again, since there are three switches, eight operating scenarios are possible. Notice here that the microgrids can be connected to Feeder-1 when switches SW-1 and SW-12 are closed, and SW-2 is open. Similarly, they can be connected to Feeder-2 when switches SW-2 and SW-12 are closed, and SW-1 is open. Furthermore, they can be connected to both the feeders when all three switches are closed. This is the most versatile configuration of a microgrid network since it can operate in grid-connected mode even when one of the feeders is down. In this configuration, the restoration of feeders following a fault is possible, where multiple feeders can be black started.

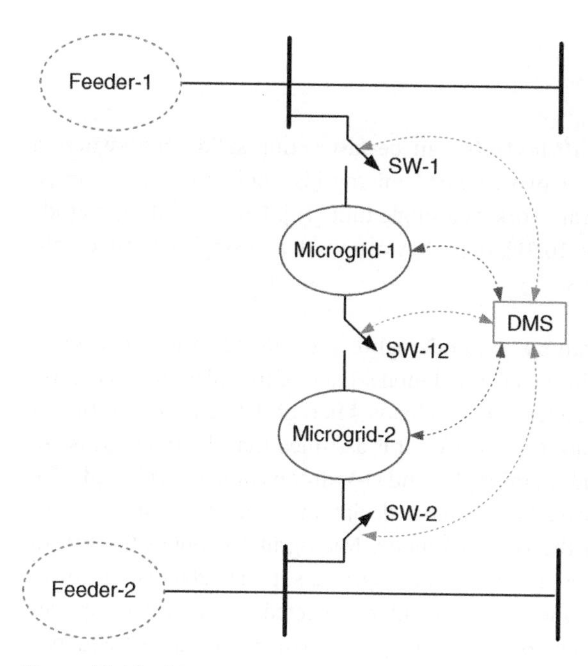

Figure 10.46 Interconnected microgrids on multiple feeders.

Table 10.4 Operating scenarios of series connected microgrids to a single feeder.

Scenario	Microgrid-1	Microgrid-2	SW-1	SW-2
1	Grid connected	Grid connected	Closed	Closed
2	Grid connected	Islanded	Closed	Open
3	Islanded	Islanded	Open	Open
4	Islanded	Connected to Microgrid-1	Open	Closed

Table 10.5 Operating scenarios of parallel connected microgrids to a single feeder.

Scenario	Microgrid-1	Microgrid-2	SW-1	SW-2	SW-12
1	Grid connected	Grid connected	Closed	Closed	Open
2	Islanded	Islanded	Open	Open	Open
3	Islanded	Grid connected	Open	Closed	Open
4	Grid connected	Islanded	Closed	Open	Open
5	Grid connected	Connected to Microgrid-1	Closed	Open	Closed
6	Connected to Microgrid-2	Grid connected	Open	Closed	Closed
7	Islanded, but connected to Microgrid-2	Islanded, but connected to Microgrid-1	Open	Open	Closed
8	Connected to both grid and Microgrid-2	Connected to both grid and Microgrid-1	Closed	Closed	Closed

The switches in Figures 10.44–10.46 can be fast acting solid state switches. Before the interconnection of two islanded microgrids, their droop gains must be normalized so that they can work as a single unit [16]. However, if microgrids select the droop gains as per (10.21), then they can work as a single cohesive unit. Example 10.16 illustrates this.

Example 10.16 In this example, two microgrids are connected together when both of them are operating in the islanded mode. Each of the microgrids is represented by an aggregation of generators and loads. Microgrid-1 has a total rating of 1 MW, while Micogrid-2's rating is 500 kW. It is assumed that the microgrids are predominantly inductive and therefore P-f and Q-V droop control is selected. The droop gains are selected based on a frequency deviation of ± 0.5 Hz from the nominal frequency of 50 Hz. In the islanded mode, Microgrid-1 supplies 960 kW of power at a frequency of 49.54 Hz, while Microgrid-2 supplies 200 kW of power at a frequency of 50.1 Hz. Once they are interconnected, Microgrid-1 supplies 790 kW and microgrid-2 supplies 395 kW of power with a collective frequency of 49.71 Hz. The results are shown in Figure 10.47, where the microgrids are synchronized when the phase angles of the voltages are nearly equal [16]. Despite this, large transients are visible, and therefore a soft synchronization scheme will have to be devised. Note that there is a slight discrepancy in the total power generation before and after the reconnection. Since the loads are passive RL, the rise in the voltage level causes a little bit of extra power being supplied to the load.

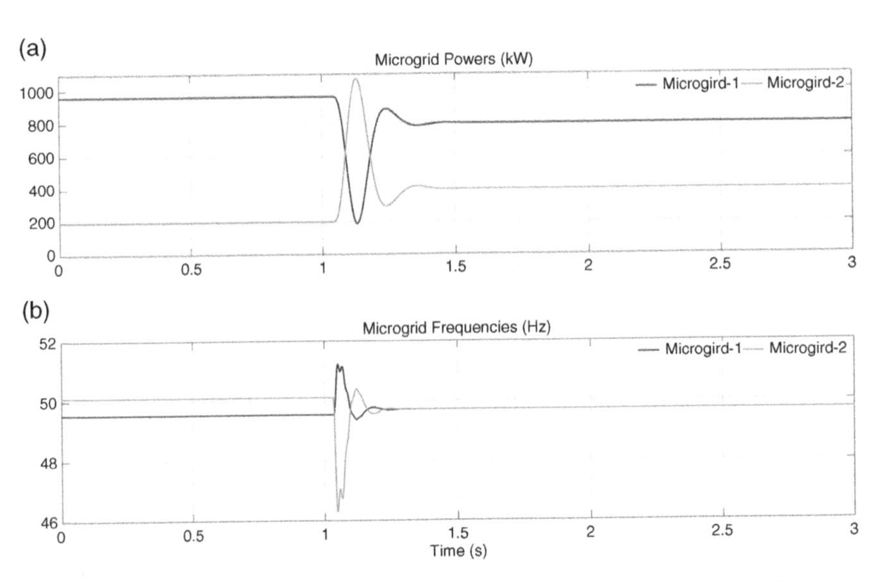

Figure 10.47 Power flow and frequency before and after the interconnection of two microgrids.

10.10 Microgrids in Cluster

Example 10.16 assumes that both microgrids follow the same principle for droop gain selection, which may not always be possible. Each microgrid is designed based on the available resources and they may have different operating principles. Therefore, it is desirable that each microgrid maintains its autonomy as far as possible, i.e. their operations should be relatively independent of each other. To explain the principle, let us consider the microgrid–utility connection through a back-to-back converter system shown in Figure 10.32. It is assumed that the back-to-back converters connect the two systems through two feeders. The main function of VSC-1 is to hold the DC voltage V_{dc} across the capacitor C_{dc} by drawing power from the AC utility through angle control. On the other hand, VSC-2 facilitates power flow between the utility and the microgrid.

Now VSC-2 must synchronize with the microgrid, which may operate at a different frequency and voltage level than the AC grid. Let us assume that the microgrid voltages at the output are given by

$$\begin{aligned} v_{MGa} &= V_{MG}\sin(\omega_{MG}t + \delta_{MG}) \\ v_{MGb} &= V_{MG}\sin(\omega_{MG}t + \delta_{MG} - 120^\circ) \\ v_{MGc} &= V_{MG}\sin(\omega_{MG}t + \delta_{MG} + 120^\circ) \end{aligned} \tag{10.86}$$

where ω_{MG} is the operating frequency of the microgrid. Noting that power flow over an AC line is mainly dependent on the relative angle difference, VSC-2's output voltage v_{C2} will be synthesized such that the required phase shift is introduced depending on the power flow between them. Let the VSC-2 output voltages be defined by

$$\begin{aligned} v_{C2a} &= V_{C2}\sin(\omega_{MG}t + \delta_{C2}) \\ v_{C2b} &= V_{C2}\sin(\omega_{MG}t + \delta_{C2} - 120^\circ) \\ v_{C2c} &= V_{C2}\sin(\omega_{MG}t + \delta_{C2} + 120^\circ) \end{aligned} \tag{10.87}$$

A balanced system operation is assumed. Then, from the measurement of the instantaneous voltages v_{MGa}, v_{MGb}, and v_{MGc} at any given instant, the instantaneous symmetrical component transformation given in (2.8) of Chapter 2 is used to get the positive sequence vector as

$$\begin{aligned} v_{MG1} &= \frac{1}{3}\left[v_{MGa} + a v_{MGb} + a^2 v_{MGc}\right] \\ &= \frac{V_{MG}}{2}\left[\sin(\omega_{MG}t + \delta_{MG}) - j\cos(\omega_{MG}t + \delta_{MG})\right] = \frac{V_{MG}}{2}(\alpha + j\beta) \end{aligned} \tag{10.88}$$

where $\alpha = \sin(\omega_{MG}t + \delta_{MG})$ and $\beta = \cos(\omega_{MG}t + \delta_{MG})$. The following three terms are defined from (10.88)

$$|v_{MG1}| = \frac{V_{MG}}{2}, \quad \alpha = \frac{\text{Re}\,(v_{MG1})}{|v_{MG1}|}, \quad \beta = \frac{\text{Im}(v_{MG1})}{|v_{MG1}|} \tag{10.89}$$

Similarly, the positive sequence of the converter side voltage can be written as

$$v_{C21} = \frac{V_{C2}}{2}\left[\sin\left(\omega_{MG}t + \delta_{C2}\right) - j\cos\left(\omega_{MG}t + \delta_{C2}\right)\right] \tag{10.90}$$

Now, for the power flow over the link (P_{Link}), the two angles must differ. Defining

$$\delta = \delta_{C2} - \delta_{MG} \tag{10.91}$$

and substituting (10.79) in (10.78), we get

$$\begin{aligned} v_{C21} &= \frac{V_{C2}}{2}\left[\sin\left(\omega_{MG}t + \delta_{MG} + \delta\right) - j\cos\left(\omega_{MG}t + \delta_{MG} + \delta\right)\right] \\ &= \frac{V_{C2}}{2}\left[(\alpha + j\beta)\cos\delta - (\beta - j\alpha)\sin\delta\right] \end{aligned} \tag{10.92}$$

From the instantaneous measurements of (10.86), the terms $|v_{MG1}|$, α, and β given in (10.89) are computed. Assuming $V_{C2} = V_{MG}$, the positive sequence component vector v_{C21} can be computed if the value of δ can be determined.

Let us assume that the microgrid needs to draw a fixed amount of power from the utility grid. The power reference is denoted by P^*_{Link}. Then this is compared with the measured link power and used in a PI controller to obtain δ through

$$\delta = K_{PP}\left(P^*_{Link} - P_{Link}\right) + K_{IP}\int\left(P^*_{Link} - P_{Link}\right)dt \tag{10.93}$$

Once δ is calculated, the vector v_{C21} is calculated from (10.92). Note that the negative sequence vector v_{C22} is the complex conjugate of v_{C21}. Therefore, with these vectors and assuming that the zero-sequence component is zero, the reference signals in the abc-frame can be computed from the inverse symmetrical component transform. If VSC-2 can follow the signals accurately, it will guarantee that the output voltages are synchronized with the microgrid. Example 10.17 shows the limits on the performance of an islanded microgrid.

Example 10.17 Let us consider a microgrid that is supplied by two identical diesel generator sets. Each set consists of a 4-stroke internal combustion engine coupled to a synchronous generator. The internal combustion engine's speed is controlled by a governor, which is a PID controller that maintains output speed by changing the fuel rate. The synchronous generator contains an exciter and an automatic voltage regulator (AVR). The AVR controls the field supply of the generator to maintain the required terminal voltage. The maximum rating of each diesel generator is 480 kW. The droop gains are selected based on ±0.5 Hz voltage drop from the nominal frequency of 50 Hz. The DGs supply a constant P-type load. At the beginning the load is 850 kW, which is shared equally between the DGs. The steady state frequency is 49.57 Hz. With the system operating in the steady state, the load is changed to 1.05 MW at 1 second. Obviously, this is beyond the capacity of the DGs in the microgrid, and it needs to draw power from the utility grid to supply this load; otherwise, load shedding may be required.

The question is how much power should be drawn from the grid. Here we stipulate that the microgrid should draw as much power as is required to meet the shortfall in its generation. Notice that when the DGs in the microgrid supply their maximum rated power, their frequency will be 49.5 Hz. A drop in the frequency below this value is indicative of a shortfall in power. Corrective actions will be required when the frequency falls below 49.5 Hz. However, to differentiate between a large transient due to load change and power shortfall, a deadtime must be provided between the time when the frequency falls below the threshold and the time of taking corrective action. This deadtime will depend on the type of generators present in the microgrid. If all the generators are power electronic interfaced, the time will be less as they cannot provide inertia. This can lead to a voltage collapse as soon as their currents saturate. On the other hand, a synchronous generator, can hold the voltage by releasing the rotational kinetic energy. However, this can only be temporary. Therefore, once the frequency crosses the threshold and stays there for a stipulated amount of time, a one-shot Schmitt trigger is used to generate a pulse "Trig." This will start the process of the microgrid drawing power from the utility. The amount of power should be such that the microgrid frequency stabilizes at 49.5 Hz. To accomplish this, another PI controller is employed based on the frequency, which is given by

$$\begin{aligned} \Delta f &= 49.5 - f_{MG} \\ P_{Link}^{*} &= K_{Pf}\Delta f + K_{If}\int \Delta f \, dt \end{aligned} \tag{10.94}$$

where f_{MG} is the measured microgrid frequency at the connection point. The VSC-2 control scheme is shown in Figure 10.48. Note that the two PI controllers get activated through the Trig signal, which is generated only when the frequency falls below the threshold frequency of 49.5 Hz. Once the power demand from the microgrid is removed, the Trig signal is removed, and the PI controllers are reset and given an input of 0 so that large starting transients do not occur when they are called into action next time the power shortfall occurs.

The results are shown in Figure 10.49. Since the two DGs are identical, their power overlap cannot be differentiated. However, it is obvious that the scheme works perfectly as the frequency, when the initial transients after load change are over, settles to 49.5 Hz and the DGs supply their maximum rated power.

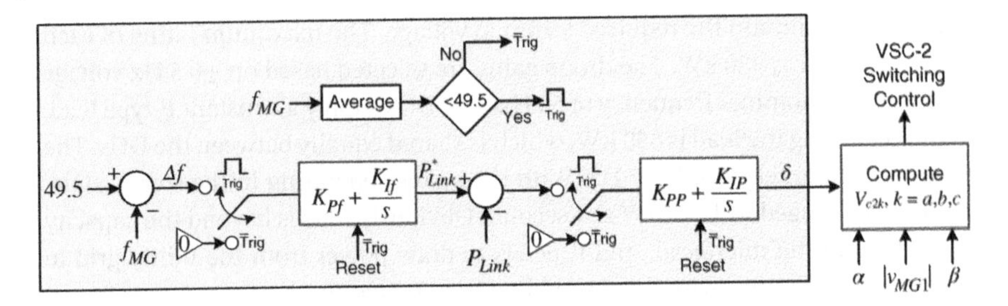

Figure 10.48 VSC-2 control scheme.

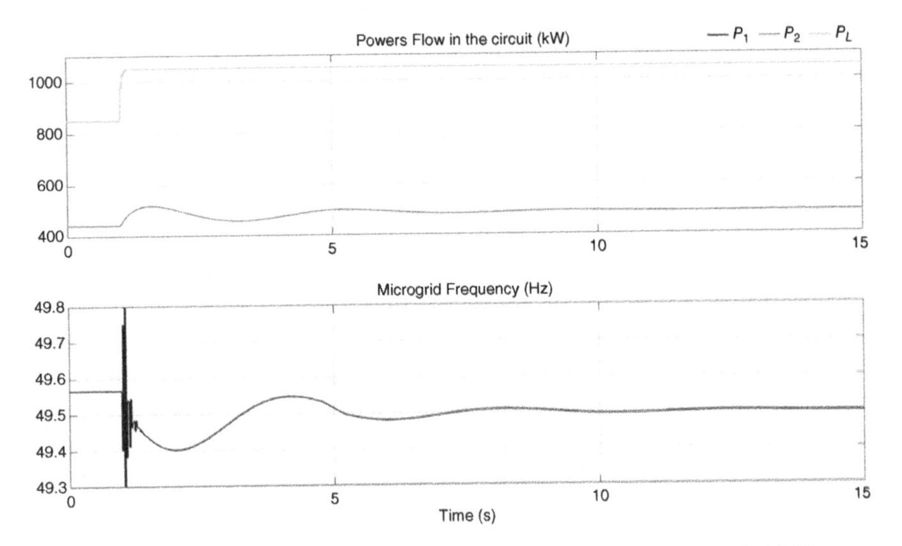

Figure 10.49 Microgrid power shortfall being supplied by utility in Example 10.15.

The only drawback of this scheme is that the converter losses must be supplied continuously even when no power flow occurs between the two systems in order to hold the DC capacitor voltage V_{dc} constant. However, using this scheme, the microgrid can operate independently of the AC system. This configuration forms the basis of microgrid clusters, which is discussed in Section 10.11.1.

10.10.1 The Concept of Power Exchange Highway (PEH)

An interesting concept of coupling microgrids is proposed in [35]. This is shown in Figure 10.50. In this scheme, the power exchange takes place through a dedicated line (i) between the microgrids and (ii) between the microgrids and the utility

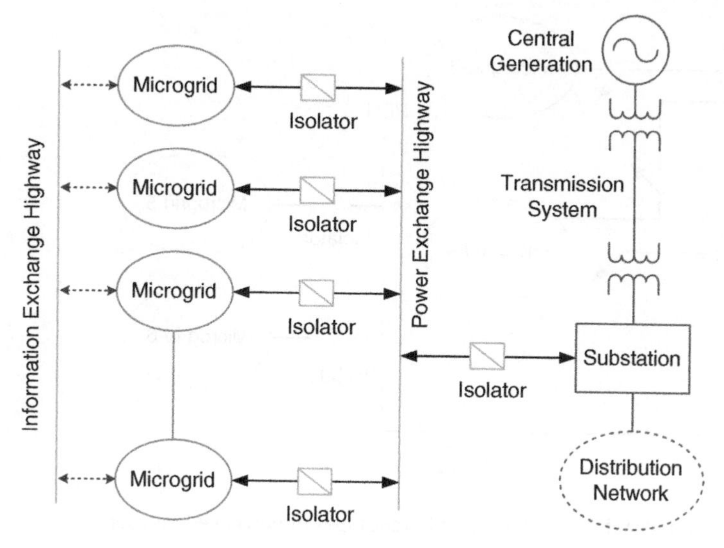

Figure 10.50 Microgrid cluster formation through the PEH.

system. This line is called the power exchange highway (PEH). In parallel with the PEH, there is an information exchange highway through which the microgrids communicate with each other. The advantage of these two highways is that all microgrids can exchange power and information with each other, irrespective of their locations vis-à-vis the other microgrids. It is also possible to have the microgrid cluster connected through the PEH with other networked microgrids, as shown in Figure 10.51. In this scheme, Microgrid-5 and 6 are connected through a PEH and can operate independently of the rest of the system. However, they can participate in energy trading as and when it is suitable. The other four microgrids are connected through switches. Obviously, when Microgrid-1 to 4 are connected to the utility grid, they must operate at the same frequency as the grid, while Microgrid-5 and 6 can operate under their respective droop control regimes.

In Figures 10.50 and 10.51, the isolators are converters. The operation of back-to-back converter connection of a microgrid with an AC grid is discussed in Example 10.17. This can also be applied to the clusters shown in the figures. However, the PEH can be a three-phase AC line or a DC line. The following conditions can be stipulated for the two configurations:

- AC-PEH: If a microgrid is AC, the isolator connecting it to the PEH is back-to-back VSCs. There need not be any isolator between the PEH and the utility substation. However, if one of the microgrids is DC, the isolator is a VSC plus DAB combination.

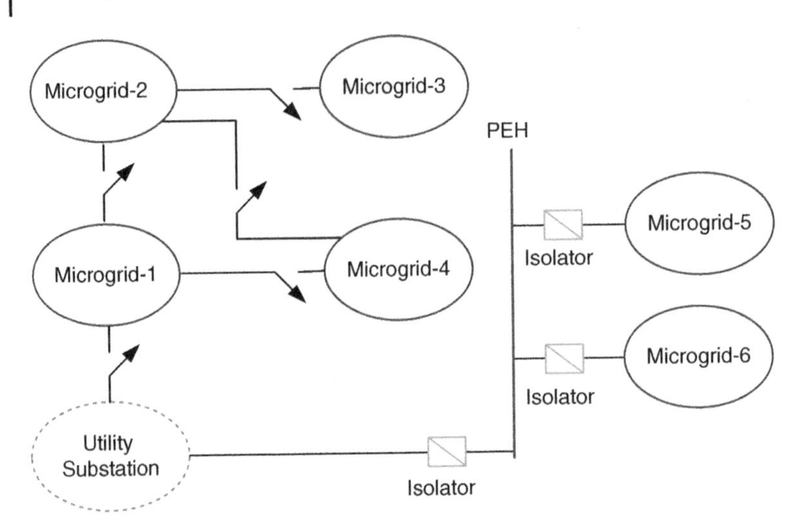

Figure 10.51 Future distribution grid with different types of networked microgrids.

- DC-PEH: If the microgrid is AC, the isolator is an AC-DC converter, whereas for a DCMG the isolator is a DC-DC converter. The isolator connecting the PEH with the utility substation is an AC-DC converter. The advantage of having a DC PEH is that the losses are less as only one converter is needed for each isolator.

The operating principle of an AC PEH is discussed in [36], and that of a DC PEH is presented in [37]. Some of their principles of operation are the same. Since the interconnection of a microgrid with a utility system through a back-to-back converter is discussed elsewhere, we discuss the power exchange through a DC PEH in the remainder of this section.

10.10.2 Operation of DC Power Exchange Highway (DC-PEH)

Let us consider an islanded microgrid cluster, as shown in Figure 10.52a. It has a total number of n AC microgrids that are connected together through a DC-PEH. No load is assumed to be connected to the DC-PEH. Each microgrid is connected to the DC-PEH through a VSC. The DC link of each VSC is connected with a storage capacitor. The capacitors are connected in parallel through the DC-PEH, which is represented by line resistances.

Nominally, each microgrid, while supplying its local load, holds the voltage across the DC capacitor connected to the VSC constant at a predefined level as well. However, when a power shortfall occurs in one (or more) microgrid, the other microgrids supply a part of their excess available power to support it.

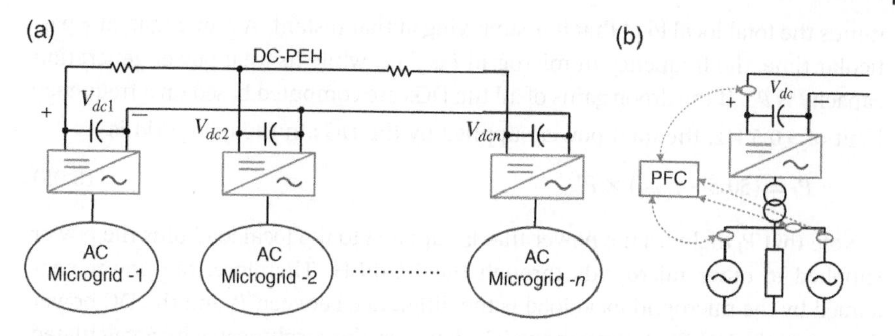

Figure 10.52 (a) Structure of islanded microgrid cluster and (b) configuration of one of the microgrids.

The excess power available in a microgrid can vary depending on the consumption of its local load. Hence, it is not possible to determine a priori how much support, in terms of real power, a microgrid can provide to others. Therefore, the surplus available power in a microgrid must be continually updated. Based on this information, the microgrid with excess available power will support the microgrids with a power shortfall.

The equivalent circuit of each of the microgrids in the cluster is shown in Figure 10.52b. It contains several DGs and loads. The VSC is connected to the microgrid bus through a transformer. A power flow controller (PFC) controls the bidirectional power flow between the microgrid and the DC-PEH, by determining either the available excess power that can be supplied to the DC-PEH or the exact amount of power shortfall that is required to be replenished for the safe operation of the microgrid. As shown in Figure 10.52b, the PFC needs information about breaker status, microgrid frequency, and DC power to generate the requisite control signals.

10.10.3 Overload Detection and Surplus Power Calculation

It is assumed that, when a microgrid is overloaded, it should draw the exact amount of power that will maintain the stability in the system, as shown in Example 10.17. The overload detection scheme is based on the frequency and is triggered when the frequency falls below the set threshold value of the microgrid. Thereafter, the power required by the microgrid can be drawn from the DC-PEH using the VSC control scheme of (10.94).

When a microgrid is overloaded, other microgrids with surplus power will be able to supply the required power depending on their load consumptions at that time. Thus, each microgrid must have a surplus power calculation scheme.

The power surplus capacity (P_{SC}) of a microgrid is defined as the difference between the total generation capacity the microgrid has at a given instant of time minus the total local load that it is supplying at that instant. Assume that, at a particular time, the frequency in microgrid-i is f_{MGi}, while its total power generation capacity is P^*_{gi}. If the droop gains of all the DGs are computed based on a frequency limit of ± 0.5 Hz, the total power supplied by the DGs in the microgrid is

$$P_i = (50.5 - f_{MGi}) \times P^*_{gi} \tag{10.95}$$

Note that P_i includes the power that is supplied to the local load plus the power supplied to other microgrids through the DC-PEH. Therefore, the power consumed by the microgrid local load is the difference between P_i and the DC power flowing to the DC-PEH through the IC. However, the surplus capacity is calculated based on only the microgrid local load. It is defined by the difference between the total power generation capacity (P^*_{gi}) and local load, i.e.

$$P_{SCi} = P^*_{gi} - (P_i - V_{dci} \times I_{dci}) \tag{10.96}$$

It is interesting to note that the PFC of Figure 10.52b does not need to know about the status of each DG for overload prevention; it just draws power from the DC-PEH to stabilize the frequency to 49.5 Hz irrespective of how many DGs are connected to the system. On the other hand, the PFC needs to know the DG status for surplus power calculation. This is because the maximum generation capacity of a microgrid changes with the availability of the DGs. It might so happen that one of the DGs is out of commission due to the required maintenance work. Also, the microgrid generation capacity can vary when plug-and-play type DGs are included. Suppose microgrid-i has n number of DGs, with their maximum power being denoted by P^*_k. The total power consumed by the microgrid is P_i. The microgrid surplus power generation capacity is then given by modifying (10.96) as

$$P_{SCi} = \left[\sum_{k=1}^{n} s_k P^*_k\right] - (P_i - V_{dci} \times I_{dci}) \tag{10.97}$$

where the $s_k = 1$ if the DG is connected; otherwise, $s_k = 0$. Therefore, a communication link between the DGs and the PFC will be required to update the DG status. Note that, since this link is only used for the status update, a low bandwidth communication channel will be sufficient for this purpose.

Example 10.18 Suppose a mostly inductive microgrid has four DGs, rated at 200, 400, 300, and 100 kW. Thus, the total power generation capacity of the microgrid (P^*_g) is 1 MW. The droop gains, selected based on a frequency deviation of ± 0.5 Hz, are 5, 2.5, 3.33, and 10 Hz/MW respectively. Therefore, from (10.21), $\alpha = 1$ Hz.

Now, suppose all the DGs in the microgrid are operational and the microgrid is operating at a frequency of 49.9 Hz. Then, from (10.95), the total power supplied by the microgrid is

$$P_i = (50.5 - f_{MGi}) \times P_{gi}^* = (50.5 - 49.9) \times 1 = 0.6 \text{ MW}$$

Let us suppose that the microgrid is supplying $V_{dci} \times I_{dci} = 150$ kW of power to the DC-PEH. Then, the total local load plus losses in the microgrid itself is 450 kW, and the microgrid surplus capacity is given from (10.96) as

$$P_{SCi} = P_{gi}^* - (P_i - V_{dci} \times I_{dci}) = 1 - (0.6 - 0.15) = 0.55 \text{ MW}$$

This implies that, even though the microgrid supplies 600 kW of power at this given instant, since its local load is 450 kW, it has a surplus capacity of 550 kW.

If the DG with the maximum capacity of 200 kW trips, then the microgrid surplus power generation capacity is modified from (10.97) as

$$P_{SCi} = (0.4 + 0.3 + 0.1) - (0.6 - 0.15) = 0.35 \text{ MW}$$

However, since the microgrid is supplying 600 kW of power to its load and to the DC-PEH, its frequency will drop to 49.75 Hz, as per (10.95), i.e.

$$P_i = 0.6 = (50.5 - f_{MGi}) \times 0.8 \Rightarrow f_{MGi} = 50.5 - \frac{0.6}{0.8} = 49.75 \text{ Hz}$$

10.10.4 Operation of DC-PEH

It is assumed that the DC-PEH operates in the I-V droop discussed in Subsection 10.6.3. To explain its operation, let us consider the three-microgrid cluster shown in Figure 10.53. Nominally, all the microgrids only supply their local loads. Therefore, the voltage references for the capacitor voltages are set as per (10.65) as

$$\begin{aligned} V_{dc1}^* &= V^* - n_1 I_{dc1} \\ V_{dc2}^* &= V^* - n_2 I_{dc2} \\ V_{dc3}^* &= V^* - n_3 I_{dc3} \end{aligned} \tag{10.98}$$

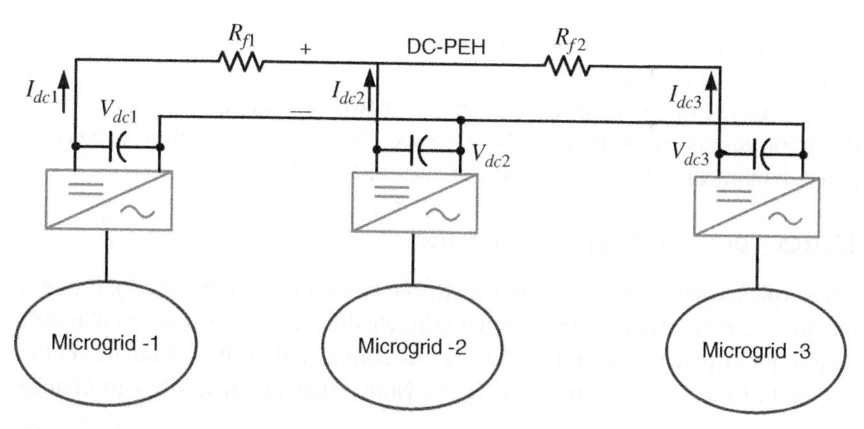

Figure 10.53 Schematic diagram of a three-microgrid cluster.

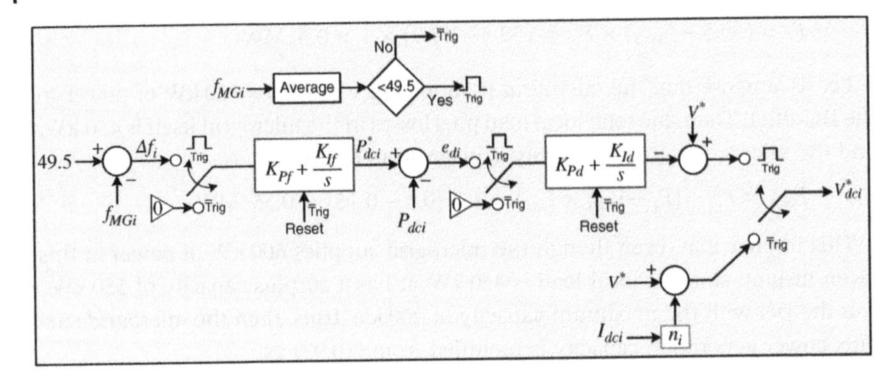

Figure 10.54 Overload prevention scheme through controlled DC voltage generation.

When there is no power flow from any of the microgrids to the DC-PEH, all the currents are zero, i.e. $I_{dc1} = I_{dc2} = I_{dc3} = 0$ and therefore $V^*_{dc1} = V^*_{dc2} = V^*_{dc3} = V^*$.

Consider the case where Microgrid-2 needs power from the DC-PEH. The power requirement (P^*_{dc2}) is obtained from the PI controller of (10.94). The DC capacitor voltage reference for Microgrid-2 is set such that this amount of power is drawn from the DC-PEH. This is accomplished through another PI controller, given by

$$\begin{aligned} e_{d2} &= P^*_{dc2} - P_{dc2} \\ V^*_{dc2} &= V^* - K_{Pd}\, e_{d2} - K_{Id} \int e_{d2}\, dt \end{aligned} \tag{10.99}$$

where $P_{dc2} = -V_{dc2} \times I_{dc2}$. The overload prevention scheme is shown in Figure 10.54. From this figure, the desired voltage of the overloaded Microgrid-i is obtained. This voltage is then regulated using the PI controller

$$\delta_i = K_{P\delta}\left(V^*_{dci} - V_{dci}\right) + K_{I\delta} \int \left(V^*_{dci} - V_{dci}\right) dt \tag{10.100}$$

The angle is then used to form the reference voltage for the IC, just like a DSTATCOM in voltage control mode.

Note that the reference voltages of the other two microgrids are obtained using the droop equations given in (10.98). However, their droop gains need to be computed dynamically. This is discussed in Section 10.11.5.

10.10.5 Dynamic Droop Gain Selection

The surplus capacity that a microgrid has can be estimated from (10.97). It might not, however, be necessary that a microgrid schedules the entire amount of power surplus for dispatch through DC-PEH. Let us assume that the maximum power the microgrid-i schedules for dispatch is P^*_i. Notice that this quantity can change

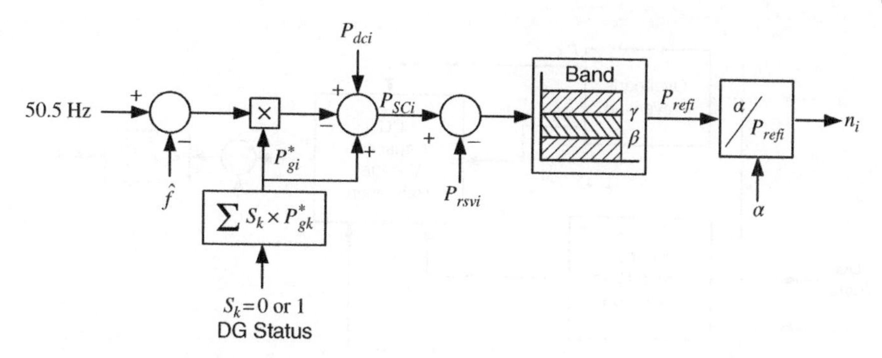

Figure 10.55 Dynamic droop gain selection scheme.

depending on the local load of each microgrid. Therefore, the droop gains need to be adjusted dynamically. In the I-V droop control discussed in Section 10.6.3, it is stipulated that (10.55), i.e. $P_i^* \times n_1 = P_i^* \times n_2 = \cdots = P_M^* \times n_M = \alpha$, remains valid.

The schematic diagram of the dynamic droop gain selection scheme is shown in Figure 10.55. Based on the estimated frequency and the status of the DGs in a microgrid, its surplus capacity (P_{SCi}) is calculated from (10.97). A certain amount of power (P_{rsvi}) is kept as reserve in the microgrid to cater for a sudden change in its local load. Thereafter, the desired maximum power output of the microgrid P_i^* is selected based on the difference ($P_{SCi} - P_{rsvi}$). Note that P_{SCi} changes for every change in the microgrid local load. This will cause a continuous fluctuation in the droop gain if P_i^* is simply chosen as $P_{SCi} - P_{rsvi}$. To prevent this, P_i^* is changed in discrete steps of $P_{SCi} - P_{rsvi}$ in a band that is given by

$$\text{If } \beta \le P_{SCi} - P_{rsvi} < \gamma, \text{then } P_i^* = \beta \tag{10.101}$$

where β is the lower limit and γ is the upper limit of the band and the bandwidth is selected as

$$\gamma - \beta = \lambda \times \sum_{k=1}^{n} \left[S_k P_k^*\right] - P_i \tag{10.102}$$

where λ is a constant that defines a percentage of the total generation capacity of the microgrid. Once P_i^* is obtained, the droop gain of the microgrid is computed from the preset value of α.

The block diagram of the dynamic droop gain selection in the power flow control scheme is shown in Figure 10.56. The step-by-step PFC operation is listed in Table 10.6.

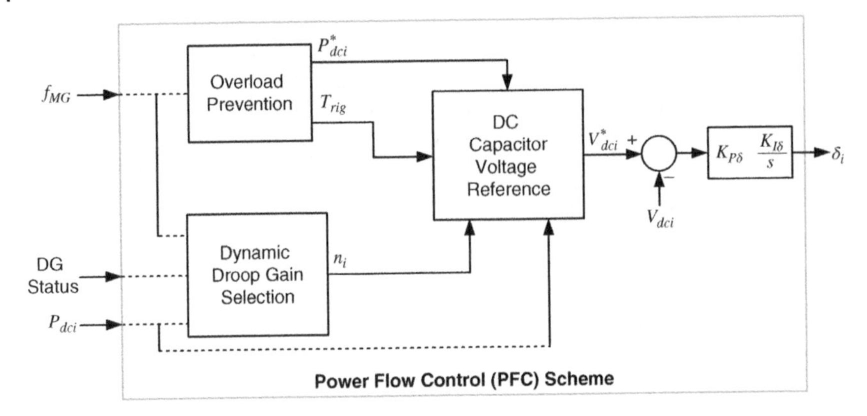

Figure 10.56 Dynamic droop gain selection in power flow control scheme.

Table 10.6 PFC operation.

At each sampling instant *k*			
Check frequency	Frequencies of all the microgrids are above 49.5 Hz	Compute V^*_{dci} from (10.98), where all the droop gains are taken as zero	
		Perform capacitor voltage control of (10.100)	
	If the frequency of a microgrid-k, ($k \neq i$) is below 49.5 Hz	Microgrid-k starts drawing power using Figure 10.54, capacitor voltage control of (10.100)	
		For microgrid-i, ($k \neq i$)	Check breaker status
			Find the surplus power P_{SCi} from (10.97)
			Find P^*_i, n_i from Figure 10.55
			Compute V^*_{dci} from (10.98)
			Perform capacitor voltage control of (10.100)

Example 10.19 (Overload in Microgrid-3)
Let us consider the microgrid cluster shown in Figure 10.53, for which the following parameters are chosen

$$R_{f1} = 0.01\,\Omega, \quad R_{f2} = 0.06\,\Omega, \quad V^* = 2.5\,\text{kV}$$

It is stipulated that the maximum power that can flow through the DC-PEH is $P_{max} = 250$ kW and the maximum allowable voltage drop $\Delta V_{max} = 100$ V. Then, from (10.72), we get

$$n = \frac{2.4 \times 10^3 \times 100}{2.5 \times 10^3} = 0.96\,\Omega$$

Hence

$$\alpha = n \times P_{max} = 2.4 \times 10^5$$

Assume that Microgrid-1 and Microgrid-2, at any given time, have surplus power of 200 kW and 100 kW respectively. Then their droop gains will be

$$n_1 = \frac{\alpha}{P_{ref1}} = 1.2\,\Omega, \quad n_2 = \frac{\alpha}{P_{ref2}} = 2.4\,\Omega$$

The total generation capacity of Microgrid-3 is assumed to be 1 MW. At the beginning, all the three microgrids in the cluster supply their local loads and no power exchange takes place through the DC-PEH. During this time, the local load in Microgrid-3 is 833 kW. At 1 second, the Microgrid-3 load increases suddenly to 1.1 MW, which is beyond its total generation capacity. This causes the Microgrid-3 frequency to drop below 49.5 Hz and hence results in the activation of the overload prevention scheme. The power reference (P^*_{dc3}) set by the PI controller and Microgrid-3 frequency are shown in Figure 10.57. The desired frequency of 49.5 Hz is reached at around 3 seconds when Microgrid-3 draws power from the DC-PEH.

The power flow through Microgrid-3 is shown in Figure 10.58a. The total generation saturates at 1 MW, while the demand increases to 1.1 MW. This excess amount of power flows from DC-PEH as shown in Figure 10.58b, where the share of power between Microgrid-1 and Microgrid-2 remains in the ratio of 2 : 1. The DC capacitor voltages of the ICs of the microgrids are shown in Figure 10.59, where it is obvious that none of the capacitor voltages violates the minimum limit of 2.4 kV.

Example 10.20 (Dynamic Droop Gain Selection)
In this example, it is assumed that Microgrid-2 requires a total power of 200 kW. Microgrid-1 has a surplus capacity of 200 kW and hence its droop gain is chosen as 1.2 Ω. To select the droop gain of Microgrid-3, λ in (10.101) is chosen as 0.05 and P_{srv3} is chosen as 50 kW, while the total generation capacity of the microgrid

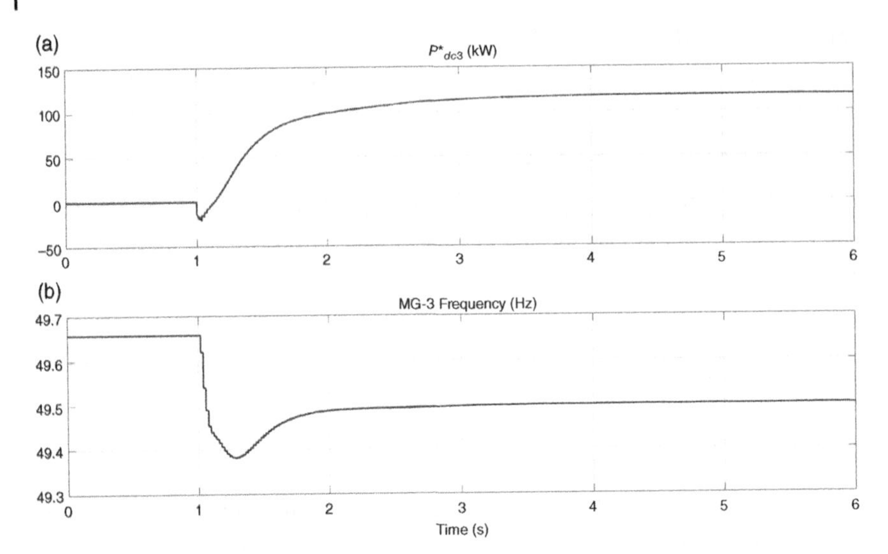

Figure 10.57 Microgrid-3 overload prevention: (a) required power to be drawn from DC-PEH and (b) Microgrid-3 frequency.

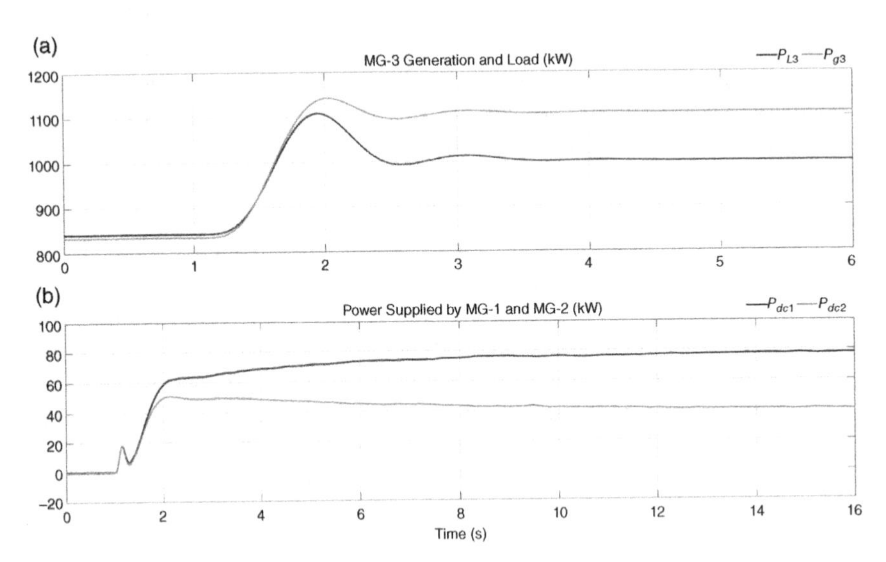

Figure 10.58 Power flow through (a) Microgrid-3 and (b) DC-PEH in Example 10.19.

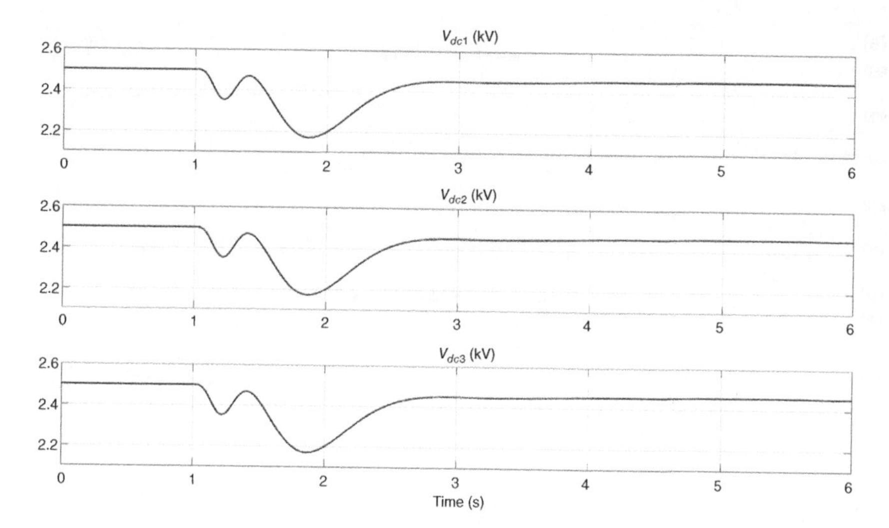

Figure 10.59 DC capacitor voltage of the ICs of the three microgrids in Example 10.19.

Table 10.7 Dynamic droop gains of Microgrid-3 in Example 10.20.

Range (kW)	P_{ref3} (kW)	n_3 (Ω)
$0 < (P_{SC3} - P_{rsv3}) \leq 50$	0	0
$50 < (P_{SC3} - P_{rsv3}) \leq 100$	50	4.8
$100 < (P_{SC3} - P_{rsv3}) \leq 150$	100	2.4
$150 < (P_{SC3} - P_{rsv3}) \leq 200$	150	1.6
$200 < (P_{SC3} - P_{rsv3}) \leq 250$	200	1.2
$250 < (P_{SC3} - P_{rsv3})$	250	0.96

is 1 MW. This implies that the dynamic droop remains constant for less than 50 kW change in the local load. The droop gain selection is given in Table 10.7, for $\alpha = 2.4 \times 10^5$.

In the beginning, Microgrid-3 is in the steady state supplying 760 kW of local load plus system losses of around 8 kW. Then $(P_{SC3} - P_{rsv3})$ is equal to 182 kW. The value of n_3 is chosen as 1.6 Ω from Table 10.7. Microgrid-3 local load changes to 780 kW at 1 second. This implies that $(P_{SC3} - P_{rsv3})$ is equal to 162 kW, including losses. Therefore, the droop gain should not change, and the power supplied by Microgrid-3 to DC-PEH should remain unchanged. Thereafter, at 6 seconds, Microgrid-3 local load changes to 830 kW, which reduces P_{ref3} to 100 kW and n_3

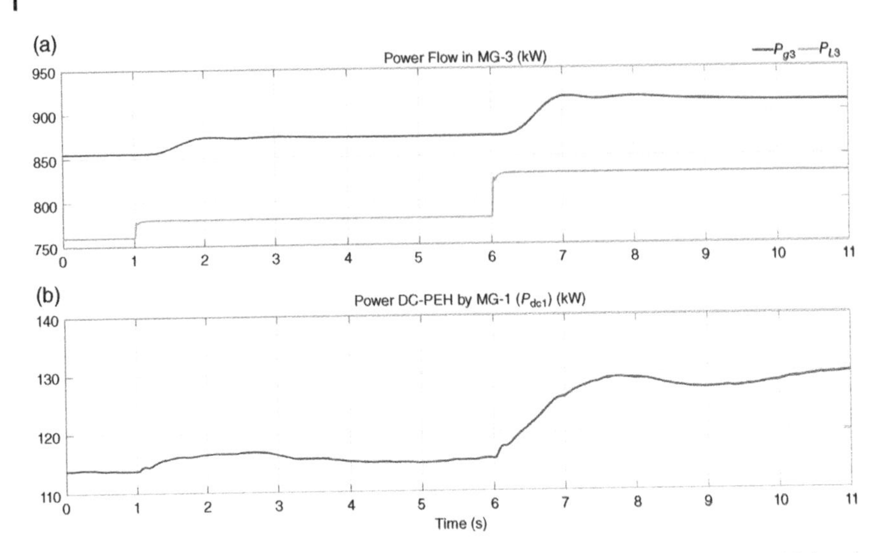

Figure 10.60 (a) Power flow in Microgrid-3 and (b) power supplied by Microgrid-1 to DC-PEH in Example 10.20.

changes to 2.4 Ω. The power flow through Microgrid-3 is shown in Figure 10.60a, while the power supplied to DC-PEH by Microgrid-1 is shown in Figure 10.60b. It is obvious that the power supplied by Microgrid-1 remains (almost) constant when the Microgrid-3 local load changes to 780 kW but increases when the load increases to 830 kW. The quantity ($P_{SC3} - P_{rsv3}$) of Microgrid-3 is shown in Figure 10.61a, while its droop gain is shown in Figure 10.61b. They follow the data given in Table 10.7.

The power sharing strategy discussed here assumes that several microgrids in a cluster can operate based purely on excess available power. However, economic aspects can also play a significant role in the trading of energy between the microgrids. Consider, for example, that a microgrid has a power shortfall and it needs to buy energy from the microgrids in a cluster. On the other hand, there may be several microgrids that can supply the power. Dynamic droop gain selection is a method that can provide an equitable distribution, provided the selling prices of the various microgrids are the same and are fixed beforehand. This, however, may not always be the case.

The basic aim of a microgrid is to cater to its local demand and make profit at the same time. These objectives can be met by drawing power from the utility when the price is low, storing power in energy storage systems when the renewable generation is high and the load demand is low, and selling power back to the grid

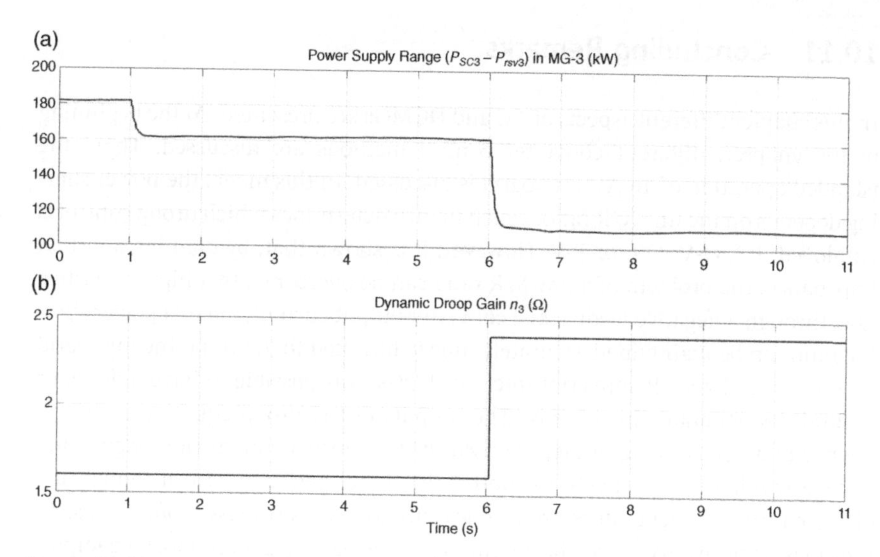

Figure 10.61 (a) Power supply range in Microgrid-3 and (b) its dynamic droop gain in Example 10.20.

during its peak. Microgrids can also exchange power between themselves without involving the grid. Therefore, energy trading schemes must be established between the microgrids, where they can work in either a cooperative or a noncooperative manner while trying to maximize their profits.

Energy trading is a very complex problem since the overall energy price at a given time must be considered. Different microgrids may have different generation costs that will depend on the types of DGs used, their fuel, and associated maintenance costs. Moreover, the utility may have different tariffs depending on the time of the day. It might so happen that the cost in the evening peak hours is significantly higher. However, during this time, several microgrids may not have sufficient reserve to sell power to the utility. Therefore, the first step in the process is to consider the cost of energy production of each microgrid vis-à-vis its load. For example, if, at a given time, a microgrid does not have sufficient excess power after supplying its load, it cannot participate in energy trading. On the other hand, if a microgrid has a power shortfall, it will have to buy from the other microgrids or from the utility, irrespective of the price of energy at the time. Therefore, a more desirable policy is to consider different operating scenarios to determine a price structure that is suitable for energy trading. Once this price structure is determined, cooperative or noncooperative game theoretic models can be implemented for the purpose of energy trading.

10.11 Concluding Remarks

In this chapter different aspects of AC and DCMGs are presented. At the beginning of the chapter, different converter control methods are discussed. Then, the islanded operation of an AC microgrid is discussed. In this mode, the power sharing depends on the line X/R ratio, which in turn determines which droop control is employed: P-f, Q-V, Q-f, or P-V. However, it is shown that, by defining a virtual impedance, the problem of a low X/R ratio can be overcome. In a highly resistive line, through a negative feedback of converter output current, the real power sharing ratio can be maintained accurately. It is to be noted that, even if the microgrid is assumed to have all converter interfaced DGs, it is possible to have a diesel or natural gas generator in the mix. The outputs of the droop equations, e.g. frequency and voltage magnitude, are used in the internal combustion engine and voltage regulator respectively to control the diesel generator. Even though the diesel/natural gas generators produce greenhouse gases (unless biofuel is used), they may be necessary in the short term to provide backup generation. One important aspect that needs to be mentioned here is that the droop gains for an islanded operation are chosen depending on the power rating of the DGs, and therefore an instability problem due to disproportionately large droop gains is not possible through this choice.

This chapter also discusses the grid-connected operation of a DCMG. A DCMG can operate in standalone mode as well. As shown using the configuration of Figure 10.36, the power exchange between the two systems can be controlled through a DAB that can control a bidirectional power flow between the systems. It is to be noted that the DCMG is still in its infancy and therefore most standards have not yet been set. However, as shown in this chapter, they can be a viable alternative, especially where most loads are DC such as in data centers.

Microgrids have progressed beyond simple systems containing plug-and-play type DERs. The concept of clustered microgrids or networked microgrids in a smart grid framework is gaining much attention currently. We therefore include discussions on coupled microgrids. These systems require different levels of control strategies. In this chapter, a method is discussed through which a microgrid can mitigate its power shortfall by borrowing power from neighboring microgrids or from utility systems. However, the inertia in a converter-dominated microgrid is almost nonexistent. Therefore, a smart load shedding strategy or battery backup may be desirable to prevent a system-wide collapse. The discussions in this chapter are mainly focused on the converter control problem that can tackle any energy shortfall problem in coupled microgrids. However, the financial aspects of energy trading need to be embedded so that an overall satisfaction level can be achieved for all participating microgrids.

Problems

10.1 Consider a predominantly inductive autonomous microgrid with three DERs. The power ratings of the DERs are:

$$\text{DER}-1: 1\ \text{MW}, \text{DER}-2: 500\ \text{kW}, \text{and DER}-3: 2\ \text{MW}$$

The droop equations are given by

$$\omega_i = \omega_r + n_i \times \left(0.5P_i^* - P_i\right)\ \text{rad/s}, \quad i = 1, 2, 3$$

where $\omega_r = 100\pi$ rad/s, ω_i is the frequency in rad/s, P_i^* is the rated power in MW, P_i is its measured power in MW, and n_i is the droop gain of microgrid-i.

(a) Determine the values of the droop gains if the frequency is to be limited within ±0.5 Hz.

(b) Suppose the DERs together are supplying 1.4 MW of power to the loads and losses, determine the operating frequency in Hz and the power supplied by each DER.

10.2 Consider again the autonomous microgrid of Problem 10.1. The droop gains in this case are given by

$$f_i = 50 + n_i \times \left(P_i^* - P_i\right)\ \text{Hz}$$

(a) Determine the droop gains if the frequency is allowed to vary between 50 Hz and 51 Hz.

(b) Assuming that the DERs are supplying 1.75 MW of loads and losses, determine the operating frequency of the microgrid.

10.3 Consider the autonomous microgrid of Figure P10.3, where DER-1 and DER-2 supply 600 kW and 300 kW respectively to a resistive load. The system parameters are:

DER ratings: $P_1^* = 1$ MW and $P_2^* = 0.5$ MW

Voltage: 11 kV (L-L), frequency: 50 Hz

Feeder parameters: $L_{f1} = 10$ mH and $L_{f2} = 20$ mH

Assuming the DER voltages are held constant at 11 kV (L-L), the DERs operate in a P-δ droop given by

$$\delta_1 = n_1 \times \left(P_1^* - P_1\right)$$
$$\delta_2 = n_2 \times \left(P_2^* - P_2\right)$$

where the droop gain n_1 is given by 0.12 rad/MW.

(a) Determine the angles δ_1 and δ_2, assuming $\delta_L = 0°$.

(b) Determine the magnitude of the load voltage V_L.

(c) Determine the load resistance R_L.

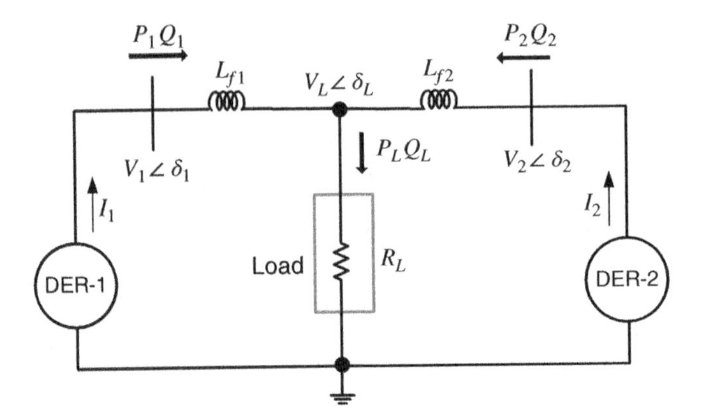

Figure P10.3 The autonomous microgrid operating under angle droop of Problem 10.3.

10.4 Consider a predominantly resistive autonomous microgrid with three DERs. The power ratings of the DERs are:

$$\text{DER}-1: 1\text{ MW}, \text{DER}-2: 500\text{ kW}, \text{and DER}-3: 2\text{ MW}$$

The reactive power ratings of the DERs in MVAr are assumed to be 50% of their respective active power ratings in MW, i.e. 500 kVAr for DER-1, etc. The DERs are controlled by the droop equations given in (10.27) and (10.28), where $V^* = 22$ kV (L-L) and $f^* = 60$ Hz. Find the droop gains so that the frequency variation is restricted to be within ± 0.5 Hz and the voltage variation is restricted to be within ± 0.06 pu.

10.5 Consider the utility connected AC microgrid, shown in Figure P10.5. Together the utility and the DERs supply a constant PQ load, where it is stipulated that DER-1 supplies twice the amount of active power supplied by DER-2. The system parameters are:
Voltage: $V_1 = V_2 = 11$ kV (L-L), 50 Hz
Feeder parameters: $L_{f1} = 20$ mH and $L_{f2} = 10$ mH
Load: $P_L = 400$ kW and $Q_L = 90$ kVAr
If the utility supplies (P_s) 100 kW:
(a) Determine the angles δ_1 and δ_2 to hold V_L (L-N) constant at 6.34 kV.
(b) Determine the reactive power Q_1 and Q_2 supplied by the DERs.
(c) Determine the reactive power supplied by the utility.

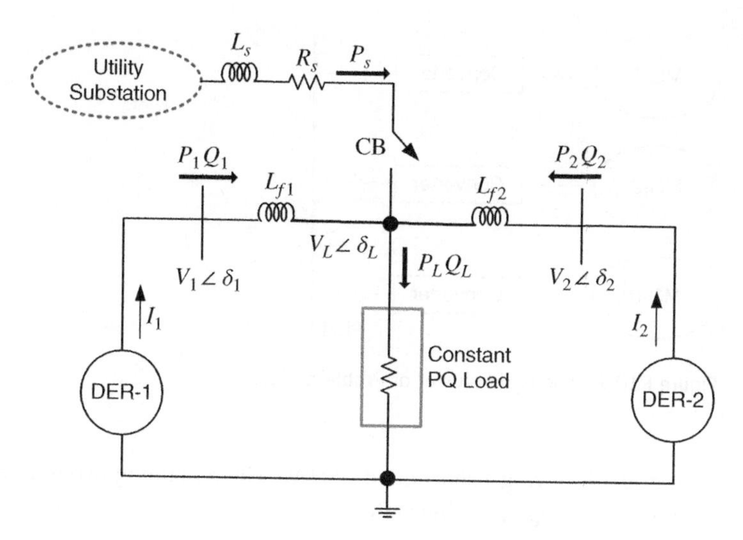

Figure P10.5 Grid-connected microgrid of Problem 10.5.

10.6 Consider the DCMG shown in Figure 10.34, where the DC-DC converters are boost converters. The system parameters are:
Voltage rating (V^*): 1 kV
Maximum voltage deviation (ΔV_m): 200 V
DC-DC Coverter-1 maximum power: 300 kW
DC-DC Coverter-2 maximum power: 100 kW
$V_{dc1} = 500$ V, $V_{dc2} = 400$ V, $R_{f1} = 0.02\ \Omega$, and $R_{f2} = 0.06\ \Omega$

The microgrid is controlled under a P-V droop control regime. Assuming that the converters together supply total load plus losses of 240 kW:

(a) Determine the duty ratios of the two boost converters.
(b) Determine the load voltage (V_L) and the load resistance (R_L).

10.7 Repeat Problem 10.6 if the microgrid operates under an I-V droop control regime.

10.8 Consider the DAB circuit shown in Figure 10.39, where the transformer is assumed to be lossless. The parameters of the DAB are:
$V_{dc} = 48$ V, $V_0 = 650$ V, $R_L = 420\ \Omega$, and frequency = 20 kHz
Transformer: 100 V: 1 kV, leakage inductance 0.79577 μH

From the power transfer relationship, determine the phase shift (ϕ).

10.9 Consider the three-microgrid cluster shown in Figure P10.9. Here, the type of converter or the type of PEH is not important. The droop gains of all the microgrids are chosen to restrict the frequency to be within ±0.5 Hz.

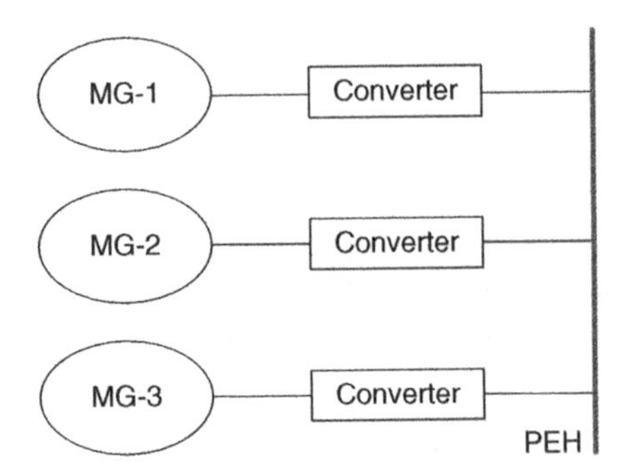

Figure P10.9 Microgrid cluster of Problem 10.9.

Microgrid-2 is having a shortfall of 200 kW. The remaining two microgrids have the following operating status:

Microgrid-1: Total capacity 2 MW. Currently supplying load at a frequency of 49.8 Hz.
Microgrid-3: Total capacity 1 MW. Currently supplying load at a frequency of 49.7 Hz.

Assuming that Microgrid-1 an Microgrid-2 share power according to their balance rating, determine how much power will be supplied by each of these microgrids. Also, determine the frequencies of these two microgrids after they start supplying power to Microgrid-2. Assume the converter and line losses are negligible.

Notes and References

Different converter control modes are very well presented in [3] and [4]. Since the power flow in a mostly inductive line depends on the relative angle difference, P-δ based droop control is presented in [6, 7]. This method, however, is reliant on the output inductance ratios and a global clock is necessary to synchronize the angles in a common timeframe. However, the controller has a superior dynamic response for converter interfaced DGs. The Virtual impedance concept has gained much attention as a tool to overcome the impedance ratio problem [11–13]. However, the stability of the system will depend on the choice of the virtual impedance. Therefore, eigenvalue analysis may be required before these parameters can be properly tuned.

The V-I droop sharing for DCMG is discussed in [38] and [39]. In [38], the droop gain selection based on voltage drop is also highlighted. In [40], the autonomous operation of AC and DC subgrids is proposed, where these are linked through an IC that normalizes the droop equations of these two subgrids. There are several papers that discuss the integrated operation of AC-DC systems: see [41–43]. Finally, a DC system does not have the natural zero-crossing of current, unlike an AC system. Therefore, the protection of a DC system is very challenging, as is discussed in [17]. A comprehensive review of DCMG protection is presented in [44].

Game theory [45, 46], which has gained significant importance in recent years, has been used in power markets for several years. Different power energy market related subjects are covered in several papers, e.g. profit allocation [47], the planning of a hybrid grid using central generation, and power converter interfaced generators [48] and energy trading in microgrids [49–51].

1 Lasseter, R., Akhil, A. Marnay, C. et al. Integration of distributed energy resources: the CERTS micro-grid concept. White Paper prepared by Consortium for Electric Reliability Technology Solutions (CERTS). https://escholarship.org/content/qt9w88z7z1/qt9w88z7z1_noSplash_394a7c3b903aebec68fdce8d79e53708.pdf, accessed 22 May 2022.

2 Navigant Research (2017). Microgrids: a global view. International Symposium on Microgrids. Newcastle, Australia (29 November 2017). http://microgrid-symposiums.org/wp-content/uploads/2017/06/Plenary_1_Asmus_v01_20171101.pdf, accessed 22 May 2022.

3 Rocabert, J., Luna, A., Blaabjerg, F., and Rodrıguez, P. (2012). Control of power converters in AC microgrids. *IEEE Trans. Power. Electron.* 27 (11): 4734–4749.

4 Levron, Y., Belikov, J., and Baimel, D. (2018). A tutorial on dynamics and control of power systems with distributed and renewable energy sources based on the dq0 transformation. *Appl. Sci.* 8 (9): 1661. https://doi.org/10.3390/app8091661.

5 Glover, J.D., Sarma, M.S., and Overbye, T.J. (2012). *Power System Analysis and Design*, 5e. Stamford, CT: Cengage Learning.

6 Majumder, R., Chaudhuri, B., Ghosh, A. et al. (2010). Improvement of stability and load sharing in an autonomous microgrid using supplementary droop control loop. *IEEE Trans. Power Sys.* 25 (2): 796–808.

7 Majumder, R., Ghosh, A., Ledwich, G., and Zare, F. (2010). Power management and power flow control with back-to-back converters in a utility connected microgrid. *IEEE Trans. Power Sys.* 25 (2): 821–834.

8 De Brabandere, K., Bolsens, B., Van den Keybus, J. et al. (2007). A voltage and frequency droop control method for parallel inverters. *IEEE Trans. Power Electron.* 22 (4): 1107–1115.

9 Majumder, R., Ledwich, G., Ghosh, A. et al. (2010). Droop control of converter interfaced micro sources in rural distributed generation. *IEEE Trans. Power Delivery* 25 (4): 2768–2778.

10 Guerrero, J.M., de Vicuña, L.G., Castilla, M., and Miret, J. (2004). A wireless controller to enhance dynamic performance of parallel inverters in distributed generation systems. *IEEE Trans. Power Electron.* 19 (5): 1205–1213.

11 Guerrero, J.M., Matas, J., de Vicuña, L.G. et al. (2007). Decentralized control for parallel operation of distributed generation inverters using resistive output impedance. *IEEE Trans. Indust. Electr.* 54 (2): 994–1004.

12 Yu, X., Khambadkone, A.M., Wang, H., and Terence, S.T.S. (2010). Control of parallel-connected power converters for low-voltage microgrid: part I: a hybrid control architecture. *IEEE Trans. Power Electron.* 25 (12): 2962–2970.

13 Pogaku, N., Prodanovic, M., Green, T.C. et al. (2007). Modeling, analysis and testing of autonomous operation of an inverter-based microgrid. *IEEE Trans. Power Electron.* 22 (2): 613–625.

14 Rocabert, J., Azevedo, G.M.S., Luna, A. et al. (2011). Intelligent connection agent for three-phase grid-connected microgrids. *IEEE Trans. Power Electron.* 26 (10): 2993–3005.

15 Goyal, M., Fan, Y., Ghosh, A., and Shahnia, F. (2016). Techniques for a wind energy system integration with an islanded microgrid. *Int. J. Emerg. Electr. Power Syst.* 17 (2): 191–203.

16 Pashajavid, E., Ghosh, A., and Zare, F. (2018). A multimode supervisory control scheme for coupling remote droop-regulated microgrids. *IEEE Trans. Smart Grid* 9 (5): 5381–5392.

17 Kumar, D., Zare, F., and Ghosh, A. (2017). DC microgrid technology: system architectures, AC grid interfaces, grounding schemes, power quality, communication networks, applications, and standardizations aspects. *IEEE Access* 5: 12230–12256.

18 Jin, Z., Savaghebi, M., Vasquez, J.C. et al. (2016). Maritime DC microgrids: a combination of microgrid technologies and maritime onboard power system for future ships. *IEEE 8th Power Electronics and Motion Control Conference (IPEMC-ECCE Asia).* Hefei, China (22–26 May 2016).

19 Hansen, J.-F., Lindtjørn, J.O., Myklebust, T.A., and Vanska, K. (2012). Onboard DC grid. *ABB Review* 2 (12): 29–33. https://library.e.abb.com/public/4b6d4b3cfd353a88c1257a25002696a8/ABB%20Review%202-2012_72dpi.pdf, accessed 22 May 2022.

20 Datta, A.J. (2019). Operation and control of DC microgrids. PhD thesis. Curtin University, Perth, Australia.

21 Zhao, B., Song, Q. et al. (2014). Overview of dual active bridge isolated bidirectional DC-DC converter for high- frequency-link-power-conversion system. *IEEE Trans. Power Electron.* 29 (8): 4091–4106.

22 De Doncker, R.W.A.A., Divan, D.M., and Kheraluwala, M.H. (1991). A three-phase soft-switched high-power-density DC/DC converter for high-power applications. *IEEE Trans. Ind. Appl.* 27 (1): 63–73.

23 Krismer, F. and Kolar, J.W. (2009). Accurate small-signal model for the digital control of an automotive bidirectional dual active bridge. *IEEE Trans. Power Electron.* 24 (12): 2756–2768.

24 Datta, A.J., Ghosh, A., Zare, F., and Rajakaruna, S. (2019). Bidirectional power sharing in an AC/DC system with a dual active bridge converter. *IET Gener. Transm. Distrib.* 13 (4): 495–501.

25 Olivares, D.E., Mehrizi-Sani, A., Etemadi, A.H. et al. (2014). Trends in microgrid control. *IEEE-PES Task Force on Microgrid. Cont. IEEE Trans. Smart Grid* 5 (4): 1905–1919.

26 Majumder, R., Ghosh, A., Ledwich, G., and Zare, F. (2009). Load sharing and power quality enhanced operation of a distributed microgrid. *IET Renew. Power Gener.* 3 (2): 109–119.

27 Lopes, J.A.P., Moreira, C.L., and Madureira, A.G. (2006). Defining control strategies for microgrids islanded operation. *IEEE Trans. Power Sys.* 21 (2): 916–924.

28 Shafiee, Q., Guerrero, J.M., and Vasquez, J.C. (2014). Distributed secondary control for islanded microgrids: a novel approach. *IEEE Trans. Power Electron.* 29 (2): 1018–1031.

29 Han, Y., Zhang, K., Li, H. et al. (2018). MAS-based distributed coordinated control and optimization in microgrid and microgrid clusters: a comprehensive overview. *IEEE Trans. Power Electron.* 33 (8): 6488–6508.

30 Shawon, M.H., Muyeen, S.M., Ghosh, A. et al. (2019). Multi-agent systems in ICT enabled smart grid: a status update on technology framework and applications. *IEEE Access* 7: 97959–97973.

31 Lasseter, R. (2011). Smart distribution: coupled microgrids. *Proc. of IEEE.* 99 (6): 1074–1082.

32 Li, Z., Shahidehpour, M., Aminifar, F. et al. (2017). Networked microgrids for enhancing the power system resilience. *Proc. IEEE* 105 (7): 1289–1310.

33 Liu, G., Starke, M.R., Ollis, B. and Xue, Y. (2016). Networked microgrid scoping study. *Technical Report ORNL/TM-2016/294.* Oak Ridge National Laboratory. https://info.ornl.gov/sites/publications/Files/Pub68339.pdf, accessed 22 May 2022.

34 Alam, M.N., Chakrabarti, S., and Ghosh, A. (2019). Networked microgrids: state-of-the-art and future perspectives. *IEEE Trans. on Indust. Informat.* 15 (3): 1238–1250.

35 Farhangi, H. (2010). The path of the smart grid. *IEEE Power Energ. Mag.* 19–28.

36 Goyal, M. (2016). Reliability improvement of autonomous microgrids through interconnection and storage. PhD thesis, Curtin University, Perth, Australia.

37 John, B., Ghosh, A., Goyal, M., and Zare, F. (2019). A DC power exchange highway based power flow management for interconnected microgrid clusters. *IEEE Syst. J.* 13 (3): 3347–3357.

38 Ito, Y., Zhongqing, Y., and Akagi, H. (2004). DC microgrid based distribution power generation system. *IEEE 4th International Power Electronics and Motion Control Conference* 3: 1740–1745.

39 Kurohane, K., Senjyu, T., Yona, A. et al. (2010). A hybrid smart AC/DC power system. *IEEE Trans. Smart Grid* 1 (2): 199–204.

40 Loh, P.C., Li, D., Chai, Y.K., and Blaabjerg, F. (2013). Autonomous operation of hybrid microgrid with AC and DC subgrids. *IEEE Trans. Power Electron.* 28 (5): 2214–2223.

41 dos Santos Neto, P.J., Barrosc, T.A.S., Silveira, J.P.C. et al. (2020). Power management techniques for grid-connected DC microgrids: a comparative evaluation. *Appl. Ener.* 269: https://doi.org/10.1016/j.apenergy.2020.115057.

42 Eghtedarpour, N. and Farjah, E. (2018). Power control and management in a hybrid AC/DC microgrid. *IEEE Trans. Smart Grid* 5 (3): 1494–1505.

43 Khorsandi, A., Ashourloo, M., and Mokhtari, H. (2014). A decentralized control method for a low-voltage DC microgrid. *IEEE Trans. Energ. Convers.* 29 (4): 793–801.

44 Beheshtaein, S., Cuzner, R.M., Forouzesh, M. et al. DC Microgrid protection: a comprehensive review. *IEEE Journal of Emerging and Selected Topics in Power Electronics* https://doi.org/10.1109/JESTPE.2019.2904588.

45 Barron, E.N. (2013). *Game Theory: An Introduction.* New York: Wiley.

46 Osbourne, M.J. (2004). *An Introduction to Game Theory.* Oxford: Oxford University Press.

47 Xia, N.X. and Yokoyama, R. (2003). Profit allocation of independent power producers based on cooperative game theory. *Elect. Power and Ener. Sys.* 25: 633–641.

48 Mei, S., Wang, Y., Liu, F. et al. (2012). Game approaches for hybrid power system planning. *IEEE Trans. Sustainable Energy* 3 (3): 506–517.

49 Ali, L., Muyeen, S.M., Bizhani, H., and Ghosh, A. (2020). Optimal planning of clustered microgrid using a technique of cooperative game theory. *Electr. Power Syst. Res.* 183, https://doi.org/10.1016/j.epsr.2020.106262.

50 Dehghanpour, K. and Nehrir, H. (2019). An agent-based hierarchical bargaining framework for power management of multiple cooperative microgrids. *IEEE Trans. Smart Grid.* 10 (1): 514–522.

51 Lee, J., Guo, J., Choi, J.K., and Zukerman, M. (2015). Distributed energy trading in microgrids: a game-theoretic model and its equilibrium analysis. *IEEE Trans. Indust. Elect.* 62 (6): 3524–3533.

11

Harmonics in Electrical and Electronic Systems

In an ideal electrical and electronic system, pure sinusoidal voltage sources provide power and energy to the different parts of the system to support single-phase or three-phase loads. The frequency of the voltage source is expected to be constant at 50 or 60 Hz and the phase voltage amplitude is expected to be below 240 V rms in low-voltage distribution networks, depending on a national grid code. There are many nonlinear loads and grid equipment such as power electronics converters and distribution transformers which can generate current and voltage signals with different frequencies: integer and noninteger multiples of the grid fundamental frequency. These signals are known as harmonics and interharmonics. These are defined in this chapter, along with their effects on power systems. Also, a review of different power quality regulations and standards is included at the end of the chapter.

11.1 Harmonics and Interharmonics

Harmonics in power systems occur in the form of a current or voltage waveform and it is an integer multiple of the power system's fundamental frequency. Conventional power electronic products, such as single-phase and three-phase diode rectifiers, generate significant low-order harmonics because of the operating modes of the rectifier (AC-DC converter) and a DC link filter to control the DC link voltage.

Interharmonics in power systems come in the form of a current or voltage waveform and are a noninteger multiple of the power system fundamental frequency. Interharmonics are mainly considered for 0–2 kHz in power system applications and can be generated by:

- Interaction between two systems operating at different frequencies.
- A rapid change of current in equipment.

Control of Power Electronic Converters with Microgrid Applications, First Edition.
Arindam Ghosh and Firuz Zare.

Published 2023 by John Wiley & Sons, Inc.

- Pulse width modulation (PWM) patterns utilized in power converters such as random and hysteresis modulations.
- Controlled rectifiers with variable firing angles.
- Multicycle control of AC systems.

Consider, for example, the following load current

$$i_{Load}(t) = 10\sin(\omega_0 t) + 3\sin(5\omega_0 t) - 0.5\sin(2.5\omega_0 t)\ \text{A} \tag{11.1}$$

where $\omega_0 = 2\pi f_0$, f_0 being the fundamental frequency, which is taken as 50 Hz. In (11.1), the load current has a fundamental component of amplitude 10 A, a harmonic component at 250 Hz with 3 A amplitude, and an interharmonic component at 125 Hz with the amplitude of 0.5 A. The instantaneous load current is shown in Figure 11.1a, while its harmonic spectrum is shown in Figure 11.1b.

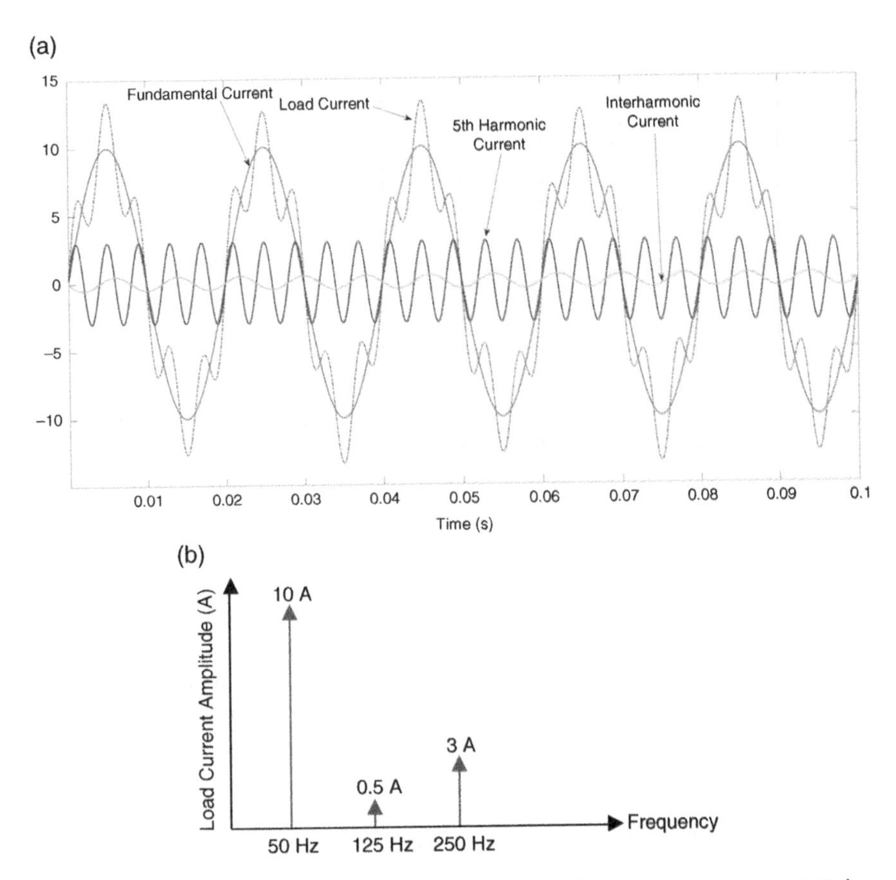

Figure 11.1 A load current waveform with harmonic and interharmonic components in (a) time domain and (b) frequency domain.

According to the International Electrotechnical Commission (IEC) 61 000 standards, the frequency range of harmonics is defined as up to 9 kHz. The method of measurement and harmonic limits are classified based on several factors such as the frequency range of 0–2 kHz or 2–9 kHz, load types and power levels, and distribution network configurations and types. Measurement methods and harmonic limits in the range of 9 kHz and above are covered by International Special Committee on Radio Interference (CISPR) standards. These essential concepts are explained in Section 11.4.

11.1.1 High-frequency Harmonics (2–150 kHz)

Two new frequency ranges, 2–9 kHz and 9–150 kHz, have been identified by international standardization committees as new disturbing frequency ranges which affect communication and control signals in smart meters, power line carriers (PLC), and ripple control signals in distribution networks. Conventional power converters and modern power electronics systems based on active front end (AFE) and PWM-based converters with fast switching operations have increased harmonics emissions above 2 kHz in low- and medium-voltage networks.

The levels of the harmonics and their spectral contents depend on power electronics technology, design, and application, and their interaction with the grid and grid-connected equipment. The switching operation of these power electronic devices causes nonsinusoidal current and voltage waveforms to affect the power quality of the grids. The utilization of power electronics systems has increased significantly in many applications, such as rooftop solar inverters and compact fluorescent lamps, with significant impacts on high-frequency and high-energy harmonics within the frequency range of 2–150 kHz.

Harmonics have short- and long-term adverse effects on grids and grid-connected electronics and power electronics equipment, such as malfunction, failure, and losses. These issues reduce the reliability, lifetime, and efficiency of the electricity networks. There are no comprehensive standards for harmonic emission within the frequency range of 2–150 kHz, to protect the current and future electricity networks and smart grids. IEC Technical Subcommittee SC 77A, Working Group 1 is the world-leading authority to prepare technical documents for international standards. It has requested international experts to define standards for harmonics within the frequency range of 2–150 kHz. Conducted emission and immunity limits and measurement methods are being developed by several IEC and CISPR standardization committees.

The new challenging issues for future grids concerning these new frequency ranges are classified as:

- Generation of high-frequency harmonics.
- Creation of new resonant frequencies.

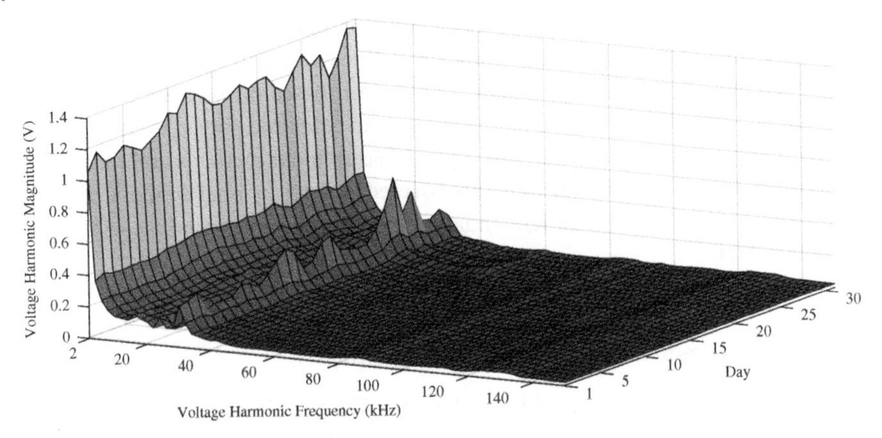

Figure 11.2 Daily and hourly average values for 2–150 kHz voltage harmonics in a distribution network.

- Harmonic interactions between different types of grid-connected systems.
- Propagation of high-frequency harmonics in medium-voltage networks.

These issues depend on many factors such as load types and profiles, the length and types of feeders, distribution transformers, and system configurations. Hence utility companies, as well as renewable energy and power electronics manufactures, have been facing new challenges to solve the harmonic issues in these frequency ranges. The current trend to manage the high-frequency harmonic concerns is to define proper harmonics standards and compatibility levels for grid connected electronic and power electronic equipment.

Figure 11.2 presents daily average values of voltage harmonics in 2–150 kHz in a low-voltage distribution network. As shown in this figure, there are some peaks on voltage harmonics at higher frequencies around 30 kHz, which indicate the high penetration of grid-connected power electronics systems with high switching frequency and the interaction with the grid and grid impedance.

11.1.2 EMI in the Frequency Range of 150 kHz–30 MHz

Electromagnetic interference (EMI) is a high-frequency noise and disturbance generated by external sources or by another part of a system. EMI is also known as radiofrequency interference (RFI). The main phenomena occur due to the conducted or radiated emission noise caused by electromagnetic and electrostatic couplings between the systems, components, enclosure, and interconnections. The disturbance may be severe enough to damage sensitive devices or significant enough to reduce the lifetime of components or degrade the regular operation

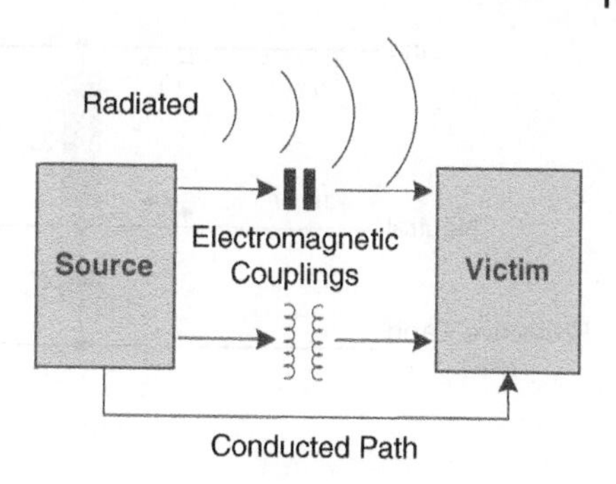

Figure 11.3 Conducted and radiated EMI noise emission and propagation.

of the system. The main EMI sources in distribution networks are lightnings, fast high voltage and high current switching, ignition systems, pulsed width modulated signals in power converters, and many other types of voltage and current transitions in electrical and electronic devices.

As shown in Figure 11.3, high-frequency noise is classified as conducted (low-impedance loop) and radiated (electromagnetic coupling), which can be propagated and transferred from one part of the circuit to another part of the system. CISPR standards are used for the high-frequency noise measurement and with specific noise limits. The frequency range depends on the standard. For example, for the conducted emission, it can be from 150 kHz to 30 MHz and for the radiated emission it can be from 30 MHz to 1 GHz. The high-frequency conducted noise is classified into common mode and differential mode noises, which are explained in Section 11.1.1.2.

11.1.3 Common Mode and Differential Mode Harmonics and Noises

In a low-voltage power system, neutral and protective earth (PE) wires provide two different current paths for electrical equipment. A PE wire is connected to a frame or an enclosure of equipment for protection and safety purposes, while a neutral wire is a main current path for single-phase systems or a zero-sequence path in three-phase systems. As shown in Figure 11.4, in an electrical system, the currents through the line and neutral are defined based on the common mode and differential mode components as

$$i_{Line}(t) = i_{COM}(t) + i_{DM}(t) \tag{11.2}$$

$$i_{Neutral}(t) = i_{COM}(t) - i_{DM}(t) \tag{11.3}$$

If the sum of the line and neutral currents is zero at any time, it means that there is no leakage or common mode current through the PE wire.

Due to stray capacitive couplings in AC motors, cables, heatsink, enclosures, and transformers and voltage stress (dv/dt) in PWM voltage, the leakage current

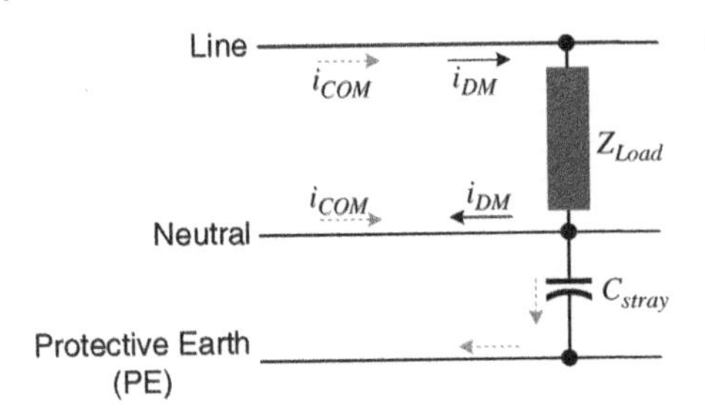

Figure 11.4 Common mode and differential mode current paths.

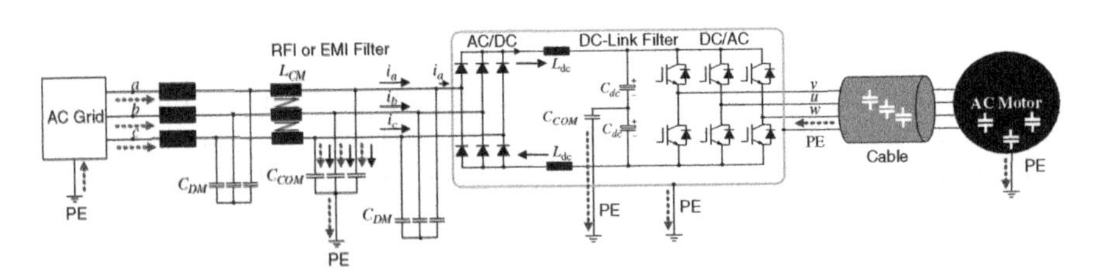

Figure 11.5 Common mode and differential mode currents in a motor drive system.

(common mode current) is a major concern for most power electronics converters. The amplitude, frequency, and waveform of the common mode current depend on the power converter topology, interconnections and PCB layout, control and modulation method, load, and filter configurations.

Figure 11.5 shows a motor drive application with a front side EMI and a DC link filter including all interconnections and parasitic couplings with respect to the PE. There are several common mode and differential mode loops that generate low- and high-frequency (common mode and differential mode) currents at the load side and the grid side.

11.1.4 Stiff and Weak Grids

A weak power system can be susceptible to a sudden change in load or operating conditions, which can cause deviation in the grid voltage and frequency. Strong grids are robust and have a high capability to handle sudden changes in power demand, load variations and fault conditions – keeping the grid voltage and frequency within the limits in which the system can be stable.

For power quality and harmonic analysis, we consider a weak AC grid as a system with a high impedance and a low power transfer capability, whereas a stiff grid has a lower impedance and higher power transfer capability. When the grid impedance is low compared to a load impedance, the load variation has less impact on the fundamental and harmonic voltages. As shown in Figure 11.6, the

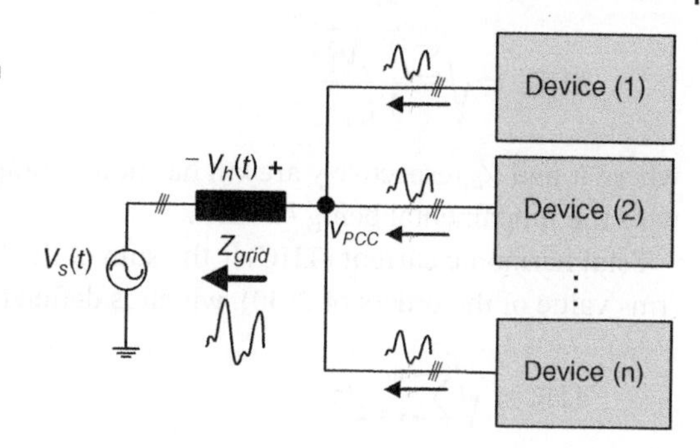

Figure 11.6 Grid impedance impact on voltage harmonics in a distribution network.

nonlinear devices in the low-voltage grid generate current harmonics which can affect the power quality of the voltage at the point of common coupling due to the voltage drop across the grid impedance $v_h(t)$ (as discussed in Chapter 9).

The grid impedance for power quality analysis should be measured and characterized for the frequency range of 0–9 kHz as the impedance at any point of common coupling varies due to the load impedances, feeders, and transformers. According to IEC standards [1, 2], the short circuit ratio (SCR) is defined as the ratio of the short circuit power of a grid to an apparent rated power of an interconnecting device. In general, a weak grid has a low SCR compared to a stiff grid. For power quality and harmonic analysis, the grid impedance varies with respect to frequency due to capacitive couplings in cables and transformers (nonlinear permittivity), as well as the nonlinear property of the magnetic core in transformers. The impact of the grid impedance is more significant at higher-frequency ranges as the voltage drop across the grid impedance depends on the order of the harmonics.

11.2 Power Quality Factors and Definitions

We briefly present harmonic distortion in Chapter 2. In this section, we present elaborate definitions of different power quality factors that impact a power grid.

11.2.1 Harmonic Distortion

As mentioned in Chapter 2, the total harmonic distortion (THD) is the ratio of the sum of the harmonic components (rms value of the orders of 2–40) to the rms value of the fundamental signal that can be defined for current and voltage signals as follows [1, 2]

$$\mathrm{THD}_i = \sqrt{\frac{\sum_{k=2}^{40} I_k^2}{I_1^2}} \tag{11.4}$$

$$\mathrm{THD}_v = \sqrt{\frac{\sum_{k=2}^{40} V_k^2}{V_1^2}} \tag{11.5}$$

where I_k and V_k respectively are kth harmonic component of current and voltage, with the fundamental being $k = 1$.

Total harmonic current (THC) is the sum of the harmonic current components (rms value of the orders of 2–40), which is defined as

$$\mathrm{THC} = \sqrt{\sum_{k=2}^{40} I_k^2} \tag{11.6}$$

Therefore, we have

$$\mathrm{THD}_i = \frac{\mathrm{THC}}{I_1}$$

Partial odd harmonic current (PHOC) is the sum of the harmonic current components (rms value of the orders of 21–39), which is defined as

$$\mathrm{POHC} = \sqrt{\sum_{k=21,23,\ldots}^{40} I_k^2} \tag{11.7}$$

Partial harmonic current (PHC) is the sum of the harmonic current components (rms value of the orders of 14–40), which is defined as

$$\mathrm{PHC} = \sqrt{\sum_{k=14}^{40} I_k^2} \tag{11.8}$$

Example 11.1 Consider the instantaneous current

$$i(t) = 10\sqrt{2}\left\{\sum_{k=1}^{40} \frac{\sin(k\omega t)}{k}\right\} \mathrm{A}$$

where $\omega = 100\pi$ rad/s. This means that the rms value of the current at the fundamental frequency of 50 Hz is 10 A, while the rms values decrease significantly to the harmonic numbers, e.g. it is 5 A for the second harmonic and 3.33 A for the third harmonic, and so on. Then from (11.4) and (11.6)–(11.8), we have

$$\mathrm{THD}_i = 0.7876, \mathrm{THC} = 7.876\ \mathrm{A}, \mathrm{PHOC} = 1.117\ \mathrm{A}, \text{and } \mathrm{PHC} = 2.222\ \mathrm{A}$$

This implies that the current THD is 78.76%.

Figure 11.7 depicts the current, when the fundamental and different harmonic components are present. For comparison, the fundamental component is also plotted in these figures. When all even and odd harmonic components ride the fundamental, the waveform almost becomes a sawtooth with a frequency that is

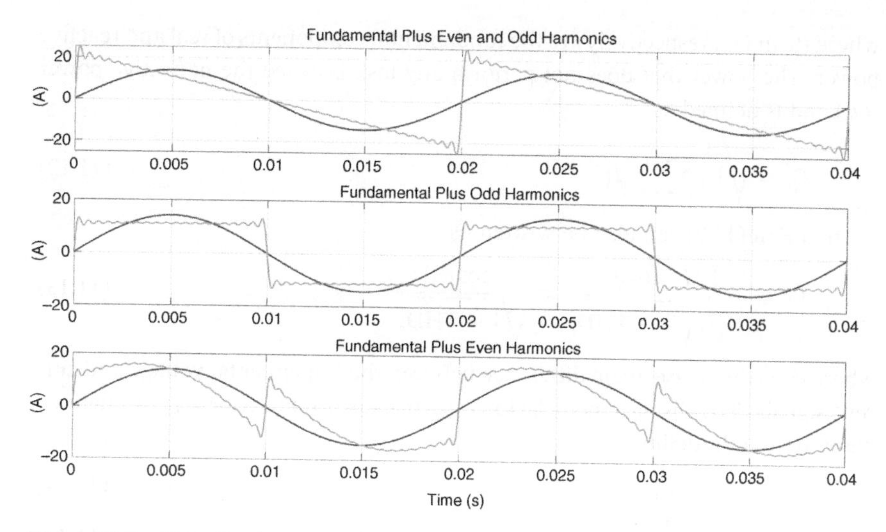

Figure 11.7 Distortion in current due to different harmonic components.

half of the fundamental frequency, as shown in Figure 11.7a. When the waveform contains the odd harmonic, it almost becomes a square wave having the same frequency as the fundamental waveform (Figure 11.7b). However, when only the even harmonics are present with the fundamental, the waveform is irregular but periodic, as shown in Figure 11.7c.

11.2.2 Power and Displacement Factors

The power factor (PF) of an electrical system is defined as the ratio of the active power (absolute value) to apparent power as

$$\mathrm{PF} = \frac{|P|}{S} \tag{11.9}$$

where P is the active power and S is the apparent power. Assuming that the voltage harmonics are negligible but load current harmonics are present, the apparent power can be written as

$$S = \sqrt{V_1^2 I^2} = \sqrt{V_1^2\left(I_1^2 + \sum\nolimits_{k=2}^{40} I_k^2\right)} = \sqrt{V_1^2 I_1^2 + V_1^2 \sum\nolimits_{k=2}^{40} I_k^2} \tag{11.10}$$

The fundamental apparent power is given by

$$S = \sqrt{V_1^2 I_1^2} = \sqrt{P_1^2 + Q_1^2} \tag{11.11}$$

where P_1 and Q_1 respectively are the fundamental components of real and reactive power. The power that does not perform any task is called the nonactive power (Q_h) and is defined as

$$Q_h = \sqrt{V_1^2 \sum_{k=2}^{40} I_k^2} \tag{11.12}$$

The PF in (11.9) can also be written as

$$\text{PF} = \frac{V_1 I_1 |\cos\phi_1|}{V_1 I_1 \sqrt{1 + \text{THD}_i^2}} = \frac{|\cos\phi_1|}{\sqrt{1 + \text{THD}_i^2}} \tag{11.13}$$

where ϕ_1 is the phase angle difference between the fundamental voltage and current signals. The displacement factor λ_{disp} and distortion factor λ_{dist} are defined based on the PF definition as

$$\lambda_{disp} = |\cos\phi_1| \tag{11.14}$$

$$\lambda_{dist} = \frac{1}{\sqrt{1 + \text{THD}_i^2}} \tag{11.15}$$

Therefore, the PF of the electrical system consists of two factors, λ_{disp} and λ_{dist}, which have two different effects on grids. The reactive power is due to the displacement factor λ_{disp}, which can affect the grid voltage level, the line current, and additional losses through cables and transformers. The current harmonic distortion is represented in the distortion factor λ_{dist}, and it has an impact on the power quality and energy efficiency of the grid, including additional core and copper losses in the power systems.

Example 11.2 Consider the instantaneous current of Example 11.1. Let us assume that the fundamental component of this current lags the supply voltage by 20°, i.e. $\phi_1 = -20°$. Then we have

$$\lambda_{disp} = |\cos\phi_1| = 0.9397, \lambda_{dist} = \frac{1}{\sqrt{1 + (0.7876)^2}} = 0.8863, \text{and}$$

$$\text{PF} = \lambda_{disp} \times \lambda_{dist} = 0.8329$$

11.3 Harmonics Generated by Power Electronics in Power Systems

In the past, one of the main harmonic contaminations in power system was the low-order harmonic currents caused by saturated iron in transformers and AC motor drive systems. Today, there are several nonlinear loads: conventional and modern power electronics converters based on semiconductor switching devices such as diodes, insulated-gate bipolar transistors (IGBTS), metal oxide silicon field effect transistors (MOSFETs) or thyristors which generate harmonics or interharmonics at different amplitudes and frequency ranges.

Power electronics technology in distribution networks allows the transfer of electrical power from renewable energy sources to grids, while regulating frequency and/or voltage for different loads such as variable speed drives and battery chargers. New demands for (i) cost and size reduction, (ii) performance and quality improvement, and (iii) flexibility on power management have promoted power electronics applications extensively in industrial, commercial, and residential sectors.

With the recent development and advancement in power electronics applications, the cost of these devices has been competitively reduced due to an increase in the demand and the presence of many competitors in the market. The more recent evolution in modern power electronics devices with large power handling capability provides a significant contribution in energy saving and the efficient use of electricity.

The penetration of grid-connected renewable energy sources based on power electronics technology has been increasing in low- and medium-voltage distribution networks. Harmonics are the main drawback of modern power electronics converters. Harmonic distortion causes unnecessary heat in power system and grid-connected equipment due to overloading of neutrals, overheating of transformers, nuisance tripping of circuit breakers, and overstressing of PF correction capacitors. Harmonic currents together with grid and system impedance lead to harmonic voltage distortion, which results in poor power quality and instability of the grids. Therefore, harmonic mitigation techniques have become an important topic in distribution networks. There are various harmonics mitigation techniques to improve line current waveform.

A classification of the different harmonic mitigation techniques is shown in Figure 11.8 and presented here:

- Passive methods based on inductor and LC filters.
- Multipulse rectifier techniques.
- Active harmonic cancelation techniques.

The pros and cons of each harmonic mitigation technique have already been reported in various literature.

Passive filters are cost-effective solutions and can provide acceptable current harmonic cancelation. However, they have the following drawbacks:

- Bulky and heavy.
- Performance depends on the load profile.
- Creation of resonances in power systems.

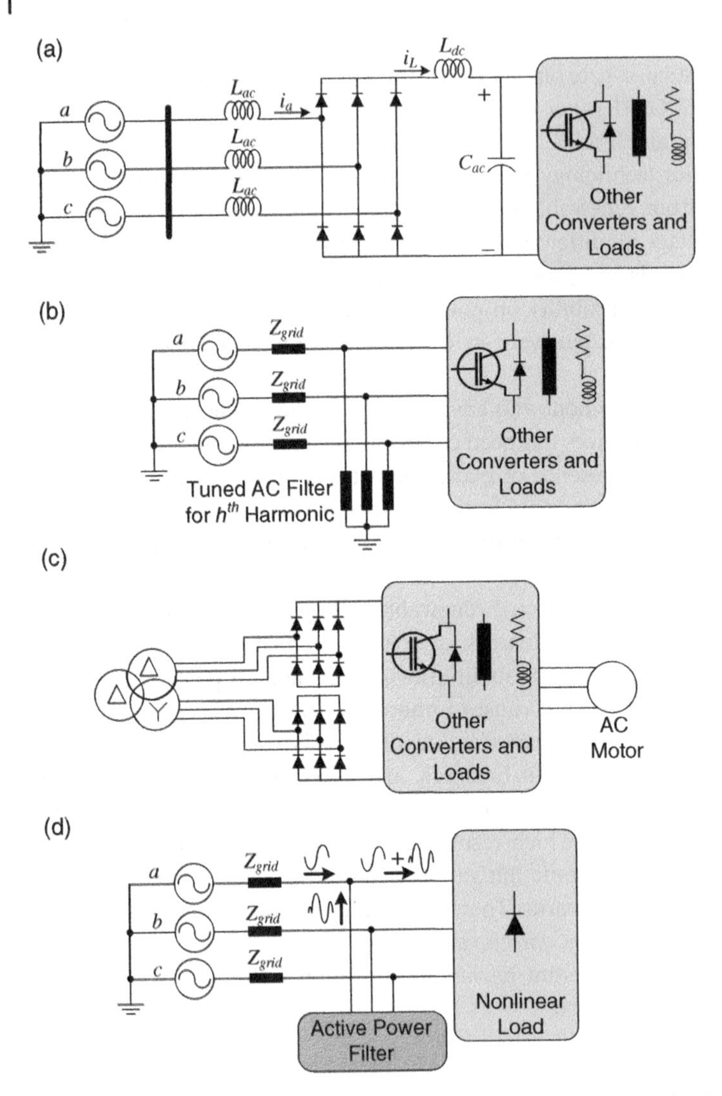

Figure 11.8 Several harmonic mitigation techniques: (a) passive inductive filter, (b) tuned LC filter, (c) 12-pulse rectifier, and (d) active power filter.

Harmonic cancelation methods using multipulse rectifiers are based on phase-shifting line current and are the preferred solutions to eliminate harmonics in high-power converters, such as large motor drive applications. However, they require bulky and expensive phase-shifting transformers, which are mostly used in step-down systems (medium- to-low voltage grids) with passive filters to eliminate the remaining harmonics.

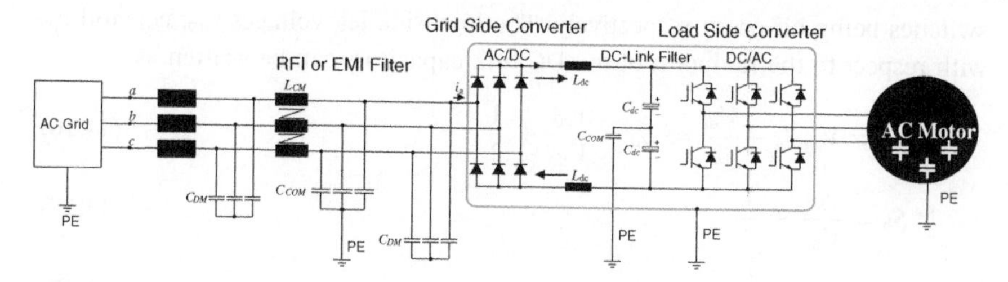

Figure 11.9 Circuit diagram of a typical power electronic system.

Active harmonic cancelation methods are based on AFE, which can reduce low-order harmonic emission significantly. These systems are not a cost-effective solution due to the utilization of semiconductor switches, sensors, advanced control systems, and gate drives.

The main focus of the discussion in this section is to identify harmonic sources at the load and grid sides. As shown in Figure 11.9, harmonics are generated in different parts of a system based on system topology, application, and load. In Sections 11.3.1–11.3.4, different power electronics converters generating harmonics are explained.

11.3.1 Harmonic Analysis at a Load Side (a Three-phase Inverter)

Figure 11.10 shows a three-phase motor drive application where a three-phase inverter is connected to a three-phase load. Let us focus on an analytical study to find the spectral contents of the load current harmonics at the DC side (i_{inv}) based on the Fourier series and switching patterns of the load side inverter.

The inverter operates based on a PWM method such as a generic sine-triangular PWM strategy and the switching functions of each leg. Let us assume that the DC link voltage is equally shared across the DC link capacitors (C_{dc}). The switching functions of the legs S_a, S_b, and S_c have the value of 0 or 1 depending on the

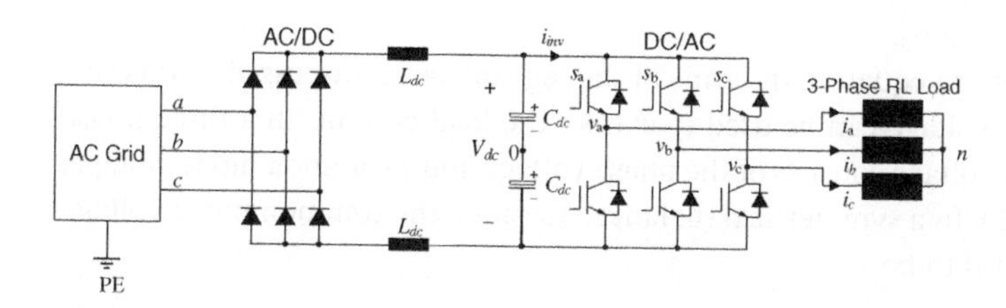

Figure 11.10 A general circuit diagram of a three-phase motor drive application with a three-phase R-L load.

switches being off or on respectively. The load-side leg voltages v_{a0}, v_{b0}, and v_{c0} with respect to the midpoint of the DC link capacitors can be written as

$$v_{a0} = V_{dc}S_a - \frac{V_{dc}}{2} \Rightarrow S_a = \frac{v_{a0}}{V_{dc}} + \frac{1}{2} \tag{11.16}$$

$$S_b = \frac{v_{b0}}{V_{dc}} + \frac{1}{2} \tag{11.17}$$

$$S_c = \frac{v_{c0}}{V_{dc}} + \frac{1}{2} \tag{11.18}$$

The double Fourier series of the leg voltage is given by [3]

$$\begin{aligned} v_{a0} &= M\frac{V_{dc}}{2}\cos(\omega_0 t) \\ &+ \frac{2V_{dc}}{\pi}\sum_{m=1}^{\infty}\sum_{k=-\infty}^{\infty}\left[J_k\left(m\frac{\pi}{2}M\right)\sin\left(\{m+k\}\frac{\pi}{2}\right) + \cos(m\omega_c t + k\omega_0 t)\right] \end{aligned} \tag{11.19}$$

where

- $\omega_c = 2\pi f_c$ is the carrier frequency.
- $\omega_0 = 2\pi f_0$ is the fundamental frequency.
- M is the modulation index.
- J_k is the Bessel function of order k.

The switching function equation of S_a can be extracted from the above leg as

$$\begin{aligned} S_a &= \frac{M}{2}\cos(\omega_0 t) \\ &+ \frac{2}{\pi}\sum_{m=1}^{\infty}\sum_{k=-\infty}^{\infty}\left[J_k\left(m\frac{\pi}{2}M\right)\sin\left(\{m+k\}\frac{\pi}{2}\right) + \cos(m\omega_c t + k\omega_0 t)\right] + \frac{1}{2} \end{aligned} \tag{11.20}$$

A similar switching function can also be extracted for S_b and S_c. The above methodology can be applied to any PWM strategy to extract the leg voltage and switching function equations.

The phase voltage equation in terms of the leg voltages is discussed in this section. The phase voltage can be used to extract the load current. In a three-phase inverter, the leg voltage consists of the phase voltage and a common mode voltage, as given in (7.15). In a symmetrical (balanced) system, the common mode voltage is shown in (7.16) to be

$$v_{n0} = \frac{v_{a0} + v_{b0} + v_{c0}}{3}$$

Now since $v_{an} = v_{a0} - v_{n0}$, the above equation can be rewritten as

$$v_{an} = \frac{2v_{a0} - v_{b0} - v_{c0}}{3} \tag{11.21}$$

The load current harmonics can be defined based on the load model and its applications. If the load is a series combination of an inductor and a resistor, the output current can be extracted as

$$i_a(t) = I_0 \cos(\omega_o t - \theta_0) + \sum_{h=1}^{\infty} I_h \sin(\omega_h t + \theta_h) \tag{11.22}$$

where I_0 is the maximum amplitude of the fundamental frequency load current and θ_0 is its phase angle. The second term on the right-hand side of (11.22) defines the spectral contents of the load current based on the load model, PWM method, and switching frequency. As there are several trigonometric terms in the phase voltage and current equations, the harmonic currents are defined in terms of a sinewave where ω_h and θ_h are the angular frequency and the phase angle of the hth harmonics, where h can have noninteger values as well.

As shown in Figure 11.10, the inverter current at the DC link side, i_{inv}, can be expressed in terms of the load currents and the inverter switching functions as

$$i_{inv} = S_a i_a + S_b i_b + S_c i_c \tag{11.23}$$

Due to the symmetry in a three-phase system, the switching functions S_b and S_c are phase shifted from S_a by 120° and therefore can be written by substituting $\omega_o t$ by $\omega_o t - 120°$ and $\omega_o t + 120°$ respectively. In a similar way, the load currents of phases b and c can also be obtained by phase shifting i_a by 120°. However, if the load impedance is unbalanced, the phase current will be unbalanced as well. The load side current harmonics are modelled in terms of the load side inverter switching function and load parameters. Thus, the harmonic emission to the grid side depends on the parameters of the entire system such as the low- and high-frequency models of the power converter, including the passive DC link filter, the EMI filter, and the grid impedance.

11.3.2 Harmonic Analysis at a Grid Side (a Three-phase Rectifier)

Three-phase AC-DC converters are utilized in many low- and high-power applications, such as motor drive systems and battery chargers. The low-order harmonics generated by the three-phase diode rectifier are mitigated using a passive filter, such as an AC or a DC inductor or a combination of these two inductors. In a three-phase balance system, there is no third and triplen harmonics due to the symmetrical operation of the diode rectifier in each leg, i.e. 120° phase shift of the conduction time instants.

Figure 11.11a shows a three-phase diode rectifier with DC link inductors connected to an inverter at the load side. Under a full load condition, it is expected

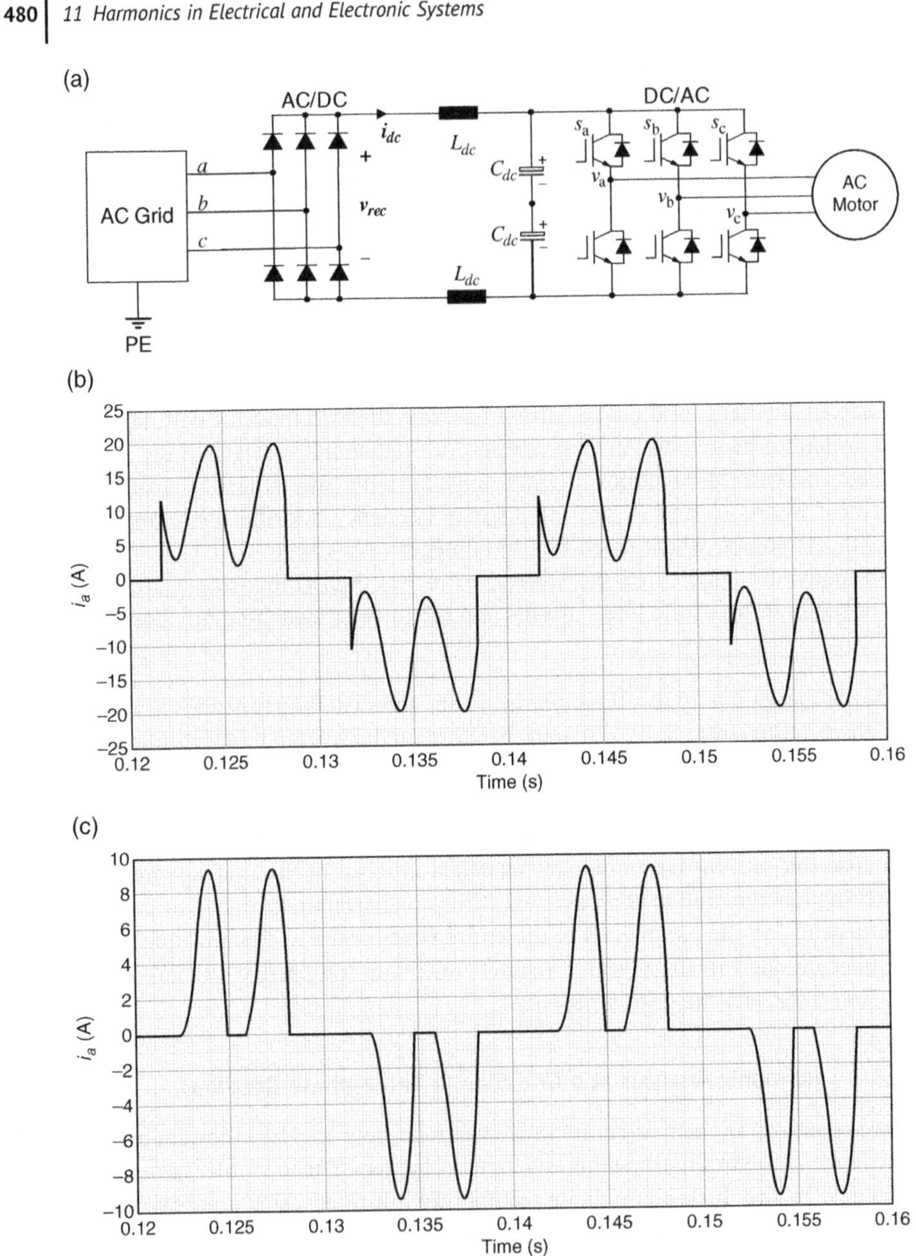

Figure 11.11 (a) A three-phase diode rectifier with (a) DC link filter and its inductor current (b) continuous conduction mode, and (c) discontinuous conduction mode.

that the DC link current is continuous (Figure 11.11b), and each diode conducts for 120° based on the grid voltage, while at a partial power condition the inductor current can be discontinuous (Figure 11.11c), and the conduction times of the diodes depend on the system and load parameters.

Mathematical equations can be extracted to find the DC and harmonic values of the rectified voltage and the DC link current for a three-phase diode rectifier. Let us assume a set of three-phase balance voltages as

$$\begin{aligned} v_a &= V_m \sin(\omega_o t) \\ v_b &= V_m \sin(\omega_o t - 120°) \\ v_c &= V_m \sin(\omega_o t + 120°) \end{aligned} \tag{11.24}$$

where V_m is the peak amplitude of the grid voltage.

The upper and lower diodes in a three-phase diode rectifier are turned on and off in which the rectified voltage can be expressed in terms of the line-to-line voltage, depending on which phases are connected to the DC link. In an ideal case, if the DC link current i_{dc} is ripple free and constant, the rectified voltage $V_{rec}(t)$ consists of six pulses, representing the line-to-line voltage for each period of $\pi/3$, as shown in Figure 11.12a. Using Fourier series, the DC and harmonic values of the rectified voltage can be extracted as

$$V_{rec}(t) = \frac{3\sqrt{3}V_m}{\pi}\left[1 - \sum_{n=1}^{\infty} \frac{2}{(36n^2 - 1)} \cos(n\pi)\cos(6n\omega_0 t)\right] \tag{11.25}$$

As shown in Figure 11.12b, and assuming the current through the inductor is continuous and the system operates in continuous conduction mode, the impedance seen from the rectifier Z_{rec} is given by

$$Z_{rec} = j2\omega L_{dc} + \frac{2Z_{inv}}{2 + j\omega C_{dc} Z_{inv}} \tag{11.26}$$

where Z_{inv} represents the impedance of the load side inverter and its load. As the rectified voltage has a DC part and harmonics at the frequencies of $6n\omega_0$, the impedance in (11.26) can be simplified in terms of the grid frequency as

$$Z_{rec} = j12n\omega_0 L_{dc} + \frac{Z_{inv}}{1 + j3n\omega_0 C_{dc} Z_{inv}} \tag{11.27}$$

The DC link consists of two capacitors in series and two inductors.

The inverter impedance can be divided into two parts: the DC impedance Z_{inv_dc} representing the active power consumed by the inverter and the load and the harmonic impedance $Z_{inv}(6n\omega_0)$. Thus, the DC link current has two parts: the DC and harmonic parts as

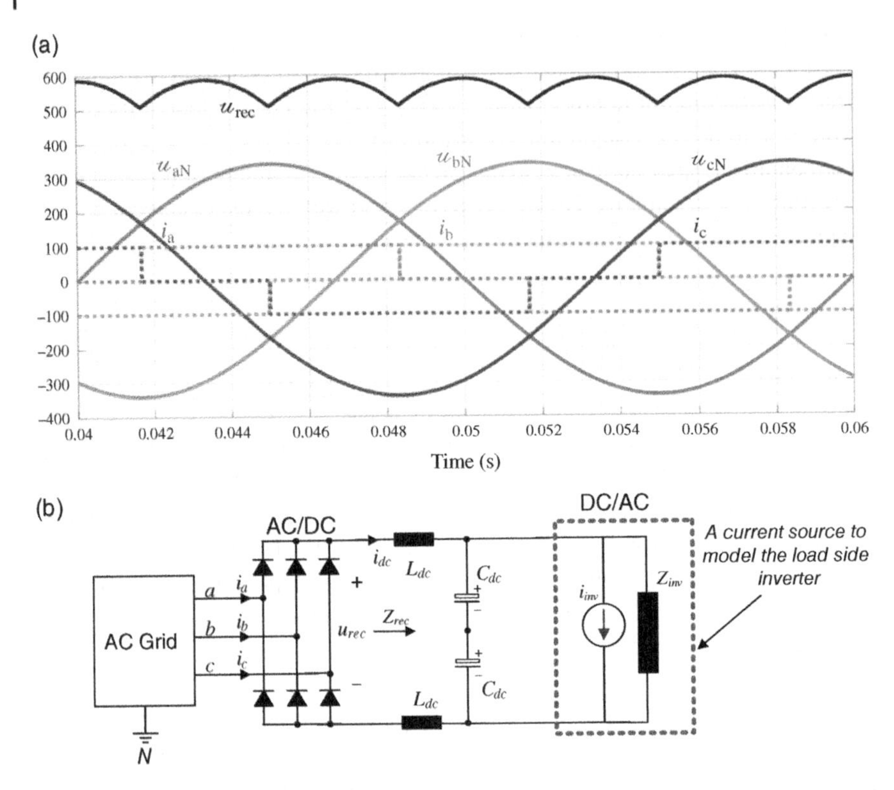

Figure 11.12 (a) The voltage output of a three-phase diode rectifier and (b) the simplified model of the inverter as a current source and an internal impedance.

$$i_{dc}(t) = \frac{3\sqrt{3}V_m}{\pi}\left[\frac{1}{Z_{inv\_dc}} - \sum_{n=1}^{\infty}\frac{1}{Z_{inv}(6n\omega_0)} \times \frac{2}{(36n^2-1)}\cos(n\pi)\cos(6n\omega_0 t)\right] \tag{11.28}$$

Based on the DC link current model in terms of the converter parameters, the grid current can be developed in terms of the switching function of the diode rectifier. Let us assume that the diodes in the three-phase AC-DC rectifier system operate without any commutations and the switching function of each phase is symmetrical with 120° conduction, as shown in Figure 11.13. The mathematical equation of the grid current can be calculated for phase-a of the grid current and similar approaches can be utilized for the other two phases. As shown in Figure 11.13, the grid current $i_{grid\_a}(t)$ is defined as the DC link current $i_{dc}(t)$ times of the diode rectifier switching function $S_{diode\_a}(t)$, i.e.

$$i_{grid\_a}(t) = S_{diode\_a}(t) \times i_{dc}(t) \tag{11.29}$$

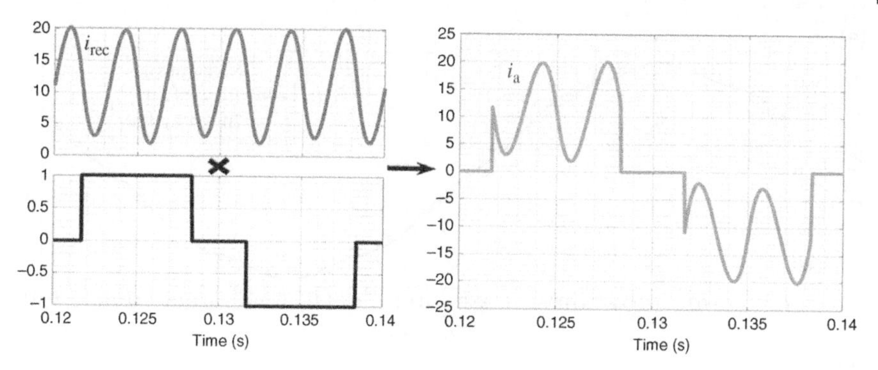

Figure 11.13 The extraction of the grid current based on the DC link current and the switching function of the diode rectifier.

The grid current can be extracted when the Fourier series of these two time domain functions $S_{diode\_a}(t)$ and $i_{dc}(t)$ are multiplied together. From the expression of the DC link current given in (11.28), the Fourier series of the switching function $S_{diode\_a}(t)$ is given by

$$S_{diode\_a}(t) = \frac{4}{\pi}\sum_{k=1,3,5,\ldots}^{\infty} \frac{\cos\left(k\pi/6\right)}{k} \sin\left(k\omega_0 t\right) \tag{11.30}$$

Thus, the grid current is expressed based on (11.28–11.30) as

$$i_{grid\_a}(t) = \frac{4}{\pi}\sum_{k=1,3,5,\ldots}^{\infty} \left[\frac{\cos\left(k\pi/6\right)}{k} \sin\left(k\omega_0 t\right)\right] \times \frac{3\sqrt{3}V_m}{\pi}\left[\frac{1}{Z_{inv\_dc}} - \sum_{n=1}^{\infty} \frac{1}{Z_{inv}(6n\omega_0)} \times \frac{2}{(36n^2-1)} \cos\left(n\pi\right)\cos\left(6n\omega_0 t\right)\right] \tag{11.31}$$

The grid current harmonics given in (11.31) represent the current harmonics generated by the rectified voltage. The total grid current harmonics are the effects of both harmonics generated by the diode rectifier (11.31) and the switching pattern of the inverter (11.23) through the entire circuit, including the DC and AC filters.

Voltage harmonics at the PCC can affect the DC link voltage at the rectifier side $V_{rec}(t)$ and consequently the DC link and grid currents. The main influence of the grid voltage harmonics is on the conductivity of the diode rectifier and the phase shifts in their current conduction times. Thus, the grid voltage background harmonics can affect the current harmonics generated by the three-phase diode rectifiers under continuous conduction mode [4].

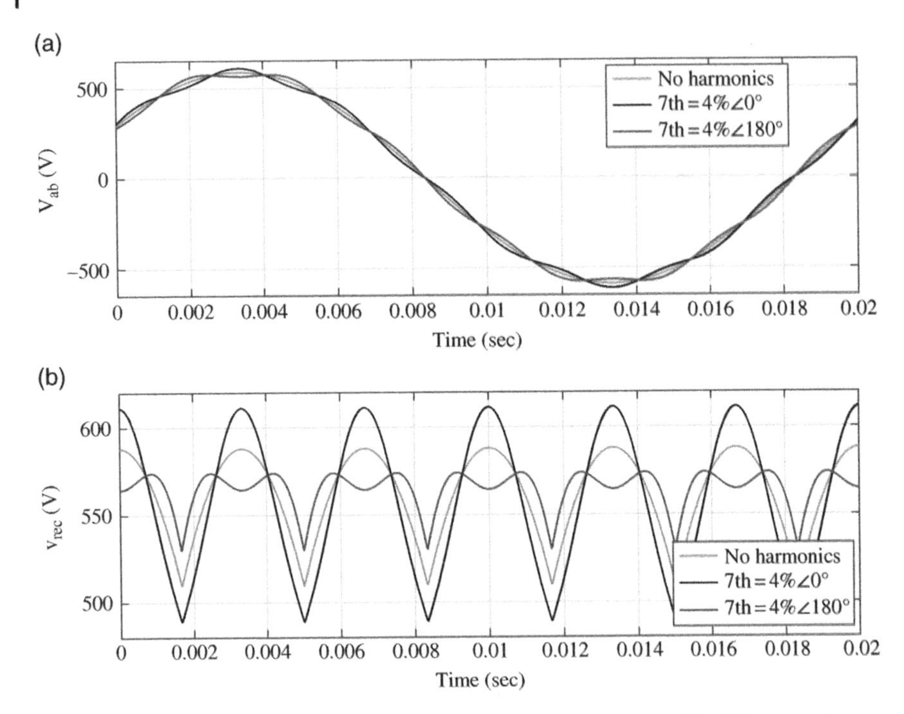

Figure 11.14 (a) The grid and (b) the rectifier voltage waveforms with different voltage harmonics.

Figure 11.14 shows the grid and the rectified voltage waveforms with different voltage harmonics at the PCC. It is obvious that the DC link voltage ripple (the rectified voltage) is significantly varied in terms of the grid voltage harmonic amplitude and phase angle. These phenomena have big impacts on the DC link current harmonic levels and the grid current as well. Similar mathematical equations can be derived for the rectified voltage and the DC link current considering the impacts of the grid voltage harmonics on the conduction times of each diode in the AC-DC rectifier.

11.3.3 Harmonic Analysis at Grid Side (Single-phase Rectifier with and without PF Correction System)

In low-power applications, such as lighting systems and electronic equipment, a single-phase diode rectifier with DC-DC converters is the most common system topology. The AC-DC diode rectifier can be cascaded with a boost converter to improve line current harmonics and PF. The combination of the single-phase diode rectifier with a boost converter is known as a power factor correction

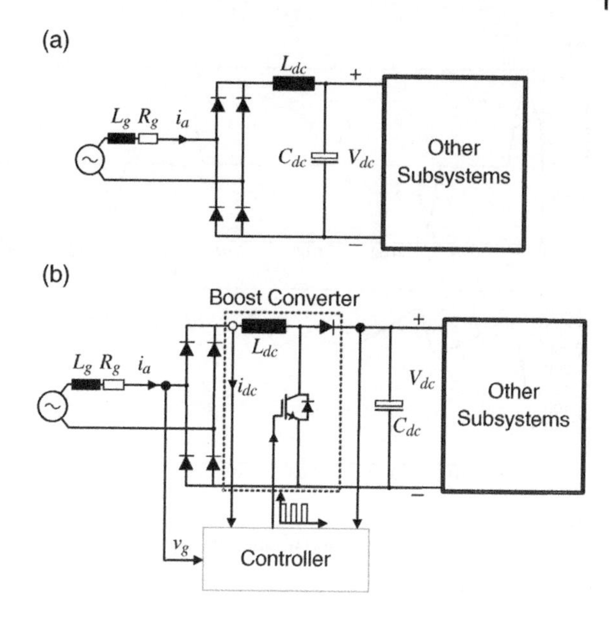

Figure 11.15 A single-phase diode rectifier (a) without a boost converter and (b) with a boost converter.

(PFC) system. The schematic diagrams of a single-phase rectifier without and with a boost converter are shown in Figure 11.15.

The single-phase diode rectifier generates 100 Hz rectified signal at the DC side of the converter and the instantaneous rectified voltage varies from zero to the maximum peak value of the grid voltage. Thus, a large storage element, a DC link capacitor, is required to control the DC link voltage at a high voltage level in such a way that the DC voltage can be utilized as a voltage source for other parts of the system. Without the boost converter, the single-phase diode rectifier operates mainly in a discontinuous conduction mode as the large DC link capacitor keeps the voltage higher than the grid voltage for some periods and the diodes will be switched off. Under this operating mode, the line current is discontinuous, as shown in Figure 11.16a. The harmonic spectrum of the current is shown in Figure 11.16b.

Therefore, the single-phase topology without a boost converter generates a distorted line current with significant low-order harmonics, mainly below 2 kHz. A main harmonic problem of the single-phase rectifier is the generation of a very high third-order harmonic content. The third harmonic and triplen harmonic current components circulate in the neutral line of a three-phase system, causing concerns of overloading the neutral conductor and the transformer overheating. One of the methods to mitigate current harmonics is to use a passive filter, such as a DC or an AC inductor. However, the performance of the converter depends on the load power (damping factor) and grid parameters.

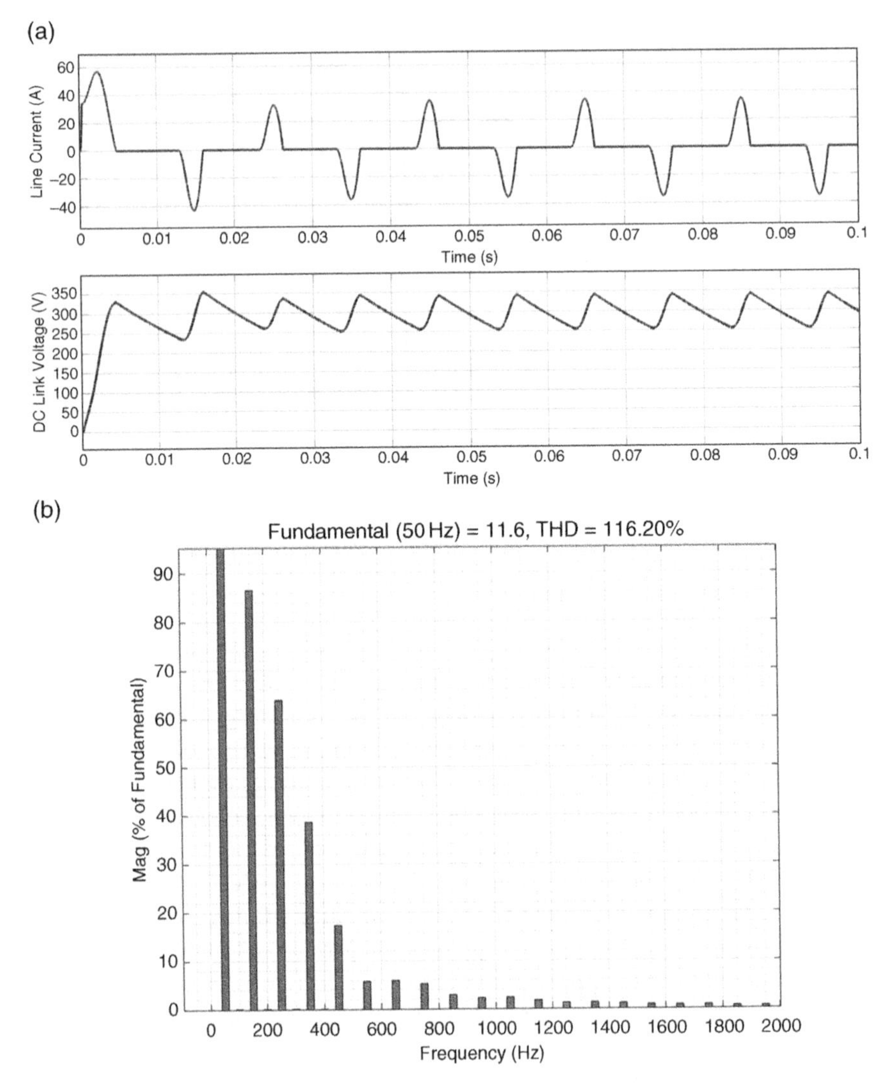

Figure 11.16 (a) Current waveform of a single-phase diode rectifier without a boost converter and (b) its harmonic spectrum.

In recent years, the use of the PFC systems has increased in many single-phase appliances where low-harmonic emissions and a high PF are required. The main advantages of the PFC system are that its line current is synchronized with the grid voltage, and it is shaped to a sinewave using a current controller. As shown in Figure 11.15b, a boost converter is cascaded with a diode rectifier at the DC link side to control the inductor current to a rectified sinewave based on a modulated high-frequency switching pattern. The control system requires current and voltage measurement to monitor and control the power flow and DC link voltage. A phase lock loop (PLL) or a direct grid voltage measurement is utilized to synchronize the PFC line current with the grid voltage to generate a unity PF.

As shown in Figure 11.17, a single-phase rectifier with a PFC system has a very low THD and low-order harmonic emission. The levels of the DC current and

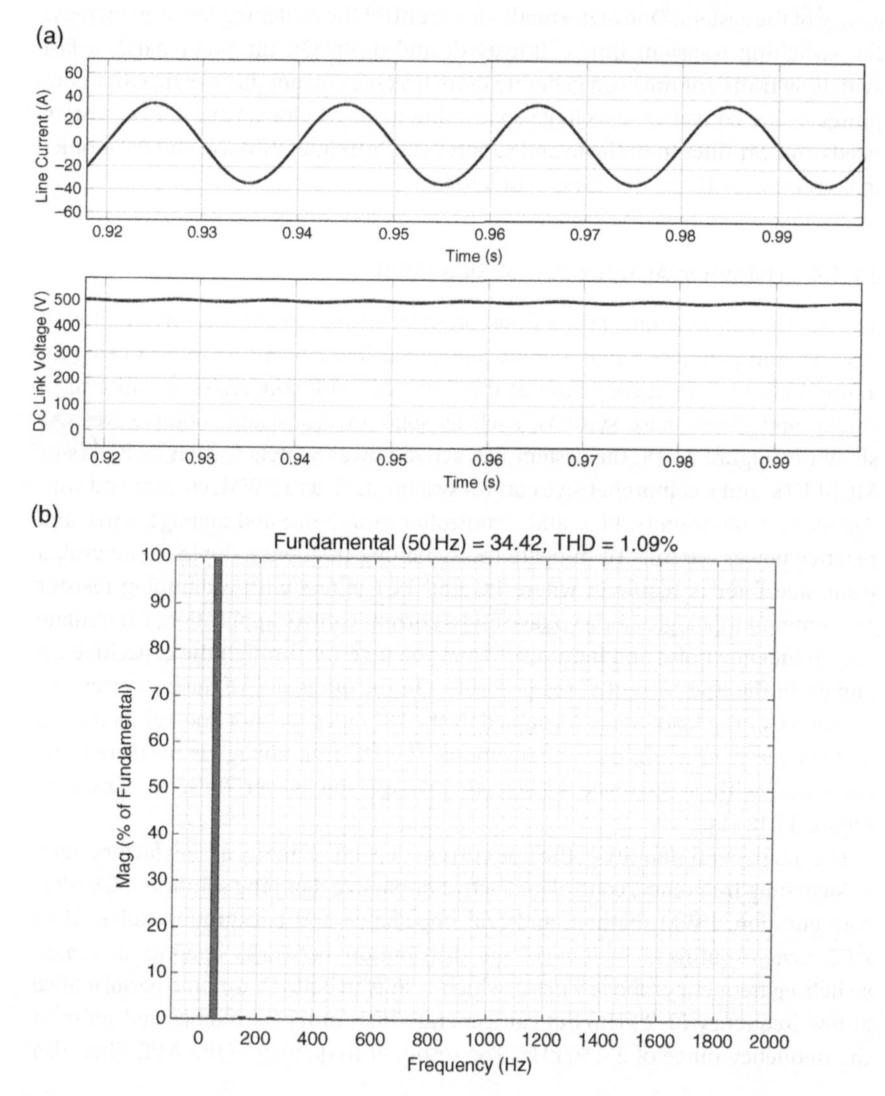

Figure 11.17 (a) Current waveform of a single-phase diode rectifier with a boost converter and (b) its harmonic spectrum.

specific low-order harmonics, i.e. the third and fifth orders, are limited according to international standards, such as IEC 61000-3-2.

The main drawbacks of the PFC system are the higher system cost for more active and passive components, high-frequency leakage, and ripple currents due to the PWM signal and voltage stress (dv/dt) across the inductor. The current ripple amplitude can be reduced if the switching frequency is increased. However, the higher switching frequency can affect the switching losses and reduce the efficiency of the system. One of the methods to control the switching loss is to increase the switching transient time – fast dv/dt and di/dt. On the other hand, a fast switching transient time can generate more leakage current due to capacitive couplings in the converter or voltage spikes due to stay inductances. A PFC system needs an EMI filter to mitigate and suppress high-frequency noise and harmonics to comply with IEC and CISPR standards.

11.3.4 Harmonic Analysis at Grid Side (AFE)

The AFE is a bidirectional power flow converter with a PWM-based inverter topology that can generate a high-quality sinusoidal line current waveform using an appropriate high-frequency filter at the grid side. AFE converters are utilized in single- and three-phase systems, such as solar inverters and motor drives. As shown in Figure 11.18, the system has active power switches, such as IGBTs or MOSFETs, and a comprehensive control system, such as a PWM, current and voltage measurement units, PLL, and a controller to stabilize and manage active and reactive power control. To mitigate the switching frequency ripple in the grid, a front side filter is required, where LC and LCL filters with a damping resistor are common topologies for a voltage and a current control application, to circulate high-frequency noise and harmonics from the inverter through the capacitive leg and clean the line current at the grid side. On the other side of the converter, the PV source or the load side is decoupled by the DC link capacitor and has its control and converter units. In most applications, the DC link voltage is measured and compared with a reference voltage to manage the power flow, as shown in Figure 11.18 [5].

The quality and stability of the line current depend on many design factors, such as switching frequency, control and active or passive damping solution, LCL filter configuration, PWM method, and grid impedance and background noise. Most AFE systems utilized in low-voltage distribution networks operate at a high switching frequency (above 9 kHz), which results in better harmonic performance at low frequency (0–2 kHz) but can generate high-order harmonics and noise in the frequency range of 2–150 kHz. The resonant frequency of the AFE filter also

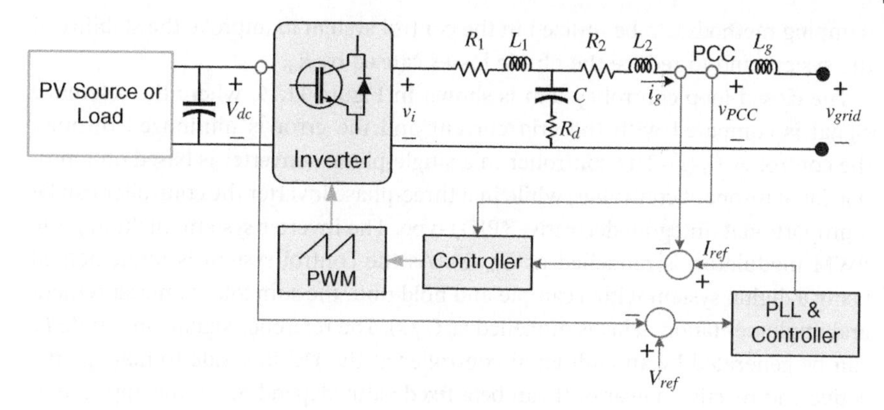

Figure 11.18 An AFE topology with current control and measuring units.

has a big impact on harmonic generation in distribution networks. The resonant frequency of the filter is around one-third of the switching frequency of the AFE converter.

The increased use of AFE topology in solar inverters and motor drive systems has been causing a major concern in the frequency range of 2–150 kHz. This has attracted the attention of international standardization organizations, such as IEC committees, to develop new standards for this frequency range.

Let us model a single-phase AFE system for stability and harmonic analysis. The block diagram of the control system is shown in Figure 11.19, where the inductors L_1 and L_2 with parasitic resistors r_1 and r_2 (ohmic losses) are a part of the AFE filter. The DC link voltage V_{dc} is modelled as an ideal voltage source with no ripple, while in practical cases, such as solar inverters or motor drive systems, the DC link voltage can be defined as $v_{dc}(t) = V_{dc} + v_{noise}(t)$, where $v_{noise}(t)$ represents the DC link voltage variation or ripple.

The capacitor branch has a resistor R_d that can improve the stability of the system. In some applications, a virtual damping resistor can be created by adding the grid side voltage or current measurement in the control loop, which is known as an active damping method. In some applications, a combination of active and passive

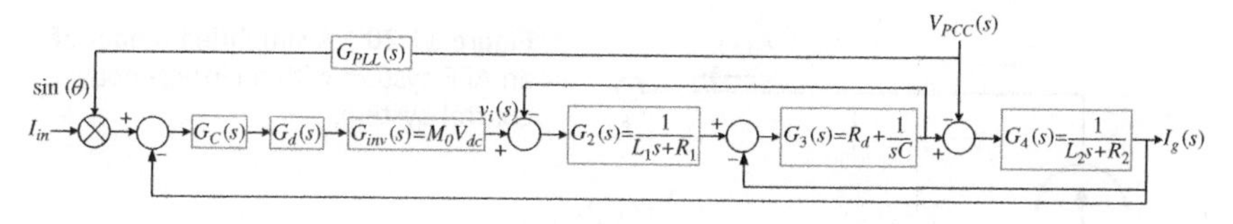

Figure 11.19 Block diagram of the closed-loop AFE control system.

damping methods can be utilized in the control system to improve the stability of the system and to reduce the ohmic losses caused by R_d.

The closed-loop control system is shown in Figure 11.19, where the reference signal is compared with the grid current and the error is minimized through the controller $G_C(s)$. The controller in a single-phase converter is based on a proportional resonant controller, while in a three-phase inverter the controller can be a proportional–integral–derivative (PID) type. The inverter system, including the PWM modulator, is modelled as $G_{inv}(s)$. As the control system is implemented using a digital system with a sample and hold unit, the sum total of measurement and implementation delay is modelled as $G_d(s)$. The reference signal amplitude I_{in} can be generated by an additional controller at the DC link side to manage the active and reactive power or it can be a fixed value depending on the application parameters. The reference signal $I_{ref}(s)$ is generated by a PLL unit to synchronize the grid current with the grid voltage. The transfer function of the PLL unit is defined as $G_{PLL}(s)$.

A simplified model of the AFE system with a closed-loop control system is shown in Figure 11.20 and can be modelled as

$$I_g(s) = I_{inv}(s) - Y_0(s)V_{PCC}(s) \tag{11.32}$$

where $Y_0(s)$ is the system admittance, $I_{inv}(s)$ is the internal current of the AFE, and $I_g(s)$ is the output current of the grid-connected AFE system. As shown in Figure 11.20, the AFE output current is influenced by two main factors:

- The inverter current $I_{inv}(s)$ and admittance $Y_0(s)$ which depend on the control parameters and filter design parameters.
- The grid voltage background noise and harmonics.

In an ideal case, if the admittance (impedance) value is zero (infinity) over the entire frequency range, the impact of the grid background noise on the converter performance is minimized and the AFE can be modelled as a current source.

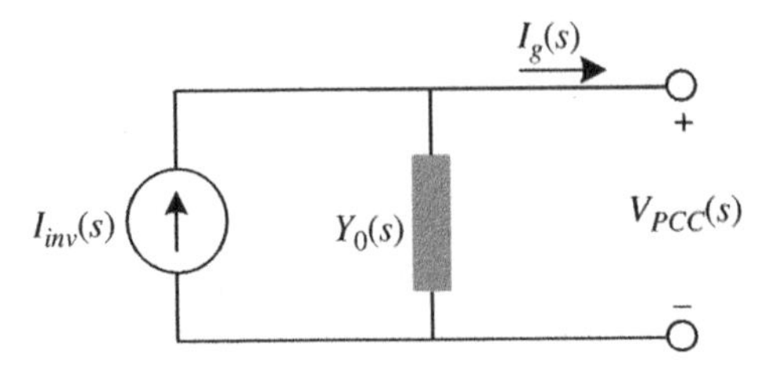

Figure 11.20 A simplified model of an AFE system with a closed-loop control system.

11.4 Power Quality Regulations and Standards

There are several standards for power quality. These are briefly discussed here.

11.4.1 IEEE Standards

IEEE societies and technical committees have a board of the IEEE Standards Association (IEEE-SA) that have developed different standards approved by the American National Standards Institute. The technical committees consist of volunteers who need not be members of the IEEE. The IEEE standards are not mandatory, while some countries, utilities, and manufacturers can refer to these standards for specific regulations or quality controls of products or services. There are several IEEE standards for power quality and harmonics, such as:

- IEEE 519 for harmonic control in electric power systems [6].
- IEEE 1159 for the monitoring and characterization of electric power quality power quality [7].
- IEEE 1409 is a guide for the application of power electronics for power quality improvement on distribution systems rated 1 through 38 kV [8].
- IEEE 1547 for interconnecting distributed resources with electric power systems [9].
- IEEE 1560 discusses methods of measurement of radiofrequency power line interference filters (100 Hz to 10 GHz) [10].
- IEEE 1662-2008 is a guide for the design and application of power electronics in electrical power systems on ships [11].
- IEEE 1826-2012 for power electronics open systems interfaces in zonal electrical distribution systems rated above 100 kW [12].
- IEEE 1709-2010 is the recommended practice for 1–35 kV medium-voltage DC power systems on ships [13].

In this section, some important IEEE standards for microgrids and low-order harmonic emissions generated by nonlinear loads and distributed energy resources are explained in detail.

11.4.2 IEEE 519

Power electronics converters, such as AC-DC rectifiers or AFE converters connected to DC-DC or DC-AC converters, are the main harmonic sources in low-voltage distribution networks. These converters are utilized in several applications, such as power supplies, motor drives, solar inverters, and electrical and electronic appliances. In medium- and high-voltage applications, other types of power electronics systems such as flexible AC transmission systems (FACTS), renewable energy systems, and other power system compensators are the harmonic sources. These power converters are defined as nonlinear loads as they generate fundamental and harmonic currents at the grid side while they are connected to an AC voltage source with a fundamental frequency of 50 or 60 Hz.

IEEE 519 evaluates harmonic emission at the system level. Therefore, a demand load current is defined for harmonic calculation as a percentage. The harmonic measurement method is based on IEC 61000-4-7 and IEC 61000-4-10 standards, where discrete Fourier transform (DFT) techniques are used. For the fundamental frequency of 50 Hz, the time domain measurement window is 200 ms to capture 10 cycles with 5 Hz resolution. When the grid frequency is 60 Hz, the measurement window is almost 200 ms (199.999 ms). Harmonics and interharmonics at 50, 55, 60, 65 Hz, etc., can be identified and the rms value of the signal at each harmonic is calculated based on these harmonic values.

The maximum and average values of the harmonics can be defined based on the measurement time, where 3-second and 10-minute time intervals are used in many power quality meters. For example, in the 3-second (3000 ms) time interval method, the rms value of each current harmonic component is calculated based on 15 samples which are measured with a 200 ms window method. The aggregated harmonic value is given by

$$I_n = \sqrt{\frac{1}{15}\sum_{k=1}^{15} I_{nk}^2} \tag{11.33}$$

Table 11.1 shows current harmonic limits for a distribution network up to 69 kV. The limits are defined based on the percentage of the maximum demand current. According to IEEE 519, the maximum demand current is calculated by the average value of the maximum current in the 12 previous months.

Table 11.1 Odd harmonics limits in percentage of demand load current (I_L) and short circuit current (I_{SC}) based limits on IEEE 519; even harmonics are limited to 25% of the odd harmonic limits.

Current Ratio	$3 \leq h < 11$	$11 \leq h < 17$	$17 \leq h < 23$	$23 \leq h < 35$	$35 \leq h < 50$	TDD
$I_{sc}/I_L < 20$	4.0	2.0	1.5	0.6	0.3	5.0
$20 < I_{sc}/I_L < 50$	7.0	3.5	2.5	1.0	0.5	8.0
$50 < I_{sc}/I_L < 100$	10.0	4.5	4.0	1.5	0.7	12.0
$100 < I_{sc}/I_L < 1000$	12.0	5.5	5.0	2.0	1.0	15.0
$1000 < I_{sc}/I_L$	15.0	7.0	6.0	2.5	1.4	20.0

In IEEE 519, the total demand distortion (TDD) is defined as

$$\mathrm{TDD}_i = \sqrt{\frac{\sum_{k=2}^{40} I_k^2}{I_{demand}^2}} \tag{11.34}$$

The main difference between THD and TDD is how the percentage of the distortion is calculated. In the THD calculation, the measured value of the fundamental current of a product is used, while in TDD the maximum demand current based on average over a period of time (for example 12 months) is utilized in the calculation. Therefore, IEEE 519 standard evaluates a system based on the demand current rather than a product based on the measured reference or fundamental current.

Example 11.3 Let us consider a network where two sets of loads (linear loads and power converters) are connected to the point of common coupling as shown in Figure 11.21. The system data considered are

$$I_a = 200\ \mathrm{A\ at\ 50\ Hz}$$

$$I_b = \begin{cases} 200\ \mathrm{A} & \mathrm{at\ 50\ Hz} \\ 70\mathrm{A} & \mathrm{at\ 250\ Hz} \\ 50\ \mathrm{A} & \mathrm{at\ 350\ Hz} \\ 20\ \mathrm{A} & \mathrm{at\ 550\ Hz} \end{cases}$$

$$I_{demand} = 1000\ \mathrm{A}$$

The THD and TDD values of the system are calculated as

$$\mathrm{THD}_{i_a} = \sqrt{\frac{0}{200^2}} = 0$$

$$\mathrm{THD}_{i_b} = \sqrt{\frac{70^2 + 50^2 + 20^2}{200^2}} = \sqrt{\frac{7800}{40000}} = 44.1\%$$

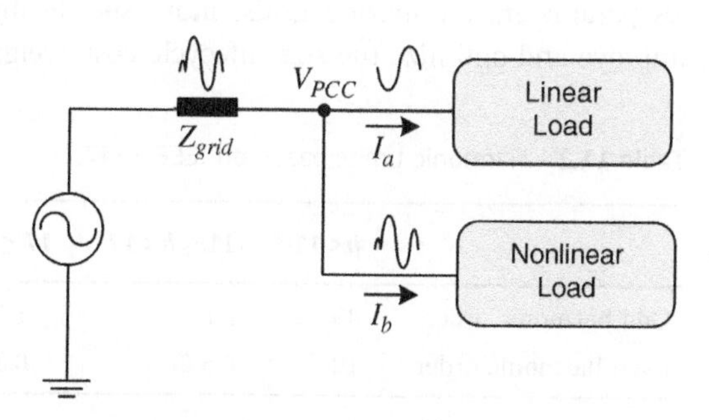

Figure 11.21 A simplified model of a network with two different loads.

$$\mathrm{THD}_{i\_total} = \sqrt{\frac{70^2 + 50^2 + 20^2}{400^2}} = \sqrt{\frac{7800}{160000}} = 22.1\%$$

$$\mathrm{TDD}_i = \sqrt{\frac{\sum_{k=2}^{40} I_k^2}{I_{demand}^2}} = \sqrt{\frac{70^2 + 50^2 + 20^2}{1000^2}} = 8.8\%$$

Thus, when the demand current or the total fundamental current is increased, the THD or the TDD value is decreased. IEEE 519 can be used to evaluate a system rather than a product.

11.4.3 IEEE 1547

One of the standards for technical specifications of distributed energy resource interconnection is IEEE 1547, which provides information regarding the operation, measurement, and safety of the system. This standard covers all technologies utilized in distributed energy resources with a total capacity of 10 MVA or less at the point of common coupling.

The TDD is calculated based on the maximum demand load current over a period of 15 or 30 minutes that is different from the maximum demand current defined in IEEE 519.

At any point of connection where a distributed energy resource is connected, the injected DC current should not be greater than 0.5% of the full rated output current. Current harmonics generated by distributed energy resources should not exceed the maximum harmonic current limits given in Table 11.2. The limits are defined based on a maximum load current demand without energy resources or the maximum current capacity of the distributed energy resource unit. The TDD value should be less than 5%.

11.4.4 IEEE 1662-2008

This standard is developed to address methods and how to analyze power electronics parameters for marine grids, more specifically technical design factors to improve and optimize the size, lifecycle cost, weight, energy efficiency, and risk

Table 11.2 Harmonic limits based on IEEE 1547.

	$h < 11$	$11 \leq h < 17$	$17 \leq h < 23$	$23 \leq h < 35$	$35 \leq h$
Odd harmonic order	4%	2%	1.5%	0.6%	0.3%
Even harmonic order	1%	0.5%	0.375%	0.15%	0.075

of the system. This document covers all power electronics topologies and related power system applications such as inverter, converter, and rectifiers (DC-DC, DC-AC, AC-DC), protecting devices such as electronic fuses, current limiters and circuit breakers, power quality compensators (active harmonic and reactive power filter), and energy storage systems.

This standard also provides technical information, which covers a wide range of power electronics equipment to design and integrate equipment, system, and shipyard with power ratings above 100 kW. Examples of these designs are:

- Single propulsor system fed from two main buses.
- Single propulsor system with a dual winding propulsion motor fed from two main buses.
- Dual propulsor system with propulsion motors fed from separate main buses.
- Dual propulsor system with dual winding propulsion motors.
- Zonal distribution with conventional propulsors.
- Integrated power system.

The AC and DC system parameters such as voltage and frequency (AC systems) fluctuations depend on the subsystem design. For example, a high voltage above 1 kV is suggested for systems with a power rating of 5 MW and above. The frequency variation for AC systems can be up to $\pm 5\%$, while the voltage amplitude variation can be from $+6$ to -10%. For DC systems, the voltage ripple or variation is recommended to be within 10%. The electrical power system grounding in ship networks is classified into three cases: (i) safety, (ii) power system, and (iii) signal, which might be tied together in different applications.

To address the power quality and interconnection of electrical systems and distributed energy resources in marine applications, two IEEE standards (IEEE Std 519TM-1992 and IEEE Std 1547TM-2003) are recommended. These standards provide technical information and references to mitigate the harmonic voltage and current generated by grid-connected electrical and electronics equipment.

11.4.5 IEEE 1826-2012

In this standard, a zonal electrical distribution system is defined as a system with a set of loads, and it can be as a part of a larger grid. If a zonal electrical distribution system has distributed energy resources or electrical storages, it can operate for periods of time. This document can be used for different applications, including maritime vessels and platforms.

The system has a limited number of electrical and control interfaces to prevent any fault to be propagated into another system. The interfaces can have three operating control states, such as (i) a central system to control the entire system,

(ii) a distributed control with an independent communication unit, and (iii) autonomous control without communication with other devices.

Any electrical and electronic device connected to a zonal electrical distribution system and a larger system should have the quality of power and operation by a customer–supplier agreement. IEEE 519 and IEEE 1547 standards can be used for interconnection and power quality analysis and design evaluation.

11.4.6 IEEE 1709-2010

This standard can be used to analyze electrical equipment connected to a DC grid on ships at the voltage level of 1–35 kV; more specifically, the standard provides design parameters to optimize the size, lifecycle cost, weight, energy efficiency, and risk issues in medium-voltage DC (MVDC) systems. The International Association of Classification Societies UR E11 provides technical reports and standards for mostly AC systems and very limited standards on DC systems. MVDC grids with voltage levels above 3 kV have no specific standards.

Similar to AC systems, in a shipboard MVDC bus system, different equipment, such as distributed energy sources, linear loads, and power electronics converter and storage devices, are connected to the DC bus. The power electronics converters play an interface between the DC bus and loads to protect the DC bus against any severe load faults. Tables 11.3 and 11.4 list the recommended rated and withstand voltages in MVDC grids.

Grounding systems, short circuit faults, and DC arc-fault mitigation techniques are some of the design concerns in MVDC grids. IEEE Std 1628-2009, IEEE Std

Table 11.3 Recommended MVDC voltage classes.

	MVDC Class kV	Nominal MVDC class rated voltage (kV)	Maximum MVDC class rated voltage (kV)
Already established classes	1.5	1.5 or ± 0.75	2 or ± 1
	3	3 or ± 1.5	5 or ± 2.5
Future design classes	6	6 or ± 3	10 or ± 5
	12	12 or ± 6	16 or ± 8
	18	18 or ± 9	22 or ± 11
	24	24 or ± 12	28 or ± 14
	30	30 or ± 15	34 or ± 17

Table 11.4 Proposed rated withstand voltages for MVDC voltage classes.

MVDC Case	Rated short duration withstand voltage to ground in kV for 1 minute	Rated lightning impulse withstand voltage to ground in kV (peak value)
1	10	45
3	20	60
6	27	75
12	35	95
18	50	110
24	70	150
30	95	200

142-2007, and IEEE Std 1100-2005 are the main standards to address grounding systems and faults.

As a DC grid has no fundamental frequency, harmonic emission is not the main concern in DC grids. However, the DC bus voltage fluctuation and noise, including current ripples, are considered in the DC networks. The main issue is related to the switching frequency of power electronics converters which generate low- and high-frequency ripples on the load current. The non-DC component of the load current can affect the quality of the DC bus voltage. Thus, short- and long-term power quality factors to measure DC link voltage spikes, sags, high-frequency noises, and surges are significantly important to evaluate the reliability and safety of the DC network. As an example, the rms value of ripple and noise voltage should not exceed 5% per unit. Some of the power quality factors which are defined in the DC grids are:

- maximum nonrepetitive peak
- maximum repetitive peak
- maximum repetitive peak-to-peak.

11.4.7 IEC Standards

The IEC, founded in 1906, is the world's leading organization with several technical committees and subcommittees to develop international standards for all electrical, electronic, and related technologies. The subcommittee 77A (SC 77A) is responsible to develop and prepare standards in the field of electromagnetic compatibility for the frequency ranges of 0–2 and 2–9 kHz. SC 77A consists of several projects, joint working groups. and the following five working groups:

- WG1: Harmonics and other low-frequency disturbances.
- WG2: Voltage fluctuations and other low-frequency disturbances.
- WG6: Low-frequency immunity tests.
- WG8: Description of the electromagnetic environment associated with the disturbances present on electricity supply networks.
- WG9: Power quality measurement methods.

IEC 61000-3-2 and IEC 61000-3-12 are general standards that define harmonic current emission limits for equipment with input current less than 16 A per phase and between 16 and 75 A per phase respectively [1, 2]. These standards control grid voltage distortion by limiting the maximum value for harmonic currents from 2nd to 40th order. The limits are defined based on several types of equipment (single-phase, three-phase, and low and high power) and grid impedance or SCR.

IEC 61000-3-16 is a similar standard to IEEE 1547, which is being developed by the IEC SC 77A, WG1, and the first version of the standard is finalized in June 2021 [14]. IEC 61000-3-16 is about the limits for currents produced by inverter-type electrical energy supplying equipment with a reference current less than or equal to 75 A per phase connected to public low-voltage systems [15].

IEC committees and working groups have published many standards for DC systems, such as IEC 62040-5-3 and IEC 61643-3 for existing DC applications. Recently, several activities in the area of low-voltage DC applications in information and communication technologies (ICT), residential and commercial buildings, etc., have led the IEC to establish a new strategic group (SG) to study the standardization of DC distribution, in which SG4 has been approved for low-voltage DC (LVDC) distribution systems up to 1500 V DC in relation to energy efficiency [16].

Some standards or guidelines for the onboard DC system, such as the voltage level, new safety regulations, and suitable protection solution, are addressed in the following standards:

- IEC 63108: Electrical installations in ships-primary DC distribution-system design architecture: first edition published in 2017.
- IEC 61660: Short-circuit currents in DC auxiliary installations in power plants and substations, Part 1: Calculation of short-circuit currents.
- IEC 60092-507: Electrical installations in ships, Part 507: Small vessels (including some information about DC distribution system): third edition published in 2014.
- IEC 61892-1: Mobile and fixed offshore units – Electrical installations – Part 1: General requirements and conditions (including DC installations up to and including 1500 V): third edition published in 2015.

11.5 Concluding Remarks

This chapter addresses harmonic emissions generated by different power electronics converters within the frequency range of 0–150 kHz. The conventional power quality and harmonic analysis is limited to harmonics up to 2 kHz, while the current standardization committees (IEC SC77A, WG1, WG8, and WG9) and utility companies around the world have reported some severe high-frequency interferences within the 2–150 kHz range in low-voltage distribution networks. Different IEEE and IEC standards for DC and microgrids systems are explained in this chapter. The generic harmonic standards to define harmonic limits for grid-connected electrical and electronic equipment are IEEE 519, IEEE 1547, IEC 61000-3-2, IEC 61000-12, and IEC 61000-3-16.

Notes and References

There are several power qualities, DC grids, and microgrid standards developed by IEEE and IEC committees which are listed in the reference section [1, 2, 5–13]. Two review papers [17, 18] presents different aspects of microgrids and marine networks, such as the classification of power system architectures, power electronics converters topologies, control and protection architecture, and energy efficiency indicators.

1 International Electrotechnical Commission IEC 61000-3-2. 2014. Electromagnetic compatibility (EMC): Part 3–2: Limits: Limits for harmonic current emissions (equipment input current ≤ 16 A per phase), Geneva, Switzerland.

2 International Electrotechnical Commission IEC 61000-3-12, Electromagnetic compatibility (EMC): Part 3–12: Limits: Limits for harmonic currents produced by equipment connected to public low voltage systems with input current > 16 A and ≤ 75 A per phase. Geneva, Switzerland. 2011

3 Yaghoobi, J., Zare, F., and Rathnayake, H. (2021). Current harmonics generated by motor-side converter: new standardizations. *IEEE Trans. Emerg. Sel. Top. Power Electron.* 9 (3): 2868–2880.

4 Alduraibi, A., Yaghoobi, J., and Zare, F. (2021). Impacts of grid voltage harmonics amplitude and phase angle values on power converters in distribution networks. *IEEE Acc.* 9: 92017–92029.

5 Khajeh, K.G., Solatialkaran, D., Zare, F., and Nadarajah, M. (2021). An enhanced full-feedforward strategy to mitigate output current harmonics in grid-tied inverters. *IET Gener. Transm. Distrib.* 15: 827–835.

6 IEEE Standard 519. IEEE Recommended Practice and Requirements for Harmonic Control in Electric Power Systems. 2014.

7 IEEE 1159-2009. IEEE Recommended Practice for Monitoring Electric Power Quality. 2009.
8 IEEE 1409-2012. IEEE Guide for Application of Power Electronics for Power Quality Improvement on Distribution Systems Rated 1 kV Through 38 kV. 2012.
9 IEEE 1547-2018, *IEEE Standard for Interconnection and Interoperability of Distributed Energy Resources with Associated Electric Power Systems Interfaces*, 2018.
10 IEEE 1560-2005. IEEE Standard for Methods of Measurement of Radio Frequency Power Line Interference Filter in the Range of 100 Hz to 10 GHz. 2005.
11 IEEE Standard 1662-2008. Guide for the Design and Application of Power Electronics in Electrical Power Systems on Ships. 2008.
12 IEEE Standard 1826-2012. IEEE Standard for Power Electronics Open System Interfaces in Zonal Electrical Distribution Systems Rated Above 100 kW. 2012.
13 IEEE Standard 1709. Recommended Practice for 1 kV to 35 kV Medium-Voltage DC Power Systems on Ships. 2010.
14 International Electrotechnical Commission SC 77A. (2022). WG 1, Harmonics and Other Low Disturbances. https://www.iec.ch/ords/f?p=103:38:501910248699179::::FSP_ORG_ID,FSP_APEX_PAGE,FSP_PROJECT_ID:1384,23,23070#, accessed 22 May 2022.
15 International Electrotechnical Commission IEC TS 61000-3-16 ED1. Electromagnetic Compatibility (EMC): Part 3-16: Limits: Limits for currents produced by the inverter of inverter-type electrical energy-supplying equipment with a reference current less than or equal to 75 A per phase connected to public low-voltage systems.
16 International Electromagnetic Commission IEC SEG4. Implementing the Standardization Framework to Support the Development of Low Voltage Direct Current and Electricity Access. Geneva, Switzerland. 2016.
17 Kumar, D., Zare, F., and Ghosh, A. (2017). DC microgrid technology: system architectures, AC grid interfaces, grounding schemes, power quality, communication networks, applications, and standardizations aspects. *IEEE Acc.* 5: 12230–12256.
18 Kumar, D. and Zare, F. (2019). A comprehensive review of maritime microgrids: system architectures, energy efficiency, power quality, and regulations. *IEEE Acc.* 7: 67249–67277. https://doi.org/10.1109/ACCESS.2019.2917082.

IEEE Press Series on Power and Energy Systems

Series Editor: Ganesh Kumar Venayagamoorthy, Clemson University, Clemson, South Carolina, USA.

The mission of the IEEE Press Series on Power and Energy Systems is to publish leading-edge books that cover a broad spectrum of current and forward-looking technologies in the fast-moving area of power and energy systems including smart grid, renewable energy systems, electric vehicles and related areas. Our target audience includes power and energy systems professionals from academia, industry and government who are interested in enhancing their knowledge and perspectives in their areas of interest.

1. *Electric Power Systems: Design and Analysis, Revised Printing*
   Mohamed E. El-Hawary
2. *Power System Stability*
   Edward W. Kimbark
3. *Analysis of Faulted Power Systems*
   Paul M. Anderson
4. *Inspection of Large Synchronous Machines: Checklists, Failure Identification, and Troubleshooting*
   Isidor Kerszenbaum
5. *Electric Power Applications of Fuzzy Systems*
   Mohamed E. El-Hawary
6. *Power System Protection*
   Paul M. Anderson
7. *Subsynchronous Resonance in Power Systems*
   Paul M. Anderson, B.L. Agrawal, J.E. Van Ness
8. *Understanding Power Quality Problems: Voltage Sags and Interruptions*
   Math H. Bollen
9. *Analysis of Electric Machinery*
   Paul C. Krause, Oleg Wasynczuk, and S.D. Sudhoff
10. *Power System Control and Stability, Revised Printing*
    Paul M. Anderson, A.A. Fouad

11. *Principles of Electric Machines with Power Electronic Applications*, Second Edition
    Mohamed E. El-Hawary
12. *Pulse Width Modulation for Power Converters: Principles and Practice*
    D. Grahame Holmes and Thomas Lipo
13. *Analysis of Electric Machinery and Drive Systems*, Second Edition
    Paul C. Krause, Oleg Wasynczuk, and S.D. Sudhoff
14. *Risk Assessment for Power Systems: Models, Methods, and Applications*
    Wenyuan Li
15. *Optimization Principles: Practical Applications to the Operations of Markets of the Electric Power Industry*
    Narayan S. Rau
16. *Electric Economics: Regulation and Deregulation*
    Geoffrey Rothwell and Tomas Gomez
17. *Electric Power Systems: Analysis and Control*
    Fabio Saccomanno
18. *Electrical Insulation for Rotating Machines: Design, Evaluation, Aging, Testing, and Repair*
    Greg C. Stone, Edward A. Boulter, Ian Culbert, and Hussein Dhirani
19. *Signal Processing of Power Quality Disturbances*
    Math H. J. Bollen and Irene Y. H. Gu
20. *Instantaneous Power Theory and Applications to Power Conditioning*
    Hirofumi Akagi, Edson H.Watanabe and Mauricio Aredes
21. *Maintaining Mission Critical Systems in a 24/7 Environment*
    Peter M. Curtis
22. *Elements of Tidal-Electric Engineering*
    Robert H. Clark
23. *Handbook of Large Turbo-Generator Operation and Maintenance,* Second Edition
    Geoff Klempner and Isidor Kerszenbaum
24. *Introduction to Electrical Power Systems*
    Mohamed E. El-Hawary
25. *Modeling and Control of Fuel Cells: Distributed Generation Applications*
    M. Hashem Nehrir and Caisheng Wang

26. *Power Distribution System Reliability: Practical Methods and Applications*
Ali A. Chowdhury and Don O. Koval
27. *Introduction to FACTS Controllers: Theory, Modeling, and Applications*
Kalyan K. Sen, Mey Ling Sen
28. *Economic Market Design and Planning for Electric Power Systems*
James Momoh and Lamine Mili
29. *Operation and Control of Electric Energy Processing Systems*
James Momoh and Lamine Mili
30. *Restructured Electric Power Systems: Analysis of Electricity Markets with Equilibrium Models*
Xiao-Ping Zhang
31. *An Introduction to Wavelet Modulated Inverters*
S.A. Saleh and M.A. Rahman
32. *Control of Electric Machine Drive Systems*
Seung-Ki Sul
33. *Probabilistic Transmission System Planning*
Wenyuan Li
34. *Electricity Power Generation: The Changing Dimensions*
Digambar M. Tagare
35. *Electric Distribution Systems*
Abdelhay A. Sallam and Om P. Malik
36. *Practical Lighting Design with LEDs*
Ron Lenk, Carol Lenk
37. *High Voltage and Electrical Insulation Engineering*
Ravindra Arora and Wolfgang Mosch
38. *Maintaining Mission Critical Systems in a 24/7 Environment*, Second Edition
Peter Curtis
39. *Power Conversion and Control of Wind Energy Systems*
Bin Wu, Yongqiang Lang, Navid Zargari, Samir Kouro
40. *Integration of Distributed Generation in the Power System*
Math H. Bollen, Fainan Hassan
41. *Doubly Fed Induction Machine: Modeling and Control for Wind Energy Generation Applications*
Gonzalo Abad, Jesús López, Miguel Rodrigues, Luis Marroyo, and Grzegorz Iwanski

42. *High Voltage Protection for Telecommunications*
Steven W. Blume

43. *Smart Grid: Fundamentals of Design and Analysis*
James Momoh

44. *Electromechanical Motion Devices*, Second Edition
Paul Krause, Oleg Wasynczuk, Steven Pekarek

45. *Electrical Energy Conversion and Transport: An Interactive Computer-Based Approach*, Second Edition
George G. Karady, Keith E. Holbert

46. *ARC Flash Hazard and Analysis and Mitigation*
J.C. Das

47. *Handbook of Electrical Power System Dynamics: Modeling, Stability, and Control*
Mircea Eremia, Mohammad Shahidehpour

48. *Analysis of Electric Machinery and Drive Systems, Third Edition*
Paul C. Krause, Oleg Wasynczuk, S.D. Sudhoff, Steven D. Pekarek

49. *Extruded Cables for High-Voltage Direct-Current Transmission: Advances in Research and Development*
Giovanni Mazzanti, Massimo Marzinotto

50. *Power Magnetic Devices: A Multi-Objective Design Approach*
S.D. Sudhoff

51. *Risk Assessment of Power Systems: Models, Methods, and Applications*, Second Edition
Wenyuan Li

52. *Practical Power System Operation*
Ebrahim Vaahedi

53. *The Selection Process of Biomass Materials for the Production of Bio-Fuels and Co-Firing*
Najib Altawell

54. *Electrical Insulation for Rotating Machines: Design, Evaluation, Aging, Testing, and Repair*, Second Edition
Greg C. Stone, Ian Culbert, Edward A. Boulter, and Hussein Dhirani

55. *Principles of Electrical Safety*
Peter E. Sutherland

56. *Advanced Power Electronics Converters: PWM Converters Processing AC Voltages*
Euzeli Cipriano dos Santos Jr., Edison Roberto Cabral da Silva

57. *Optimization of Power System Operation*, Second Edition
Jizhong Zhu

58. *Power System Harmonics and Passive Filter Designs*
J.C. Das

59. *Digital Control of High-Frequency Switched-Mode Power Converters*
Luca Corradini, Dragan Maksimoviæ, Paolo Mattavelli, Regan Zane

60. *Industrial Power Distribution*, Second Edition
Ralph E. Fehr, III

61. *HVDC Grids: For Offshore and Supergrid of the Future*
Dirk Van Hertem, Oriol Gomis-Bellmunt, Jun Liang

62. *Advanced Solutions in Power Systems: HVDC, FACTS, and Artificial Intelligence*
Mircea Eremia, Chen-Ching Liu, Abdel-Aty Edris

63. *Operation and Maintenance of Large Turbo-Generators*
Geoff Klempner, Isidor Kerszenbaum

64. *Electrical Energy Conversion and Transport: An Interactive Computer-Based Approach*
George G. Karady, Keith E. Holbert

65. *Modeling and High-Performance Control of Electric Machines*
John Chiasson

66. *Rating of Electric Power Cables in Unfavorable Thermal Environment*
George J. Anders

67. *Electric Power System Basics for the Nonelectrical Professional*
Steven W. Blume

68. *Modern Heuristic Optimization Techniques: Theory and Applications to Power Systems*
Kwang Y. Lee, Mohamed A. El-Sharkawi

69. *Real-Time Stability Assessment in Modern Power System Control Centers*
Savu C. Savulescu

70. *Optimization of Power System Operation*
Jizhong Zhu

71. *Insulators for Icing and Polluted Environments*
Masoud Farzaneh, William A. Chisholm

72. *PID and Predictive Control of Electric Devices and Power Converters Using MATLAB®/Simulink®*
Liuping Wang, Shan Chai, Dae Yoo, Lu Gan, Ki Ng

73. *Power Grid Operation in a Market Environment: Economic Efficiency and Risk Mitigation*
Hong Chen

74. *Electric Power System Basics for Nonelectrical Professional*, Second Edition
Steven W. Blume

75. *Energy Production Systems Engineering*
Thomas Howard Blair

76. *Model Predictive Control of Wind Energy Conversion Systems*
Venkata Yaramasu, Bin Wu

77. *Understanding Symmetrical Components for Power System Modeling*
J.C. Das

78. *High-Power Converters and AC Drives*, Second Edition
Bin Wu, Mehdi Narimani

79. *Current Signature Analysis for Condition Monitoring of Cage Induction Motors: Industrial Application and Case Histories*
William T. Thomson, Ian Culbert

80. *Introduction to Electric Power and Drive Systems*
Paul Krause, Oleg Wasynczuk, Timothy O'Connell, Maher Hasan

81. *Instantaneous Power Theory and Applications to Power Conditioning*, Second Edition
Hirofumi, Edson Hirokazu Watanabe, Mauricio Aredes

82. *Practical Lighting Design with LEDs*, Second Edition
Ron Lenk, Carol Lenk

83. *Introduction to AC Machine Design*
Thomas A. Lipo

84. *Advances in Electric Power and Energy Systems: Load and Price Forecasting*
Mohamed E. El-Hawary

85. *Electricity Markets: Theories and Applications*
Jeremy Lin, Jernando H. Magnago

86. *Multiphysics Simulation by Design for Electrical Machines, Power Electronics and Drives*
Marius Rosu, Ping Zhou, Dingsheng Lin, Dan M. Ionel, Mircea Popescu, Frede Blaabjerg, Vandana Rallabandi, David Staton

87. *Modular Multilevel Converters: Analysis, Control, and Applications*
Sixing Du, Apparao Dekka, Bin Wu, Navid Zargari

88. *Electrical Railway Transportation Systems*
Morris Brenna, Federica Foiadelli, Dario Zaninelli

89. *Energy Processing and Smart Grid*
James A. Momoh

90. *Handbook of Large Turbo-Generator Operation and Maintenance*, 3rd Edition
Geoff Klempner, Isidor Kerszenbaum

91. *Advanced Control of Doubly Fed Induction Generator for Wind Power Systems*
Dehong Xu, Dr. Frede Blaabjerg, Wenjie Chen, Nan Zhu

92. *Electric Distribution Systems*, 2nd Edition
Abdelhay A. Sallam, Om P. Malik

93. *Power Electronics in Renewable Energy Systems and Smart Grid: Technology and Applications*
Bimal K. Bose

94. *Distributed Fiber Optic Sensing and Dynamic Rating of Power Cables*
Sudhakar Cherukupalli, and George J Anders

95. *Power System and Control and Stability,* Third Edition
Vijay Vittal, James D. McCalley, Paul M. Anderson, and A. A. Fouad

96. *Electromechanical Motion Devices: Rotating Magnetic Field-Based Analysis and Online Animations,* Third Edition
Paul Krause, Oleg Wasynczuk, Steven D. Pekarek, and Timothy O'Connell

97. *Applications of Modern Heuristic Optimization Methods in Power and Energy Systems*
Kwang Y. Lee and Zita A. Vale

98. *Handbook of Large Hydro Generators: Operation and Maintenance*
Glenn Mottershead, Stefano Bomben, Isidor Kerszenbaum, and Geoff Klempner

99. *Advances in Electric Power and Energy: Static State Estimation*
Mohamed E. El-hawary

100. *Arc Flash Hazard Analysis and Mitigation*, Second Edition
J.C. Das

101. *Maintaining Mission Critical Systems in a 24/7 Environment,* Third Edition
Peter M. Curtis

102. *Real-Time Electromagnetic Transient Simulation of AC-DC Networks*
Venkata Dinavahi and Ning Lin

103. *Probabilistic Power System Expansion Planning with Renewable Energy Resources and Energy Storage Systems*
Jaeseok Choi and Kwang Y. Lee

104. *Power Magnetic Devices: A Multi-Objective Design Approach*, Second Edition
Scott D. Sudhoff

105. *Optimal Coordination of Power Protective Devices with Illustrative Examples*
Ali R. Al-Roomi

106. *Resilient Control Architectures and Power Systems*
Craig Rieger, Ronald Boring, Brian Johnson, and Timothy McJunkin

107. *Alternative Liquid Dielectrics for High Voltage Transformer Insulation Systems: Performance Analysis and Applications*
Edited by U. Mohan Rao, I. Fofana, and R. Sarathi

108. *Introduction to the Analysis of Electromechanical Systems*
Paul C. Krause, Oleg Wasynczuk, and Timothy O'Connell

109. *Power Flow Control Solutions for a Modern Grid using SMART Power Flow Controllers*
Kalyan K. Sen and Mey Ling Sen

110. *Power System Protection: Fundamentals and Applications*
John Ciufo and Aaron Cooperberg

111. *Soft-Switching Technology for Three-phase Power Electronics Converters*
Dehong Xu, Rui Li, Ning He, Jinyi Deng, and Yuying Wu

112. *Power System Protection*, Second Edition
Paul M. Anderson, Charles Henville, Rasheek Rifaat, Brian Johnson, and Sakis Meliopoulos

113. *High Voltage and Electrical Insulation Engineering*, Second Edition
Ravindra Arora and Wolfgang Mosch.

Index

Control of Power Electronic Converters with Microgrid Applications, First Edition.
Arindam Ghosh and Firuz Zare.

Published 2023 by John Wiley & Sons, Inc.

Printed and bound by CPI Group (UK) Ltd, Croydon, CR0 4YY

Printed and bound by CPI Group (UK) Ltd, Croydon, CR0 4YY

23/08/2023

08103885-0003